Introduction to Regression Methods for Public Health Using R

Introduction to Regression Methods for Public Health Using R teaches regression methods for continuous, binary, ordinal, and time-to-event outcomes using R as a tool. Regression is a useful tool for understanding the associations between an outcome and a set of explanatory variables, and regression methods are commonly used in many fields, including epidemiology, public health, and clinical research. The focus of this book is on understanding and fitting regression models, diagnosing model fit, and interpreting and writing up results. Examples are drawn from public health and clinical studies. Designed for students, researchers, and practitioners with a basic understanding of introductory statistics, this book teaches the basics of regression and how to implement regression methods using R, allowing the reader to enhance their understanding and begin to grasp new concepts and models.

The text includes an overview of regression (Chapter 2); how to examine and summarize the data (Chapter 3), simple (Chapter 4) and multiple (Chapter 5) linear regression; binary, ordinal, and conditional logistic regression, and log-binomial regression (Chapter 6); Cox proportional hazards regression (survival analysis) (Chapter 7); handling data arising from a complex survey design (Chapter 8); and multiple imputation of missing data (Chapter 9). Each chapter closes with a comprehensive set of exercises.

Key Features:

- Comprehensive coverage of the most commonly used regression methods, as well as how to use regression with complex survey data or missing data
- Accessible to those with only a first course in statistics
- Serves as a course textbook, as well as a reference for public health and clinical researchers seeking to learn regression and/or how to use R to do regression analyses
- Includes examples of how to diagnose the fit of a regression model
- Includes examples of how to summarize, visualize, table, and write up the results
- Includes R code to run the examples

Ramzi W. Nahhas teaches Biostatistics in the Department of Population and Public Health Sciences, Boonshoft School of Medicine at Wright State University in Dayton, Ohio. In addition to teaching, he is actively involved in research collaborations with faculty, residents, and students, primarily in his own department and the Department of Psychiatry.

Introduction to Regression Methods for Public Health Using R

Ramzi W. Nahhas

CRC Press is an imprint of the
Taylor & Francis Group, an **informa** business
A CHAPMAN & HALL BOOK

Designed cover image: © Ramzi W. Nahhas

First edition published 2025
by CRC Press
2385 NW Executive Center Drive, Suite 320, Boca Raton FL 33431

and by CRC Press
4 Park Square, Milton Park, Abingdon, Oxon, OX14 4RN

CRC Press is an imprint of Taylor & Francis Group, LLC

ISBN: 978-1-032-20307-2 (hbk)
ISBN: 978-1-032-20318-8 (pbk)
ISBN: 978-1-003-26319-7 (ebk)

DOI: 10.1201/9781003263197

Typeset in Latin Modern font
by KnowledgeWorks Global Ltd.

Publisher's note: This book has been prepared from camera-ready copy provided by the authors.

Contents

Preface

This text was written to be used in a second biostatistics course for Master of Public Health students; however, students in any field will find it useful. Students in many disciplines take an introductory statistics course, providing foundational competencies but perhaps not enough to use more advanced methods without additional training. There are a plethora of textbooks covering topics such as linear regression, logistic regression, and survival analysis aimed at those with a background in mathematical statistics and/or without a focus specifically on public health and/or without a focus on using R statistical software. The goal of this text is to provide a gentle introduction to regression methods, using R, that covers all the basics and a bit more, with examples drawn from public health data.

The text began in 2016 as course notes and evolved over time into what you see here. My hope is that what you learn from this book will give you the knowledge and skills to understand and carry out appropriate basic regression analyses, as well as the foundation and confidence to go deeper. When you are ready to go deeper, there are excellent texts that cover each of the methods covered herein, as well as R programming, in much greater detail (e.g., Faraway, 2016; Fox, 2015; Fox and Weisberg, 2019; Harrell, 2015; Klein and Moeschberger, 2010; Kleinman and Horton, 2014; Lohr, 2021; Lumley, 2010; van Buuren, 2018; Weisberg, 2014; Wickham et al., 2023; Wickham, 2019). Additionally, improvements over standard regression methods are available using hierarchical (multilevel, random coefficient) and related shrinkage estimation procedures such as parametric empirical-Bayes/semi-Bayes and penalized-likelihood methods (Efron, 2013; Greenland, 2000; Harrell, 2015; Efron, 2023).

An online version of this text may be found at bookdown.org/rwnahhas/RMPH[1].

About the Author

Ramzi W. Nahhas[2] teaches biostatistics at Wright State University, Dayton, Ohio, where he is Professor in the Department of Population and Public Health Sciences[3]. In addition to teaching, he is actively involved in research collaborations with faculty, residents, and students, primarily in his own department and the Department of Psychiatry. If you have any comments or suggestions, you may contact the author at ramzi.nahhas@wright.edu[4].

[1] https://bookdown.org/rwnahhas/RMPH
[2] https://people.wright.edu/ramzi.nahhas
[3] https://medicine.wright.edu/population-and-public-health-sciences
[4] mailto:ramzi.nahhas@wright.edu

Acknowledgments

Thank you to the reviewers who took the time to provide very helpful feedback: Ashna Jagtiani, Chinwe C. Uzodinma-Jones, Jianrong Wu, Scott J. Pearson, Pratyaydipta Rudra, Sander Greenland (specifically, Section 4.5, "Interpreting p-values"), and Paul von Hippel (specifically, Chapter 9, "Multiple Imputation of Missing Data"). I have tried to incorporate your suggestions, and any errors that remain are solely mine. Thank you to the faculty, staff, and students of the Department of Population and Public Health Sciences, Boonshoft School of Medicine, Wright State University. It is a pleasure to interact each day with such a talented, dedicated, and wonderful group of individuals. Thank you to my editor, Lara Spieker, for skillfully shepherding me through the publication process. Finally, thank you to my family for your continual support and encouragement; you mean the world to me.

1

Introduction

Regression, in general, is a useful tool for understanding the association between one variable (the outcome, or dependent variable) and a set of other variables (the predictors, or independent variables, or explanatory variables). Some users of regression may treat it like a "black box" – a method that can be used without any need to understand the inner workings. They might identify an outcome and a set of predictors, throw them all into a regression model as-is, and see what results the computer outputs. Using regression in this way can lead to incorrect conclusions. It is important to understanding the model, the meaning of the terms in the model, and how to properly fit, check, and interpret the results.

Knowing the basics of regression may give a researcher a false sense of confidence in their results. If you know you have no idea how to use regression, you will seek help. If you know the basics but do not realize that there is a lot more, you may unknowingly produce incorrect or misleading results. The goal of this text is to introduce you to the various important concepts of regression analyses and the tools available in R. Until you have a lot of practice using regression, it is wise to consult with a statistician who can guide you and give you feedback; but you will be able to carry out the analyses yourself, in many cases, using the tools you learn here. After working your way through this text, you will know a lot more than you did, you will know your limits, and you will have the foundation to expand those limits over time with experience and further study.

The text assumes a level of statistical knowledge obtainable in a typical introductory statistics course. Briefly, the aspects of regression covered are fitting the model, visualizing aspects of the model fit, testing assumptions, possible modifications when the assumptions are not met, testing associations between predictors and the outcome, and interpreting and presenting the results.

The text begins with some preliminaries: an overview of regression in general (Chapter 2) and how to examine and summarize the data (Chapter 3). Following are chapters describing in detail how to carry out specific regression methods: simple (Chapter 4) and multiple (Chapter 5) linear regression; binary, ordinal, and conditional logistic regression, and log-binomial regression (Chapter 6); and Cox proportional hazards regression (survival analysis) (Chapter 7). The final two chapters cover handling data arising from a complex survey design (Chapter 8) and multiple imputation of missing data (Chapter 9), methods that can be applied in conjunction with all these regression methods. Each chapter closes with a comprehensive set of exercises.

1.1 R and RStudio

This book is about using R (R Core Team, 2022) to do regression, not a book about learning R itself. The text assumes you have some basic familiarity with R and RStudio (an integrated development environment for R, making using R much easier), although you do not need to be an expert by any means. See RStudio Education for Beginners[1] for links to installing R and RStudio and for excellent learning resources.

After you install R and RStudio, install some packages that will be used in this text.

```
install.packages(
  c("tidyverse",
    "Rcpp", "gtsummary", "flextable", "gt",
    "car", "gmodels", "Hmisc", "MASS",
    "sjPlot", "sjmisc", "effects",
    "ResourceSelection", "logbin",
    "survival", "bshazard",
    "survey",
    "mice", "miceadds", "mitml", "howManyImputations")
)
```

1.2 Datasets

To seamlessly use the code in the text, create an R project and a folder called "Data" in the same location as your R project. For more information on R projects, see "Workflow: scripts and projects" in R for Data Science[2] (Wickham et al., 2023). In the Data folder, place the teaching datasets listed below. Descriptions, including instructions for downloading and processing these datasets, can be found in Appendix A. In some cases, the datasets are provided as-is; in other cases, you will need to download and run some R code to create them.

- NHANES (2017-2018)
- United Nations Human Development Data (2020)
- U.S. Natality (2018)
- COVID-19 county-level data
- NSDUH (2019)
- Framingham Heart Study (BioLINCC teaching dataset)
- CAMP (BioLINCC teaching dataset)
- Digitalis (BioLINCC teaching dataset)
- Opioid

NOTE: The datasets are meant for teaching, not research. The analyses herein are meant solely as teaching examples illustrating the use of regression methods using R. Results found in this text, and datasets provided with this text, should not be used to draw conclusions about any health conditions or relationships between variables.

[1] https://education.rstudio.com/learn/beginner/
[2] https://r4ds.hadley.nz/workflow-scripts

1.3 Functions

The file `Functions_rmph.R` contains custom functions written for use with this text and can be downloaded from the book's github site, RMPH Resources[3]. Place this file in the same location as your R project. To load the functions, run the following code once per R session.

```
source("Functions_rmph.R")
```

1.4 Software information and conventions

The **knitr** package (Xie, 2015) and the **bookdown** package (Xie, 2023) were used to compile this text. Package names and inline code are formatted in a typewriter font (e.g., `knitr`, `lm(Y ~ X)`), and function names are followed by parentheses (e.g., `lm()`).

```
# Code and output
```

[3]https://github.com/rwnahhas/RMPH_Resources

2

Overview of Regression Methods

In this chapter, you will learn:

- A general definition of regression;
- Reasons to use regression; and
- How to distinguish between commonly used regression methods.

2.1 Introduction

Introductory statistics classes often cover methods for testing and assessing the magnitude of an association between two variables, such as correlation, the two-sample t-test, analysis of variance (ANOVA), the chi-square test, and others. **Regression** is a general term used for methods that assess the association between an outcome and one or more predictors by positing a model for the outcome as a function of the predictors. Many basic statistical methods are actually special cases of regression with a single predictor.

Unlike, for example, correlation, in regression you must designate one variable to be the outcome, also known as the dependent variable. The other variables are predictors, or independent variables. In some contexts, predictors may be referred to as covariates or confounders.

A **simple linear regression** model has the following equation.

$$Y = \beta_0 + \beta_1 X + \epsilon \tag{2.1}$$

Chapter 4 will explain what all these symbols represent, but for now just know that Equation 2.1 describes a model in which the values of an outcome Y are linearly related to the values of a predictor X – pairs of points (X, Y) are scattered randomly around a line. Figure 2.1 illustrates data generated from a simple linear regression model. At each X value, the corresponding Y value is the sum of the height of the line at X and some random noise.

In practice, Y is some outcome of interest (e.g., fasting glucose) and X is candidate predictor of that outcome (e.g., body fat). If the line has a positive slope (as in Figure 2.1), we say that the predictor has a positive association with the outcome – individuals with larger X values tend to have larger Y values. If the line has a negative slope, then the predictor has a negative association with the outcome – individuals with larger X values tend to have smaller Y values.

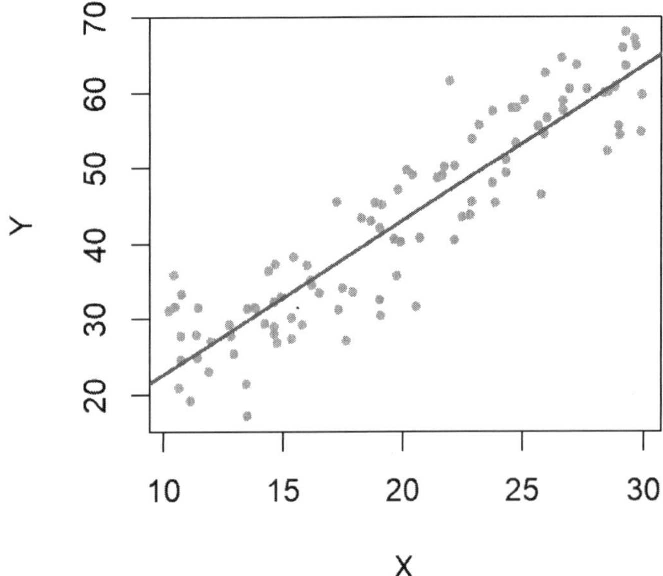

FIGURE 2.1 Linear regression example

More complicated regression settings have equations that are more complicated than Equation 2.1, including perhaps having more than one predictor. For example, a **multiple linear regression** model with K predictors is described by Equation 2.2.

$$Y = \beta_0 + \beta_1 X_1 + \beta_2 X_2 + \ldots + \beta_K X_K + \epsilon \tag{2.2}$$

2.2 Why use regression?

Regression can be used for any or all of the following purposes:

- **Testing a theory:** A theory that implies a certain functional relationship between the outcome Y and a predictor X can be tested by comparing the hypothesized model to simpler or more complex models and seeing which model fits best.
- **Estimation and prediction**: After fitting the model to observed data, one can estimate the average outcome value at a specified value of a predictor, as well as the distribution of outcome values for individuals with that predictor value.
- **Machine learning:** The fields of artificial intelligence and machine learning (AI/ML), sometimes placed under the umbrella of "data science", "data mining", "analytics", or "statistical learning" are, essentially, attempts to predict an outcome based on a set of predictors ("features"). Regression is one method among many of making such a prediction. For a gentle introduction to these methods in R, see James et al. (2021). For a more in-depth treatment, see Hastie et al. (2016).
- **Testing an association:** Is there a significant association between Y and X? In the case of simple linear regression, this question is answered by testing the null hypothesis

$H_0 : \beta_1 = 0$. Under the null hypothesis, the outcome does not depend on the predictor. If there is enough evidence to reject the null hypothesis, we conclude that there is a significant association.

- **Estimating a rate of change:** How does Y change as X changes? In the case of simple linear regression, this question is answered by estimating the magnitude of β_1, the regression slope.
- **Controlling for confounding:** Is there an association between Y and X_1 after adjusting for $X_2, \ldots, X_K$? When the data arise from an observational study, an observed association between a single predictor and the outcome may be spurious due to confounding. A third variable may actually be associated with each, and those associations induce an association between the predictor of interest and the outcome. Alternatively, the estimate of a real association may be biased due to confounding. Regression adjustment for confounding is a powerful tool for attempting to isolate the effect of one predictor on the outcome from the effects of other potentially confounding variables.
- **Effect modification:** Does the association between Y and X depend on the value of another variable, Z? If so, then Z is an effect modifier. Ignoring an effect modifier may lead to estimating an effect that may not apply to any individual cases, as it will be averaged over the modifier-level-specific effects. Regression methods can be used to estimate the extent of effect modification by including an "interaction" term.

2.3 Taxonomy of regression methods

The nature of the outcome variable determines which regression method is appropriate. Below is a list of outcome types, along with at least one example of each.

- A **continuous** variable has a range of possible values and can take on any value in an interval (e.g., waist circumference, age, body mass index). In practice, continuous variables take on a finite set of values due to measurement instrument limitations and/or rounding, but the underlying variable can in theory take on any numeric value.
- A **categorical** variable has a small number of possible values (e.g., smoking status, race/ethnicity, presence/absence of a condition).
- A **binary** variable is a categorical variable with exactly two levels (e.g., presence/absence of disease).
- An **ordinal** variable is a categorical variable in which the levels have a natural ordering (e.g., depression categorized as none, mild, moderate, and severe).
- A **nominal** variable is a categorical variable in which the levels do not have a natural ordering (e.g., marital status with levels married, living with partner, separated, divorced, never married, and widowed).
- A **count** variable takes on non-negative integer values (0, 1, 2, ...) and describes a number of items or events (e.g., days of hospitalization in a year).
- An **event time** variable describes the time from a time origin to an event (e.g., time from surgery to cancer remission).

Table 2.1 identifies the appropriate regression method for each outcome type, common statistical methods that are special cases of each regression method, and the predictor type associated with each special case. For example, "correlation" is a special case of simple linear regression with a continuous predictor.

TABLE 2.1 Taxonomy of regression methods

Outcome	Regression Method	Special Case	Type of Predictor Variable(s) for Special Case
Continuous	Simple linear	Correlation	Continuous
		Two-sample t-test	Binary
		One-way ANOVA	Categorical with three or more levels
	Multiple linear	Two-way ANOVA	Two categorical
		ANCOVA	One categorical, one continuous
Binary	Binary logistic	Chi-square test	Categorical
Ordinal	Ordinal logistic	Spearman's rho	Ordinal
		Kendall's tau-b	Ordinal
Nominal	Multinomial logistic	Chi-square test	Categorical
Count	Poisson		
Event time	Cox (survival analysis)	Mantel-Haenszel	Binary
		Log-rank test	Categorical

The above methods all assume each observation is independent of the others, as is the case, for example, with cross-sectional data that are a simple random sample from a population. If you have **repeated measures** on the same individuals (e.g., a cohort study with follow-up over time) or **clustered observations** (e.g., in a study of hospital patients from a group of hospitals, patients within a hospital are more similar to each other than are patients from different hospitals), then the independence assumption does not hold and other methods are needed. The following methods extend certain methods to handle dependent data.

- **Generalized least squares** and **linear mixed models** extend linear regression to handle dependent data.
- **Generalized estimating equations (GEE)** and **generalized linear mixed models** extend logistic and Poisson regression to handle dependent data.

A few other terms to be familiar with:

- **Simple** or **univariate** regression models have only one predictor. This is also referred to as **bivariate** regression (referring to the fact that there are two variables involved – one outcome and one predictor).
- **Multiple** or **multivariable** regression models have more than one predictor. The term **multivariate** regression is often used for this scenario, as well; however, it technically refers to the case of multiple *outcomes*.

The following are examples of research questions for which you could use each of the regression methods listed in Table 2.1.

- **Simple linear regression:** Is there an association between (the outcome) triglycerides and (the predictor) body mass index?

- **Multiple linear regression:** Is there an association between triglycerides and body mass index, adjusted for age, sex, and race/ethnicity?
- **Binary logistic regression:** What is the magnitude of association between obesity status (BMI < 30 kg/m^2 vs. ≥ 30 kg/m^2) and hours of weekly physical activity?
- **Ordinal logistic regression:** What is the magnitude of association between overweight/obesity category (1 = "normal", BMI < 25; 2 = "overweight", $25 \leq$ BMI < 30; 3 = "obese", BMI ≥ 30 kg/m^2) and hours of weekly physical activity?
- **Multinomial logistic regression:** Is there an association between the most-consumed non-alcoholic beverage (regular soda, diet soda, tea, coffee, energy drink, or water) in the month after an educational seminar on the topic of cardiovascular risk and the content of the seminar (standard content vs. standard content + experimental content to be evaluated for effectiveness)?
- **Poisson regression:** Is there a difference in mean length of stay (number of days from hospital admission to discharge) between those with different insurance types (private, Medicare, Medicaid, uninsured)? Is the number of cases of COVID-19 per 100,000 associated with the proportion of individuals in a county who report wearing a mask most of the time? Note: Another form of regression, negative binomial regression, may be more appropriate depending on the magnitude of spread in the data.
- **Cox proportional hazards regression:** Is there a difference in time from cancer remission to recurrence between patients given a new therapy and those given the standard of care?

Of these methods, this text covers simple (Chapter 4) and multiple (Chapter 5) linear regression, binary and ordinal logistic regression (Chapter 6), and Cox proportional hazards regression (survival analysis) (Chapter 7). However, before getting into the details of each regression method, the following chapter describes the necessary preliminary step of examining and summarizing the analysis variables (Chapter 3).

2.4 Exercises

1. What is a general term used for methods that assess the association between an outcome and one or more predictors?

a. Correlation
b. T-test
c. Regression
d. Analysis of variance

2. True or false? A simple linear regression model has only one predictor.

3. True or false? The outcome in a binary logistic regression model can take on only two possible values (e.g., presence or absence of a disease).

4. True or false? Regression methods can estimate the extent of effect modification by controlling for confounding.

5. Which of the following are valid purposes for regression?

a. Controlling for confounding
b. Estimating the magnitude of effect modification
c. Testing an association
d. All of the above

6. Match the outcome type with the appropriate definition.

Outcome Type	Definition
Nominal	Has a small number of possible values.
Binary	A categorical variable in which the levels have a natural ordering.
Continuous	A categorical variable in which the levels do not have a natural ordering.
Event time	Takes on non-negative integer values (0, 1, 2, ...) and describes a number of items or events.
Ordinal	Has a range of possible values and can take on any value in an interval.
Count	Describes the time from a time origin to an event.
Categorical	A categorical variable with exactly two levels.

7. Match the outcome type with the appropriate regression method.

Outcome Type	Regression Method
Nominal	Linear
Binary	Cox
Continuous	Poisson
Event time	Multinomial logistic
Ordinal	Binary logistic
Count	Ordinal logistic

For the following research questions, the wording assumes the outcome is mentioned first. For example, "Is (outcome) associated with (predictor)"? For the research question "Is the prevalence of Type 2 diabetes mellitus (T2DM) associated with median household income?", the outcome is prevalence of T2DM and the predictor is median income.

For each of the following research questions:

- What is the outcome type (continuous, binary, ordinal, nominal, count, or event time)?
- What is the appropriate regression method?

8. Is total cholesterol associated with weight?

9. Is time to ischemic heart disease associated with binge drinking? (Degerud et al., 2021)

10. Is seropositivity for COVID-19 (positive/negative) associated with occupation? (Alsuwaidi et al., 2021)

11. Is hospital length of stay in days following surgery associated with age?

12. Is body mass index category (underweight, normal, overweight, obese) associated with hours of vigorous activity per week?

3

Data Summarization

In this chapter, you will learn how to:

- Numerically and visually examine the distributions of continuous and categorical variables;
- Create a complete-case analysis dataset; and
- Summarize a dataset using a "Table 1" of descriptive statistics.

Some of the R programming code used in this chapter uses elements of the `tidyverse` library (Wickham, 2023), in particular the pipe operator `%>%` and functions such as `select()`. This chapter also uses the `gtsummary` library to create tables. Load these libraries before proceeding.

```
library(tidyverse)
library(gtsummary)
```

3.1 Examining the data

Before fitting a regression model, it is a good idea to examine each of the variables individually. One reason to do this is to look for anomalous values that may require you to look more closely at your data source. Additionally, a common element in the presentation of regression analysis results is a table containing descriptive statistics for each analysis variable.

How a variable is examined depends on whether it is continuous or categorical (as defined in Chapter 2).

- For continuous variables, create a numerical summary and plot a histogram using `summary()` and `hist()`.
- For categorical variables, create a frequency table and plot a bar chart using `table()` and `barplot()`.

NOTE: Throughout this text, we assume that categorical variables are coded in R as `factor` variables (for more information, see "Factors" in R for Data Science[1] (Wickham et al., 2023) or Chapter 3 in An Introduction to R for Research[2] (Nahhas, 2024)).

Example 3.1: Using data from a random subset of 1,000 adults from the NHANES 2017-2018 examination teaching dataset (see Appendix A.1), summarize the continuous variables systolic blood pressure (`sbp`) and age (`RIDAGEYR`) and the categorical variables gender (`RIAGENDR`) and annual household income category (`income`).

[1] https://r4ds.hadley.nz/factors
[2] https://bookdown.org/rwnahhas/IntroToR/managevars.html

First, load the NHANES examination teaching dataset using `load()`.

```
load("Data/nhanes1718_adult_exam_sub_rmph.Rdata")
# For convenience, give the dataset a shorter name
nhanes <- nhanes_adult_exam_sub
```

Next, use `summary()` to look at some basic descriptive statistics for the continuous variables. None of these values seem out of the ordinary, although note that there are 42 missing values for SBP (`NA's`).

```
# Continuous variables
summary(nhanes$sbp)
```

```
##    Min. 1st Qu.  Median    Mean 3rd Qu.    Max.    NA's
##      83     111     121     124     134     234      42
```

```
summary(nhanes$RIDAGEYR)
```

```
##    Min. 1st Qu.  Median    Mean 3rd Qu.    Max.
##    20.0    32.0    47.0    47.7    61.0    80.0
```

Use `hist()` to create a visualization of the entire distribution for each variable. As shown in Figure 3.1, SBP is a bit skewed to the right, which is typical for many health-related measures, and there are more younger than older individuals. There is a spike at the upper end of the age distribution because NHANES, for privacy reasons, reports ages $\geq$ 80 years as exactly 80 years (see the NHANES documentation for `RIDAGEYR`[3], accessed May 20, 2022).

```
par(mfrow=c(1,2))
hist(nhanes$sbp,
     main = "",
     xlab = "Systolic Blood Pressure (mmHg)")
hist(nhanes$RIDAGEYR,
     main = "",
     xlab = "Age (years)")
```

Computations of common descriptive statistics are demonstrated below. The option `na.rm = T` is needed if there are any missing values; otherwise, the functions will return `NA` (indicating "missing" or "unknown").

```
# Mean
mean(nhanes$sbp, na.rm = T)
```

```
## [1] 123.5
```

```
# Standard deviation
sd(nhanes$sbp, na.rm = T)
```

```
## [1] 17.65
```

[3]https://wwwn.cdc.gov/Nchs/Nhanes/2017-2018/DEMO_J.htm#RIDAGEYR

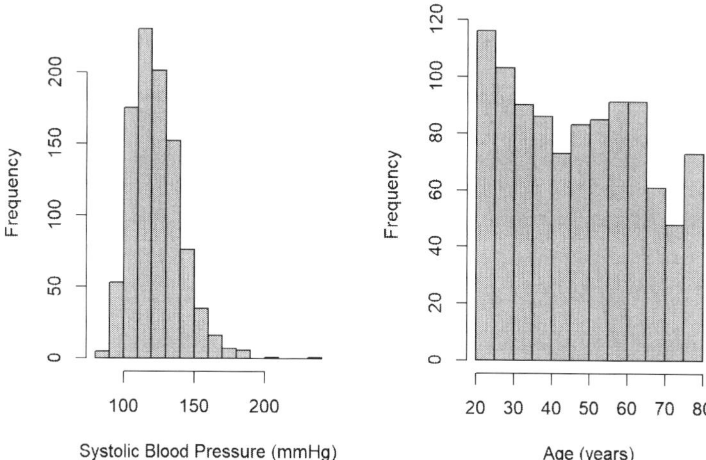

FIGURE 3.1 Histograms of continuous variables

```
# Median
median(nhanes$sbp, na.rm = T)
```

```
## [1] 121
```

```
# Interquartile range
IQR(nhanes$sbp, na.rm = T)
```

```
## [1] 23
```

```
# 25th and 75th percentile
# (sometimes also referred to as the IQR)
quantile(nhanes$sbp,
         probs = c(0.25, 0.75),
         na.rm = T)
```

```
## 25% 75%
## 111 134
```

```
# Minimum
min(nhanes$sbp, na.rm = T)
```

```
## [1] 83
```

```
# Maximum
max(nhanes$sbp, na.rm = T)
```

```
## [1] 234
```

```
# Number of non-missing values for a single variable
sum(complete.cases(nhanes$sbp))
```

```
## [1] 958
```

```
# Number of missing values for a single variable
sum(is.na(nhanes$sbp))
```

```
## [1] 42
```

For categorical variables, use `table()` and `prop.table()` to examine the frequency and proportion of observations at each **level** (each possible value of a categorical variable). The `useNA = "ifany"` option tells R to include the number of missing values in the frequency table. In `prop.table()`, `useNA = "ifany"` is omitted here, resulting in proportions of non-missing cases.

```
# Categorical variables
table(nhanes$income, useNA = "ifany")
```

```
##
##          <$25,000 $25,000 to <$55,000           $55,000+               <NA>
##               156                 254                480                110
```

```
prop.table(table(nhanes$income))
```

```
##
##          <$25,000 $25,000 to <$55,000           $55,000+
##            0.1753              0.2854             0.5393
```

When a variable has no missing values, the output of `table(, useNA = "ifany")` simply excludes the `<NA>` column. To force it to appear, use `table(, useNA = "always")`, as demonstrated below.

```
table(nhanes$RIAGENDR, useNA = "always")
```

```
##
##   Male Female   <NA>
##    482    518      0
```

```
prop.table(table(nhanes$RIAGENDR))
```

```
##
##   Male Female
##  0.482  0.518
```

In general, we would like to know if there are missing values. Therefore, omitting the `useNA` option altogether is not recommended, as that would result in exclusion of the `<NA>` column whether there are missing values or not. An exception is when using `prop.table()` to obtain the proportion of non-missing values.

The upper income group is most common and there are 110 individuals with missing income values. Missing income values are common in survey data, as some individuals are reluctant to disclose their income, even when the response options are ranges of values. Also, NHANES has gender response options "Male" and "Female" and, in this subset of the data, there are more females than males.

Options for visualizing the distribution of categorical variables include vertical and horizontal bar charts created with `barplot()`, as shown in Figure 3.2.

```
par(mfrow=c(1,2))
# barplot() expects frequencies, not the raw data, so use table inside barplot()
barplot(table(nhanes$RIAGENDR),
        ylab = "Frequency", xlab = "Gender")
barplot(table(nhanes$income), horiz=T, cex.names = 0.65,
        xlab = "Frequency", ylab = "Annual Household Income")
```

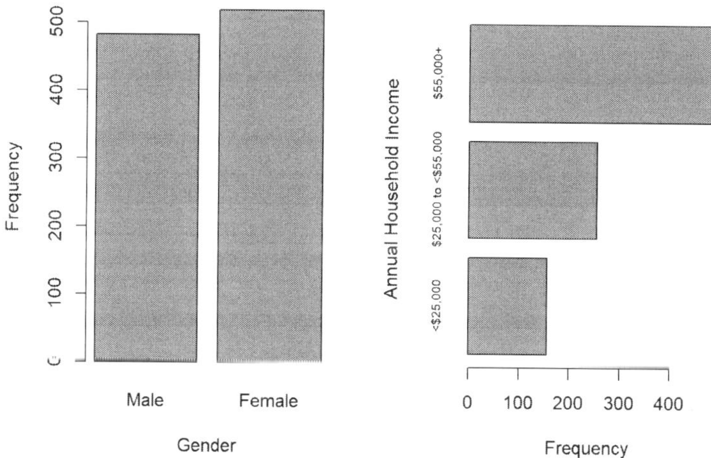

FIGURE 3.2 bar charts of categorical variable frequencies

Substituting `prop.table(table())` for `table()` results in plotting proportions instead of frequencies, as shown in Figure 3.3.

```
barplot(prop.table(table(nhanes$income)),
        ylab = "Proportion", xlab = "Annual Household Income")
```

3.1.1 Detailed description of all variables in a dataset

`summary()` can be used on multiple variables all at once, as can `complete.cases()`.

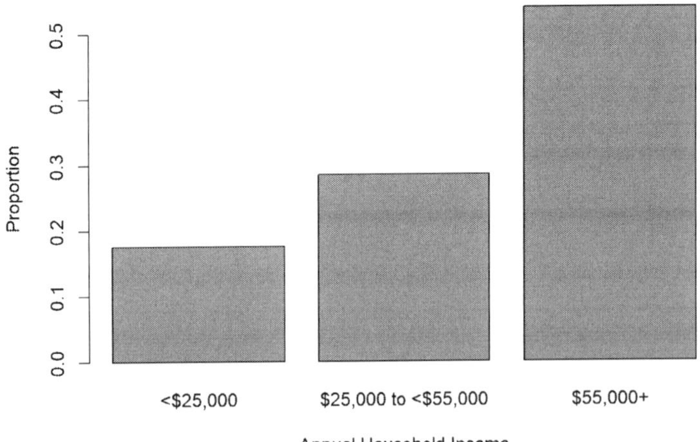

FIGURE 3.3 Bar chart of categorical variable proportions

```
# Summary of the data
summary(nhanes[, c("sbp", "RIDAGEYR", "RIAGENDR", "income")])
```

```
##       sbp          RIDAGEYR       RIAGENDR                    income
##  Min.   : 83   Min.   :20.0   Male  :482   <$25,000            :156
##  1st Qu.:111   1st Qu.:32.0   Female:518   $25,000 to <$55,000:254
##  Median :121   Median :47.0                $55,000+            :480
##  Mean   :124   Mean   :47.7                NA's                :110
##  3rd Qu.:134   3rd Qu.:61.0
##  Max.   :234   Max.   :80.0
##  NA's   :42
```

```
# Number of cases
nrow(nhanes[, c("sbp", "RIDAGEYR", "RIAGENDR", "income")])
```

```
## [1] 1000
```

```
# Number of cases with no missing data
sum(complete.cases(nhanes[, c("sbp", "RIDAGEYR", "RIAGENDR", "income")]))
```

```
## [1] 855
```

```
# Number of cases with any missing data
sum(!complete.cases(nhanes[, c("sbp", "RIDAGEYR", "RIAGENDR", "income")]))
```

```
## [1] 145
```

The `describe()` function in the `Hmisc` library (Harrell, 2023) also summarizes multiple variables at once and, additionally, provides more detail than `summary()`.

```
# Access the describe function without loading
# the entire Hmisc library using the :: syntax
nhanes %>%
  select(sbp, RIDAGEYR, RIAGENDR, income) %>%
  Hmisc::describe()
# (results not shown)
```

If desired, use `write()` to export the results to an external file.

```
write(Hmisc::html(Hmisc::describe(nhanes)), "nhanes_description.html")
```

3.2 Missing data options

When examining the data, you will often find there are variables with missing values. Standard regression analysis algorithms require all values to be non-missing, so the default option of regression software, when faced with data that contain missing outcome and/or predictor values, is to remove any case that has any missing values, resulting in what is called a complete case analysis. A more rigorous approach is to use the method of multiple imputation to randomly fill in missing values. Complete case analysis is easier, but "easier" does not mean "better". For the purpose of learning regression methods it does, however, make sense to stick to the easier missing data method first and then later, after becoming familiar with regression, learn the more rigorous method.

3.2.1 Complete case analysis

A **complete case analysis** is one in which all cases with missing data on any variable are removed from the analysis. In R, complete case analysis is the default when using a regression function – that is, R automatically removes cases with any missing data before fitting the regression model. However, if you want to first describe the data used in a complete case analysis, you need to explicitly remove the cases with missing values yourself in order to compute descriptive statistics that are based on the same sample as your regression analysis.

If you instead use a dataset that contains missing values to compute descriptive statistics for each variable, the results for a specific variable will be based on all available cases with non-missing values for that variable regardless of whether any other variables have missing values for those cases. Thus, descriptive statistics for different variables might be based on different subsets of cases. To ensure that the descriptive statistics for all variables are based on the same sample as each other, and the same sample as will be used when fitting the regression model, remove cases that have missing data for *any* variable in the analysis.

Example 3.1 (continued): For the same data and variables summarized previously, the following code demonstrates two methods for creating a complete case analysis dataset. Both methods result in the correct cases, but the first retains all the variables in the dataset, including those not being used in the analysis. The second method only retains the analysis variables.

```
# Remove cases with missing data and keep all variables
SUB <- complete.cases(nhanes[, c("sbp", "RIDAGEYR", "RIAGENDR", "income")])
complete.dat <- subset(nhanes, subset = SUB)

# Number of cases and variables in the original dataset
dim(nhanes)
```

```
## [1] 1000   85
```

```
# Number of cases and variables in the complete case dataset
dim(complete.dat)
```

```
## [1] 855  85
```

```
# Alternative that does the same thing
complete.dat <- nhanes %>%
  drop_na(sbp, RIDAGEYR, RIAGENDR, income)

# Number of cases and variables in the complete case dataset
dim(complete.dat)
```

```
## [1] 855  85
```

```
# Remove cases with missing data and only keep variables of interest
complete.dat <- nhanes %>%
  select(sbp, RIDAGEYR, RIAGENDR, income) %>%
  drop_na()

# Number of cases and variables in the complete case dataset
dim(complete.dat)
```

```
## [1] 855   4
```

```
# Summary of the complete data
summary(complete.dat)
```

```
##       sbp          RIDAGEYR       RIAGENDR                          income
##  Min.   : 89   Min.   :20.0   Male  :425   <$25,000              :148
##  1st Qu.:111   1st Qu.:32.0   Female:430   $25,000 to <$55,000:248
##  Median :121   Median :48.0                $55,000+              :459
##  Mean   :124   Mean   :48.1
##  3rd Qu.:134   3rd Qu.:62.0
##  Max.   :234   Max.   :80.0
```

3.2.2 Multiple imputation

Multiple imputation (MI) of missing data is a method of filling in missing values based on the observed associations between the analysis variables. The filling in (imputation) is done randomly and multiple times to avoid pretending the missing values are known with certainty. Following MI, fit the regression model multiple times, once using each imputed dataset, and combine the results according to certain mathematical rules (Rubin, 1987).

In the example we have been using, the original dataset has 1000 cases, of which only 855 have complete data on all the variables of interest. A complete case analysis uses just those 855 cases. An MI analysis, however, uses all 1000 cases. However, due to the missing data, we do not *actually* have a sample size of 1000 complete cases. The mathematical rules used to combine the results appropriately account for the fact that we have more information than just the 855 complete cases, but less information than a dataset with 1000 complete cases. Multiple imputation will be discussed in more detail in Chapter 9.

3.3 Creating a "Table 1"

In most published articles, there is a "Table 1" containing descriptive statistics for the sample. This may include, for example, the mean and standard deviation for continuous variables, the frequency and proportion for categorical variables, and perhaps also the number of missing values.

The brute force method of creating such a table would be to compute each statistic for each variable of interest and then copy and paste the results into a table. Having done this even once, you will wish for an easier method! There are many possible solutions, but one that is quick and easy to use is demonstrated here – the `gtsummary` package (Sjoberg et al., 2021, 2023). Many examples can be found on the `gtsummary` package GitHub site[4], in tbl_summary() tutorial[5], and in Table Gallery[6].

Other R packages not covered here that facilitate table creation include `flextable` (Gohel and Skintzos, 2023), `tableone` (Yoshida and Bartel, 2022), and `table1` (Rich, 2023).

3.3.1 Overall

Example 3.1 (continued): Create a table of summary statistics for all the variables we have been summarizing in this chapter for the entire sample ("overall").

The code below loads the `gtsummary` library and uses `tbl_summary()` with default settings to generate Table 3.1.

The default settings produce a table with the frequency and proportion for categorical variables, the median and interquartile range (IQR) for continuous variables (here, the 25th and 75th percentiles, not their difference), and the number of missing values (if any) (show in rows headed by "Unknown").

```
nhanes %>%
  # Select the variables to be included in the table
  select(sbp, RIDAGEYR, RIAGENDR, income) %>%
  tbl_summary()
```

[4]https://github.com/ddsjoberg/gtsummary
[5]https://www.danieldsjoberg.com/gtsummary/articles/tbl_summary.html
[6]https://www.danieldsjoberg.com/gtsummary/articles/gallery.html

TABLE 3.1 Table 1 using tbl_summary() with default settings

Characteristic	**N = 1,000**
sbp	121 (111, 134)
Unknown	42
RIDAGEYR	47 (32, 61)
RIAGENDR	
Male	482 (48%)
Female	518 (52%)
income	
<$25,000	156 (18%)
$25,000 to <$55,000	254 (29%)
$55,000+	480 (54%)
Unknown	110

[1] Median (IQR); n (%)

The default is to include missing values, and this can be overridden by setting the `missing` option to "no", as in the code below. The resulting table is not shown but is identical to the table above other than the "Unknown" rows are omitted. In particular, it is still based on the full sample size (1000 observations).

```
nhanes %>%
  # Select the variables to be included in the table
  select(sbp, RIDAGEYR, RIAGENDR, income) %>%
  tbl_summary(
    missing = "no"
  )
```

To create a table for a complete case analysis sample, start with a complete case analysis dataset, as in the code below (which uses `complete.dat`, the dataset created in Section 3.2.1). Additional options are demonstrated below. See `?gtsummary::tbl_summary` for more options.

- `statistic`: The default for continuous variables is the median and IQR. The default for categorical variables is the frequency and proportion. Below, this option is used to instead compute the mean and standard deviation for continuous variables (and the default for categorical variables is coded explicitly).
- `digits`: `tbl_summary()` guesses the number of digits to which to round. Use this option to set the number yourself. Since two statistics are specified in the `statistic` option, two numbers are specified here, one for each statistic.
- `type`: `tbl_summary()` guesses whether a variable is continuous or categorical based on its distribution. Use this option to specify the types yourself, in particular if you need to override the default.
- `label`: The default row labels are the variable names or labels (if the dataset has been labeled, for example, using the `Hmisc` library `label()` function). Use this option to change the row headers.
- `modify_header`: The default column header is "Characteristic". Use this option to change the column header. Surrounding text by `**` results in a bold font.
- `modify_caption`: Use this option to add a table caption.
- `bold_labels`: Use this option to display the row labels in a bold font.

The results are shown in Table 3.2.

```
complete.dat %>%
  select(sbp, RIDAGEYR, RIAGENDR, income) %>%
  tbl_summary(
    statistic = list(all_continuous()  ~ "{mean} ({sd})",
                     all_categorical() ~ "{n}    ({p}%)"),
    digits = list(all_continuous()  ~ c(2, 2),
                  all_categorical() ~ c(0, 1)),
    type = list(sbp      ~ "continuous",
                RIDAGEYR ~ "continuous",
                RIAGENDR ~ "categorical",
                income   ~ "categorical"),
    label  = list(sbp      ~ "SBP (mmHg)",
                  RIDAGEYR ~ "Age (years)",
                  RIAGENDR ~ "Gender",
                  income   ~ "Annual Income")
  ) %>%
  modify_header(label = "**Variable**") %>%
  modify_caption("Participant characteristics  (complete case analysis)") %>%
  bold_labels()
```

TABLE 3.2 Participant characteristics (complete case analysis)

Variable	**N = 855**
SBP (mmHg)	123.57 (17.57)
Age (years)	48.11 (17.43)
Gender	
Male	425 (49.7%)
Female	430 (50.3%)
Annual Income	
<$25,000	148 (17.3%)
$25,000 to <$55,000	248 (29.0%)
$55,000+	459 (53.7%)

[1] Mean (SD); n (%)

NOTE: If a variable is a factor with exactly two levels labeled "Yes" and "No", then `tbl_summary()` by default will only include the row corresponding to "Yes". The same applies with variables that have values TRUE/FALSE or 1/0. Use the `value` option to change the row displayed (see `?gtsummary::tbl_summary` for details). Alternatively, set `type` to "categorical" to display both rows.

3.3.2 By outcome or exposure

In many published research articles, descriptive statistics are presented not only "overall" (over the entire sample) but also by the outcome or exposure. If the "by" variable is continuous then, for the purpose of the descriptive table only, create a categorical version with as many levels as you would like "by" columns, where each level corresponds to a range of values. A common method, demonstrated here, is to use a **median split** in which a binary variable is created based on whether the value of the continuous variable is below or at least as large as the median value. This results in a "by" variable with two levels and approximately equal sample sizes in each level.

To create a table of descriptive statistics by outcome or exposure, use the `by` argument. To also include a column with the "overall" summaries, use `add_overall()`. To stratify by more than one variable, use `tbl_strata()` (see `?gtsummary::tbl_strata` for more information).

Categorical outcome or exposure

Example 3.1 (continued): Create a table of summary statistics overall and by gender.

The following code produces Table 3.3 displaying descriptive statistics by gender.

NOTES:

- The `all_stat_cols()` option in `modify_header()` adds the frequency and proportion of the "by" variable in the column header.
- The "by" variable (`RIAGENDR`) was omitted from the `type` and `label` options since leaving it in results in an error.
- The code below illustrates how to assign the table to an object (`TABLE1`), and then view it by typing the name of the object. This is not actually necessary for this example, but will facilitate exporting the table to an external file (see Section 3.3.3).

```
TABLE1 <- complete.dat %>%
  select(sbp, RIDAGEYR, RIAGENDR, income) %>%
  tbl_summary(
    # The "by" variable
    by = RIAGENDR,
    statistic = list(all_continuous()  ~ "{mean} ({sd})",
                     all_categorical() ~ "{n}    ({p}%)"),
    digits = list(all_continuous()  ~ c(2, 2),
                  all_categorical() ~ c(0, 1)),
    type = list(sbp      ~ "continuous",
                RIDAGEYR ~ "continuous",
                income   ~ "categorical"),
    label  = list(sbp      ~ "SBP (mmHg)",
                  RIDAGEYR ~ "Age (years)",
                  income   ~ "Annual Income")
  ) %>%
  modify_header(
    label = "**Variable**",
    # The following adds the % to the column total label
    # <br> is the location of a line break
    all_stat_cols() ~ "**{level}**<br>N = {n} ({style_percent(p, digits=1)}%)"
  ) %>%
  modify_caption("Participant characteristics, by gender") %>%
  bold_labels() %>%
  # Include an "overall" column
  add_overall(
    last = FALSE,
    # The ** make it bold
    col_label = "**All Participants**<br>N = {N}"
  )
```

```
TABLE1
```

TABLE 3.3 Participant characteristics, by gender

Variable	All Participants N = 855	Male N = 425 (49.7%)	Female N = 430 (50.3%)
SBP (mmHg)	123.57 (17.57)	124.55 (16.01)	122.60 (18.96)
Age (years)	48.11 (17.43)	47.26 (17.38)	48.94 (17.45)
Annual Income			
<$25,000	148 (17.3%)	65 (15.3%)	83 (19.3%)
$25,000 to <$55,000	248 (29.0%)	127 (29.9%)	121 (28.1%)
$55,000+	459 (53.7%)	233 (54.8%)	226 (52.6%)

[1] Mean (SD); n (%)

Median split for a continuous outcome or exposure

Example 3.1 (continued): Create a table of summary statistics overall and by SBP, using a median split to create two SBP groups.

The code below creates a dichotomous version of `sbp` based on a median split and then uses this new variable as the `by` variable to produce Table 3.4.

```
MEDIAN <- median(complete.dat$sbp)
LABEL0 <- paste("SBP <", MEDIAN,  sep = "")
LABEL1 <- paste("SBP >=", MEDIAN, sep = "")

complete.dat <- complete.dat %>%
  mutate(sbp_median_split = as.numeric(sbp >= MEDIAN),
         sbp_median_split = factor(sbp_median_split,
                                   levels = 0:1,
                                   labels = c(LABEL0, LABEL1)))
# Checking derivation
tapply(complete.dat$sbp, complete.dat$sbp_median_split, range)
```

```
## $`SBP <121`
## [1]  89 120
##
## $`SBP >=121`
## [1] 121 234
```

```
# Create table
TABLE1 <- complete.dat %>%
  # Select the median split variable, not the original variable
  select(sbp_median_split, RIDAGEYR, RIAGENDR, income) %>%
  tbl_summary(
    # Use the median split variable as the "by" variable
    by = sbp_median_split,
    statistic = list(all_continuous()  ~ "{mean} ({sd})",
                     all_categorical() ~ "{n}    ({p}%)"),
    digits = list(all_continuous()  ~ c(2, 2),
                  all_categorical() ~ c(0, 1)),
    type = list(RIDAGEYR ~ "continuous",
                RIAGENDR ~ "categorical",
                income   ~ "categorical"),
    label  = list(RIDAGEYR ~ "Age (years)",
                  RIAGENDR ~ "Gender",
                  income   ~ "Annual Income")
```

```
  ) %>%
  modify_header(
    label = "**Variable**",
    all_stat_cols() ~ "**{level}**<br>N = {n} ({style_percent(p, digits=1)}%)"
  ) %>%
  modify_caption("Participant characteristics, by SBP") %>%
  bold_labels() %>%
  add_overall(last = FALSE,
              col_label = "**All Participants**<br>N = {N}")
```

```
TABLE1
```

TABLE 3.4 Participant characteristics, by SBP

Variable	All Participants N = 855	SBP <121 N = 412 (48.2%)	SBP >=121 N = 443 (51.8%)
Age (years)	48.11 (17.43)	41.57 (15.84)	54.19 (16.63)
Gender			
Male	425 (49.7%)	180 (43.7%)	245 (55.3%)
Female	430 (50.3%)	232 (56.3%)	198 (44.7%)
Annual Income			
<$25,000	148 (17.3%)	67 (16.3%)	81 (18.3%)
$25,000 to <$55,000	248 (29.0%)	120 (29.1%)	128 (28.9%)
$55,000+	459 (53.7%)	225 (54.6%)	234 (52.8%)

[1] Mean (SD); n (%)

3.3.3 Exporting to an external file

To export a `gtsummary` table to a Microsoft Word or HTML file, use the following syntax which starts with the `tbl_summary` object (called `TABLE1` above) and then uses the `flextable` (Gohel and Skintzos, 2023) or `gt` (Iannone et al., 2023) package to do the exporting.

```
# Make sure these are installed:
# install.packages(c("Rcpp", "gtsummary", "flextable", "gt"))

TABLE1 %>%
  as_flex_table() %>%
  flextable::save_as_docx(path = "MyTable1.docx")

TABLE1 %>%
  as_gt() %>%
  gt::gtsave(filename = "MyTable1.html")
```

3.3.4 Adding p-values to Table 1

Often, in a published research article, a Table 1 that displays descriptive statistics by the outcome or an exposure includes p-values that test, for each variable, the null hypothesis that the variable has the same mean (or median or proportion) across all groups in the population. P-values can be easily added to a `tbl_summary` table using `add_p` (see `?gtsummary::add_p.tbl_summary` for all the options, including the various statistical tests available).

Example 3.1 (continued): Create a table of descriptive statistics, by SBP, including t-tests for continuous variables and chi-square tests for categorical variables.

The code below loads the `gtsummary` library and uses `tbl_summary()` with default settings to generate Table 3.5.

```
complete.dat %>%
  select(sbp_median_split, RIDAGEYR, RIAGENDR, income) %>%
  tbl_summary(
    by = sbp_median_split,
    statistic = list(all_continuous()   ~ "{mean} ({sd})",
                     all_categorical() ~ "{n}    ({p}%)"),
    digits = list(all_continuous()  ~ c(2, 2),
                  all_categorical() ~ c(0, 1)),
    label  = list(RIDAGEYR ~ "Age (years)",
                  RIAGENDR ~ "Gender",
                  income   ~ "Annual Income")
  ) %>%
  add_p(
    test = list(all_continuous()  ~ "t.test",
                all_categorical() ~ "chisq.test"),
    pvalue_fun = function(x) style_pvalue(x, digits = 3)
  )
```

TABLE 3.5 Participant characteristics, by SBP, including p-values

Characteristic	**SBP <121**, N = 412	**SBP >=121**, N = 443	**p-value**
Age (years)	41.57 (15.84)	54.19 (16.63)	<0.001
Gender			<0.001
Male	180 (43.7%)	245 (55.3%)	
Female	232 (56.3%)	198 (44.7%)	
Annual Income			0.728
<$25,000	67 (16.3%)	81 (18.3%)	
$25,000 to <$55,000	120 (29.1%)	128 (28.9%)	
$55,000+	225 (54.6%)	234 (52.8%)	

[1] Mean (SD); n (%)

[2] Welch Two Sample t-test; Pearson's Chi-squared test

3.3.5 Should p-values be added to a Table 1?

It is very easy to add p-values to a Table 1, but are they recommended? In general, no, regardless of whether the data arose from an observational study (Vandenbroucke et al., 2007) or a randomized trial (Moher et al., 2010).

If displaying sample characteristics by the outcome for the purpose of providing crude (unadjusted) tests of association between a set of predictors and the outcome, then including p-values in Table 1 does make some sense. However, typically the subsequent regression analysis will provide adjusted tests of association, and conclusions will be drawn from that adjusted analysis, so unadjusted tests may not be relevant.

A potential reason for displaying p-values in a Table 1 of participant characteristics by the primary exposure of interest is to attempt to demonstrate the extent to which the characteristics differ between exposure groups and may therefore confound the outcome-exposure relationship. But what matters for confounding is not if the groups differ in the population (which is what the p-values are testing) but how much they differ in *this sample*. P-values, in this context, are not relevant; what matters are the magnitudes of differences in characteristics between the exposure groups, as well as the magnitude of association between characteristics and the outcome (Vandenbroucke et al., 2007).

Yes, p-values are related to the magnitude of differences, but they are very dependent on the sample size, as well. In a small sample, even a meaningfully large difference might not lead to a small p-value, resulting in an incorrect conclusion of "no confounding". Conversely, in a large sample, even a small, non-meaningful, difference might result in small p-value, resulting in an incorrect conclusion of "confounding". Yet another error can occur in the case of a small, seemingly non-meaningful, difference that is not statistically significant but which is for a predictor that is very strongly associated with the outcome. The p-value would lead to a conclusion of "no confounding" yet even a small difference between exposure groups in a predictor strongly associated with the outcome can result in meaningful confounding (Dales and Ury, 1978; Vandenbroucke et al., 2007). Even worse, if the exposure groups were determined using randomization (e.g., a randomized clinical trial) then we already know the null hypothesis is true so p-values are irrelevant (Moher et al., 2010; Altman, 1985; Senn, 1994). Under randomization, any differences observed between groups in the sample are entirely due to randomness, not to any underlying difference between the groups.

Thus, using p-values to provide evidence for or against confounding can be misleading or even nonsensical. In a confirmatory analysis (see Section 5.23), potential confounders are identified using subject-matter knowledge based on prior research and included in a regression model regardless of their observed associations.

In summary, including p-values in a Table 1 is easy to do but may not be relevant and can, at times, be misleading or meaningless.

3.4 Exercises

1. True or false? A bar chart is an appropriate visualization for a continuous variable.

2. True or false? An appropriate numerical summary for a categorical variable is a frequency table.

3. For descriptive statistical functions such as `mean()` and `sd()`, if R returns a missing value `NA`, what option can you use to return a non-missing value (assuming the variable you are describing has some non-missing values)?

4. What is the default method of handling missing data when using a regression function in R?

5. Give a reason for removing cases with missing values before doing a regression analysis.

6. Using the COVID dataset (`covid_20210908_rmph.rData`, see Appendix A.4), numerically and visually examine the continuous variable `hospitals_per_100k` (number of hospitals per 100,000 persons) and the categorical variable `CensusRegionName`.

7. Using the United Nations Human Development Data (`unhdd2020.rmph.rData`, see Appendix A.2), create a "Table 1" of descriptive statistics (mean and standard deviation for continuous variables, frequency and proportion for categorical variables), overall and by Human Development Index group (`hdi_group`). Use a complete case analysis that includes the following variables:

- `hdi`: Human Development Index (HDI)
- `life`: Life expectancy at birth (years)
- `educ_expected`: Expected years of schooling (years)
- `gii`: Gender Inequality Index
- `urban`: Urban population (%)

8. Using the COVID dataset (`covid_20210908_rmph.rData`, see Appendix A.4), create a "Table 1" of descriptive statistics, overall and by the number of hospitals per 100,000 persons (`hospitals_per_100k`) (use a median split to create a binary version of this "by" variable). Describe the following variables in this table:

- `pop.usafacts`: County population
- `cases.usafacts.20210908`: COVID-19 cumulative cases as of 2021-09-08
- `deaths.usafacts.20210908`: COVID-19 cumulative deaths as of 2021-09-08
- `MedianAge2010`: Median age of county in 2010
- `CensusRegionName`: Name of census region

Hint: Remove the `statistic` option in `tbl_summary()` to display the default statistics (median and interquartile range) which are more appropriate for data that are skewed, such as county population. This is equivalent to replacing `"{mean} ({sd})"` with `"{median} ({p25}, {p75})"`.

4

Simple Linear Regression

In this chapter, you will learn how to:

- Write and interpret a simple linear regression equation;
- Fit and visualize a simple linear regression model;
- Interpret the estimated regression coefficients for a model with a continuous predictor;
- Interpret the estimated regression coefficients for a model with a categorical predictor;
- Test the significance of a single regression coefficient;
- Use a multiple degree of freedom test to test the significance of multiple coefficients simultaneously;
- Compute predictions from the model;
- Compute confidence intervals for the regression coefficients;
- Compute and visualize confidence intervals for the mean outcome;
- Compute and visualize prediction intervals for individual observations; and
- Fit and visualize a curve using a polynomial function.

Some of the R programming code used in this chapter uses elements of the `tidyverse` library (Wickham, 2023), in particular the pipe operator `%>%` and functions such as `mutate()`. Load this library before proceeding.

```
library(tidyverse)
```

4.1 Introduction

In general, a **regression** analysis estimates the association between an **outcome** (Y) and one or more **predictors** ($X_1, X_2, \ldots, X_K$). **Linear regression** is used when Y is continuous. The term **simple linear regression (SLR)** is typically used for the setting where there is only one predictor (X) and that predictor is continuous. In this text, the term SLR refers to *any* linear regression with just one predictor, whether that predictor is continuous, categorical, or modeled as a curve (e.g., a quadratic polynomial). Chapter 5 extends SLR to the case of multiple predictors (multiple linear regression, MLR). This chapter is not meant to be an exhaustive treatment of SLR. Rather, it is meant to give just enough of an introduction to the concepts of linear regression to set the stage for MLR.

Two examples are used throughout this chapter to demonstrate the use of R for SLR. The remainder of this introduction presents some basic analyses to set the stage, along with some example R code for visualizing the relationship between the outcome and predictor.

Example 4.1: Levels of fasting glucose, measured using a blood draw following a fast, are one criteria used to diagnose Type 2 diabetes. An individual is classified as "normal", "pre-diabetic", or "diabetic" if their fasting glucose is <5.6 mmol/L, 5.6–6.9 mmol/L, or 7 mmol/L or greater on two separate tests, respectively (https://www.mayoclinic.org/diseases-conditions/type-2-diabetes/diagnosis-treatment/drc-20351199, Accessed September 17, 2021). The code below investigates the distribution of fasting glucose (mmol/L) using data from adults age $\geq$ 20 years in NHANES 2017-2018. The teaching dataset `nhanes1718_adult_fast_sub_rmph.Rdata` contains a random subsample of 1,000 adults who had blood drawn after fasting (see Appendix A.1). After investigating the distribution, investigate how the mean fasting glucose varies with waist circumference (a continuous predictor).

After loading the data, use a histogram to examine the distribution of fasting glucose (`LBDGLUSI`) and add a vertical line to visualize the location of the average value (Figure 4.1).

```
load("Data/nhanes1718_adult_fast_sub_rmph.Rdata")
# For convenience, give the dataset a shorter name
nhanesf <- nhanes_adult_fast_sub
```

```
hist(nhanesf$LBDGLUSI, xlab = "Fasting Glucose (mmol/L)", main =  "", breaks = 20)
MEAN <- mean(nhanesf$LBDGLUSI, na.rm = T)
abline(v = MEAN, lwd = 3, col = "red")
```

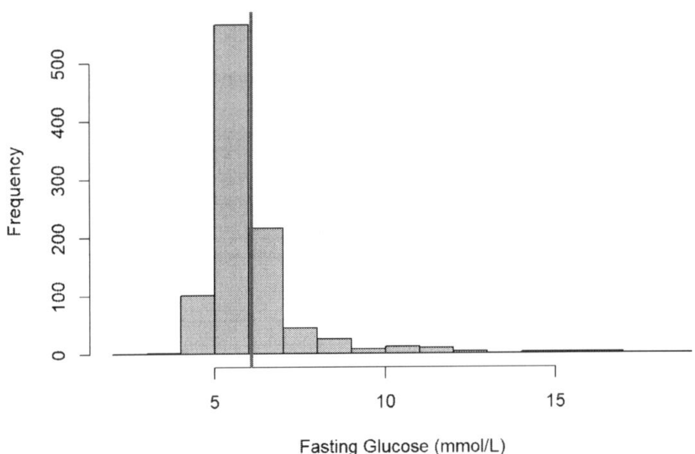

FIGURE 4.1 Histogram of fasting glucose (vertical bar indicates the mean value)

The notation $E(Y)$ refers to the "expected value" or mean of a random variable Y. In this sample, the average fasting glucose (FG) is 6.1 mmol/L; thus, $E(FG) = 6.1$. This is the average value, but the histogram demonstrates that there is a lot of variation between individuals. Is there some other variable that might explain some of that variation? We might hypothesize that groups of people with different amounts of body fat have different average values of fasting glucose.

Below, we examine how fasting glucose is associated with body fat, using waist circumference (`BMXWAIST`) as our measure of adiposity. If you are *only* interested in a measure of association, then estimating the correlation would be enough, and we could compute that the correlation between fasting glucose and waist circumference is .295.

```
cor(nhanesf$LBDGLUSI, nhanesf$BMXWAIST, use = "complete.obs")
```

```
## [1] 0.2952
```

But SLR can do more than just estimate a measure of association; it can find the **best fitting line** describing the linear relationship between fasting glucose and waist circumference, as shown in Figure 4.2.

```
plot(LBDGLUSI ~ BMXWAIST, data = nhanesf,
     col = "gray",
     ylab = "Fasting Glucose (mmol/L)",
     xlab = "Waist Circumference (cm)",
     las = 1,  # Rotate the axis labels so they are all horizontal
     pch = 20, # Change the plotting symbol to solid circles
     font.lab = 2, font.axis = 2) # Bold font
# The lm() function fits a linear regression (discussed more later)
abline(lm(LBDGLUSI ~ BMXWAIST, data = nhanesf), lwd = 2, col = "red")
```

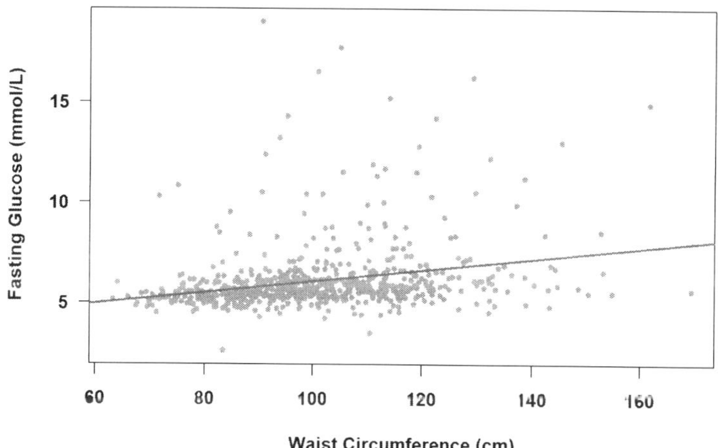

FIGURE 4.2 Simple linear regression of fasting glucose on waist circumference

With just the mean all we had was $E(FG)$; now we have $E(FG|WC)$, which is read as "the expected value of fasting glucose given waist circumference". The line in Figure 4.2 shows us how fasting glucose differs, on average, between individuals with different waist circumference, which is much more informative than just the single, overall, mean value.

In Example 4.1, the predictor variable is continuous. When the predictor is categorical, taking on only a few values, then rather than fit a straight line SLR estimates the mean outcome at each discrete level of the predictor.

Example 4.2: How does average fasting glucose differ between never, past, and current smokers (`smoker`)?

```
# NOTE: plot.default is used because plot results in boxplots
#       instead of the actual data points.
#       plot.default does not have a data argument so we must
#       extract y and x using the $ notation.
plot.default(nhanesf$LBDGLUSI ~ nhanesf$smoker,
             col = "gray",
             ylab = "Fasting Glucose (mmol/L)",
             xlab = "Smoking Status",
             las = 1, pch = 20, font.lab = 2, font.axis = 2,
             xaxt = "n") # Suppresses the x-axis
# Add custom x-axis
axis(1, at = 1:3, labels = levels(nhanesf$smoker), font = 2)
# Compute means
MEANS <- tapply(nhanesf$LBDGLUSI, nhanesf$smoker, mean, na.rm = T)
# Add means to the plot
# NOTE: For points() and lines() the arguments are x, y not y ~ x
points(1:3, MEANS, pch = 20, cex = 3)
lines( 1:3, MEANS, col = "red", lwd=2)
```

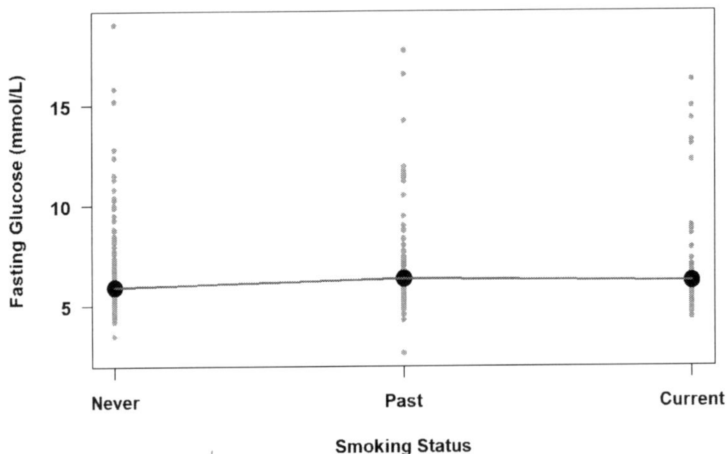

FIGURE 4.3 Simple linear regression of fasting glucose on smoking status

The line connecting the mean fasting glucose at different levels of smoking status in Figure 4.3 is not estimated by SLR; it is just added here to demonstrate that with a categorical predictor the means are not constrained to fall on a straight line.

NOTE: This is not a book about R programming, but rather how to use R to do regression analyses. If you are still relatively new to R and want to learn more about how a code chunk presented in this text works, then play around with changing different options or investigating pieces of the code. For example, change the value of `font.lab` in the previous code chunk and see what changes in the plot; or type `MEANS` in the R console to see what output `tapply()` produces; or type `?tapply` to learn more about that function.

4.2 Notation and interpretation

While we are going to try in this text to not get into too much mathematical detail, we do need some notation to help keep everything straight. If you have never seen this sort of notation before, do not worry if it is a lot to digest at first. Over time, the notation will become more familiar. The initial descriptions that follow are dense, but we will unpack the details as we go.

The data are n independent pairs of observed values of the outcome Y and predictor X. The data for the ith case (or individual, or observation) are denoted (y_i, x_i) – each case has an associated outcome value and predictor value. Equation 4.1 describes the simple linear regression model for Y with a **continuous** predictor X.

$$Y = \beta_0 + \beta_1 X + \epsilon \tag{4.1}$$

where β_0 and β_1 are the intercept and slope, respectively, of the best fit line, and ϵ is the **error** term, which is *assumed* to have a normal distribution with the same variance at all values of X. We denote the assumption about the error term using the notation $\epsilon \sim N(0, \sigma^2)$. The model described by Equation 4.1 *assumes* that the relationship between Y and X is linear and the error term captures random variation between individuals and random measurement error.

For a **continuous predictor**, the **intercept** (β_0) is $E(Y|X = 0)$, the mean (average) outcome when the predictor is zero; and the **slope** (β_1) is the difference in mean outcome associated with a one-unit difference in X.

For a **categorical predictor**, however, the notation is very different. Equation 4.2 describes the simple linear regression model for Y with a **categorical** predictor X with L levels.

$$Y = \beta_0 + \beta_1 I(X = 2) + \beta_2 I(X = 3) + ... + \beta_{L-1} I(X = L) + \epsilon \tag{4.2}$$

where the **indicator function** I(logical condition) equals 1 when the logical condition is true and 0 when it is false. For example, $I(X = 2) = 1$ when $X = 2$ and 0 when $X \neq 2$. One of the levels ($X = 1$) is left out of Equation 4.2 and is referred to as the **reference level.**

Suppose X has $L = 3$ levels. If we plug each of 1, 2, and 3 in for X in Equation 4.2 we get the following three equations:

$$\begin{aligned} E(Y|X = 1) &= \beta_0 \\ E(Y|X = 2) &= \beta_0 + \beta_1 \\ E(Y|X = 3) &= \beta_0 + \beta_2 \end{aligned}$$

Solving these equations for the βs results in:

$$\begin{aligned} \beta_0 &= E(Y|X = 1) \\ \beta_1 &= E(Y|X = 2) - E(Y|X = 1) \\ \beta_2 &= E(Y|X = 3) - E(Y|X = 1) \end{aligned}$$

So, for a categorical predictor, the individual β terms do *not* correspond to the intercept and slope of a line. Instead, β_0 corresponds to the mean outcome at the reference level, and the other βs correspond to differences in the mean outcome between other levels and the reference level.

The interpretations of regression coefficients, and the distinctions in their interpretations between continuous and categorical predictors, are very important to keep in mind when fitting and interpreting regression models. In summary:

- β_0 is the mean outcome when the predictor is 0 (if the predictor is continuous) or at its reference level (if the predictor is categorical).
- Each other β is the difference in mean outcome between two groups of individuals. If the predictor X is continuous, β_1 is the difference in mean outcome between groups that differ by one-unit in X. If the predictor X is categorical, each β other than the intercept is the difference in mean outcome between those at one of the non-reference levels of X and those at the reference level.

The following examples help to make these concepts more clear.

Example 4.1 (continued): For the regression of fasting glucose on waist circumference (a continuous predictor), the regression equation is written as follows.

$$\text{FG} = \beta_0 + \beta_1 \text{WC} + \epsilon$$

β_0 is the intercept of the best fit line and corresponds to the mean fasting glucose among those with a waist circumference of 0 cm. Clearly, this is not a meaningful quantity! However, that does not mean there is anything wrong with the regression fit. Later, we will discuss *centering* a predictor so the value of the intercept is more meaningful (see Section 4.3.4).

β_1 is the slope of the best fit line and corresponds to the difference in mean fasting glucose between those who differ by 1 cm in waist circumference. In Figure 4.2, for every 1 cm you move to the right on the horizontal axis, the line "slopes" upwards by an amount equal to our best estimate of β_1. The nature of the equation, specifically that fasting glucose is assumed to be linearly related to waist circumference, implies that this difference between groups is the same regardless of the level of waist circumference (the slope is the same at all points on the line). The two groups could have waist circumferences of 90 and 91 or 120 and 121; in either case the pair of groups are assumed to differ in mean fasting glucose by the same amount (the slope). Relaxing this assumption can be accomplished by fitting a curve instead of a straight line (see Section 4.8).

Example 4.2 (continued): Using "Never" as the reference level, the regression of fasting glucose on smoking status (a categorical predictor) is written:

$$\text{FG} = \beta_0 + \beta_1 I(X = \text{Past}) + \beta_2 I(X = \text{Current}) + \epsilon$$

β_0 corresponds to the mean fasting glucose among never smokers (the reference level) (see Figure 4.3), β_1 corresponds to the difference in mean fasting glucose between past and never smokers, and β_2 corresponds to the difference in mean fasting glucose between current and never smokers. Once again, for emphasis, the non-intercept βs are *not* mean outcomes at non-reference levels; they are *differences* in the mean outcome between non-reference levels and the reference level.

The indicator functions $I()$ are also referred to as **dummy variables**. If a variable is coded as a factor, then the creation of dummy variables is done for you automatically when you run a regression. For a categorical variable with L levels, $L-1$ dummy variables (each a 0/1 variable) are created, one for each level other than the reference level, and automatically included in the regression model. For example, for smoking status, two dummy variables are created for you by R, one that is 1 when smoking status is Past and 0 otherwise, and another that is 1 when smoking status is Current and 0 otherwise.

4.3 SLR model with a continuous predictor

Now you are ready to use R to fit your first regression model! To fit a linear regression, use the `lm()` function ("lm" stands for "linear model"). The syntax is `lm(y ~ x, data = dat)` where `dat` is a `data.frame` containing variables with the names `y` and `x`. If your dataset, outcome, and/or predictor have different names, then use those names instead of `dat`, `y`, and/or `x`, respectively. After fitting the model, use `summary()` to view the output.

Example 4.1 (continued): Regress fasting glucose on the continuous predictor waist circumference. Typically, the output of `lm()` is assigned to an object, in this case `fit.ex4.1`. The name you give the object is arbitrary. Assigning the output to an object allows you to use `summary()` to later view formatted output and to use `$` notation to extract elements of the output.

```
fit.ex4.1 <- lm(LBDGLUSI ~ BMXWAIST, data = nhanesf)
summary(fit.ex4.1)
```

```
##
## Call:
## lm(formula = LBDGLUSI ~ BMXWAIST, data = nhanesf)
##
## Residuals:
##     Min      1Q Median      3Q     Max
## -3.012 -0.685 -0.280  0.168 13.184
##
## Coefficients:
##             Estimate Std. Error t value            Pr(>|t|)
## (Intercept)   3.3041     0.2950   11.20 <0.0000000000000002 ***
## BMXWAIST      0.0278     0.0029    9.59 <0.0000000000000002 ***
## ---
## Signif. codes:  0 '***' 0.001 '**' 0.01 '*' 0.05 '.' 0.1 ' ' 1
##
## Residual standard error: 1.52 on 963 degrees of freedom
##   (35 observations deleted due to missingness)
## Multiple R-squared:  0.0871, Adjusted R-squared:  0.0862
## F-statistic: 91.9 on 1 and 963 DF,  p-value: <0.0000000000000002
```

4.3.1 Estimated regression coefficients

The `Coefficients:` section of the output contains estimates of and hypothesis tests for the intercept (β_0), labeled `(Intercept)`, and the coefficient for waist circumference (the slope, β_1), labeled `BMXWAIST`. The `Estimate` column displays the estimated regression coefficients. Thus, the best fit regression line has an intercept of 3.304 and a slope of 0.0278. If you want to extract just the `Coefficients` table, use `summary(fit.ex4.1)$coef`. Additionally, you can extract any element of the table using bracket `[]` notation or row and column headers, as shown below.

```
summary(fit.ex4.1)$coef
```

```
##             Estimate Std. Error t value                       Pr(>|t|)
## (Intercept)  3.30409   0.295020  11.200 0.000000000000000000000000001873
## BMXWAIST     0.02776   0.002895   9.588 0.000000000000000000007392885480
```

```
# Extract the slope using [] notation
summary(fit.ex4.1)$coef[2,1]
```

```
## [1] 0.02776
```

```
# Extract the slope using row and column headers
summary(fit.ex4.1)$coef["BMXWAIST", "Estimate"]
```

```
## [1] 0.02776
```

```
# Extract the p-value for the slope
summary(fit.ex4.1)$coef["BMXWAIST", "Pr(>|t|)"]
```

```
## [1] 0.000000000000000000007393
```

The `Coefficients` table also displays the **standard error** of each `Estimate`, its **t value** (= `Estimate/Std. Error`), and its **p-value** (labeled `Pr(>|t|)`). The t value tells you how many standard deviations away from 0 the estimate falls. The p-value for the intercept is usually not of interest, but the p-value for the predictor tests the null hypothesis that the outcome has no association with the predictor or, equivalently, that the slope is zero. The null hypothesis is that the best fit line is a horizontal line, indicating that the expected mean outcome is the same at all values of the predictor (that is, there is no association between the outcome and the predictor). Using the expected value notation, if fasting glucose and waist circumference are unrelated, then $E(\text{FG}|\text{WC}) = E(\text{FG})$, that is, the average fasting glucose does not vary with waist circumference.

4.3.2 Other outputs of summary()

- **Residual standard error (SE):** The `Residual standard error` is 1.52. This is an estimate of σ, the standard deviation of the model error term ϵ. To extract this estimate from the output, use `summary(fit.ex4.1)$sigma`.
- **Residual SE degrees of freedom (df):** The residual SE df is 963 and corresponds to the sample size minus the number of regression parameters. In this example, there are two regression parameters (the intercept and the slope). In general, the number of regression parameters is the number of rows in the `Coefficients` section. Therefore, the sample size with no missing data is $963 + 2 = 965$. To extract the residual SE df from the output, use `fit.ex4.1$df.residual`.

- **Number of observations excluded:** A note that 35 observations were deleted due to "missingness" – they had a missing value for at least one of the variables in the regression model. To extract the sample size used in the analysis (the number of observations in the dataset with no missing values on any analysis variable), use `length(fit.ex4.1$residuals)`.
- **Multiple R^2:** The `Multiple R-squared` value (.0871), which is a measure of goodness-of-fit ranging from 0 (no association) to 1 (perfect linear association), and is the square of the Pearson correlation between the outcome and predictor. It is interpreted as the proportion of variation in the outcome explained by the model. To extract R^2 from the output, use `summary(fit.ex4.1)$r.squared`.
- **Adjusted R^2:** The `Adjusted R-squared` (.0862) is an alternative that penalizes R^2 based on how many predictors are in the model. To extract adjusted R^2 from the output, use `summary(fit.ex4.1)$adj.r.squared`.
- **Global F test:** The `F-statistic:` section provides a global F test of the null hypothesis that all the coefficients other than the intercept are 0, and this section can typically be ignored. For SLR with a continuous predictor, this provides no new information – the F statistic is the square of the slope's t value, and the F statistic p-value is identical to the slope's p-value. When there is more than one predictor, as there will be in Chapter 5, the global F test is usually not of interest. Typically, we are interested in tests of individual predictors, not a comparison of a model with all the predictors to a model with none. To extract the F statistic and corresponding df values, use `summary(fit.ex4.1)$fstatistic`.

Use `confint()` to get confidence intervals (CI) for the intercept and slope. It is always a good idea to report a CI along with any estimate.

```
confint(fit.ex4.1)
```

```
##               2.5 %  97.5 %
## (Intercept) 2.72514 3.88305
## BMXWAIST    0.02208 0.03344
```

4.3.3 Writing it up

Below is an example of how to write up these results for a report or publication.

Example 4.1 (continued): Linear regression was used to test the association between fasting glucose (mmol/L) and waist circumference (cm) using data from 965 adults from NHANES 2017-2018. 8.71% of the variation in fasting glucose was explained by waist circumference (R^2 = .0871). There was a significant positive association between fasting glucose and waist circumference (B = 0.0278; 95% CI = 0.0221, 0.0334; p <.001). On average, for every 1-cm difference in waist circumference, adults differ in mean fasting glucose by 0.0278 mmol/L. Table 4.1 summarizes the regression results.

In Chapter 3, `tbl_summary()` from the `gtsummary` library (Sjoberg et al., 2021, 2023) was used to produce a table of descriptive statistics. Similarly, the `tbl_regression()` function easily creates nice regression tables (see Tutorial:tbl_regression[1] and Table Gallery[2] for more information).

[1]https://www.danieldsjoberg.com/gtsummary/articles/tbl_regression.html
[2]https://www.danieldsjoberg.com/gtsummary/articles/gallery.html

```
library(gtsummary)
TABLE2 <- fit.ex4.1 %>%
  tbl_regression(intercept = T,
                 estimate_fun = function(x) style_sigfig(x, digits = 4),
                 pvalue_fun   = function(x) style_pvalue(x, digits = 3),
                 label        = list(BMXWAIST ~ "Waist circumference (cm)")) %>%
  modify_caption("Regression of fasting glucose (mmol/L) on waist circumference")
```

```
TABLE2
```

TABLE 4.1 Regression of fasting glucose (mmol/L) on waist circumference

Characteristic	Beta	95% CI	p-value
(Intercept)	3.304	2.725, 3.883	<0.001
Waist circumference (cm)	0.0278	0.0221, 0.0334	<0.001

[1] CI = Confidence Interval

To save the table to a Microsoft Word document, use the following code.

```
TABLE2 %>%
  as_flex_table() %>%
  flextable::save_as_docx(path = "MyTable2.docx")
```

NOTES:

- This is cross-sectional data, so the write-up was carefully worded to not make any kind of causal claim. We did not write that a "change" or "increase" in waist circumference "causes" or "leads to" greater fasting glucose. We specifically use the words "differ" or "difference" because in cross-sectional data we can only draw conclusions about differences in the mean outcome between individuals with different values of the predictor, not what happens as an individual changes over time.
- As described in Appendix A.1, the NHANES data we are using are a subsample of the full data and intended only for use in teaching regression methods. No conclusions should be drawn from these analyses about the outcomes we are analyzing. Also, the NHANES complex survey design was not accounted for in these analyses; we will discuss how to account for complex survey designs in Chapter 8.
- Even if we had used the full NHANES data and appropriately accounted for the complex survey design, we would still generally not want to draw any substantive conclusions from a simple linear regression, as this is observational data. This is an **unadjusted** analysis and may be biased due to confounding. We will use multiple linear regression later to carry out **adjusted** analyses in Chapter 5. This is one of the main features of regression that makes it useful in epidemiological research – the ability to control for confounding by adjusting for potential confounders in a multiple regression model.

4.3.4 Centering a continuous predictor

"Centering" refers to replacing a continuous predictor with its value after subtracting off some value, typically the mean or median (although you can center at any number). The primary purpose of centering is to make the intercept interpretable as the mean outcome when the predictor is at its centering value.

To center a continuous predictor X at its mean, compute $cX = X - \bar{X}$ where $\bar{X}$ is the mean of X and enter cX into the model rather than X. To center at a different value, replace $\bar{X}$ with that value (e.g., the median).

Example 4.1 (continued): Re-fit the regression of fasting glucose on waist circumference after centering waist circumference at its mean.

```
MEAN <- mean(nhanesf$BMXWAIST, na.rm = T)
nhanesf$cBMXWAIST <- nhanesf$BMXWAIST - MEAN
fit.ex4.1.centered <- lm(LBDGLUSI ~ cBMXWAIST, data = nhanesf)
MEAN
```

```
## [1] 100.5
```

The uncentered output was:

```
round(summary(fit.ex4.1)$coef, 6)
```

```
##             Estimate Std. Error t value Pr(>|t|)
## (Intercept)  3.30409   0.295020  11.200        0
## BMXWAIST     0.02776   0.002895   9.588        0
```

and the new, centered output is:

```
round(summary(fit.ex4.1.centered)$coef, 6)
```

```
##             Estimate Std. Error t value Pr(>|t|)
## (Intercept)  6.09342   0.049041 124.251        0
## cBMXWAIST    0.02776   0.002895   9.588        0
```

The only difference between these two outputs is the intercept. Without centering, the intercept was the mean fasting glucose at waist circumference = 0 cm, which was not meaningful since no such humans exist (or could exist!). After centering, the intercept is the mean fasting glucose at waist circumference = 100.5 cm. Centering waist circumference made the intercept interpretable but had no effect on the regression coefficient for waist circumference.

4.4 SLR model with a categorical predictor

Earlier we showed that when you have a categorical predictor linear regression is essentially comparing the mean outcome between levels of that predictor. This is exactly what two-sample t-tests and one-way analysis of variance (ANOVA) do – compare means between independent samples. These methods are equivalent to a linear regression with a single categorical predictor. The advantage of using linear regression, however, is that it is easy to adjust for additional variables by adding more predictors to the model (as we will do in Chapter 5).

Example 4.2 (continued): Regress fasting glucose on the categorical predictor smoking status, which has three levels: Never, Past, and Current (as seen by using the `levels()` function below). The syntax is the same as for a continuous predictor, but the interpretation of the estimated regression coefficients differs.

```
levels(nhanesf$smoker)
```

```
## [1] "Never"   "Past"    "Current"
```

```
fit.ex4.2 <- lm(LBDGLUSI ~ smoker, data = nhanesf)
summary(fit.ex4.2)
```

```
##
## Call:
## lm(formula = LBDGLUSI ~ smoker, data = nhanesf)
##
## Residuals:
##     Min      1Q Median      3Q     Max
## -3.752 -0.722 -0.367  0.168 13.058
##
## Coefficients:
##               Estimate Std. Error t value             Pr(>|t|)
## (Intercept)     5.9423     0.0666   89.23 < 0.0000000000000002 ***
## smokerPast      0.4199     0.1190    3.53              0.00044 ***
## smokerCurrent   0.2551     0.1442    1.77              0.07722 .
## ---
## Signif. codes:  0 '***' 0.001 '**' 0.01 '*' 0.05 '.' 0.1 ' ' 1
##
## Residual standard error: 1.6 on 997 degrees of freedom
## Multiple R-squared:  0.0131, Adjusted R-squared:  0.0111
## F-statistic: 6.62 on 2 and 997 DF,  p-value: 0.00139
```

```
confint(fit.ex4.2)
```

```
##                 2.5 % 97.5 %
## (Intercept)    5.8116 6.0730
## smokerPast     0.1864 0.6535
## smokerCurrent -0.0279 0.5380
```

The primary way in which this differs from the output for a continuous predictor is that the `Coefficients` table now has, in addition to the `(Intercept)` row, $L - 1$ rows (where L = the number of levels of the categorical predictor), each one corresponding to the difference in the estimated mean outcome between a level and the reference level. **By default, `lm()` sets the reference level to be the first level of the categorical predictor.**

This output states that:

- The estimated mean fasting glucose among never smokers is 5.94 mmol/L (95% CI = 5.81, 6.07) (for a categorical predictor the intercept is the mean at the reference level).
- Past smokers have significantly greater mean fasting glucose than never smokers (B = 0.420; 95% CI = 0.186, 0.653; p <.001).
- Current smokers have greater mean fasting glucose than never smokers (B = 0.255; 95% CI = -0.028, 0.538; p = .077), but this difference is not statistically significant.

4.4.1 Recoding as a factor

In the previous example, smoker was coded as a factor variable. If it were coded instead as a numeric variable with levels, say, 1, 2, and 3, then lm() would treat it, incorrectly, as a continuous variable and assume that the difference in mean outcome between levels 2 and 1 is the same as between levels 3 and 2 (this is what it means to assume a linear relationship). But we do not want to make this assumption for a categorical predictor. To query whether or not a variable is coded as a factor, use is.factor().

```
is.factor(nhanesf$smoker)
```

```
## [1] TRUE
```

If is.factor() returned FALSE then we would need to convert the variable to a factor before running lm(). Our dataset contains a made-up variable cat_not_factor that has three levels (1, 2, and 3) but is not coded as a factor. The following code converts this variable to a factor with labels A, B, and C.

```
table(nhanesf$cat_not_factor, useNA = "ifany")
```

```
##
##   1   2   3
## 100 700 200
```

```
is.factor(nhanesf$cat_not_factor)
```

```
## [1] FALSE
```

```
nhanesf <- nhanesf %>%
  mutate(cat_factor =
           factor(cat_not_factor,
                  # Enter the existing values
                  levels = c(1,2,3),
                  # Enter new values or omit the labels option
                  # if you want the old values to be the factor level labels
                  labels = c("A", "B", "C")))
# Check
table(nhanesf$cat_not_factor,
      nhanesf$cat_factor, useNA = "ifany")
```

```
##
##       A   B   C
##   1 100   0   0
##   2   0 700   0
##   3   0   0 200
```

```
is.factor(nhanesf$cat_factor)
```

```
## [1] TRUE
```

4.4.2 What happens if you fit the model without coding a categorical variable as a factor?

To illustrate the difference between treating the predictor in the previous section as continuous vs. categorical, fit the regression of fasting glucose on each and see how the results differ.

```
fit.not_factor <- lm(LBDGLUSI ~ cat_not_factor, data = nhanesf)
fit.factor     <- lm(LBDGLUSI ~ cat_factor,     data = nhanesf)

round(summary(fit.not_factor)$coef, 4)
```

```
##                Estimate Std. Error t value Pr(>|t|)
## (Intercept)      5.9014     0.2052 28.7656   0.0000
## cat_not_factor   0.0913     0.0946  0.9651   0.3347
```

```
round(summary(fit.factor)$coef, 4)
```

```
##             Estimate Std. Error t value Pr(>|t|)
## (Intercept)   5.9611     0.1612 36.9736   0.0000
## cat_factorB   0.1321     0.1724  0.7662   0.4438
## cat_factorC   0.1985     0.1975  1.0053   0.3150
```

If (incorrectly) treating the predictor as continuous, we get just one regression coefficient for the predictor (0.0913) which implies that the difference in estimated mean outcome between individuals at levels 2 and 1, or between those at levels 3 and 2, each a 1-unit difference, is 0.0913. It also implies that the difference in estimated mean outcome between those at levels 3 and 1 (a 2-unit difference) is $2 \times 0.0913 = 0.1826$. Treating a predictor as a continuous variable *assumes* that the relationship between the outcome and predictor is linear.

Treating the predictor (correctly) as categorical, however, places no restrictions on the differences in mean outcome between levels of the predictor. When treating the predictor as a categorical variable (by coding it as a factor), the model implies that the difference in mean outcome between those at levels B and A is 0.1321, between levels C and B is $0.1985 - 0.1321 = 0.0664$, and between levels C and A is 0.1985. Comparing this to the previous paragraph, we see that failing to code a categorical variable as a factor leads to overly restrictive assumptions about the differences in mean outcome between levels.

4.4.3 Re-leveling

To compare the mean outcome between two non-reference levels of a categorical predictor, you could simply subtract their two regression coefficients (as was done in the previous paragraph when comparing levels C and B). However this only provides an estimate, not a CI or p-value. A method which yields all three is to **re-level** the predictor to change the reference level to one of those levels, followed by refitting the model.

Example 4.2 (continued): To compare current and past smokers, set one of them to be the reference level. For example, setting "Past" as the reference level moves it to be first in the ordering of the levels, leading `lm()` to treat it as the reference level when fitting the model.

```
nhanesf.releveled <- nhanesf %>%
  mutate(smoker = relevel(smoker, ref = "Past"))

# Check before (rows) vs. after (columns)
table(nhanesf$smoker, nhanesf.releveled$smoker, useNA = "ifany")
```

```
##
##           Past Never Current
##   Never      0   579       0
##   Past     264     0       0
##   Current    0     0     157
```

```
# Re-fit the model using the modified dataset
fit.ex4.2.releveled <- lm(LBDGLUSI ~ smoker, data = nhanesf.releveled)

# Use cbind() to combine the coef and confint output columns
cbind(round(summary(fit.ex4.2.releveled)$coef, 4),
      round(confint(fit.ex4.2.releveled),       4))
```

```
##                Estimate Std. Error t value Pr(>|t|)   2.5 %  97.5 %
## (Intercept)      6.3623     0.0986  64.510   0.0000  6.1687  6.5558
## smokerNever     -0.4199     0.1190  -3.529   0.0004 -0.6535 -0.1864
## smokerCurrent   -0.1649     0.1615  -1.021   0.3075 -0.4818  0.1520
```

Re-leveling does not change the overall fit of the model, as indicated by the residual standard error, R^2, and F statistic (compare `summary(fit.ex4.2)` and `summary(fit.ex4.2.releveled)` to verify this). However, it does change the interpretation of the coefficients. The intercept is now the mean fasting glucose among past smokers (the new reference level) and the other coefficients correspond to differences in mean outcome between each level and past smokers. Thus, we conclude that current smokers were observed to have lower fasting glucose than past smokers, but this difference was not statistically significant (B = -0.165; 95% CI = -0.482, 0.152; p = .308).

4.4.4 Multiple DF test for a categorical predictor

P-values for categorical predictor non-reference levels give us tests of significance for specific pairwise comparisons between levels. But what if you want a single overall test of significance for the categorical predictor? The null hypothesis of no association corresponds to the mean outcome being equal at all levels of the predictor or, equivalently, that all differences in mean outcome between each level and the reference level are 0. For example, for "smoking status", which has three levels, the null hypothesis is that both $\beta_1 = 0$ and $\beta_2 = 0$ simultaneously. $\beta_1 = 0$ corresponds to Past and Never smokers having equal means, $\beta_2 = 0$ corresponds to Current and Never smokers having equal means, and if both are true then Current, Past, and Never smokers have equal means implying no association between smoking status and the outcome. Since these are two tests, this would be a "2 df" (two degree of freedom) test. In general, for a categorical predictor with L levels, the corresponding test will be a $L - 1$ df test.

Obtain the p-value corresponding to this **multiple degree of freedom (df) test** by calling the `Anova()` (upper case `A`) function in the `car` library (Fox et al., 2023; Fox and Weisberg, 2019). The `type = 3` option specifies that we want a **Type III test**, which is a test of a predictor after adjusting for all other terms in the model (there are no other terms in SLR, but there will be when we get to multiple linear regression). A Type III test for a predictor X compares the regression model to a reduced model in which that predictor has been removed.

Example 4.2 (continued): Is the predictor smoking status significantly associated with the outcome fasting glucose? We already fit the regression model and assigned it to the object `fit.ex4.2` so enter that object into the `car::Anova()` function.

```
car::Anova(fit.ex4.2, type = 3)
```

```
## Anova Table (Type III tests)
##
## Response: LBDGLUSI
##              Sum Sq  Df F value                Pr(>F)
## (Intercept)   20445   1 7961.81 <0.0000000000000002 ***
## smoker           34   2    6.62                0.0014 **
## Residuals      2560 997
## ---
## Signif. codes:  0 '***' 0.001 '**' 0.01 '*' 0.05 '.' 0.1 ' ' 1
```

The Type III test states that the 2 degree of freedom test of smoker is statistically significant (p = .0014). In other words, we reject the null hypothesis that mean fasting glucose is the same for never, past, and current smokers. If reporting this test formally, we would write "F[2, 997] = 6.62, p = .001" where the 2 and 997 are the `smoker` and `Residuals` degrees of freedom, respectively.

In the case of a single categorical predictor, the Type III test is equivalent to the global F test shown in the `lm()` output above since both tests compare a model with one predictor to a model without any predictors. This will not be the case in multiple linear regression. `car::Anova()` is used for tests of individual predictors adjusted for all others, whereas the global F-test in the `lm()` output is used to test all predictors simultaneously by comparing the regression model to a model with just an intercept.

4.4.5 Special case: binary predictor

If a categorical predictor has only two levels (binary, dichotomous), it will only have one indicator function in the regression and there is no need for a Type III test – simply read the p-value from the `Coefficients` table. The comparison of mean outcome between the two levels is the same as the overall test of a binary predictor. For example, suppose we regress fasting glucose on gender (for which NHANES only reports responses of male and female) (`RIAGENDR`).

```
fit.binary <- lm(LBDGLUSI ~ RIAGENDR, data = nhanesf)
summary(fit.binary)$coef
```

```
##                Estimate Std. Error t value  Pr(>|t|)
## (Intercept)      6.2973    0.07491  84.067 0.0000000
## RIAGENDRFemale  -0.3758    0.10165  -3.697 0.0002304
```

```
car::Anova(fit.binary, type = 3)
```

```
## Anova Table (Type III tests)
##
## Response: LBDGLUSI
##              Sum Sq  Df F value                Pr(>F)
## (Intercept)  18123    1  7067.3 < 0.0000000000000002 ***
## RIAGENDR        35    1    13.7              0.00023 ***
## Residuals     2559 998
## ---
## Signif. codes:  0 '***' 0.001 '**' 0.01 '*' 0.05 '.' 0.1 ' ' 1
```

The `RIAGENDRFemale` p-value in the regression coefficient table is identical to the `RIAGENDR` Type III test p-value in the `Anova Table`.

4.4.6 Writing it up

Below is an example of how to write up these results for a report or publication. Notice how this differs from when the predictor was continuous – for a categorical predictor we report the overall association plus the pairwise differences.

Example 4.2 (continued): We used linear regression to test the association between fasting glucose (mmol/L) and smoking status (Never, Past, Current) using data from 1000 adults from NHANES 2017-2018. 1.31% of the variation in fasting glucose was explained by smoking status (R^2 = .0131). The association between fasting glucose and smoking status was statistically significant (F[2, 997] = 6.62, p = .001). Mean fasting glucose was significantly greater among past smokers than among never smokers (B = 0.420; 95% CI = 0.186, 0.653; p <.001), but not significantly different between current and never smokers (B = 0.255; 95% CI = -0.028, 0.538; p = .077) or between current and past smokers (B = -0.165; 95% CI = -0.482, 0.152; p = .308).

Table 4.2 summarizes the regression results.

```
library(gtsummary)
TABLE2 <- fit.ex4.2 %>%
  tbl_regression(intercept = T,
                 estimate_fun = function(x) style_sigfig(x, digits = 3),
                 pvalue_fun   = function(x) style_pvalue(x, digits = 3),
                 label        = list(smoker ~ "Smoking Status")) %>%
  # add_global_p() adds the multiple df Type III p-values from car::Anova
  # keep = T keeps the individual p-values comparing levels, as well
  add_global_p(keep = T) %>%
  modify_caption("Regression of fasting glucose (mmol/L) on smoking status")
```

```
TABLE2
```

TABLE 4.2 Regression of fasting glucose (mmol/L) on smoking status

Characteristic	Beta	95% CI	p-value
(Intercept)	5.94	5.81, 6.07	<0.001
Smoking Status			0.001
Never	—	—	
Past	0.420	0.186, 0.653	<0.001
Current	0.255	-0.028, 0.538	0.077

[1] CI = Confidence Interval

The p-value in the "Smoking Status" row corresponds to the 2 df Type III test of smoking status obtained using `car::Anova()`.

4.5 Interpreting p-values

In the examples we have discussed, the **p-value** for a regression coefficient β tests the null hypothesis $H_0 : \beta = 0$ vs. the alternative hypothesis $H_1 : \beta \neq 0$. If the associated regression term is a continuous predictor, the null hypothesis is that the outcome has no association with the predictor. If the associated regression term is an indicator function for the level of a categorical predictor, then the null hypothesis is that the mean outcome is the same at that level and the reference level.

Typically, the **level of significance** is set at .05 (for arbitrary historical reasons) and we "reject the null hypothesis" if $p < .05$ and "fail to reject the null hypothesis" if $p \geq .05$. Importantly, rejecting the null hypothesis does not mean we have proved the null hypothesis to be false or the alternative to be true. Similarly, failing to reject the null does not mean we have proved the null to be true or the alternative to be false.

The p-value is computed under two assumptions: (1) Equation 4.1 or 4.2 and the associated assumptions constitute the model that is generating the data and (2) the regression coefficient (β) is zero (the null hypothesis). Under these assumptions, the p-value is the probability of observing a sample that results in a regression coefficient with a magnitude at least as large as what was observed. For a continuous predictor, for example, the p-value answers the following question: "If there really is no association between the outcome and predictor, and the regression model really is correct, how likely is it that we would observe an association at least this strong?" In Example 4.1, p <.001 which means that if the true model really is a horizontal line, and the regression assumptions are true, there is very little chance we would observe as steep a slope or steeper (as large an association or larger).

In epidemiology, the p-value is often said to be used to attempt to "rule out chance" as an explanation for an observed association. **The p-value, however, is not the probability that random chance produced the observed association (a common misconception).** More specifically, it is *not* the probability that the null hypothesis of no association is true. Rather, the p-value is the probability of observing a slope so large *assuming* the null is true (assuming chance, or assuming no association). You cannot compute the probability of something you are assuming to be true, but you can assume it is true and then compute the probability of observing an extreme result.

A key distinction is between the estimated regression coefficient, which is a measure of the strength of association between the predictor and the outcome, and the p-value, which is a measure of the compatibility of the data with the null hypothesis (Greenland et al., 2016b). In a small sample, one could observe a meaningfully large estimated regression coefficient yet have a large p-value and fail to reject the null. Why? Because, assuming the null is true, it is not unusual in a small sample to end up with a large regression coefficient. Conversely, in a large sample, one could observe a small and not meaningfully far from zero estimated regression coefficient, yet have a small p-value and reject the null. Why? Because, assuming the null is true, only small deviations from no effect are likely in large sample sizes.

Conclusion: While p-values are commonly reported in research studies, they are far less important than estimates of the magnitude of the estimated regression coefficients and their confidence intervals. In particular, because p-values are so dependent on sample size, pay more attention to the size of regression coefficients than to p-values when deciding if an association or effect is meaningfully large.

NOTES:

- When working with very large samples, for example when conducting a secondary analysis of data from a large, national survey, it is common to find that many of the p-values for regression coefficients are very small. Rather than just declare these effects to be "significant," base your conclusions primarily on the sizes of the regression coefficients.
- When working with a designed study with a sample size determined by a power analysis, this scenario is less likely to occur. A different problem arises, however. The sample size required to achieve the desired power is often underestimated, resulting in an underpowered study. "The reason is well known among those who have served as grant and research reviewers: The final effective sample size and thus the final standard error is most determined not by the total size of the trial, but rather by the number of highly informative cases or events (e.g., deaths, remissions, clinically large reductions in biomarkers, etc.) which, in practice, are almost inevitably less frequent in the study than was promised in the proposal, even when the total sample was very large. This is unsurprising because the frequency estimates in proposals are usually based on population data, but the study will impose restrictions that reduce baseline event frequencies below those in population estimates (e.g., by excluding patients in poorest health to minimize liability concerns and drop-out)" (Sander Greenland, personal communication, 2024). For example, Olivier et al. (2024) carried out a systematic review of 344 cardiovascular trials and found that, when compared to the observed event rates, 61% overestimated the event rate during trial design (mean relative deviation = 12.3%; 95% CI = 5.6%, 16.4%; $p < .001$).
- P-values can also be computed for alternative hypotheses in which the tested coefficient value(s) are not zero. This is often advisable when the goal is to test non-inferiority, superiority, or equivalence (Greenland et al., 2016b; Rafi and Greenland, 2020a,b).
- P-values have generated a lot of debate. See Wasserstein and Lazar (2016) for the American Statistical Association's official stance on p-values, Greenland et al. (2016b) for a consensus supplementary statement about misinterpretations of p-values, confidence intervals, and power, the supplementary comments following Wasserstein and Lazar (2016) for additional commentary and differing opinions, and Wasserstein et al. (2019) for a follow-up article.

4.6 Predictions from the model

The term **prediction** is used here to refer to the process of making a best guess for the outcome at a given predictor value. Technically, this is *estimation* of the mean outcome $E(Y|X = x)$. As we will see later, whether you are estimating the mean outcome or predicting the outcome for an individual, the point estimate will be the same; however, the interval estimate will differ.

For a model with a continuous predictor, each point on the regression line is the estimated mean outcome at a given value of the predictor ($E(Y|X)$). For a categorical predictor, the estimates are the mean outcome values at the levels of the predictor. To compute these estimates, you *could* manually enter various predictor values into the model and compute the result, as in the following examples.

Example 4.1 (continued):

```
round(summary(fit.ex4.1)$coef, 4)
```

```
##             Estimate Std. Error t value Pr(>|t|)
## (Intercept)   3.3041     0.2950  11.200        0
## BMXWAIST      0.0278     0.0029   9.588        0
```

The estimated mean fasting glucose for those with a waist circumference of 100 cm is approximately $3.3041 + 0.0278 \times 100 = 6.0841$ mmol/L.

Example 4.2 (continued):

```
round(summary(fit.ex4.2)$coef, 4)
```

```
##               Estimate Std. Error t value Pr(>|t|)
## (Intercept)     5.9423     0.0666  89.229   0.0000
## smokerPast      0.4199     0.1190   3.529   0.0004
## smokerCurrent   0.2551     0.1442   1.769   0.0772
```

The estimated mean fasting glucose for Current smokers is $5.9423 + 0.4199 \times 0 + 0.2551 \times 1 = 6.1974$ mmol/L.

However, neither of the above are exact because we rounded each coefficient. Better to let R do the computation for you by using the `predict()` function. Repeat the examples above, this time using `predict()`, along with `interval = "confidence"` which will compute a 95% CI for each estimated mean outcome. In each case, use the `newdata` argument to supply a `data.frame` with the value at which we want a prediction.

```
# Example 4.1
predict(fit.ex4.1,
        newdata = data.frame(BMXWAIST = 100),
        interval = "confidence")
```

```
##    fit   lwr   upr
## 1 6.08 5.984 6.177
```

```
# Example 4.2
predict(fit.ex4.2,
        newdata = data.frame(smoker = "Current"),
        interval = "confidence")
```

```
##     fit   lwr   upr
## 1 6.197 5.946 6.448
```

The manual calculation may turn out to be spot on; but this will not always be true due to rounding. If you want an *exact* answer, and a CI, use `predict()`.

4.7 Confidence intervals and prediction intervals

When reporting a parameter estimate, also report a confidence interval for the parameter. **A 95% confidence interval (CI) for a population parameter is a random interval that has 95% probability of containing the true parameter.** For example, if (5.1%, 7.3%) is a 95% CI for the prevalence of some disease in a population, that means that there is a 95% chance that this interval contains the true prevalence.

We are going to look at three different kinds of intervals that you can create in simple linear regression: **(1)** CIs for regression coefficients, **(2)** a CI for the mean outcome, and **(3)** a prediction interval (PI) for an individual observation.

- The **regression coefficients** are the βs – the intercept and slope (for a continuous predictor) or the intercept and differences in mean outcome between levels and the reference level (for a categorical predictor).
- The **estimated mean outcome** is the value of the outcome on the regression line which we computed in Section 4.6.
- The **predicted outcome for an individual observation** is identical to the estimated mean outcome, but the PI for the outcome for an individual observation will be wider than the CI for the estimated mean outcome because individual observations are more variable than the mean.

4.7.1 CIs for regression coefficients

Earlier, we computed CIs for each of the individual regression coefficients using `confint(fit)`.

```
# Example 4.1 (continuous predictor)
confint(fit.ex4.1)
```

```
##                2.5 %  97.5 %
## (Intercept) 2.72514 3.88305
## BMXWAIST    0.02208 0.03344
```

Based on this output, we conclude that a 95% CI for the population slope (β_1) of the regression of fasting glucose on waist circumference is 0.0221, 0.0334. Combining this with the regression coefficient and p-value from the earlier regression output, we would report this as follows: Waist circumference is significantly associated with fasting glucose (B = 0.0278; 95% CI = 0.0221, 0.0334; p <.001).

For a categorical predictor,

```
# Example 4.2 (categorical predictor)
confint(fit.ex4.2)
```

```
##                  2.5 % 97.5 %
## (Intercept)     5.8116 6.0730
## smokerPast      0.1864 0.6535
## smokerCurrent  -0.0279 0.5380
```

Recall that the intercept (β_0) is the mean outcome at the reference level, and the other parameters are differences in mean between those levels and the reference level. Thus, a 95% CI for the population mean fasting glucose among never smokers is 5.8116, 6.0730. To get a CI for the mean at another level, re-level the predictor to make that level the reference level (see Section 4.4.3), re-fit the model, and re-compute the CI for the intercept.

A 95% CI for the population difference in mean fasting glucose between past and never smokers is 0.1864, 0.6535. We would report this as follows: Mean fasting glucose is significantly different between past and never smokers (B = 0.42; 95% CI = 0.19, 0.65; p <.001). To get a CI for a pairwise comparison between levels where neither is the reference level, re-level the predictor (see Section 4.4.3), re-fit the model, and re-compute the CI for the appropriate regression coefficient.

4.7.2 CI for the mean outcome

We already computed a CI for the mean outcome earlier when we used `predict()` along with `interval = confidence`. By way of reminder,

```
# Example 4.1 (continuous predictor)
predict(fit.ex4.1,
        newdata = data.frame(BMXWAIST = 100),
        interval = "confidence")
```

```
##    fit   lwr   upr
## 1 6.08 5.984 6.177
```

CIs for the mean over the entire range of X values are referred to as a **confidence band** since they visually form a band around the regression line. To **visualize the confidence band** for a continuous predictor, compute the CIs over the range of possible X values.

```
X <- seq(min(nhanesf$BMXWAIST, na.rm=T),
         max(nhanesf$BMXWAIST, na.rm=T), by = 1)
mypred.cont <- predict(fit.ex4.1,
                       newdata = data.frame(BMXWAIST=X),
                       interval="confidence")
```

This produces a matrix with one row for each X value and three columns – the estimated mean and the lower and upper confidence limits. View the X values along with the predictions by using `data.frame` to create a dataset with X and the values in `mypred.cont`.

```
head(data.frame(X, mypred.cont))
```

```
##      X   fit   lwr   upr
## 1 63.2 5.059 4.826 5.291
## 2 64.2 5.086 4.859 5.314
## 3 65.2 5.114 4.892 5.336
## 4 66.2 5.142 4.925 5.359
## 5 67.2 5.170 4.957 5.382
## 6 68.2 5.197 4.990 5.405
```

In Figure 4.4, we plot `fit` vs. `X` to see the regression line (solid line), and each of `lwr` and `upr` vs. `X` to see the confidence band (dashed lines). With a reasonably large sample size, the confidence band will be quite narrow, so below it is plotted both on the original scale and zoomed in (using the `ylim` option to restrict the y-axis limits) to see the band more clearly.

```
par(mfrow=c(1,2))
plot(LBDGLUSI ~ BMXWAIST,    data = nhanesf,
     col="gray",
     ylab = "Fasting Glucose (mmol/L)",
     xlab = "Waist Circumference (cm)",
     las = 1, pch = 20, font.lab = 2, font.axis = 2, cex.axis = 0.75)
# NOTE: For lines() the arguments are x, y not y ~ x
lines(X, mypred.cont[, "fit"], col = "red",   lty = 1, lwd = 2)
lines(X, mypred.cont[, "upr"], col = "blue", lty = 5, lwd = 2)
lines(X, mypred.cont[, "lwr"], col = "blue", lty = 5, lwd = 2)

plot(LBDGLUSI ~ BMXWAIST,    data = nhanesf, ylim = c(4.5, 8.5),
     col="gray",
     ylab = "(zoomed in)",
     xlab = "Waist Circumference (cm)",
     las = 1, pch = 20, font.lab = 2, font.axis = 2, cex.axis = 0.75)
# NOTE: For lines() the arguments are x, y not y ~ x
lines(X, mypred.cont[, "fit"], col = "red",   lty = 1, lwd = 2)
lines(X, mypred.cont[, "upr"], col = "blue", lty = 5, lwd = 2)
lines(X, mypred.cont[, "lwr"], col = "blue", lty = 5, lwd = 2)
```

In general, the confidence band will be narrower in the middle of the predictor range and wider at the edges because we are more confident in the location of the middle of the line than we are about the location of the ends of the line. This could be shown mathematically but, just conceptually, consider that a slightly different sample of points would result in a slightly shifted and rotated line and that any rotation is going to have a greater impact on the ends of the line than the middle.

If the predictor is categorical, then rather than plotting over a continuous range of X values, we plot at the set of discrete levels of X (using `levels()`), as shown in Figure 4.5.

```
# Example 4.2 (categorical predictor)
X <- levels(nhanesf$smoker)
mypred.cat <- predict(fit.ex4.2,
                      newdata = data.frame(smoker=X),
                      interval="confidence")
data.frame(X, mypred.cat)
```

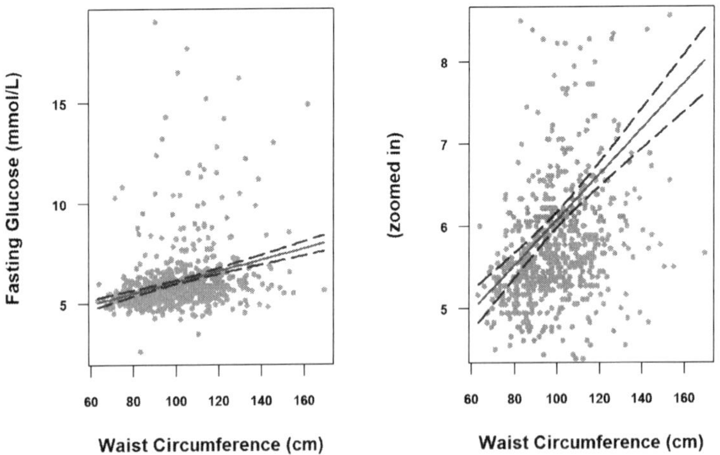

FIGURE 4.4 Confidence band for a continuous predictor

```
##          X   fit   lwr   upr
## 1   Never 5.942 5.812 6.073
## 2    Past 6.362 6.169 6.556
## 3 Current 6.197 5.946 6.448
```

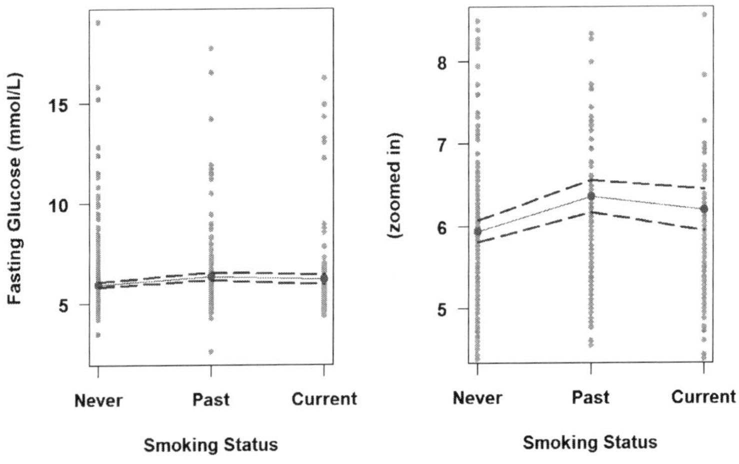

FIGURE 4.5 Confidence band for a categorical predictor

```
par(mfrow=c(1,2))
plot.default(nhanesf$LBDGLUSI ~ nhanesf$smoker,
             col="gray",
             ylab = "Fasting Glucose (mmol/L)",
             xlab = "Smoking Status",
             las = 1, pch = 20,
             font.lab = 2, font.axis = 2, xaxt = "n")
axis(1, at = 1:3, labels = levels(nhanesf$smoker), font = 2)
# NOTE: For points() and lines() the arguments are x, y not y ~ x
points(1:3, mypred.cat[, "fit"], col = "red", pch = 20, cex = 1.5)
lines( 1:3, mypred.cat[, "fit"], col = "red")
lines( 1:3, mypred.cat[, "upr"], col = "blue", lty = 5, lwd = 2)
lines( 1:3, mypred.cat[, "lwr"], col = "blue", lty = 5, lwd = 2)

plot.default(nhanesf$LBDGLUSI ~ nhanesf$smoker, ylim = c(4.5, 8.5),
             col="gray",
             ylab = "(zoomed in)",
             xlab = "Smoking Status",
             las = 1, pch = 20,
             font.lab = 2, font.axis = 2, xaxt = "n")
axis(1, at = 1:3, labels = levels(nhanesf$smoker), font = 2)
# NOTE: For points() and lines() the arguments are x, y not y ~ x
points(1:3, mypred.cat[, "fit"], col = "red", pch = 20, cex = 1.5)
lines( 1:3, mypred.cat[, "fit"], col = "red")
lines( 1:3, mypred.cat[, "upr"], col = "blue", lty = 5, lwd = 2)
lines( 1:3, mypred.cat[, "lwr"], col = "blue", lty = 5, lwd = 2)
```

Unlike for a continuous predictor, the width of the confidence band is not necessarily greater on the "ends" when you have a categorical predictor. The width of the band now is instead related only to the sample size at each level of X – this dataset contains more never smokers so the CI for the prediction is narrower at that level.

4.7.3 PI for an individual observation

The third type of interval of interest is a **prediction interval** (PI) for the outcome for an *individual observation* at a specific predictor value. It turns out that the estimated value of the mean outcome and the predicted value for an individual's outcome are identical, but their intervals are not. The PI for an individual's outcome will be wider than the CI for the mean outcome – individuals are more variable than the average of individuals.

A 95% prediction interval for the outcome Y for an individual with a specific predictor value $X = x$ is interpreted as follows: *Among all individuals in the population with* $X = x$, we expect 95% of them to have outcome values within the 95% prediction interval.

In `predict()`, add `interval = "prediction"` to compute the PI for an individual observation.

```
# Example 4.1 (continuous predictor)
predict(fit.ex4.1,
        newdata = data.frame(BMXWAIST = 100),
        interval = "prediction")
```

```
##    fit   lwr   upr
## 1 6.08 3.089 9.071
```

The estimated average fasting glucose for individuals in the population with a waist circumference of 100 cm is 6.08 mmol/L, and we expect 95% of such individuals to have fasting glucose between 3.09 and 9.07 mmol/L.

```
# Example 4.2 (categorical predictor)
predict(fit.ex4.2,
        newdata = data.frame(smoker = "Current"),
        interval = "prediction")
```

```
##      fit   lwr   upr
## 1 6.197 3.043 9.352
```

The average fasting glucose for individuals who are current smokers is 6.20 mmol/L, and we expect 95% of current smokers to have a fasting glucose between 3.04 and 9.35 mmol/L.

The PIs for individual observations over a range of X values form a **prediction band**. To visualize the prediction band, use the same code as in Section 4.7.2 but with `interval ="prediction"` instead of `interval="confidence"` in the call to `predict()`. The results for Examples 4.1 and 4.2 are shown in Figure 4.6 and Figure 4.7, respectively.

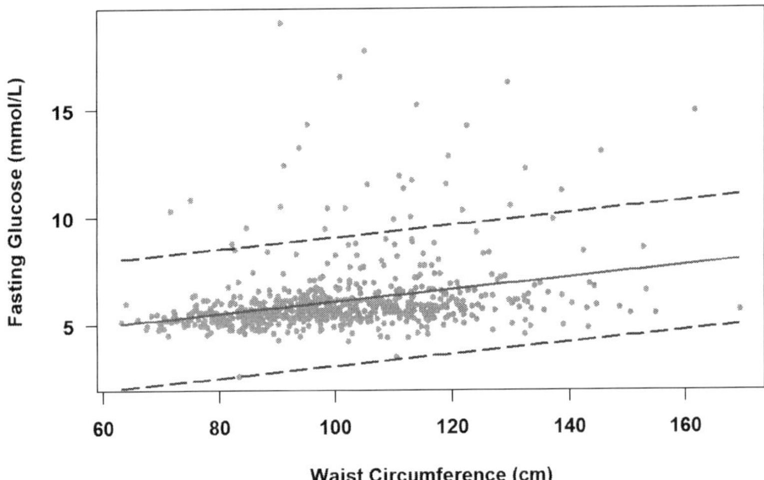

FIGURE 4.6 Prediction band for a continuous predictor

There is no need to zoom in this time because the prediction band is quite wide. Since we are less confident in the outcome for an individual observation than for the mean of individual observations, a prediction band will always be wider than a confidence band.

NOTE: As the sample size increases, a prediction band will become more accurate but will always be wide since it captures 95% of the observations. A confidence band, however, will get narrower as the sample size increases.

In summary, you can obtain three kinds of intervals from a regression – for the population regression coefficients, for the population mean outcome, and for the outcome for an individual observation. For the mean and individual the centers of the intervals are the same, but the PI for an individual observation will always be wider than the CI for the mean.

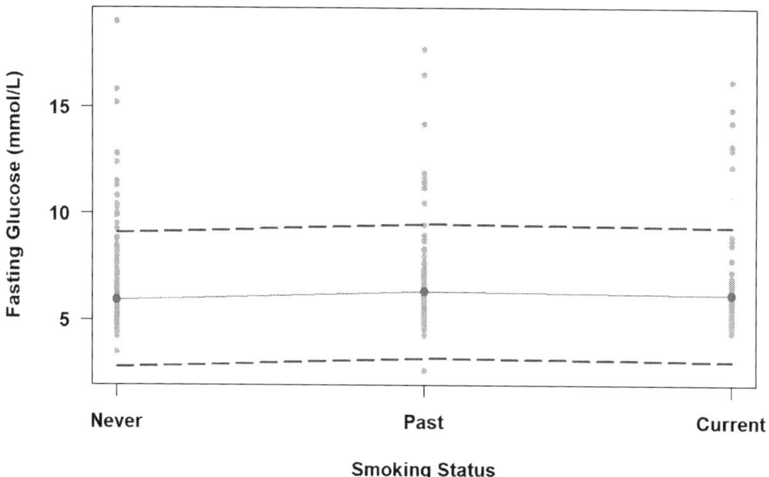

FIGURE 4.7 Prediction band for a categorical predictor

4.8 Fitting curves using polynomials

If the relationship between the outcome and a continuous predictor is non-linear, a curve may fit better than a straight line. A common way to fit a curve is to use a polynomial function, like a quadratic or cubic. Such a curve does not have a single slope – the rate of change in Y is not constant and depends on X.

The equations for linear, quadratic, and cubic models are, respectively:

$$\begin{array}{lrcl} \text{Linear:} & Y & = & \beta_0 + \beta_1 X + \epsilon \\ \text{Quadratic:} & Y & = & \beta_0 + \beta_1 X + \beta_2 X^2 + \epsilon \\ \text{Cubic:} & Y & = & \beta_0 + \beta_1 X + \beta_2 X^2 + \beta_3 X^3 + \epsilon \end{array}$$

The βs are not the same between the models (e.g., β_0 is not the same for all three models), we are just using the same notation for all three models. Also, higher order polynomials are possible, as well.

Example 4.3: Use nation-level data compiled by the United Nations Development Programme (United Nations Development Programme, 2020) (see Appendix A.2) to regress Y = Adult female mortality (2018, per 1,000 people) (`mort_adult_f`) on X = Female population with at least some secondary education (2015-2019, % ages 25 and older) (`educ_f`). Try linear, quadratic, and cubic models and determine statistically and visually which fits the best.

```
load("Data/unhdd2020.rmph.rData")
# Fit linear, quadratic, and cubic models
fit.poly.1 <- lm(mort_adult_f ~ educ_f, data = unhdd)
fit.poly.2 <- lm(mort_adult_f ~ educ_f + I(educ_f^2), data = unhdd)
fit.poly.3 <- lm(mort_adult_f ~ educ_f + I(educ_f^2) + I(educ_f^3), data = unhdd)
# NOTE: You must use I(X^2) rather than X^2 in the model, but this is not an
# indicator function, it is telling lm to use the transformation inside ().
```

Test if the cubic curve fits significantly better than quadratic by testing the null hypothesis $H_0 : \beta_3 = 0$ in the cubic fit. If not, then test if the quadratic curve fits significantly better than linear by testing the null hypothesis $H_0 : \beta_2 = 0$ in the quadratic fit. If not, then a quadratic does not fit significantly better than a line.

```
round(summary(fit.poly.3)$coef, 4)
```

```
##             Estimate Std. Error t value Pr(>|t|)
## (Intercept) 263.6288    26.8635  9.8137   0.0000
## educ_f       -3.0509     2.0016 -1.5243   0.1294
## I(educ_f^2)  -0.0022     0.0412 -0.0535   0.9574
## I(educ_f^3)   0.0001     0.0002  0.5433   0.5877
```

```
round(summary(fit.poly.2)$coef, 4)
```

```
##             Estimate Std. Error t value Pr(>|t|)
## (Intercept) 274.6177    17.6385  15.569   0.0000
## educ_f       -4.0651     0.7205  -5.642   0.0000
## I(educ_f^2)   0.0199     0.0063   3.166   0.0019
```

```
round(summary(fit.poly.1)$coef, 4)
```

```
##             Estimate Std. Error t value Pr(>|t|)
## (Intercept)  229.748    10.7916   21.29        0
## educ_f        -1.838     0.1599  -11.49        0
```

Based on the output, the cubic term (`I(educ_f^3)`) in the cubic model is not statistically significant (p = .5877), but the quadratic term (`I(educ_f^2)`) in the quadratic model is (p = .0019). Thus, we conclude that the best fitting polynomial is a quadratic.

NOTE: With a large sample size, even a small deviation from linearity will have a small p-value. In that case, a visual comparison is more helpful than a statistical hypothesis test. Whatever the sample size, it is typically a good idea to judge the comparison of polynomials more on the visual difference or similarity than on p-values.

To plot the fitted polynomial curves, plot the data first and then overlay lines or curves based on predictions from the model over a range of predictor values.

```
plot(mort_adult_f ~ educ_f, data = unhdd,
     col = "gray",
     ylab = "Adult female mortality per 1000",
     xlab = "Adult female pop with secondary education (%)",
     las = 1, pch = 20, font.lab = 2, font.axis = 2)
# Get predicted values over a range for plotting
# Add na.rm=T in case there are missing values
X <- seq(min(unhdd$educ_f, na.rm = T),
         max(unhdd$educ_f, na.rm = T), by = 1)
PRED1 <- predict(fit.poly.1, data.frame(educ_f = X))
PRED2 <- predict(fit.poly.2, data.frame(educ_f = X))
PRED3 <- predict(fit.poly.3, data.frame(educ_f = X))
# Add lines to the plot
# NOTE: For lines() the arguments are x, y not y ~ x
lines(X, PRED1, lwd = 2, lty = 1)
lines(X, PRED2, lwd = 2, lty = 2)
lines(X, PRED3, lwd = 2, lty = 3)
```

```
# Add a legend
par(font = 2) # Bold
legend(70, 400, c("Linear", "Quadratic", "Cubic"),
       lwd = c(2,2,2), lty = c(1,2,3), seg.len = 4,
       bty = "n", title = "Fit")
```

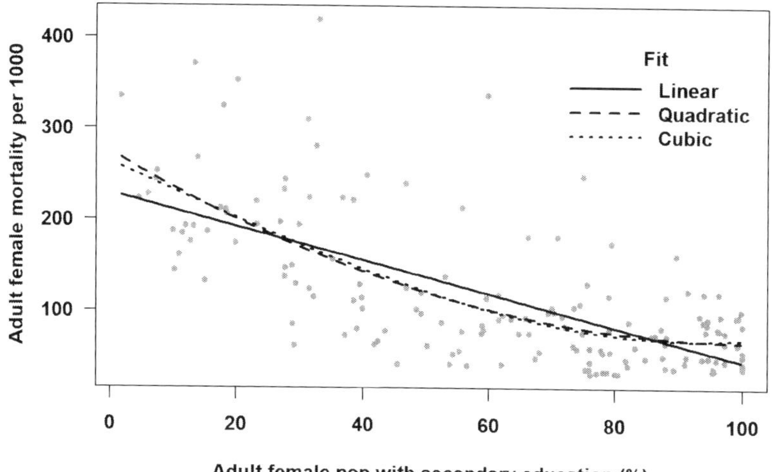

FIGURE 4.8 Using linear regression to fit a curve

In this example, Figure 4.8 confirms our conclusions from the statistical tests – the quadratic curve fits the data better than a straight line, and the cubic is not an improvement over the quadratic. So what is the relationship between nations' adult female mortality rate and secondary school attainment? Based on this data, it appears that the mortality rate is lower in nations with greater secondary school attainment, but only for those with low to moderate attainment. Among those with the highest attainment, there is no relationship (the curve levels off at the higher values of X). A linear fit underestimates the slope among countries with lower attainment and overestimates the slope among those with higher attainment.

Overall test of a predictor modeled as a polynomial

What if we want a single overall test for a predictor that is fit using a polynomial? For a linear predictor, this was easy – we simply read the p-value from the regression output. But when you have a polynomial of order two (quadratic) or greater, the predictor is modeled using more than one term. As when we tested a categorical variable with more than two levels, we need an overall multiple df test of all the terms at the same time. In general, we need to compare the model that includes X as a polynomial to a reduced model that excludes X altogether. The `anova()` (lowercase `a`) function can accomplish this goal.

```
# Reduced model (all X terms removed):
fit.reduced <- lm(mort_adult_f ~ 1, data = unhdd)
```

```
# Compare reduced and full models
anova(fit.reduced, fit.poly.2)
```

```
## Error in anova.lmlist(object, ...) :
##   models were not all fitted to the same size of dataset
```

What went wrong? `fit.poly.2` included the outcome `mort_adult_f` as well as the predictor `educ_f` whereas `fit.reduced` only included `mort_adult_f`. There were missing values in `educ_f` that resulted in the available observations being different for the two fits. Comparing the sample sizes used in each regression verifies that they are not the same.

```
length(fit.poly.2$residuals)
```

```
## [1] 162
```

```
length(fit.reduced$residuals)
```

```
## [1] 181
```

How do we solve this? `lm()` has a `subset` option which can be used to remove cases with missing values for the variable that was used in `fit.poly.2` but not `fit.reduced`. (Had we removed missing data before doing any of these analyses, this would not have been an issue.)

```
# Reduced model with correct sample
fit.reduced <- lm(mort_adult_f ~ 1, data = unhdd,
                  subset = complete.cases(educ_f))
# Compare reduced and full models
anova(fit.reduced, fit.poly.2)
```

```
## Analysis of Variance Table
##
## Model 1: mort_adult_f ~ 1
## Model 2: mort_adult_f ~ educ_f + I(educ_f^2)
##   Res.Df     RSS Df Sum of Sq    F               Pr(>F)
## 1    161 1028500
## 2    159  530058  2    498442 74.8 <0.0000000000000002 ***
## ---
## Signif. codes:  0 '***' 0.001 '**' 0.01 '*' 0.05 '.' 0.1 ' ' 1
```

The "Df" column has a 2 because this is a 2 degree of freedom test; we are testing two terms simultaneously – the linear term `educ_f` and the quadratic term `I(educ_f^2)`. According to the `Analysis of Variance Table`, we can reject the null hypothesis that both of these terms have coefficients equal to 0 and conclude that there is a significant association between secondary school attainment and mortality rate among adult females ($p < .001$). That is, the quadratic curve fits significantly better than a horizontal line at the mean outcome (no association). As before, be careful when using p-values to compare models. A visualization such as Figure 4.8 is valuable for clarifying the magnitude of differences between competing models. In this case, the figure is consistent with the hypothesis test – the quadratic model is meaningfully different from a horizontal line.

4.9 Assumptions

In Section 4.2 we discussed a number of assumptions about a linear regression model. The validity of regression coefficient estimates, confidence intervals, and significance tests based on a regression model depends on a number of assumptions: independent observations, normality of errors, linearity, and constant error variance. These assumptions will be discussed in Chapter 5, including how to assess their validity and what to do when they fail.

4.10 Exercises

1. True or false? Simple linear regression describes the setting where there is more than one predictor.

2. True or false? When the predictor is continuous, a simple linear regression estimates the best fitting line.

3. True or false? When the predictor is categorical, a simple linear regression estimates the mean outcome at each level of the predictor.

4. What is the interpretation of the intercept β_0 in a simple linear regression with an uncentered continuous predictor? With a centered continuous predictor? With a categorical predictor?

5. What is the interpretation of the slope β_1 in a simple linear regression with a continuous predictor? Does centering affect this interpretation?

6. In a simple linear regression with a categorical predictor with L levels, what is the interpretation of the coefficients $\beta_1, ..., \beta_{L-1}$?

7. For a simple linear regression with a continuous predictor, write out the null and alternative hypothesis that are tested by the p-value for that predictor. Express the null and alternative hypotheses as statements about the slope of the regression line.

8. What quantities should always be reported along with the p-value for a regression coefficient?

9. Suppose you have large sample size and a small p-value. Does this imply you have a meaningfully large regression coefficient? Why or why not?

10. Suppose you have small sample size and a large p-value. Does this imply you have a negligibly small regression coefficient? Why or why not?

For Exercises 11-18, use the Digitalis teaching dataset (`dig_rmph.rData`, see Appendix A.6) to answer questions related to the following research question: Is there a significant association between the predictor body mass index (`BMI`) and the outcome heart rate (`HEARTRTE`)? Fit the appropriate regression model and then answer the questions.

11. What kind of predictor is BMI, continuous or categorical?

12. Visualize the relationship (plot the outcome vs. the predictor with the regression line included).

13. When you fit the regression model, how many observations were removed due to missing values?

14. What proportion of the variation in heart rate is explained by BMI?

15. Provide the effect estimate (regression slope), its 95% CI, p-value, and a statement regarding whether the association is statistically significant.

16. Example 4.1 concluded "On average, for every 1-cm difference in waist circumference, adults differ in mean fasting glucose by 0.0278 mmol/L". Provide the corresponding interpretation of the regression slope in the regression of heart rate on BMI. Be careful to not imply causality, as this is cross-sectional data.

17. What is the predicted mean heart rate (and 95% CI) for those with a BMI of 20 kg/m^2? For those with a BMI of 35 kg/m^2?

18. Among those with a BMI of 20 kg/m^2, within what interval do we expect 95% of heart rate values to fall? Among those with a BMI of 35 kg/m^2?

For Exercises 19-23, use the Digitalis teaching dataset (`dig_rmph.rData`, see Appendix A.6) to answer questions related to the following research question: Does heart rate (`HEARTRTE`) differ significantly between treatment group (`TRTMT` = Placebo, Digoxin)? Fit the appropriate regression model and then answer the questions.

19. Visualize the relationship (plot the outcome vs. the predictor, along with the mean at each level and a line connecting the means).

20. Provide the p-value for the test of this association and state whether the association is statistically significant.

21. What is the mean and 95% CI for the difference in heart rate between those in the Digoxin group and those in the Placebo group? Is this difference statistically significant?

22. What is the estimated mean heart rate (and 95% CI) for each of the treatment groups?

23. For each treatment group, within what interval do we expect 95% of heart rates to fall?

For Exercises 24-25, use the UN Human Development Data and the methods in this chapter to answer each research question (`unhdd2020.rmph.rData`, see Appendix A.2).

24. What is the association between the outcome life expectancy at birth (years) (`life`) and expected years of schooling (years) (`educ_expected`)?

25. How does the Gender Inequality Index (`gii`) (GII) differ between Human Development Index (HDI) groups (`hdi_group`)?

For Exercises 26-27, use the CAMP teaching dataset `camp_0_48_rmph.rData` (see Appendix A.6).

26. Does post-bronchodilator Forced Expiratory Volume at 1 second (FEV1) (`POSFEV0`) differ significantly between treatment groups (Budesonide, Nedocromil, Placebo) (`TG`)?

27. Following up on the previous Exercise, estimate the two treatment effects (Budesonide vs. Placebo, Nedocromil vs. Placebo) along with their 95% confidence intervals and p-values.

5

Multiple Linear Regression

In this chapter, you will learn how to:

- Write and interpret a multiple linear regression equation;
- Examine variables prior to including them in a regression model;
- Fit a multiple linear regression model;
- Interpret the estimated regression coefficients;
- Distinguish between confounders, mediators, and moderators;
- Test and visualize interactions between predictors;
- Obtain predictions from the model;
- Compute three types of confidence or prediction intervals;
- Diagnose the fit of the model;
- Adjust the model to improve the fit, if needed;
- Examine collinearity of predictors and remove redundancies;
- Look for outliers and influential observations;
- Distinguish between confirmatory and exploratory analyses;
- Prevent an increase in Type I error when carrying out multiple tests;
- Carry out a sensitivity analysis to assess the robustness of results;
- Avoid overgeneralizing the implications of a regression model; and
- Appropriately summarize the methods and results of a multiple linear regression analysis.

Load `tidyverse` before proceeding. In addition, we will use a few custom functions found in the file `Functions_rmph.R` (downloadable from RMPH Resources[1]). To access these functions, `source()` the file containing their code.

```
library(tidyverse)
# Put Functions_rmph.R in same folder as your R project
source("Functions_rmph.R")
```

5.1 Introduction

In Chapter 4, we covered the basics of linear regression with just one predictor (simple linear regression, or SLR). Multiple linear regression (MLR) extends SLR to allow more than one predictor in the model. The effect of each predictor in an MLR model is "adjusted for" the others. MLR is a valuable tool for public health research, which often relies on observational data. Randomized experiments are not always possible in public health studies, resulting in the need to adjust for confounding. MLR is a common method of estimating the association between an exposure and a continuous outcome adjusted for confounding due to other variables.

[1] https://github.com/rwnahhas/RMPH_Resources

5.2 Notation and interpretation

The data are from n independent sets of observed values of an outcome Y and predictors $X_1, X_2, ..., X_K$. The data for the ith case (or individual, or observation) are denoted $(y_i, x_{i1}, x_{i2}, ..., x_{iK})$ – each case has an associated outcome value and set of predictor values. Equation 5.1 describes the multiple linear regression model.

$$Y = \beta_0 + \beta_1 X_1 + \beta_2 X_2 + \ldots + \beta_K X_K + \epsilon \tag{5.1}$$

β_0 is the **intercept**, the other β terms are the **effects** of the predictors, and ϵ is the **error** term, which is assumed to have a normal distribution with the same variance at all predictor values. We denote the assumption about the error term using the notation $\epsilon \sim N(0, \sigma^2)$, which is read "epsilon has a normal distribution with a mean of zero and a variance of sigma squared". The fact that σ^2 does not have a subscript i means that all individuals have the same error variance – individuals' true observed values vary about the values predicted by the regression, and how much they vary is assumed to not depend on the characteristics of any individual. The model described by Equation 5.1 *assumes* the relationship between Y and each $X_k, (k = 1, \ldots, K)$, is linear when all the other predictors are held constant, and the error term captures random variation between individuals, as well as outcome measurement error.

For a **continuous predictor** X_k, the corresponding regression coefficient β_k is a slope, interpreted as the difference in the mean Y associated with a one-unit difference in X_k when holding all other predictors fixed (or "controlling for" or "adjusted for" the other predictors).

As discussed in Chapter 4, if X_k is a **categorical predictor** with L levels, then instead of just one corresponding X term in the model, there are $L-1$ indicator variables. Each of the $L-1$ corresponding regression coefficients is interpreted as the difference between the mean Y at that level of X_k and the mean Y at the reference level, when holding all other predictors fixed (or "controlling for" or "adjusted for" the other predictors).

NOTE: The interpretations of the MLR regression coefficients are the same as in SLR except for the addition of "when holding all other predictors fixed". Imagine two groups of cases that have the same characteristics in all but one respect; they have the same predictor values for all Xs other than X_k and they differ in X_k by one unit. The model assumes that the mean outcome in these two groups differs by β_k. If the predictor in which they differ is categorical, then "differ by one unit" corresponds to "one group is at the reference level of X and the other is at a non-reference level".

Example 5.1: In the previous chapter, we estimated the relationship between fasting glucose (FG; mmol/L) (`LBDGLUSI`) and each of waist circumference (WC; `BMXWAIST`) and smoking status (`smoker`; Never, Past, Current) among a subset of adult participants in NHANES 2017-2018. We found that each was significantly associated with fasting glucose. However, we did not adjust for potential confounders of these relationships. In this chapter, we will use MLR to adjust each for confounding due to the other, as well as due to age (`RIDAGEYR`), gender (`RIAGENDR`; Male, Female), race/ethnicity (`RIDRETH3`; Mexican American, Other Hispanic, Non-Hispanic White, Non-Hispanic Black, Non-Hispanic Asian, Other/Multi), and `income` (<\$25,000, \$25,000 to <\$55,000, \$55,000+).

Based on Equation 5.1, the MLR model for Example 5.1 is written as follows.

$$\begin{aligned}
\text{FG} &= \beta_0 \\
&+ \beta_1 \text{WC} \\
&+ \beta_2 I(\text{Smoker} = \text{Past}) + \beta_3 I(\text{Smoker} = \text{Current}) \\
&+ \beta_4 \text{Age} \\
&+ \beta_5 I(\text{Gender} = \text{Female}) \\
&+ \beta_6 I(\text{Race} = \text{Other Hispanic}) + \beta_7 I(\text{Race} = \text{Non-Hispanic White}) \\
&+ \beta_8 I(\text{Race} = \text{Non-Hispanic Black}) + \beta_9 I(\text{Race} = \text{Non-Hispanic Asian}) \\
&+ \beta_{10} I(\text{Race} = \text{Other/Multi}) \\
&+ \beta_{11} I(\text{Income} = \$25{,}000 \text{ to } <\$55{,}000) + \beta_{12} I(\text{Income} = \$55{,}000+) + \epsilon
\end{aligned}$$

For each categorical predictor, the first level is left out, as it is the reference level.

5.3 Complete case analysis dataset

We will use a complete case analysis in this chapter, excluding all individuals (cases) that have missing values for any of the analysis variables (see Section 3.2.1). Complete case analysis is already the default for the `lm()` function, but if we want all our results, in particular the descriptive statistics in our "Table 1", to be based on the same sample as used to fit the regression model, then we must explicitly remove individuals with missing values ahead of time. See Chapter 9 for how to handle missing data using multiple imputation.

Example 5.1 (continued): Use `summary()` to assess the extent of missing data in the variables in our analysis and create a complete case analysis dataset.

```
load("Data/nhanes1718_adult_fast_sub_rmph.Rdata")
nhanesf <- nhanes_adult_fast_sub
rm(nhanes_adult_fast_sub)

nhanesf %>%
  select(LBDGLUSI, BMXWAIST, smoker, RIDAGEYR,
         RIAGENDR, RIDRETH3, income) %>%
  summary()
```

```
##     LBDGLUSI        BMXWAIST          smoker       RIDAGEYR      RIAGENDR
##  Min.   : 2.61   Min.   : 63.2   Never  :579   Min.   :20.0   Male  :457
##  1st Qu.: 5.33   1st Qu.: 88.3   Past   :264   1st Qu.:34.0   Female:543
##  Median : 5.72   Median : 97.8   Current:157   Median :47.0
##  Mean   : 6.09   Mean   :100.5                 Mean   :47.9
##  3rd Qu.: 6.22   3rd Qu.:111.0                 3rd Qu.:61.0
##  Max.   :19.00   Max.   :169.5                 Max.   :80.0
##                  NA's   :35
##                RIDRETH3                    income
##  Mexican American   :120   <$25,000            :164
##  Other Hispanic     : 71   $25,000 to <$55,000:224
##  Non-Hispanic White:602   $55,000+            :489
##  Non-Hispanic Black:115   NA's                :123
##  Non-Hispanic Asian: 48
##  Other/Multi        : 44
##
```

Waist circumference and income have missing values. The code below creates a complete case dataset using the method that removes all other variables from our dataset. See Section 3.2.1 for an alternative method that retains all the other variables.

```
nhanesf.complete <- nhanesf %>%
  select(LBDGLUSI, BMXWAIST, smoker, RIDAGEYR,
         RIAGENDR, RIDRETH3, income) %>%
  drop_na()
nrow(nhanesf)
```

```
## [1] 1000
```

```
nrow(nhanesf.complete)
```

```
## [1] 857
```

While the full dataset has 1000 observations, the complete case dataset has only 857.

5.4 Examine the data

In Chapter 3, we learned how to numerically and visually summarize variables one at a time. Here, we will apply those methods to examine the variables for a specific analysis. At this stage of the analysis, we are only looking for very obvious problems, such as values that are highly unusual, extremely skewed variables, or categorical variables with sparse levels. We also want to check that all categorical variables are coded as factors. At later stages, when we examine regression diagnostics, we may make modifications to variables in order to resolve lack of fit of the model.

5.4.1 Outcome

We need to make sure the outcome variable is appropriate for a linear regression. It should be continuous (not just a few unique values) and, ideally, normally distributed (or at least symmetric), although we will learn that normality is not essential if the sample size is large.

To be sure the outcome really is continuous, count the number of unique values. If there are only a few, then linear regression may not be appropriate; consider instead ordinal logistic regression (Section 6.22) or binary logistic regression (after dichotomizing the outcome) (Chapter 6). Next, create a numeric summary and plot a histogram of the outcome. Later, after you have learned more about regression diagnostics, you will be able to identify skew and outliers in these plots and may immediately realize that a transformation is needed. You will investigate these potential problems more closely when checking the model assumptions in Sections 5.14 to 5.17, but sometimes you can save time by noticing the need for a transformation before fitting the model.

Example 5.1 (continued): Examine a numeric summary and histogram of the outcome, fasting glucose.

```
# Count the number of unique values
length(unique(nhanesf.complete$LBDGLUSI))
```

```
## [1] 96
```

```
# Numerical summary
summary(nhanesf.complete$LBDGLUSI)
```

```
##    Min. 1st Qu.  Median    Mean 3rd Qu.    Max.
##    2.61    5.33    5.72    6.11    6.22   19.00
```

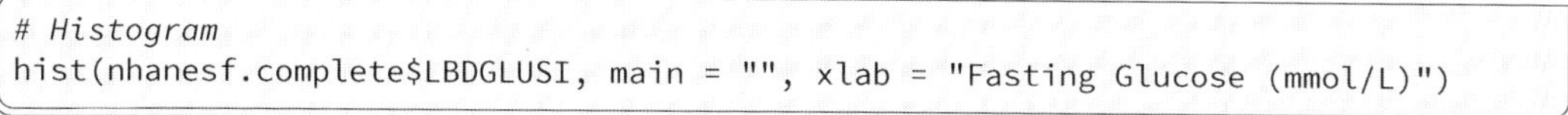

```
# Histogram
hist(nhanesf.complete$LBDGLUSI, main = "", xlab = "Fasting Glucose (mmol/L)")
```

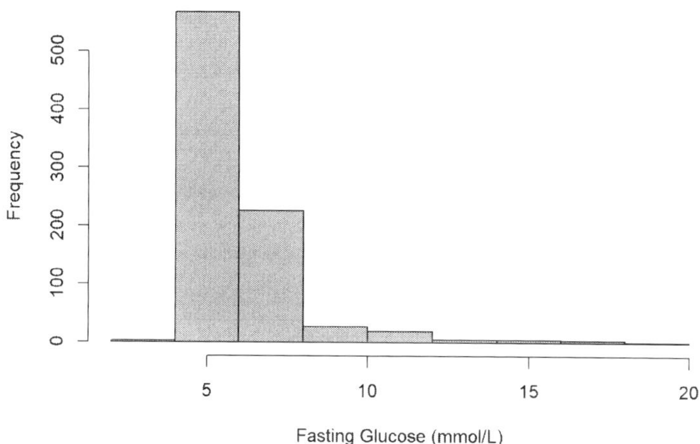

FIGURE 5.1 Histogram of fasting glucose (mmol/L)

The summary indicates that the outcome has many (96) unique values, so it is appropriate to treat it as continuous. While there are no strange values, Figure 5.1 shows that the outcome is skewed.

5.4.2 Continuous predictors

Count the number of unique values – if there are only a few, then you should consider treating the predictor as categorical instead. Create a numeric summary and plot a histogram for each continuous predictor. There is no requirement that continuous predictors be normally distributed; however, skewed distributions and extreme values can lead to a problem with influential observations (discussed in Section 5.22). What we are primarily looking for at this point are strange values and to be sure each predictor is coded as expected. For example, if what you thought should be a continuous predictor has only a few unique values, then perhaps the variable should be coded as categorical instead.

Example 5.1 (continued):

```
# Count the number of unique values
length(unique(nhanesf.complete$BMXWAIST))
```

```
## [1] 375
```

```
length(unique(nhanesf.complete$RIDAGEYR))
```

```
## [1] 61
```

```
# Numerical summary
summary(nhanesf.complete$BMXWAIST)
```

```
##    Min. 1st Qu.  Median    Mean 3rd Qu.    Max.
##    63.2    88.3    98.3   100.8   112.2   169.5
```

```
summary(nhanesf.complete$RIDAGEYR)
```

```
##    Min. 1st Qu.  Median    Mean 3rd Qu.    Max.
##    20.0    34.0    47.0    47.8    61.0    80.0
```

```
# Histograms
par(mfrow=c(1,2))
hist(nhanesf.complete$BMXWAIST, main = "", xlab = "Waist Circumference (cm)")
hist(nhanesf.complete$RIDAGEYR, main = "", xlab = "Age (years)")
```

The summaries indicate the following:

- Each predictor has many unique values indicating that each should indeed be treated as continuous.
- Figure 5.2 shows that waist circumference is slightly skewed, but neither waist circumference nor age have very extreme values. We will still, however, check for influential observations in Section 5.22 since these histograms can only show extreme values one-variable-at a time – what matters for MLR is how different a case is from the others when considering all the predictors simultaneously.
- Neither variable has any strange values

What if there were strange values? For example, if there were age values of, say, 142 or -5, then we should investigate further. Are these data entry errors that can be corrected? If they cannot be corrected with certainty, then it is reasonable to set these values to missing. Additionally, what if there were only a few unique values so you would like to code the variable as a factor instead? R code to accomplish these tasks can be found in Chapter 3 in An Introduction to R for Research[2] (Nahhas, 2024) and "Transform" in R for Data Science[3] (Wickham et al., 2023).

[2] https://bookdown.org/rwnahhas/IntroToR/managevars.html
[3] https://r4ds.hadley.nz/transform

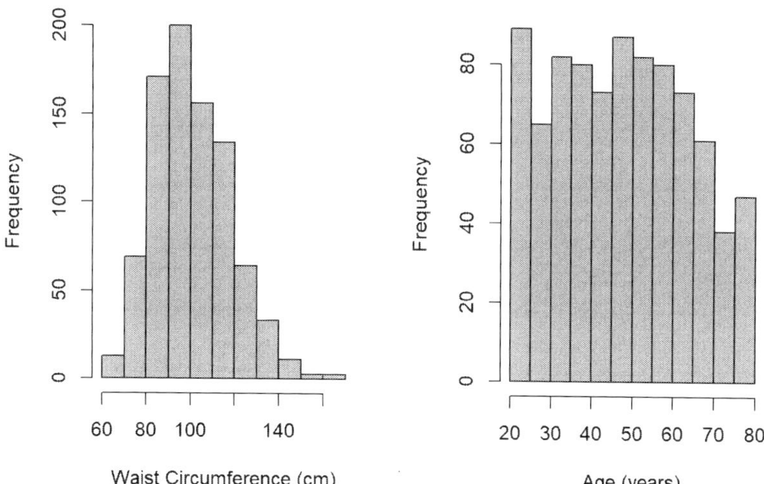

FIGURE 5.2 Histograms of continuous predictors

5.4.3 Categorical predictors

The main issue to watch out for with categorical predictors is sparse levels (levels that have a small sample size) since the estimated mean within a small group of cases will be very imprecise. Create a frequency table for each categorical predictor (optionally, also plot a bar chart). Look for levels that have only a few observations, as these may need to be collapsed with other levels. You could use, for example, 30, 50, or 100 as a cutoff, but any cutoff is subjective. If you are concerned that your regression results may depend too much on the choice of how to collapse categorical predictors, carry out a sensitivity analysis to compare the results under different scenarios (see Section 5.25). Finally, check that each categorical predictor is coded as a factor, recoding as a factor if not (see Section 4.4.1).

Example 5.1 (continued):

```
# Categorical predictors
table(nhanesf.complete$smoker, useNA = "ifany")
```

```
##
##   Never    Past Current
##     500     222     135
```

```
table(nhanesf.complete$RIAGENDR, useNA = "ifany")
```

```
##
##   Male Female
##    388    469
```

```
# table(nhanesf.complete$RIDRETH3, useNA = "ifany")
# Use count so the display is vertical
# (easier to read when there are many levels)
nhanesf.complete %>%
  count(RIDRETH3)
```

```
##              RIDRETH3   n
## 1    Mexican American  97
## 2      Other Hispanic  55
## 3 Non-Hispanic White 534
## 4 Non-Hispanic Black  98
## 5 Non-Hispanic Asian  36
## 6         Other/Multi  37
```

```
table(nhanesf.complete$income, useNA = "ifany")
```

```
##
##            <$25,000 $25,000 to <$55,000             $55,000+
##                 157                 221                  479
```

```
c(is.factor(nhanesf.complete$smoker),
  is.factor(nhanesf.complete$RIAGENDR),
  is.factor(nhanesf.complete$RIDRETH3),
  is.factor(nhanesf.complete$income))
```

```
## [1] TRUE TRUE TRUE TRUE
```

From the output we conclude the following:

- Each predictor has only a few levels, so treating each as categorical is appropriate.
- The race/ethnicity variable may have a few sparse levels.
- Each is coded as a factor.

5.4.4 Overall description of the data

As a partial alternative to the steps described above, you can use `Hmisc::describe()` (see Section 3.1.1) to get a detailed description of all the variables with one command. It does not provide all the information we need (e.g., whether or not a categorical variable is coded as a factor), but it is useful.

```
nhanesf.complete %>%
  select(LBDGLUSI, BMXWAIST, smoker, RIDAGEYR,
         RIAGENDR, RIDRETH3, income) %>%
  Hmisc::describe()
# (results not shown)
```

5.4.5 Collapsing sparse levels

This section demonstrates how to collapse a categorical predictor that has sparse levels. Here, we combine race/ethnicity levels that have 55 cases or fewer with other levels to create a new variable. By this definition, the "Mexican American" group is sparse and it makes sense to combine it with the "Hispanic" group. Additionally, "Non-Hispanic Asian" and "Other/Multi" are both small groups, so we combine them. Remember to always check any derivation by comparing observations before and after.

```
nhanesf.complete <- nhanesf.complete %>%
  mutate(race_eth = fct_collapse(RIDRETH3,
    "Hispanic"           = c("Mexican American", "Other Hispanic"),
    "Non-Hispanic Other" = c("Non-Hispanic Asian", "Other/Multi")))

# Check derivation
TAB <- table(nhanesf.complete$RIDRETH3, nhanesf.complete$race_eth,
             useNA = "ifany")
addmargins(TAB, 1) # Adds a row of column totals
```

```
##
##                       Hispanic Non-Hispanic White Non-Hispanic Black
##   Mexican American          97                  0                  0
##   Other Hispanic            55                  0                  0
##   Non-Hispanic White         0                534                  0
##   Non-Hispanic Black         0                  0                 98
##   Non-Hispanic Asian         0                  0                  0
##   Other/Multi                0                  0                  0
##   Sum                      152                534                 98
##
##                       Non-Hispanic Other
##   Mexican American                     0
##   Other Hispanic                       0
##   Non-Hispanic White                   0
##   Non-Hispanic Black                   0
##   Non-Hispanic Asian                  36
##   Other/Multi                         37
##   Sum                                 73
```

The two-way frequency table shows that the derivation code did what we intended ("Mexican American" and "Other Hispanic" were combined into "Hispanic"; "Non-Hispanic Asian" and "Other/Multi" were combined into "Non-Hispanic Other"). The "Sum" row provides the sample sizes in the new levels.

NOTE: This example is provided to illustrate how to go about collapsing a sparse predictor. In practice, think carefully about the context. By collapsing the way we did, we are assuming that the mean outcome (fasting glucose) is not meaningfully different between the groups of individuals in the levels that were combined. An alternative would be to exclude individuals in sparse levels from an analysis entirely, which would result in altering the scope of the analysis. For example, if we excluded those in the "Other/Multi" group, the results would not be generalizable to that group.

5.4.6 Visualizing the unadjusted relationships

Before fitting the model, visualize the association between the outcome and each predictor. These plots will not correspond exactly to the MLR model since they will not be adjusted for any other predictors, but they are a useful place to start just to make sure the predictors you are using are what you think they are (if the plot looks very strange, go back and look at your code to make sure you are using the variables you want to use and that they were coded correctly). Later, in Section 5.7, we will examine plots that are adjusted for the other predictors in the model.

Example 5.1 (continued): Plot the relationship between fasting glucose (`LBDGLUSI`) and each of waist circumference (`BMXWAIST`), smoking status (`smoker`), age (`RIDAGEYR`), gender

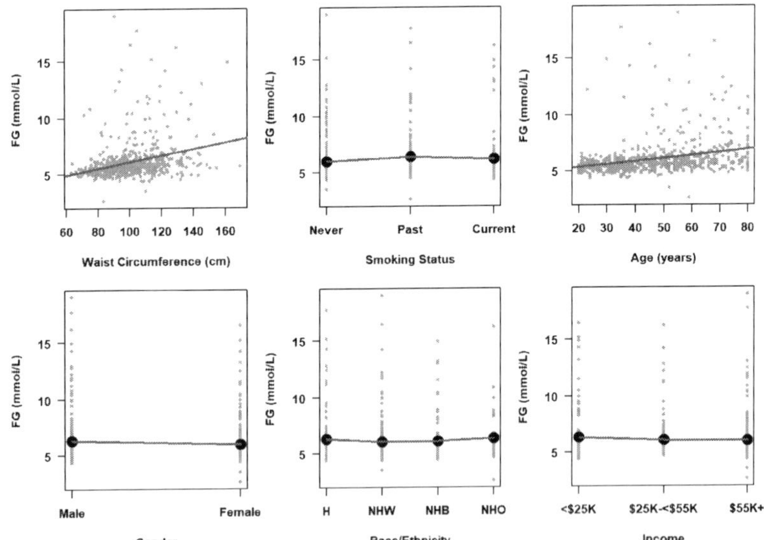

FIGURE 5.3 Unadjusted associations with fasting glucose (mmol/L)

(`RIAGENDR`), race/ethnicity (`RIDRETH3`), and `income`, one predictor at a time. The plots are shown in Figure 5.3.

You *could* use the code found in Section 4.1 to create each plot one at a time. Or, you can use the function `plotyx()` found in `Functions_rmph.R` (which you loaded at the beginning of this chapter). Each time you call `plotyx()`, pass the names of the specific outcome and predictor you are interested in plotting as arguments to the function. As a preliminary step, the code below makes copies of the race/ethnicity and income variables with shorter labels so they will fit in the plot.

```
# Rename the levels with shorter labels that fit in the plot
levels(nhanesf.complete$race_eth)
```

```
## [1] "Hispanic"           "Non-Hispanic White" "Non-Hispanic Black"
## [4] "Non-Hispanic Other"
```

```
nhanesf.complete$race_ethb <- nhanesf.complete$race_eth
levels(nhanesf.complete$race_ethb) <- c("H", "NHW", "NHB", "NHO")

levels(nhanesf.complete$income)
```

```
## [1] "<$25,000"             "$25,000 to <$55,000" "$55,000+"
```

```
nhanesf.complete$incomeb <- nhanesf.complete$income
levels(nhanesf.complete$incomeb) <- c("<$25K", "$25K-<$55K", "$55K+")
```

NOTE: The syntax is `plotyx("y", "x", data)` (outcome first, then predictor, then dataset) where `"y"` and `"x"` are character strings naming the variables you want to plot.

```
# If you have not already loaded this file...
# ...do so now to make plotyx() available
source("Functions_rmph.R")

# Change margins
# Default par(mar=c(5, 4, 4, 2) + 0.1) for c(bottom, left, top, right)
par(mar=c(4.5, 4, 1, 1))
par(mfrow=c(2,3))
plotyx("LBDGLUSI", "BMXWAIST",  nhanesf.complete,
       ylab = "FG (mmol/L)", xlab = "Waist Circumference (cm)")
plotyx("LBDGLUSI", "smoker",    nhanesf.complete,
       ylab = "FG (mmol/L)", xlab = "Smoking Status")
plotyx("LBDGLUSI", "RIDAGEYR",  nhanesf.complete,
       ylab = "FG (mmol/L)", xlab = "Age (years)")
plotyx("LBDGLUSI", "RIAGENDR",  nhanesf.complete,
       ylab = "FG (mmol/L)", xlab = "Gender")
# Using race_ethb and incomeb (the versions with shorter labels)
plotyx("LBDGLUSI", "race_ethb", nhanesf.complete,
       ylab = "FG (mmol/L)", xlab = "Race/Ethnicity")
plotyx("LBDGLUSI", "incomeb",   nhanesf.complete,
       ylab = "FG (mmol/L)", xlab = "Income")
```

5.5 Fitting the MLR model

This section demonstrates how to fit a MLR model in R. For comparison, we first look at the unadjusted models we fit in Chapter 4 (each only had one predictor variable). Second, we will fit a model adjusted for confounding due to other predictors. Since, in this chapter, we removed cases with missing values on any of a set of variables, the unadjusted results here differ slightly from those in Chapter 4.

5.5.1 Unadjusted

The unadjusted results are shown in Tables 5.1 and 5.2.

```
fit.ex4.1.cc  <- lm(LBDGLUSI ~ BMXWAIST, data = nhanesf.complete)
```

TABLE 5.1 Regression of fasting glucose (mmol/L) on waist circumference

Characteristic	Beta	95% CI	p-value
(Intercept)	3.204	2.574, 3.833	<0.001
Waist Circumference (cm)	0.0288	0.0227, 0.0350	<0.001

```
fit.ex4.2.cc  <- lm(LBDGLUSI ~ smoker,   data = nhanesf.complete)
```

TABLE 5.2 Regression of fasting glucose (mmol/L) on smoking status

Characteristic	Beta	95% CI	p-value
(Intercept)	5.96	5.81, 6.10	<0.001
Smoking Status			0.002
Never	—	—	
Past	0.451	0.192, 0.711	<0.001
Current	0.240	-0.072, 0.552	0.132

If just using the unadjusted models, we would conclude the following:

- There is a significant positive unadjusted association between fasting glucose and waist circumference (B = 0.029; 95% CI = 0.023, 0.035; p <.001).
- There is a significant association between fasting glucose and smoking status (p = .002). Past Smokers had significantly greater fasting glucose than Never Smokers (B = 0.451; 95% CI = 0.192, 0.711; p <.001). See Section 4.4 for how to re-level smoking in order to compare Past vs. Current.

However, as this is observational data, we should be wary of drawing conclusions from unadjusted models.

5.5.2 Adjusted

To fit an adjusted model, include multiple predictors in `lm()` separated by `+` signs. Then, just as with a SLR model, use `summary()`, `confint()`, and `car::Anova(, type = 3)` to summarize the results.

```
fit.ex5.1 <- lm(LBDGLUSI ~ BMXWAIST + smoker + RIDAGEYR +
      RIAGENDR + race_eth + income, data = nhanesf.complete)
summary(fit.ex5.1)
```

```
##
## Call:
## lm(formula = LBDGLUSI ~ BMXWAIST + smoker + RIDAGEYR + RIAGENDR +
##     race_eth + income, data = nhanesf.complete)
##
## Residuals:
##    Min     1Q Median     3Q    Max
## -3.694 -0.749 -0.234  0.290 13.033
##
## Coefficients:
##                             Estimate Std. Error t value             Pr(>|t|)
## (Intercept)                  2.97301    0.37665    7.89 0.0000000000000091 ***
## BMXWAIST                     0.02451    0.00306    8.00 0.0000000000000040 ***
## smokerPast                   0.21121    0.12515    1.69              0.09185 .
## smokerCurrent                0.09776    0.15063    0.65              0.51650
## RIDAGEYR                     0.02505    0.00331    7.58 0.0000000000000926 ***
## RIAGENDRFemale              -0.32795    0.10514   -3.12              0.00188 **
## race_ethNon-Hispanic White -0.50864    0.14524   -3.50              0.00049 ***
## race_ethNon-Hispanic Black -0.24764    0.19919   -1.24              0.21411
## race_ethNon-Hispanic Other  0.01644    0.21618    0.08              0.93938
## income$25,000 to <$55,000  -0.10038    0.16273   -0.62              0.53753
## income$55,000+             -0.09381    0.14724   -0.64              0.52421
## ---
```

```
## Signif. codes:  0 '***' 0.001 '**' 0.01 '*' 0.05 '.' 0.1 ' ' 1
##
## Residual standard error: 1.51 on 846 degrees of freedom
## Multiple R-squared:  0.171,  Adjusted R-squared:  0.161
## F-statistic: 17.4 on 10 and 846 DF,  p-value: <0.0000000000000002
```

```
confint(fit.ex5.1)
```

```
##                                   2.5 %   97.5 %
## (Intercept)                     2.23373  3.71228
## BMXWAIST                        0.01850  0.03052
## smokerPast                     -0.03443  0.45685
## smokerCurrent                  -0.19789  0.39342
## RIDAGEYR                        0.01856  0.03154
## RIAGENDRFemale                 -0.53431 -0.12158
## race_ethNon-Hispanic White -0.79371 -0.22356
## race_ethNon-Hispanic Black -0.63860  0.14331
## race_ethNon-Hispanic Other -0.40786  0.44075
## income$25,000 to <$55,000 -0.41978  0.21903
## income$55,000+                 -0.38281  0.19519
```

```
car::Anova(fit.ex5.1, type = 3)
```

```
## Anova Table (Type III tests)
##
## Response: LBDGLUSI
##             Sum Sq  Df F value             Pr(>F)
## (Intercept)    142   1   62.30 0.0000000000000091 ***
## BMXWAIST       146   1   64.02 0.0000000000000040 ***
## smoker           7   2    1.44            0.23638
## RIDAGEYR       131   1   57.42 0.0000000000000926 ***
## RIAGENDR        22   1    9.73            0.00188 **
## race_eth        39   3    5.68            0.00075 ***
## income           1   2    0.23            0.79082
## Residuals     1929 846
## ---
## Signif. codes:  0 '***' 0.001 '**' 0.01 '*' 0.05 '.' 0.1 ' ' 1
```

Table 5.3 combines all of the output into a single table.

```
library(gtsummary)
TABLE <- fit.ex5.1 %>%
  tbl_regression(intercept = T,
               estimate_fun = function(x) style_sigfig(x, digits = 3),
               pvalue_fun   = function(x) style_pvalue(x, digits = 3),
               label  = list(BMXWAIST ~ "Waist Circumference (cm)",
                             smoker   ~ "Smoking Status",
                             RIDAGEYR ~ "Age (years)",
                             RIAGENDR ~ "Gender",
                             race_eth ~ "Race/Ethnicity",
                             income   ~ "Annual Income")) %>%
  add_global_p(keep = T) %>%
  modify_caption("Linear regression results for fasting glucose (mmol/L) vs.
               waist circumference (cm) and smoking, adjusted for demographics")
```

```
TABLE
```

TABLE 5.3 Linear regression results for fasting glucose (mmol/L) vs. waist circumference (cm) and smoking, adjusted for demographics

Characteristic	Beta	95% CI	p-value
(Intercept)	2.97	2.23, 3.71	<0.001
Waist Circumference (cm)	0.025	0.018, 0.031	<0.001
Smoking Status			0.236
Never	—	—	
Past	0.211	-0.034, 0.457	0.092
Current	0.098	-0.198, 0.393	0.516
Age (years)	0.025	0.019, 0.032	<0.001
Gender			0.002
Male	—	—	
Female	-0.328	-0.534, -0.122	0.002
Race/Ethnicity			<0.001
Hispanic	—	—	
Non-Hispanic White	-0.509	-0.794, -0.224	<0.001
Non-Hispanic Black	-0.248	-0.639, 0.143	0.214
Non-Hispanic Other	0.016	-0.408, 0.441	0.939
Annual Income			0.791
<$25,000	—	—	
$25,000 to <$55,000	-0.100	-0.420, 0.219	0.538
$55,000+	-0.094	-0.383, 0.195	0.524

How do we interpret these regression coefficients? The regression equation is as follows.

$$\begin{aligned}
\text{FG} &= \beta_0 \\
&+ \beta_1 \text{WC} \\
&+ \beta_2 I(\text{Smoker} = \text{Past}) + \beta_3 I(\text{Smoker} = \text{Current}) \\
&+ \beta_4 \text{Age} \\
&+ \beta_5 I(\text{Gender} = \text{Female}) \\
&+ \beta_6 I(\text{Race} = \text{Non-Hispanic White}) + \beta_7 I(\text{Race} = \text{Non-Hispanic Black}) \\
&+ \beta_8 I(\text{Race} = \text{Non-Hispanic Other}) \\
&+ \beta_9 I(\text{Income} = \$25{,}000 \text{ to } <\$55{,}000) + \beta_{10} I(\text{Income} = \$55{,}000+) + \epsilon
\end{aligned}$$

The intercept is the mean outcome when all continuous predictors are 0 and all categorical predictors are at their reference level.

- In this example, the intercept (2.97 mmol/L) corresponds to the estimated fasting glucose for individuals with a waist circumference of 0 cm who have never smoked, are age = 0 years, male, Hispanic, and have <$25,000 annual household income. Waist circumference = 0 cm is, of course, not possible; also, the model was fit to data on adults so age = 0 years is far beyond the range of ages analyzed. Had we centered waist circumference and age, then the intercept would be more interpretable (see Section 4.3.4). The lack of centering here does not invalidate the model, it only makes it so the intercept does not correspond to a prediction for a plausible combination of predictor values.

For a continuous predictor, the regression coefficient is the estimated difference in the outcome associated with a 1-unit difference in the predictor, when holding all other predictors fixed.

- When holding all other predictors fixed, individuals with 1 cm greater waist circumference are estimated to have, on average, 0.025 mmol/L greater fasting glucose. "When holding all other predictors fixed" means that this statement is the same for any combination of the other predictors; it does not matter at what values you hold the other predictors fixed. Another way to say this is that individuals with the same smoking status, age, gender, race/ethnicity, and household income who differ in waist circumference by 1 cm differ in average fasting glucose by 0.025 mmol/L.
- The estimated difference in the outcome associated with an other than 1-unit difference is obtained by multiplying the coefficient by the units. For example, when holding all other predictors fixed, individuals with 5 cm greater waist circumference are estimated to have, on average, $5 \times 0.025 = 0.125$ mmol/L greater fasting glucose.

For a categorical predictor with L levels, there will be $L-1$ regression coefficients, each representing the difference in mean outcome between individuals at that level and individuals at the reference level, when holding all other predictors fixed.

- Comparing individuals who have the same values for all predictors except smoking status, past smokers have an average fasting glucose that is 0.211 mmol/L greater than never smokers.

The summaries below compare the results for each of waist circumference and smoking status before and after adjusting for the other predictors.

- Both before (B = 0.029; 95% CI = 0.023, 0.035; p <.001) and after adjusting for confounding (B = 0.025; 95% CI = 0.018, 0.031; p <.001), we found a significant association between fasting glucose and waist circumference. Additionally, the magnitude of the association did not change much, indicating that there was not much confounding.
- Before adjusting for confounding, we found a significant association between fasting glucose and smoking status (p = .002). However, after adjusting for waist circumference, age, gender, race/ethnicity, and income, there is no significant association between fasting glucose and smoking status (p = .236). Additionally, after adjusting for confounding, the magnitude of the regression coefficients were strongly attenuated (much closer to 0). Thus, for smoking status, failure to adjust for confounding was strongly biasing the results.

5.6 Residuals

Many regression concepts and diagnostic tools we will discuss use **residuals** to diagnose assumptions about the error term ϵ – the part of the outcome *not* explained by the MLR model. Based on the MRL model

$$Y = \beta_0 + \beta_1 X_1 + \beta_2 X_2 + \ldots + \beta_K X_K + \epsilon$$

we get the following formula for the predicted value for the ith observation given its predictor values:

$$\hat{y}_i = \hat{\beta}_0 + \hat{\beta}_1 x_{i1} + \hat{\beta}_2 x_{i2} + \ldots + \hat{\beta}_K x_{iK}.$$

The hat (^) on top of a term signifies that it is a *prediction* (of y_i) or *estimate* (of one of the βs), and the lowercase x's are the specific values of the predictors at which we are computing the predicted value. $\hat{y}_i$ is our best guess of the outcome given these predictor values. Notice that there is no random error term in this second equation – this is not the equation of a model but rather the equation for obtaining a prediction and, given the fitted model and a set of predictor values, the prediction is the same every time.

Every individual with the same set of predictor values has the same predicted value of the outcome. But, of course, not every individual has the same *observed* value of the outcome. For each individual, the difference between their observed and predicted outcome is their **residual**. If an observation has a residual of 0, then the prediction from the model is exactly the same as the observed value – the model captures this observation perfectly. If an observation has a large residual, however, then the prediction from the model is far from the observed value – the model does not capture this observation well.

The **prediction** for the ith observation, $\hat{y}_i$, is also called the **fitted value**. The **unstandardized residual** is the difference between the observed outcome and the fitted value. For example, if $\hat{\beta}_0$ and $\hat{\beta}_1$ are the estimates of the intercept β_0 and slope β_1 in a SLR with a continuous predictor, then the residual e_i for the ith observation is expressed as

$$e_i = y_i - \hat{y}_i = y_i - \left(\hat{\beta}_0 + \hat{\beta}_1 x_i\right).$$

Visually, a residual is the perpendicular distance from an observed value to the regression line, as shown in Figure 5.4 for two observations. Points above the line have positive residuals, and those below the line have negative residuals.

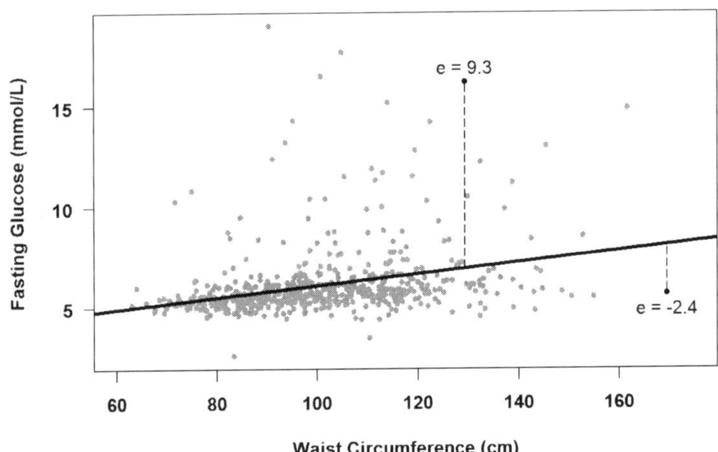

FIGURE 5.4 What is a residual?

5.6.1 Computing residuals

Many diagnostic tools that use residuals automatically compute them for you, but there may be times you need to compute them yourself. The basic definition of a residual given in

Section 5.6 is for an **unstandardized residual** – the raw difference between the observed and fitted values. Unstandardized residuals are appropriate if you want to examine the residuals on the same scale as the outcome. However, if you want to compare the magnitude of residuals to an objective standard, then you must first convert them to a standardized scale.

Standardized residuals are residuals divided by an estimate of their standard deviation, the result being that they have a standard deviation very close to 1, no matter what the scale of the outcome. **Studentized residuals** are similar to standardized residuals except that, for each case, the residual is divided by the standard deviation estimated from the regression with that case removed. For Studentized residuals, the objective standard to which they may be compared is the t distribution with degrees of freedom equal to the model residual degrees of freedom minus 1. For large sample sizes, this is very close to a standard normal distribution and Studentized residuals can be thought of as the number of standard deviations a case's outcome value is away from the regression line. For example, a Studentized residual of 2 corresponds to a point that is approximately 2 standard deviations above the regression line.

The `residuals()` (or `resid()`), `rstandard()`, and `rstudent()` functions compute unstandardized, standardized, and Studentized residuals, respectively. For example,

```
r1  <- resid(fit.ex5.1)
r2  <- rstandard(fit.ex5.1)
r3  <- rstudent(fit.ex5.1)
# Compare means and standard deviations
COMPARE <- round(data.frame(Mean = c(mean(r1), mean(r2), mean(r3)),
                            SD   = c(sd(r1),   sd(r2),   sd(r3))), 5)
rownames(COMPARE) <- c("Unstandardized", "Standardized", "Studentized")
COMPARE
```

```
##                   Mean    SD
## Unstandardized 0.00000 1.501
## Standardized   0.00004 1.002
## Studentized    0.00234 1.016
```

You can also extract the unstandardized residuals from an `lm()` object.

```
r1 <- fit.ex5.1$residuals
```

All three types of residuals have a mean of approximately 0 (for unstandardized residuals, it will be exactly 0), and standardized and Studentized residuals each have a standard deviation that is approximately 1. This makes standardized and Studentized residuals each comparable to an objective standard. As shown in Figure 5.5, the distribution of all three types of residuals will have approximately the same shape.

5.7 Visualizing the adjusted relationships

Earlier, we plotted the outcome vs. each predictor to visualize the *unadjusted* relationships. How do we visualize the *adjusted* relationships? Remember that now the regression coefficients

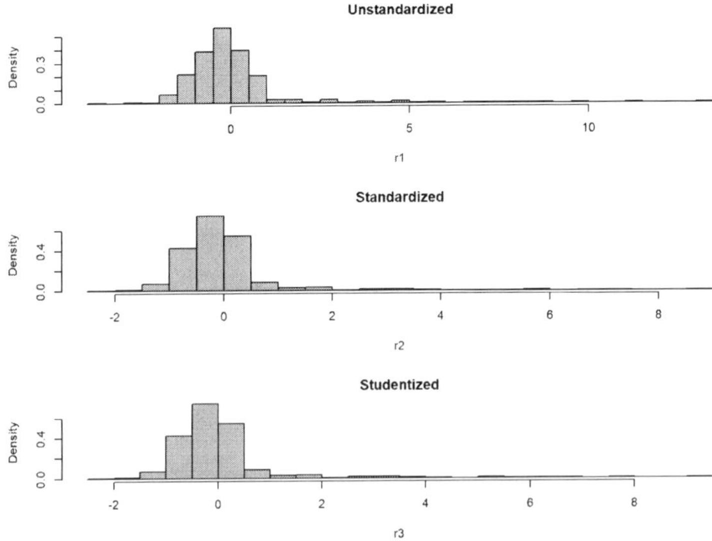

FIGURE 5.5 Three different types of residuals

are adjusted for all other terms in the model. So the relationship we would like to see is the relationship between the outcome and each predictor after removing the effect of all the other predictors. We can do this using an **added variable plot**, using the following steps.

- Denote the outcome as Y, the predictor of interest as X, and the other predictors as $Z_1, Z_2, ...$ (collectively denoted as $\mathbf{Z}$).
- Regress Y on $\mathbf{Z}$ and store the residuals (R_{yz}) (the part of the outcome not explained by the other predictors).
- Regress X on $\mathbf{Z}$ and store the residuals (R_{xz}) (the part of the predictor of interest not explained by the other predictors).
- Plot R_{yz} vs. R_{xz}.

Even though each of these steps is a linear regression, the method works even if X is binary or one of the indicator variables from a factor variable with more than two levels (as illustrated below). Fortunately, you do not have to actually do these steps yourself. They are automatically carried out by the `car::avPlots()` function (Fox et al., 2023; Fox and Weisberg, 2019). Figure 5.6 illustrates this function for a few predictors.

```
# To plot for all the predictors.
# car::avPlots(fit.ex5.1, ask = F, layout = c(2,3))

car::avPlots(fit.ex5.1, terms = . ~ BMXWAIST + RIDAGEYR + smoker)
```

Notice the axis labels: the vertical axis is labeled "LBDGLUSI | others" which corresponds to the outcome given all the predictors other than the one on the horizontal axis. The horizontal axis on each plot tells you to which predictor each added variable plot corresponds. For example, "RIDAGEYR | others" corresponds to age adjusted for all the other predictors. For a categorical predictor, there is a plot for each non-reference level. Each level of a categorical predictor has two possible values (0 or 1) so the added variable plot will have, generally, two clouds of points, one for cases where that level is 0 and one for cases where that level is 1

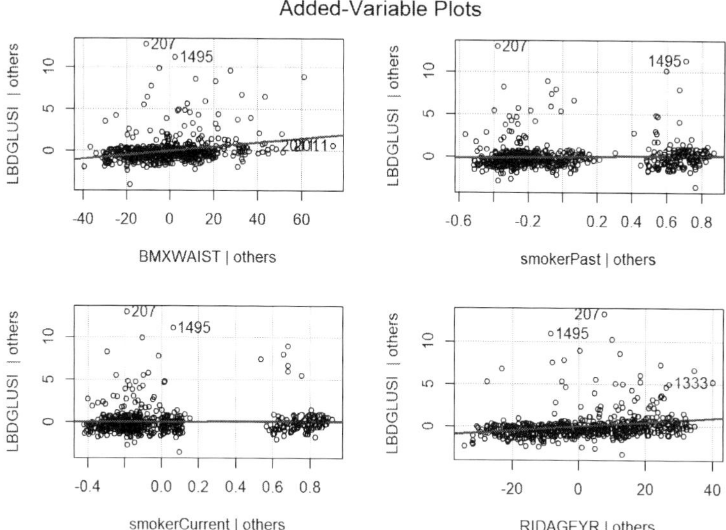

FIGURE 5.6 Added variable plots

(but sometimes a cloud could be split into multiple clouds due to the adjustment for other predictors).

The slope of each line in each panel is the regression coefficient for that term in the model. Thus, if the line is exactly horizontal, then there is no association between Y and X after adjusting for the other predictors. In our example, we see that the added-variable plot for waist circumference has a positive slope, corresponding to the positive adjusted regression coefficient in Table 5.3.

5.8 Types of predictor variables

In the discussion so far, we have included multiple predictors in a model without making any explicit distinction between their roles. This section assumes there is a single **primary predictor** of interest for which you would like to test the association with the outcome. In this context, we discuss the distinctions between the roles of other predictors, which might be confounders, mediators, or moderators, and illustrate these distinctions using **causal diagrams**.

5.8.1 Confounder

In many studies, there is a primary predictor of interest and the goal is to obtain an unbiased estimate of the effect of that predictor on the outcome. The causal diagram, ignoring all other variables, is illustrated in Figure 5.7.

FIGURE 5.7 Association between predictor and outcome

However, especially in an observational study, there may be confounders – variables that are associated with both the predictor and the outcome and are not in the causal pathway from predictor to outcome, as illustrated in Figure 5.8.

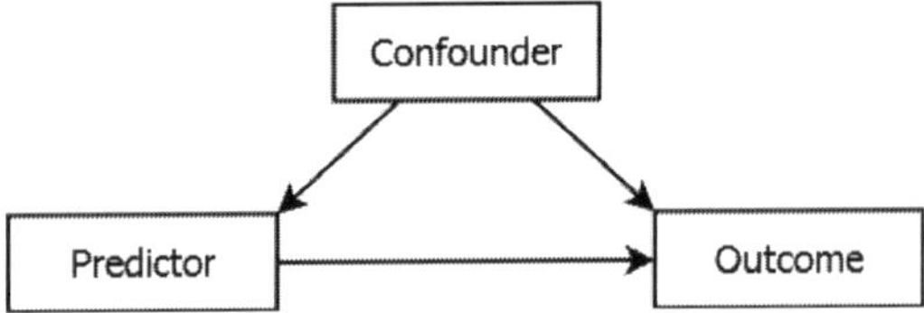

FIGURE 5.8 Confounded association

Suppose the confounder is positively associated with both the predictor and the outcome, but the predictor is not actually associated with the outcome. Failure to adjust for confounding may lead to the spurious (incorrect) conclusion that the predictor *is* associated with the outcome; differences in the outcome corresponding to differences in the predictor are actually due to differences in the confounder. When there is an association between the predictor and outcome, failure to adjust for confounding may result in not identifying that association, or over- or underestimating it. Including the confounder in the regression model along with the predictor is an attempt to adjust for confounding. Confounding can be adjusted for in the study design (e.g., matching, randomization, restriction of the scope) or analysis (e.g., stratification, standardization, regression adjustment). In this text, we focus only on regression adjustment.

For example, suppose a researcher's goal is to estimate the effect of weight loss on metabolic syndrome, defined as having at least three of the following five risk factors: (1) large waist circumference, (2) high blood pressure or taking blood pressure medication, (3) elevated triglycerides, (4) elevated fasting glucose or taking medication to lower glucose, and (5) low high-density lipoprotein (for specific cutoffs, see http://my.clevelandclinic.org/health/articles/metabolic-syndrome, accessed January 4, 2021). However, the effect of weight loss on metabolic syndrome may be confounded by income, as illustrated in Figure 5.9.

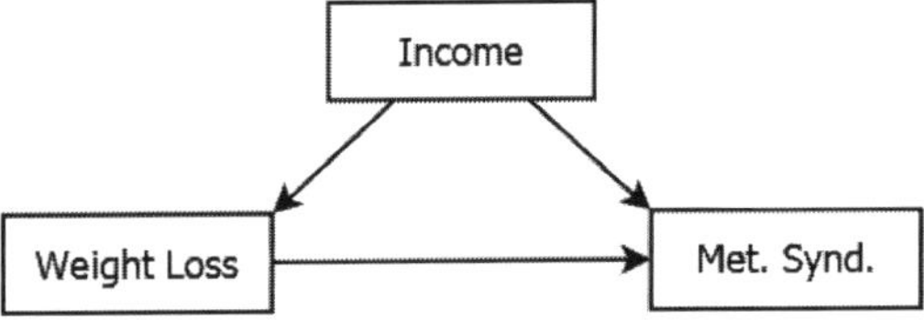

FIGURE 5.9 Income confounding the association between weight loss and metabolic syndrome

Individuals with fewer financial resources may find it more difficult to lose weight and also may be more likely to have poor metabolic characteristics. Thus, the effect of weight loss is confounded with the effect of income. In order to estimate the true (unconfounded) association between weight loss and metabolic syndrome in an observational study of individuals spanning a range of income levels, you need to adjust for potential confounding due to income. Failure to adjust for income may result in obtaining a biased estimate of association; part of the unadjusted "weight loss effect" may, in fact, be an "income effect".

5.8.2 Mediator

A mediator is like a confounder in that it is associated with both the predictor and the outcome. However, unlike a confounder, it *is* in the causal pathway. When mediation is present, the predictor typically has both a direct effect on the outcome (not through the mediator) and an indirect effect (through the mediator), as illustrated in Figure 5.10. If you are interested in the total effect of the predictor on the outcome, regardless of the pathway, then do not adjust for a mediator; doing so would bias the predictor's effect estimate by removing some of the effect you are actually interested in.

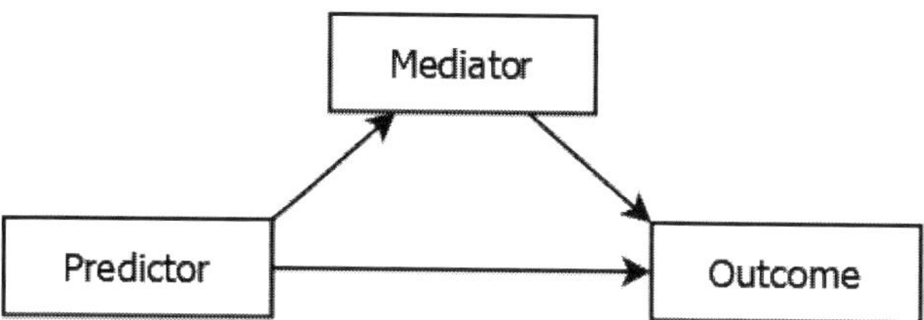

FIGURE 5.10 Mediated association

For example, adipocytokines may mediate the effect of weight loss on metabolic syndrome (Matsuzawa, 2006; Rolland et al., 2011), as illustrated in Figure 5.11.

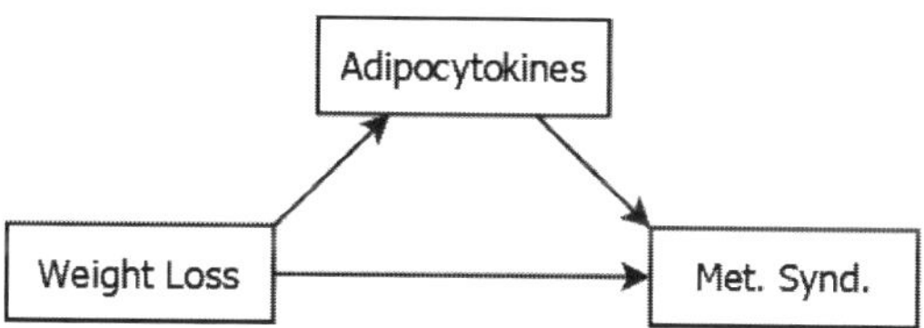

FIGURE 5.11 Adipocytokines mediating the association between weight loss and metabolic syndrome

Weight loss leads to an improvement in the characteristics that define metabolic syndrome, in part, due to its effect on levels of adipocytokines. Adjusting for adipocytokine levels would result in attenuating the estimate of the total effect of weight loss since you would be removing part of its effect (the indirect effect through adipocytokines). Therefore, if you are interested in the total effect of weight loss, do not adjust for adipocytokine levels. Compare this to the role of income as a confounder in Figure 5.9. Income *precedes* weight loss in the causal pathway – the effect of weight loss on metabolic syndrome is not through changes in income, but might be explained by differences in income between those who differ in weight loss. Thus, in these examples, income is a confounder while adipocytokine level is a mediator.

Another example is related to the study of health disparities. Should you "control for" socioeconomic status (SES) when studying racial disparities in health outcomes? In such a study, the health outcomes of individuals of different race/ethnicities are compared. "Race/ethnicity," a social not biological construct, acts as a proxy for structural racism, the actual reason for disparities (American Medical Association, 2020). In the U.S., SES is correlated with both race/ethnicity and health outcomes which may lead one to believe it is a confounder that should be adjusted for. However, SES is in the causal pathway between racism and health outcomes and therefore is a mediator of their association (see Figure 5.12).

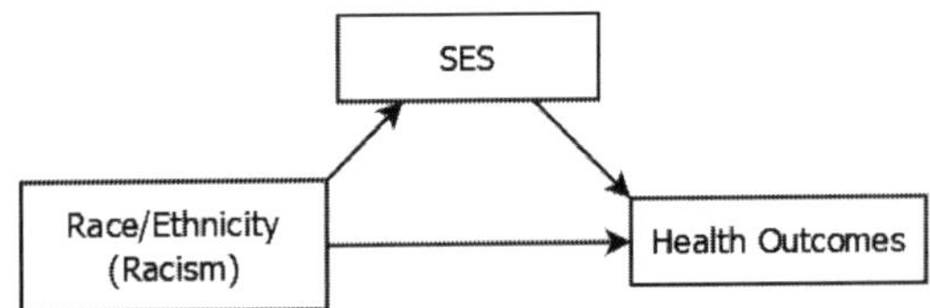

FIGURE 5.12 Socioeconomic status mediating racial disparities in health outcomes

If you "control for SES" you will remove part of the effect you are trying to estimate and underestimate disparities in health outcomes, perhaps even concluding there are no disparities. For example, Yehia et al. (2020), after adjusting for a number of demographic variables, conclude that race/ethnicity is not associated with COVID-19 mortality. However, Katikireddi et al. (2021) contend that this conclusion is in error exactly because the analysis adjusted for mediators. See also, for example, Meghani and Chittams (2015) regarding SES, and Zalla et al. (2021) and Schnake-Mahl and Bilal (2021) regarding the role of geography as a mediator of racial disparities in COVID-19 mortality.

Investigating the nature and magnitude of mediation, and decomposing the total effect into its direct and indirect components, is the realm of **mediation analysis** and is beyond the scope of this text (see, for example, Hayes (2022)). However, even if you are interested only in the total effect, it is vital to understand the distinction between mediators and confounders, and to not include mediators in a regression model — including a mediator will adjust out part of the very effect you are trying to estimate.

5.8.3 Moderator

In the above examples, there is a single predictor effect. In the case of confounding, the effect is obscured but there is still just one effect and the solution is to adjust for the confounder in the design or analysis. In the case of mediation, the effect is in part due to another variable but, again, there is still just a single effect of interest. Some variables, however, are moderators (or effect modifiers) – the effect of the predictor on the outcome *depends on* and *varies with* the level of the moderator. The predictor does not have a single effect, but rather a range of effects spanning the range of values of the moderator, as illustrated in Figure 5.13. The multiple lines going from Predictor to Outcome correspond to multiple magnitudes of association, with values that depend on the value of the moderator.

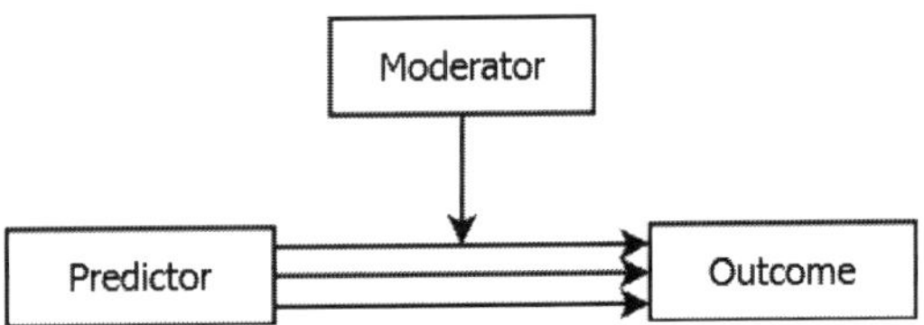

FIGURE 5.13 Moderated association

A moderator in a regression model is a term that is involved in an interaction (discussed in Section 5.9). In a regression model, include both the moderator and its interaction with the predictor.

For example, the effect of weight loss on metabolic syndrome may be moderated by baseline metabolic characteristics (see Figure 5.14) . Weight loss might have a greater impact among individuals with more room for improvement. By including the baseline measurement and a baseline $\times$ weight loss interaction in the regression model, you can estimate how the weight loss effect varies between those with different baseline metabolic characteristics.

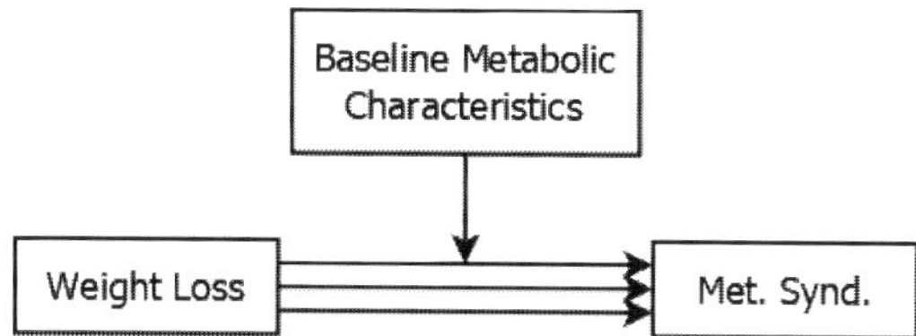

FIGURE 5.14 Baseline values moderating the association between weight loss and metabolic syndrome

Including a variable as a moderator also takes care of any confounding bias due to that variable, but in a different way than when including a confounder without an interaction. Including an interaction is similar to stratifying the analysis by another variable and estimating the effect of a variable within levels of another. If you stratify a regression analysis by a confounder that is not a moderator, then within each stratum you would get (approximately) the same effect. The regression coefficients would be approximately the same between strata (exactly the same in theory, but in practice they would vary, just not meaningfully). That single within-strata effect may be different than the effect if you ignored the confounder, but it would be the same within strata. By stratifying, you are removing the confounding – the confounder does not vary within strata so is no longer associated with the predictor or outcome within strata. When you "adjust" for a confounder in a regression, this is sort of what is happening, but mathematically it is different than stratifying. If you were to stratify your analysis by a moderator, however, then you would get *different* effects between strata – the regression coefficient for the predictor would vary between strata.

5.9 Interactions

In the MLR models we have encountered so far, each regression coefficient is interpreted as the effect of a predictor while holding other predictors at fixed values and the effect does not depend on at which specific values you hold the other predictors. How do we model a scenario in which the effect *does* depend on the level of one of the other variables? This is accomplished by including an **interaction** in the model and treats the other variable as a moderator (effect modifier) (Section 5.8). An interaction between two predictors is referred to as a **two-way interaction**. It is possible to have three-way (or more way) interactions, but this section will only discuss two-way interactions.

5.9.1 Understanding a two-way interaction using stratification

Effect modification and the use of interactions tend to be challenging for many students of regression when first encountered. Rather than diving straight into the details of including an interaction in a regression model, we start by clarifying what an interaction is through the use of stratification. As described at the end of Section 5.8.3, including an interaction is similar to stratifying the analysis by another variable and estimating the effect of a variable within levels of another.

Example 5.2: In the regression of fasting glucose on waist circumference, does the effect of waist circumference on fasting glucose depend on gender? Stratify by gender, fit a regression model for each gender, and compare the waist circumference effects between the two models. Are they the same or different? If they are different, then gender is an effect modifier (it modifies the waist circumference association with fasting glucose).

```
# Stratify by gender
nhanesf.complete.M <- nhanesf.complete %>%
  filter(RIAGENDR == "Male")
nhanesf.complete.F <- nhanesf.complete %>%
  filter(RIAGENDR == "Female")
```

Run the following to verify that the code results in datasets of the correct size.

```
# Check
table(nhanesf.complete$RIAGENDR, useNA = "ifany")
nrow(nhanesf.complete.M)
nrow(nhanesf.complete.F)
```

```
# Fit for Males
fit.ex5.2.strat.M <- lm(LBDGLUSI ~ BMXWAIST, data = nhanesf.complete.M)
# Fit for Females
fit.ex5.2.strat.F <- lm(LBDGLUSI ~ BMXWAIST, data = nhanesf.complete.F)

# Fit ignoring gender (for comparison)
# (fit earlier)
# fit.ex4.1.cc  <- lm(LBDGLUSI ~ BMXWAIST, data = nhanesf.complete)
coef(fit.ex4.1.cc)["BMXWAIST"]
```

```
## BMXWAIST
##  0.02883
```

```
# Males only
coef(fit.ex5.2.strat.M)["BMXWAIST"]
```

```
## BMXWAIST
##  0.04024
```

```
# Females only
coef(fit.ex5.2.strat.F)["BMXWAIST"]
```

```
## BMXWAIST
##   0.01988
```

The crude (unadjusted) regression coefficient for waist circumference was 0.029; however, this effect differs between males (0.040) and females (0.020).

An obvious question is, "How different do stratum-specific effects have to be before concluding there is meaningful effect modification?" There is no absolute answer to this question. In the next sections you will learn how to test if the effects significantly differ between strata. However, as with all hypothesis tests, this only clarifies statistical significance, not practical significance. Therefore, we will also use visualizations to help assess if the magnitude of effect modification is meaningful.

5.9.2 Including an interaction in a regression model

While one *could* use stratification to handle effect modification, typically it is done by including in the model both the effect modifier and the **interaction** between the effect modifier and the predictor of interest. Advantages of this method include the ability to test the significance of an interaction (are the effects significantly different between strata?) and to estimate how the effect modifier is associated with the outcome, neither of which is possible using stratification.

When including an interaction between two predictors in a model, include each of the predictors individually (the **main effects**) as well as their interaction. The syntax in R is `lm(Y ~ X + Z + X:Z)` where `X:Z` is the interaction term.

NOTES:

- Interaction is not the same as correlation. Correlation has to do with how two variables are related to each other whereas interaction has to do with how each of two variables impacts the effect of the other on the outcome.
- When there are just two predictors – the predictor of interest and a categorical effect modifier – stratification and interaction will lead to the same stratum-specific effect estimates for the predictor of interest. When there are additional predictors in the model, this may not be the case. Stratification is equivalent to interacting the effect modifier with *every* other predictor in the model whereas including an interaction only interacts the effect modifier with the predictor of interest. Even without additional predictors, stratification and interaction will not lead to exactly the same confidence intervals and p-values since the interaction model assumes the error variance is the same across all strata while the stratified model does not.
- Unless the value of 0 is meaningful, it can be helpful to center any continuous predictor that is part of an interaction, as that will make the intercept and main effects more interpretable. Centering is not required, however, and does not change the fit of the model.

Example 5.2 (continued): In the regression of fasting glucose on waist circumference, does the effect of waist circumference on fasting glucose depend on gender? Use an interaction to answer this question. Just for comparison, since this is the first model with an interaction we have fit, here we first fit a model that includes waist circumference and gender as a confounder but with no interaction, then we fit the model that includes both main effects and a waist circumference × gender interaction to answer the question.

```
# Fit with no interaction (for comparison)
fit.ex5.2.noint <- lm(LBDGLUSI ~ BMXWAIST + RIAGENDR,
                      data = nhanesf.complete)

# Fit with interaction
fit.ex5.2.int <- lm(LBDGLUSI ~ BMXWAIST + RIAGENDR +
                      BMXWAIST:RIAGENDR,
                    data = nhanesf.complete)
```

5.9.3 Regression equation with no interaction

Before explaining the output from the model with an interaction, let's review the interpretation of the output of the model without an interaction. The model with no interaction is written as

$$\text{FG} = \beta_0 + \beta_1 \text{WC} + \beta_2 I(\text{Gender} = \text{Female}) + \epsilon$$

and has the following (rounded) estimated regression coefficients.

```
round(summary(fit.ex5.2.noint)$coef, 3)
```

```
##                Estimate Std. Error t value Pr(>|t|)
## (Intercept)       3.436      0.329  10.440    0.000
## BMXWAIST          0.028      0.003   9.018    0.000
## RIAGENDRFemale   -0.314      0.108  -2.910    0.004
```

From the output above, we see that the equation for the estimated mean is:

$$E(\text{FG}|\text{WC}, \text{Gender}) = 3.436 + 0.028 \times \text{WC} - 0.314 \times I(Gender = Female)$$

The interpretation of these coefficients is as follows:

- **Intercept:** The intercept is the mean outcome when all predictors are 0 or at their reference level. I(Gender = Female) = 0 when Gender = Male (the reference level for gender). Thus, the intercept (3.436 mmol/L) is the mean fasting glucose among males with a waist circumference of 0 cm. As previously mentioned, WC = 0 cm is not a plausible value, but this does not invalidate the model; it just means the intercept does not correspond to a meaningful quantity.
- **Coefficient for WC:** For a given gender, a 1 cm difference in WC is associated with a 0.028 mmol/L difference in fasting glucose. This model *assumes* this effect is the same for males and females. The model in the next section, with an interaction, will relax this assumption.
- **Coefficient for Gender:** Females have an average fasting glucose that is -0.314 mmol/L lower (since the coefficient is negative) than males with the same WC. This model *assumes* this gender difference is the same at all WC values. The model in the next section, with an interaction, will relax this assumption.

5.9.4 Regression equation with an interaction

The model with an interaction is written as

$$\text{FG} = \beta_0 + \beta_1 \text{WC} + \beta_2 I(\text{Gender} = \text{Female}) + \beta_3 [WC \times I(Gender = Female)] + \epsilon$$

and has the following (rounded) estimated regression coefficients.

```
round(summary(fit.ex5.2.int)$coef, 3)
```

```
##                         Estimate Std. Error t value Pr(>|t|)
## (Intercept)                2.209      0.503   4.396    0.000
## BMXWAIST                   0.040      0.005   8.276    0.000
## RIAGENDRFemale             1.746      0.649   2.690    0.007
## BMXWAIST:RIAGENDRFemale   -0.020      0.006  -3.218    0.001
```

Equation 5.2 is the equation for the estimated mean.

$$\begin{aligned} E(\text{FG}|\text{WC}, \text{Gender}) &= 2.209 \\ &+ 0.040 \times \text{WC} \\ &- 1.746 \times I(\text{Gender} = \text{Female}) \\ &- 0.020 \times \text{WC} \times I(\text{Gender} = \text{Female}) \end{aligned} \tag{5.2}$$

The interpretation of these coefficients is as follows.

- **Intercept:** The estimated mean fasting glucose among males with WC = 0 cm is 2.209 mmol/L.
- **Coefficient for WC ("WC main effect"):** *Among males* (the value at which the other term in the interaction is 0 or at its reference level), a 1 cm difference in WC is associated with a 0.040 mmol/L difference in fasting glucose.
- **Coefficient for Gender ("Gender main effect"):** *Among those with WC = 0 cm* (the value at which the other term in the interaction is 0 or at its reference level), females have an average fasting glucose that is 1.746 mmol/L lower (since the coefficient is negative) than males.
- **Coefficient for WC × Gender ("interaction effect"):** This model allows the effect of each predictor in the interaction to differ depending on the level of the other. The interaction effect describes the rate at which that occurs. The interaction has two interpretations. First, the WC effect on fasting glucose is -0.020 different for females than for males. Second, the gender difference in mean fasting glucose changes by -0.020 mmol/L for each 1 cm difference in WC.

NOTE: When there is an interaction in the model, be very careful when interpreting the main effects – they are NOT the overall effects of each predictor since each predictor involved in an interaction does not have a single overall effect (see Section 5.9.11 for how to get an appropriate overall test for a predictor involved in an interaction). **A main effect is the effect of a predictor when the other predictor in the interaction is 0 or at its reference level.** The answer to "What is the effect of WC on fasting glucose?" is "it depends on gender" (and vice versa). There is no single overall effect for WC or for gender – each depends on the other.

5.9.4.1 Examining the equation for each gender

Break down the regression equation by gender to see how exactly the WC effect differs between males and females. With just two predictors, one of which is categorical, this is exactly the same as stratifying (Section 5.9.1) (at least for the coefficient estimates).

To get the regression equation for males, plug Gender = Male into Equation 5.2. Doing so results in $I(\text{Gender} = \text{Female}) = 0$ and the regression equation reduces to the following.

$$\begin{aligned} E(\text{FG}|\text{WC}, \text{Gender} = \text{Male}) &= 2.209 \\ &+ 0.040 \times \text{WC} \end{aligned}$$

For females, $I(\text{Gender} = \text{Female}) = 1$, so the regression equation reduces to the following.

$$\begin{aligned} E(\text{FG}|\text{WC}, \text{Gender} = \text{Female}) &= (2.209 + 1.746) \\ &+ (0.040 - 0.020) \times \text{WC} \\ &= 3.955 \\ &+ 0.020 \times \text{WC} \end{aligned}$$

For males:

- **Intercept:** Among males, the mean fasting glucose for those with WC = 0 cm is 2.209 mmol/L.
- **Coefficient for WC:** Among males, a 1 cm difference in WC is associated with a 0.040 mmol/L difference in fasting glucose.

For females:

- **Intercept:** Among females, the mean fasting glucose for those with WC = 0 cm is 3.955 mmol/L.
- **Coefficient for WC:** Among females, a 1 cm difference in WC is associated with a 0.020 mmol/L difference in fasting glucose.

In Section 5.9.5 we will visualize these relationships to help assess their meaningfulness and in Section 5.9.6 we will test to see if these differences are statistically significant.

5.9.5 Visualizing an interaction

Visualize the impact of including an interaction by plotting the regression lines for the models with and without an interaction. The `plot_model()` function from the `sjPlot` library (Lüdecke, 2023, Lüdecke (2021)) provides a convenient way to visualize the effect of a predictor. The syntax `terms = c("BMXWAIST", "RIAGENDR")` requests a plot in which the first variable (`BMXWAIST`) is the x-axis variable and which displays the effect of the first variable on the outcome at each level of the second (`RIAGENDR`). Using `type = "pred"` displays the predicted values assuming that all other variables in the model are at a fixed value while `type = "eff"` displays predicted values that are averaged over predictions at different levels of the other predictors. When any numeric predictors are not centered, `type = "pred"` could lead to unrealistic predictions, so we will use `type = "eff"`.

Model with no interaction: In the *left panel* of Figure 5.15, the WC effects (slopes) for males and females are *assumed* to be the same, with a common estimate of 0.028, which comes from the model with no interaction `fit.ex5.2.noint`. Also, the female line is slightly below the male line by an amount equal to the gender effect (-0.314, also from `fit.ex5.2.noint`). The assumption of no interaction is equivalent to assuming that the lines are *parallel*.

Model with an interaction: In the *right panel*, including an interaction term has the effect of fitting separate regression lines for males and females. The WC slope for males (0.040) is much steeper than for females (0.020), indicating that WC has a greater effect on fasting glucose for males than for females.

```
library(sjPlot)

# Effect of BMXWAIST at each level of RIAGENDR
# in the model with no interaction
plot_model(fit.ex5.2.noint,
           type = "eff",
           terms = c("BMXWAIST", "RIAGENDR"),
           title = "No Interaction",
           axis.title = c("Waist Circumference (cm)",
                          "Fasting Glucose (mmol/L)"))

# Effect of BMXWAIST at each level of RIAGENDR
# in the model with an interaction
plot_model(fit.ex5.2.int,
           type = "eff",
           terms = c("BMXWAIST", "RIAGENDR"),
           title = "Interaction",
           axis.title = c("Waist Circumference (cm)",
                          "Fasting Glucose (mmol/L)"))
```

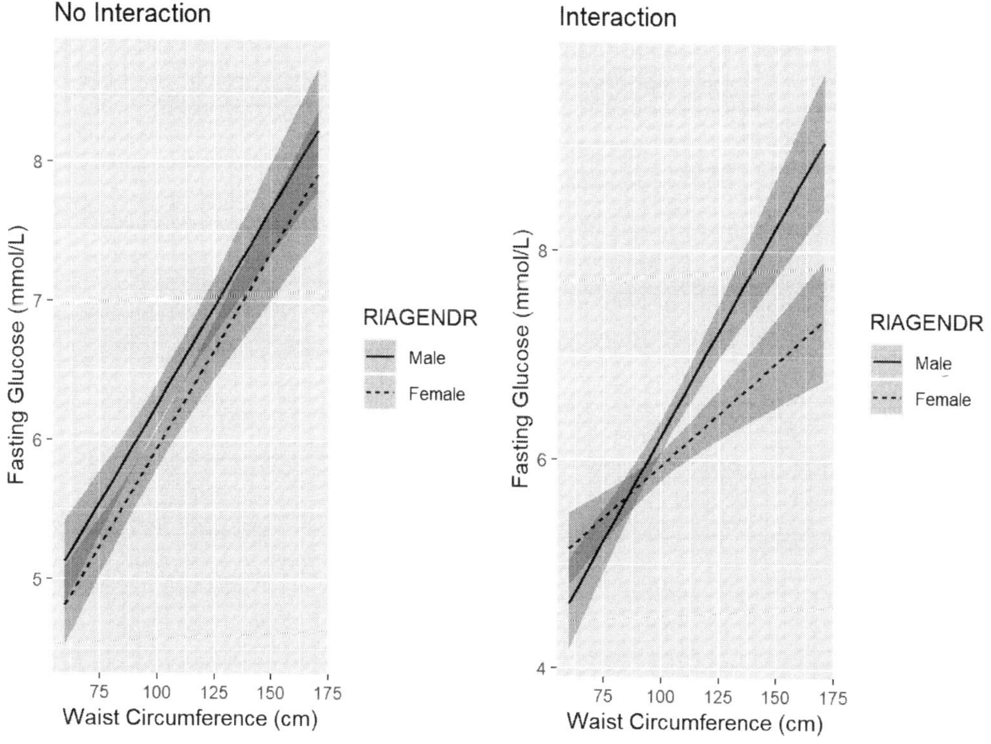

FIGURE 5.15 Linear regression interaction plot (continuous x categorical)

Although it is a subjective judgment, visually it appears that the interaction is meaningfully large. In the next section, we will test to see if the magnitude of the interaction is statistically significant.

5.9.6 Testing the difference between the slopes

In the right panel of Figure 5.15, is the difference between the male and female slopes statistically significant? This is equivalent to asking, "Are the lines in the plot significantly different from parallel?" Recall the regression equation for this interaction model.

$$\text{FG} = \beta_0 + \beta_1\text{WC} + \beta_2 I(\text{Gender} = \text{Female}) + \beta_3[WC \times I(Gender = Female)] + \epsilon$$

The p-value for the regression coefficient for the interaction tests the null hypothesis $H_0 : \beta_3 = 0$. Under the null hypothesis, the lines are parallel. Thus, to test the difference between the slopes, examine the interaction p-value.

```
round(summary(fit.ex5.2.int)$coef, 3)
```

```
##                          Estimate Std. Error t value Pr(>|t|)
## (Intercept)                 2.209      0.503   4.396    0.000
## BMXWAIST                    0.040      0.005   8.276    0.000
## RIAGENDRFemale              1.746      0.649   2.690    0.007
## BMXWAIST:RIAGENDRFemale    -0.020      0.006  -3.218    0.001
```

From the output, the p-value for the test of interaction is .001. In this case, since the interaction was between a continuous predictor and a binary predictor, the interaction is just one term, so the test can be obtained directly from the regression coefficient table. In Section 5.9.10 we will discuss what to do when the interaction involves a categorical predictor with more than two levels.

Since $p < .05$, we can make the following two inferences:

- There is a significant difference in the WC effect on fasting glucose between males and females (the lines are significantly different from parallel).
- There is a significant difference in the gender effect on fasting glucose between those with differing WC (the vertical distance between the lines depends on how large or small WC is).

5.9.7 Estimating and testing the significance of the slope at each level of a moderator

We just showed that the WC effect is significantly greater for males than for females. What if you want to estimate and test the significance of the WC effect for each gender individually (testing whether the line for males is significantly different from a horizontal line and, separately, testing whether the line for females is significantly different from a horizontal line)?

The null hypothesis for the test of the WC effect among males is a test of the slope for males, $H_0 : \beta_1 = 0$. Since this involves just one regression coefficient, you can read this right off the output in the `BMXWAIST` row of the `COEFFICIENTS` table since that row corresponds to β_1.

Thus, the WC effect for males is statistically significant (B = 0.040; 95% CI = 0.031, 0.050; p <.001).

However, the slope for females is the sum of two regression coefficients ($\beta_1 + \beta_3$). How do we conduct the test for females? You *could* re-fit the model after setting "Female" to be the reference level instead of "Male". In this new model, the WC main effect would be the slope for females. However there is a more generally applicable method that is useful to learn and does not involve re-fitting the model.

In our current model, in which "Male" was the reference level for gender, the test of the WC effect among females is $H_0 : \beta_1 + \beta_3 = 0$. The function `gmodels::estimable()` (Warnes et al., 2022) can be used to estimate and test any linear combination of regression coefficients. To use this function with a model with four regression coefficients (including the intercept), specify a vector (a, b, c, d) corresponding to the linear combination $a\beta_0 + b\beta_1 + c\beta_2 + d\beta_3$ of interest. The function will then estimate this sum and test the null hypothesis that it is 0. For females, since we want to test $\beta_1 + \beta_3$, our vector is $(0, 1, 0, 1)$.

NOTE: When using `gmodels::estimable()` to estimate the effect of X at levels of Z, always include a 1 in the slot for the main effect of X, do *not* include a 1 for the main effect of Z, and include the value of Z in the slot for the $X \times Z$ interaction. In this example, X is WC and Z is $I(\text{Gender} = \text{Female})$, so we include a 1 in the WC main effect slot (slot 2) and a 1 or 0 in the interaction slot (slot 4) depending on whether gender is female or male.

To make sure you are putting the numbers in the correct slots, look at the number, order, and spelling of the coefficients in the `lm` object. In the code below, the vector specifying the linear combination is a named vector. The names are not required but are a good idea to ensure the correct numbers are entered in the correct slots.

```
# Number of coefficients
length(coef(fit.ex5.2.int))
```

```
## [1] 4
```

```
# Select the 1st column
# Include drop=F to keep it as a matrix (which displays vertically)
summary(fit.ex5.2.int)$coef[, 1, drop=F]
```

```
##                           Estimate
## (Intercept)                2.20932
## BMXWAIST                   0.04024
## RIAGENDRFemale             1.74592
## BMXWAIST:RIAGENDRFemale -0.02036
```

```
# Test slope for females
gmodels::estimable(fit.ex5.2.int,
                   c("(Intercept)"              = 0,
                     "BMXWAIST"                 = 1,
                     "RIAGENDRFemale"           = 0,
                     "BMXWAIST:RIAGENDRFemale" = 1),
                   conf.int = 0.95)
```

```
##             Estimate Std. Error t value  DF    Pr(>|t|) Lower.CI Upper.CI
## (0 1 0 1)  0.01988    0.004051   4.907 853 0.000001106  0.01193  0.02783
```

Any terms omitted will be assumed to be multiplied by zero. Thus, the following abbreviated code produces the same result.

```
gmodels::estimable(fit.ex5.2.int,
                   c("BMXWAIST"                  = 1,
                     "BMXWAIST:RIAGENDRFemale" = 1),
                   conf.int = 0.95)
```

You can also get the same answer without naming the terms, but then you must include all the slots. However, use this method with caution as without names it is easy to enter values in the wrong slots.

```
gmodels::estimable(fit.ex5.2.int, c(0,1,0,1), conf.int = 0.95)
```

Although we already tested the slope for males above, we could use `gmodels::estimable()` instead. For males, the vector would be $(0, 1, 0, 0)$ (to test $H_0 : \beta_1 = 0$).

```
# Test slope for males
# Optionally, round the output
round(
  gmodels::estimable(fit.ex5.2.int,
                     c("(Intercept)"              = 0,
                       "BMXWAIST"                 = 1,
                       "RIAGENDRFemale"           = 0,
                       "BMXWAIST:RIAGENDRFemale" = 0),
                     conf.int = 0.95)
, 4)
```

```
##             Estimate Std. Error t value  DF Pr(>|t|) Lower.CI Upper.CI
## (0 1 0 0)    0.0402     0.0049   8.276 853        0   0.0307   0.0498
```

Conclusion: The WC effects for males (Est = 0.040; 95% CI = 0.031, 0.050; p <.001) and females (Est = 0.020; 95% CI = 0.012, 0.028; p <.001) are statistically significant.

5.9.8 A two-way interaction has two interpretations

Above, we saw how the WC effect differed by gender. We can also break down the regression equation by WC to see how the gender effect differs by WC. Let's see what the equations look like at WC = 80, 100, and 120 cm.

To get the regression equation at WC = 80, plug this value into Equation 5.2 to get the following.

$$\begin{aligned} E(\text{FG}|\text{WC} = 80, \text{Gender}) &= 2.209 \\ &+ 0.040 \times 80 \\ &- 1.746 \times I(\text{Gender} = \text{Female}) \\ &- 0.020 \times 80 \times I(\text{Gender} = \text{Female}) \end{aligned}$$

which reduces to

$$E(\text{FG}|\text{WC} = 80, \text{Gender}) = 5.429 + 0.117 \times I(\text{Gender} = \text{Female})$$

The math was carried out before rounding so you may get a different answer if you start with the values rounded to three digits. Similarly, for the other WC values, we get the following.

$$\begin{aligned} E(\text{FG}|\text{WC} = 100, \text{Gender}) &= 6.233 - 0.290 \times I(\text{Gender} = \text{Female}) \\ E(\text{FG}|\text{WC} = 120, \text{Gender}) &= 7.038 - 0.698 \times I(\text{Gender} = \text{Female}) \end{aligned}$$

As WC increases from 80 to 120 cm (a total of 40 cm), the gender effect changes by 40 $\times$ the interaction effect, or $40 \times -0.0204 = -0.8145$, going from 0.117 to -0.698. Visualize this by looking at how far apart the Male and Female regression lines are at different WC values, as shown in Figure 5.16.

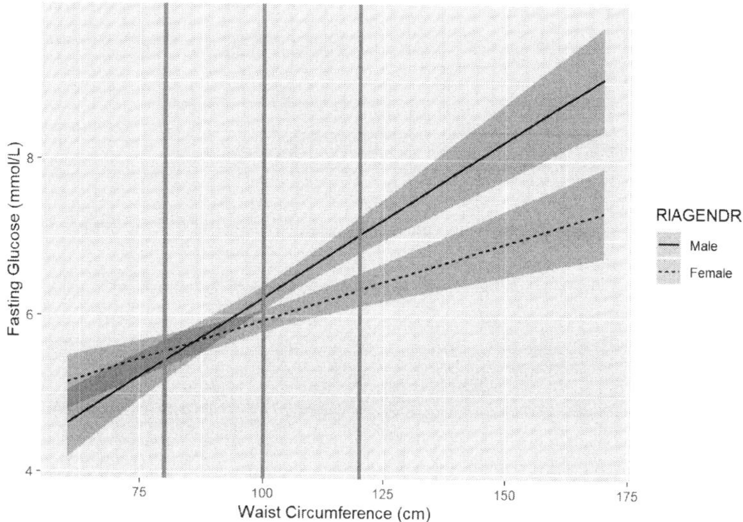

FIGURE 5.16 Gender effect depends on waist circumference

Another interaction plot that shows the effect of `RIAGENDR` at each level of `BMXWAIST` can be created using `plot_model()` with the terms reversed as `terms = c("RIAGENDR", "BMXWAIST")` (Figure 5.17). By default, if the second term is continuous, the effect of the first will be shown at the following values of the second: mean – 1 standard deviation, mean, mean + 1 standard deviation. To specify different values (e.g., 80, 100, 120), use the syntax below.

```
plot_model(fit.ex5.2.int,
           type = "eff",
           terms = c("RIAGENDR", "BMXWAIST [80, 100, 120]"),
           axis.title = c("Gender",
                          "Fasting Glucose (mmol/L)"),
           legend.title = "Waist Circumference(cm)",
           title = "")
```

Conclusion: The magnitude of the association between WC and fasting glucose differs significantly between males and females (p = .001) (this is the p-value for the interaction). The association between WC and fasting glucose is significant for both males (Slope = 0.040; 95% CI = 0.031, 0.050; p <.001) and females (Slope = 0.020; 95% CI = 0.012, 0.028; p <.001), but is significantly lower for females (B-interaction (difference in slopes) = -0.020; 95% CI = -0.033, -0.008; p = .001).

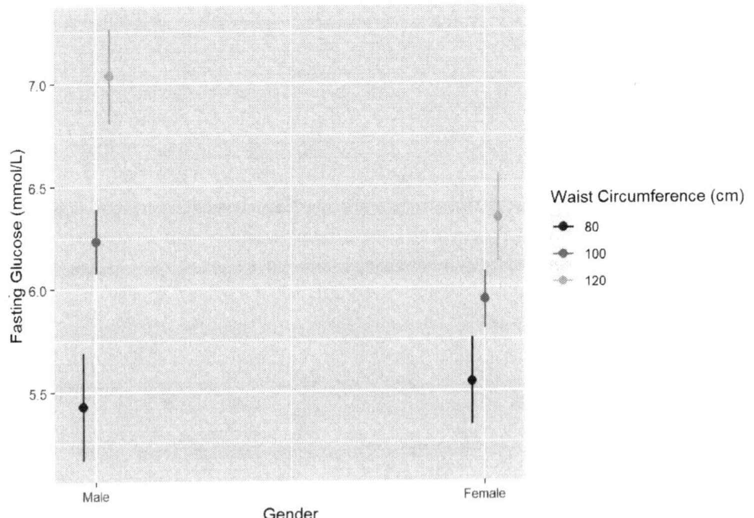

FIGURE 5.17 Another way of seeing that the gender effect depends on waist circumference

If you feel confused by interactions, you are not alone! Assuming your confusion is not due to the author's lack of clarity, take comfort in the fact that interactions are inherently tricky, but you will learn and get used to them with experience. They are easy to misunderstand and misinterpret, so if you feel confused that is a good thing in one respect – it means you know enough to be careful and ask for help until you are comfortable dealing with interactions on your own.

5.9.9 Types of two-way interactions

In regression, you can have the following different kinds of two-way interactions:

- Interaction between a continuous predictor and a categorical predictor
- Interaction between two categorical predictors
- Interaction between two continuous predictors

5.9.9.1 Continuous × categorical

This is the type of interaction discussed in Example 5.2 in Sections 5.9.1 to 5.9.8 (WC × gender).

5.9.9.2 Categorical × categorical

An interaction between two categorical predictors can be more complicated. For example, if you have a categorical predictor with 2 levels interacted with a categorical predictor with 3 levels, you do not have two regression lines but rather $2 \times 3 = 6$ means. In addition, it is possible that even if neither predictor was sparse enough to warrant collapsing levels, one or the other might be sparse within levels of the other.

Example 5.3 Regress fasting glucose on gender and income, along with their interaction, interpret the regression coefficients, and visualize the interaction.

First, check to see if any combination of these predictors is sparse.

```
table(nhanesf.complete$RIAGENDR, nhanesf.complete$income,
      useNA = "ifany")
```

```
##
##          <$25,000 $25,000 to <$55,000 $55,000+
##   Male         67                 106      215
##   Female       90                 115      264
```

All combinations have relatively large sample sizes. If some combination had a very small sample size, then one or more of the interaction terms would be estimated very imprecisely. In that case, it might be a good idea to collapse one or more of the predictors involved in the interaction (see Section 5.4.5).

The syntax for fitting this model is the same as for the continuous × categorical model.

```
fit.ex5.3.int <- lm(LBDGLUSI ~ RIAGENDR + income +
                      RIAGENDR:income,
                    data = nhanesf.complete)
round(summary(fit.ex5.3.int)$coef, 3)
```

```
##
##                                           Estimate Std. Error t value Pr(>|t|)
## (Intercept)                                  6.233      0.199  31.266    0.000
## RIAGENDRFemale                               0.216      0.263   0.820    0.413
## income$25,000 to <$55,000                    0.101      0.255   0.396    0.693
## income$55,000+                               0.102      0.228   0.447    0.655
## RIAGENDRFemale:income$25,000 to <$55,000    -0.712      0.343  -2.077    0.038
## RIAGENDRFemale:income$55,000+               -0.742      0.303  -2.448    0.015
```

Including the intercept, there are six coefficients above. These can be combined to estimate mean fasting glucose at the six possible combinations of gender × income. Let's compute the mean at each level. We *could* use `gmodels::estimable()` but for estimating the mean outcome at levels of categorical predictors (as opposed to the slopes of continuous predictors, which are not estimated means) using `predict()` is much easier.

```
# Get levels of each categorical predictors
LEVELS1 <- levels(nhanesf.complete$income)
LEVELS2 <- levels(nhanesf.complete$RIAGENDR)

# Create a data.frame to store the predictions
preddat <- data.frame(
  income   = rep(LEVELS1, length(LEVELS2)),
  RIAGENDR = rep(LEVELS2, each=length(LEVELS1))
)

# Add the predictions
preddat$pred <- predict(fit.ex5.3.int, preddat)

# View the predicted means
preddat
```

```
##                income RIAGENDR  pred
## 1             <$25,000     Male 6.233
## 2 $25,000 to <$55,000     Male 6.334
## 3             $55,000+     Male 6.335
## 4             <$25,000   Female 6.449
## 5 $25,000 to <$55,000   Female 5.837
## 6             $55,000+   Female 5.809
```

We can also visualize these means using `plot_model()`, as shown in Figure 5.18.

```
plot_model(fit.ex5.3.int,
           type = "eff",
           terms = c("income", "RIAGENDR"),
           axis.title = c("Income",
                          "Fasting Glucose (mmol/L)"),
           title = "")
```

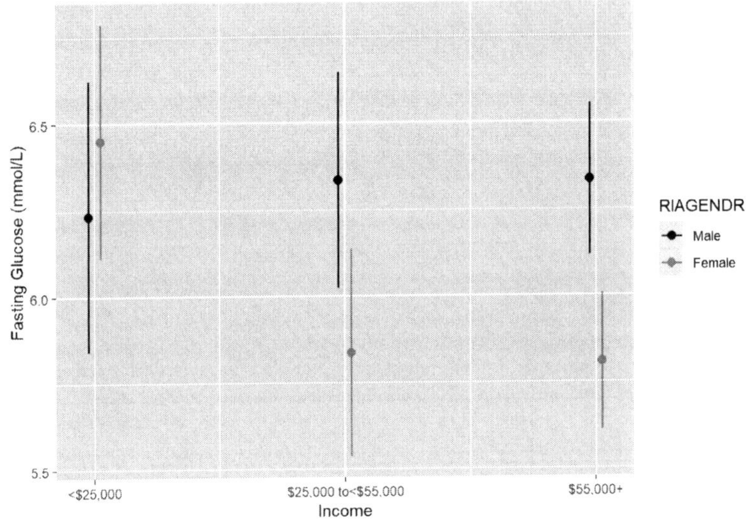

FIGURE 5.18 Interaction of two categorical predictors: Fasting Glucose vs. Income by Gender

Is the interaction statistically significant? When you have an interaction between categorical predictors with L_1 and L_2 levels, there will be $(L_1 - 1) \times (L_2 - 1)$ interaction terms. If this product is >1, which it will be if either categorical predictor has more than two levels, then you cannot get this test from the default `lm()` output since it is a multiple degree of freedom test (you are testing more than one coefficient simultaneously). But you can get it using `car::Anova()`.

```
car::Anova(fit.ex5.3.int, type = 3)
```

```
## Anova Table (Type III tests)
##
## Response: LBDGLUSI
##                  Sum Sq  Df F value                 Pr(>F)
## (Intercept)        2603   1  977.55 <0.0000000000000002 ***
## RIAGENDR              2   1    0.67                  0.413
## income                1   2    0.11                  0.898
## RIAGENDR:income      17   2    3.17                  0.042 *
## Residuals          2266 851
## ---
## Signif. codes:  0 '***' 0.001 '**' 0.01 '*' 0.05 '.' 0.1 ' ' 1
```

We conclude that there is a statistically significant interaction between gender and income (p = .042). From Figure 5.18, we see that females have a greater mean fasting glucose among those with income <$25,000 but a lower mean fasting glucose among those with income ≥$25,000.

What if you want to estimate the difference between specific levels of one predictor at a certain level of the other?

For example, estimate the difference in mean fasting glucose between males and females among those with different incomes.

- The main effect regression coefficient for `RIAGENDRFemale` is the difference in mean fasting glucose between females and males at the reference level of income, which is <$25,000.
- The difference in mean outcome between females and males among those making $25,000 to <$55,000 is the sum of the `RIAGENDRFemale` main effect and the `RIAGENDRFemale:income$25,000 to <$55,000` interaction term.
- The difference in mean outcome between females and males among those making $55,000+ is the sum of the `RIAGENDRFemale` main effect and the `RIAGENDRFemale:income$55,000+` interaction term.

Since these are sums of coefficients, use `gmodels::estimable()` to estimate these effects. First, look at the coefficient vector to confirm the number, order, and spelling of the terms in the model.

```
length(coef(fit.ex5.3.int))
```

```
## [1] 6
```

```
summary(fit.ex5.3.int)$coef[, 1, drop=F]
```

```
##                                          Estimate
## (Intercept)                                6.2328
## RIAGENDRFemale                             0.2158
## income$25,000 to <$55,000                  0.1007
## income$55,000+                             0.1021
## RIAGENDRFemale:income$25,000 to <$55,000  -0.7122
## RIAGENDRFemale:income$55,000+             -0.7418
```

There are six slots. The gender main effect is `RIAGENDRFemale` and the interaction terms are `RIAGENDRFemale:income$25,000 to <$55,000` and `RIAGENDRFemale:income$55,000+`. Since we are estimating gender effects at levels of income, always assign a 1 to the gender main effect slot, and assign a 0 or 1 to each interaction term, depending on the level of income.

```
rbind(
  # Females vs. Males, Income <$25,000
  # (income reference level, so no 1 in any interaction slot)
  gmodels::estimable(fit.ex5.3.int,
                     c("RIAGENDRFemale" = 1),
                     conf.int = 0.95),
  # Females vs. Males, Income $25,000 to <$55,000
  gmodels::estimable(fit.ex5.3.int,
                     c("RIAGENDRFemale" = 1,
                       "RIAGENDRFemale:income$25,000 to <$55,000" = 1),
                     conf.int = 0.95),
  # Females vs. Males, Income $55,000+
  gmodels::estimable(fit.ex5.3.int,
                     c("RIAGENDRFemale" = 1,
                       "RIAGENDRFemale:income$55,000+" = 1),
                     conf.int = 0.95)
)
```

```
##                    Estimate Std. Error t value  DF  Pr(>|t|) Lower.CI Upper.CI
## (0 1 0 0 0 0)      0.2158     0.2633  0.8197 851 0.4126023  -0.3010  0.73262
## (0 1 0 0 1 0)     -0.4964     0.2197 -2.2592 851 0.0241231  -0.9276 -0.06513
## (0 1 0 0 0 1)     -0.5260     0.1499 -3.5087 851 0.0004739  -0.8202 -0.23174
```

Equivalently, you could use the following code. It is a matter of preference which form you use. The former is less error-prone in that the slots are named and you do not have to keep track of the position of the slots. Also, the former is more intuitive if you understand what is needed as "main effect + an interaction term". However, the latter is nice in that it is easy to see the similarities and differences between the three vectors since they are lined up. They all have a 1 in the main effect slot, the first has no 1 in any interaction slot, and the remaining have a 1 in consecutive interaction slots.

```
rbind(
  gmodels::estimable(fit.ex5.3.int, c(0, 1, 0, 0, 0, 0), conf.int = 0.95),
  gmodels::estimable(fit.ex5.3.int, c(0, 1, 0, 0, 1, 0), conf.int = 0.95),
  gmodels::estimable(fit.ex5.3.int, c(0, 1, 0, 0, 0, 1), conf.int = 0.95)
)
```

Conclusion:

- Among those making <$25,000 per year, fasting glucose is greater in females than males, although this difference is not statistically significant (Est = 0.22; 95% CI = -0.30, 0.73; p = .413).
- Among those making $25,000 to <$55,000 per year, fasting glucose is significantly lower in females than males (Est = -0.50; 95% CI = -0.93, -0.07; p = .024).
- Among those making $55,000 or more per year, fasting glucose is significantly lower in females than males (Est = -0.53; 95% CI = -0.82, -0.23; p <.001).

5.9.9.3 Continuous × continuous

An interaction between two continuous predictors is simpler in one way but more complicated in another. It will always involve just one regression coefficient, but the relationship cannot be *completely* visualized as a set of regression lines since each predictor is continuous, with more than just a few possible values. However, to partially illustrate the interaction, you can plot a regression line for one predictor at a set of specific values of the other (like we did for the continuous × categorical interaction).

Example 5.4: Regress fasting glucose on the continuous predictors waist circumference and age, along with their interaction, and interpret the regression coefficients. After fitting the model, visualize the association between waist circumference and fasting glucose for those ages 30, 40, and 50 years.

The regression equation is

$$\text{FG} = \beta_0 + \beta_1 \text{Age} + \beta_2 \text{WC} + \beta_3 (\text{Age} \times \text{WC}) + \epsilon$$

and the interpretation of each parameter is as follows.

- β_0 is the mean fasting glucose among those age 0 years with WC = 0 cm.
- β_1 is the age effect among those with WC = 0 cm. Because there is an Age $\times$ WC interaction, the "age effect" varies with WC.
- β_2 is the WC effect among those age 0 years. Because there is an Age $\times$ WC interaction, the "WC effect" varies with age.
- β_3 is the interaction effect. It represents the difference in the age effect for every cm of WC away from 0 cm, and the difference in the WC effect for every year of age away from 0 years.

Fitting the model is no different than with other forms of interactions.

```
# Fit model
fit.ex5.4.int <- lm(LBDGLUSI ~ RIDAGEYR + BMXWAIST +
                      RIDAGEYR:BMXWAIST,
                    data = nhanesf.complete)

# View output
round(summary(fit.ex5.4.int)$coef, 4)
```

```
##                   Estimate Std. Error t value Pr(>|t|)
## (Intercept)         4.8462     0.8924  5.4307   0.0000
## RIDAGEYR           -0.0320     0.0191 -1.6790   0.0935
## BMXWAIST            0.0015     0.0089  0.1709   0.8644
## RIDAGEYR:BMXWAIST   0.0005     0.0002  2.9033   0.0038
```

As mentioned, the interaction is a single term in the model so the test of significance of the interaction can be read right off the output (interaction p = .004). That is the simple aspect of a continuous $\times$ continuous interaction. The complicated aspect is that to visualize this model we must choose a few specific values of one predictor and plot the regression lines for the other predictor at those levels. This is what we did for a continuous $\times$ categorical interaction, but in that case there were a finite number of categorical levels; here, there are an infinite number of values you can pick from, so for this example a few interesting values were chosen (ages 30, 40, and 50 years).

Visualize the interaction using `plot_model()`, as shown in Figure 5.19.

```
plot_model(fit.ex5.4.int,
           type = "eff",
           terms = c("BMXWAIST", "RIDAGEYR [30, 40, 50]"),
           axis.title = c("Waist Circumference (cm)",
                          "Fasting Glucose (mmol/L)"),
           legend.title = "Age (years)",
           title = "")
```

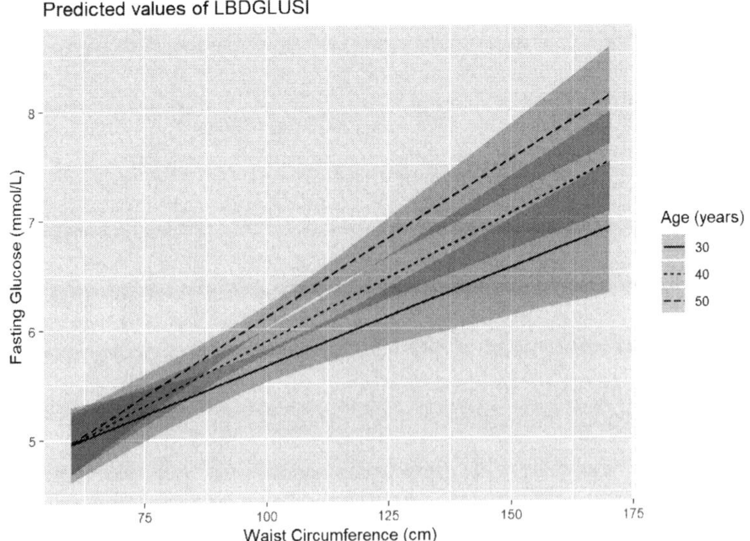

FIGURE 5.19 Interaction of two continuous predictors: fasting glucose vs. waist circumference, by age

Recall that the regression coefficient table is the following.

```
round(summary(fit.ex5.4.int)$coef, 4)
```

```
##                    Estimate Std. Error t value Pr(>|t|)
## (Intercept)          4.8462     0.8924  5.4307   0.0000
## RIDAGEYR            -0.0320     0.0191 -1.6790   0.0935
## BMXWAIST             0.0015     0.0089  0.1709   0.8644
## RIDAGEYR:BMXWAIST    0.0005     0.0002  2.9033   0.0038
```

We can interpret the numbers in this table in light of Figure 5.19.

- The coefficient for `BMXWAIST` is 0.0015. This is the slope for fasting glucose vs. WC when Age = 0 years.
- The interaction term is statistically significant (p = .004). Thus, we can reject the null hypothesis that the lines are parallel, and conclude that the age effect varies with WC, and the WC effect varies with age.
- The interaction term is positive (0.0005). This means that
  - the slopes for fasting glucose vs. WC at ages 30, 40, and 50 years are $0.0015 + 30 \times 0.0005$, $0.0015 + 40 \times 0.0005$, and $0.0015 + 50 \times 0.0005$, respectively;
  - the slope is larger among those who are older, as can be seen by the fact that lines become more steep at older age; and
  - differences in fasting glucose between those of different ages are greater among those with greater WC (the lines are further apart for larger WC values).

What if you want to estimate the WC slopes at each of these ages?

The WC slope at a given age is the `BMXWAIST` main effect plus age times the `RIDAGEYR:BMXWAIST` interaction. Use `gmodels::estimable()` to estimate these effects. First, look at the coefficient vector to confirm the number, order, and spelling of the terms in the model.

```
length(coef(fit.ex5.4.int))
```

```
## [1] 4
```

```
summary(fit.ex5.4.int)$coef[, 1, drop=F]
```

```
##                      Estimate
## (Intercept)         4.8462057
## RIDAGEYR           -0.0319988
## BMXWAIST            0.0015185
## RIDAGEYR:BMXWAIST   0.0005434
```

There are four slots. The WC main effect is in slot 3, and the interaction term is in slot 4. While we still enter a 1 in the main effect slot (as in all our previous examples), in the interaction slot we enter the age value rather than a 1 (since age is continuous).

```
round(rbind(
  gmodels::estimable(fit.ex5.4.int,
                     c("BMXWAIST"          = 1,
                       "RIDAGEYR:BMXWAIST" = 30),
                     conf.int = 0.95),
  gmodels::estimable(fit.ex5.4.int,
                     c("BMXWAIST"          = 1,
                       "RIDAGEYR:BMXWAIST" = 40),
                     conf.int = 0.95),
  gmodels::estimable(fit.ex5.4.int,
                     c("BMXWAIST"          = 1,
                       "RIDAGEYR:BMXWAIST" = 50),
                     conf.int = 0.95)
), 6)
```

```
##              Estimate Std. Error t value  DF Pr(>|t|) Lower.CI Upper.CI
## (0 0 1 30)   0.01782   0.004105   4.340 853 0.000016 0.009761  0.02588
## (0 0 1 40)   0.02325   0.003186   7.297 853 0.000000 0.016998  0.02951
## (0 0 1 50)   0.02869   0.003234   8.871 853 0.000000 0.022339  0.03503
```

Conclusion:

- Among those age 30 years, those with 1-unit greater WC have, on average, a fasting glucose that is 0.0178 greater (Est = 0.0178; 95% CI = 0.0098, 0.0259; p <.001).
- Among those age 40 years, those with 1-unit greater WC have, on average, a fasting glucose that is 0.0233 greater (Est = 0.0233; 95% CI = 0.0170, 0.0295; p <.001).
- Among those age 50 years, those with 1-unit greater WC have, on average, a fasting glucose that is 0.0287 greater (Est = 0.0287; 95% CI = 0.0223, 0.0350; p <.001).

5.9.10 Test of interaction

This section reviews the methods for carrying out a test of significance of the interaction term. We already carried out this test for each type of interaction in the previous sections, but the method is summarized here.

For a regression with outcome Y and an interaction between X and Z, a statistical test of interaction is a test of the null hypothesis that the magnitude of the effect of X on Y does not depend on Z, and that the magnitude of the effect of Z on Y does not depend on X.

Visually, the null hypothesis corresponds to parallel lines in plots such as Figures 5.15 and 5.19. Statistically, the null hypothesis is that all the regression coefficients that define the interaction are zero.

When the interaction involves just one term in the model (one row in the regression coefficient table) you can get the p-value for the interaction test by looking at the regression table. For example, if the regression formula is $Y = \beta_0 + \beta_1 X + \beta_2 Z + \beta_3 XZ + \epsilon$, the null hypothesis is $H_0 : \beta_3 = 0$. In our example of an interaction between two continuous predictors, we would look in the row with the term that contains a : (`RIDAGEYR:BMXWAIST`).

```
round(summary(fit.ex5.4.int)$coef, 4)
```

```
##                    Estimate Std. Error t value Pr(>|t|)
## (Intercept)          4.8462     0.8924  5.4307   0.0000
## RIDAGEYR            -0.0320     0.0191 -1.6790   0.0935
## BMXWAIST             0.0015     0.0089  0.1709   0.8644
## RIDAGEYR:BMXWAIST    0.0005     0.0002  2.9033   0.0038
```

When at least one of the terms in the interaction is a categorical predictor with more than two levels, the interaction will consist of multiple terms. For example, if Z is categorical with levels 1 (the reference level), 2, and 3, the regression formula is

$$Y = \beta_0 + \beta_1 X + \beta_2 I(Z = 2) + \beta_3 I(Z = 3) + \beta_4 X \times I(Z = 2) + \beta_5 X \times I(Z = 3) + \epsilon$$

and the null hypothesis of no interaction is $H_0 : \beta_4 = 0$ and $\beta_5 = 0$. In our example of an interaction between two categorical predictors, one of which had more than two levels, the rows with the : terms do *not* provide the interaction test, as they are tests of each of the interaction terms individually. The estimated regression coefficients were:

```
round(summary(fit.ex5.3.int)$coef, 4)
```

```
##                                          Estimate Std. Error t value Pr(>|t|)
## (Intercept)                                6.2328     0.1994 31.2658   0.0000
## RIAGENDRFemale                             0.2158     0.2633  0.8197   0.4126
## income$25,000 to <$55,000                  0.1007     0.2547  0.3956   0.6925
## income$55,000+                             0.1021     0.2283  0.4472   0.6549
## RIAGENDRFemale:income$25,000 to <$55,000  -0.7122     0.3429 -2.0768   0.0381
## RIAGENDRFemale:income$55,000+             -0.7418     0.3030 -2.4483   0.0146
```

Recall that to test the main effect for a categorical predictor with more than two levels, we used `car::Anova()` to carry out the multiple df test. We can do the same for a multiple df interaction:

```
car::Anova(fit.ex5.3.int, type = 3)
```

```
## Anova Table (Type III tests)
##
## Response: LBDGLUSI
##                 Sum Sq   Df F value                Pr(>F)
## (Intercept)       2603    1  977.55 <0.0000000000000002 ***
## RIAGENDR             2    1    0.67                 0.413
```

```
## income            1   2   0.11          0.898
## RIAGENDR:income  17   2   3.17          0.042 *
## Residuals      2266 851
## ---
## Signif. codes:  0 '***' 0.001 '**' 0.01 '*' 0.05 '.' 0.1 ' ' 1
```

We conclude for this example that there is a significant gender × income interaction (F[2, 851] = 3.17, p = .042). The same method (`car::Anova()`) would be used for a continuous × categorical interaction if the categorical predictor has more than two levels.

By way of reminder, the tests of each of the two main effects are *not* tests of the overall effect of each of the corresponding predictors. They are, rather, tests of significance of each predictor at the reference level of the other predictor in the interaction. The following subsection will discuss how to obtain an appropriate overall test for each predictor involved in an interaction.

5.9.11 Overall test of a predictor involved in an interaction

In Section 5.9.10, we assessed the significance of an interaction with a test of the null hypothesis that one predictor effect does not vary with another predictor. A test of interaction answers a question such as, "Does the effect of waist circumference on fasting glucose depend on gender?"

But what if we want to answer the question, "Is waist circumference significantly associated with fasting glucose?" in a model which contains WC interacted with another predictor? This is a comparison of two models – one that includes WC and one that does not. When there is no interaction, this would simply be a test of the WC regression coefficient. When there is an interaction, however, this is a simultaneous test of the WC main effect and the WC interaction with the other predictor.

For example, if the model is

$$\text{FG} = \beta_0 + \beta_1 \text{WC} + \beta_2 I(\text{Gender} = \text{Female}) + \beta_3 [\text{WC} \times I(\text{Gender} = \text{Female})] + \epsilon$$

the null hypothesis we would like to test is $H_0 : \beta_1 = 0$ and $\beta_3 = 0$. We cannot use `car::Anova()` for this test because these coefficients are from different terms in the model. Instead, to carry out this test, use `anova()` (lowercase a), which compares two nested models: a full model compared to a reduced model with some terms from the full model removed. In this example, we want to compare the full model (with both main effects and the interaction) to the model that does not have WC at all (excludes both the WC main effect and the interaction, but retains the gender main effect). Even though there are no cases with missing values in `nhanesf.complete`, the code `subset = complete.cases(BMXWAIST)` is added here in case, in the future, you imitate this code for an example in which you have not removed cases with missing values. This option ensures that `fit0` is fit to the same sample as was used when fitting the full model (those with no missing values for any of the analysis variables).

```
# Fit the reduced model
fit0 <- lm(LBDGLUSI ~  RIAGENDR, data = nhanesf.complete,
           subset = complete.cases(BMXWAIST))
# Compare reduced and full models
anova(fit0, fit.ex5.2.int)
```

```
## Analysis of Variance Table
##
## Model 1: LBDGLUSI ~ RIAGENDR
## Model 2: LBDGLUSI ~ BMXWAIST + RIAGENDR + BMXWAIST:RIAGENDR
##   Res.Df  RSS Df Sum of Sq    F              Pr(>F)
## 1    855 2295
## 2    853 2071  2       225 46.3 <0.0000000000000002 ***
## ---
## Signif. codes:  0 '***' 0.001 '**' 0.01 '*' 0.05 '.' 0.1 ' ' 1
```

Thus, we conclude that the model with both WC and a WC × gender interaction is significantly better than the model with just gender (F[2,853] = 46.3, p <.001), or that WC is significantly associated with fasting glucose.

5.9.12 When to include an interaction

If a research question is of the form "Does the association between the outcome Y and predictor X depend on (some other variable) Z?", then the regression model used to answer the question must include an interaction between X and Z. In that case, there is an *a priori* interaction of interest. In general, when carrying out a confirmatory analysis (see Section 5.23), use subject-matter knowledge to *a priori* decide which interactions to include, and then leave them in the model regardless of their statistical significance. Testing candidate interactions and only including those that are statistically significant can result in biased conclusions about statistical significance unless one uses advanced techniques to obtain valid inferences that account for the method of data exploration (Faraway, 1992; Harrell, 2015).

What if, rather than having specific interactions you want to test, you are interested in exploring many possible interactions, and finding a model that includes only interactions that seem meaningful? In an exploratory analysis (see Section 5.23), it is acceptable to try different interactions and include or not include them based on statistical significance and/or the magnitude of the interaction effects. It is important, if using this approach, to explicitly state you are carrying out an exploratory analysis, state what criteria you used to determine which interactions to include in your final model, and note as a limitation that the p-values for all terms in the model (both main effects and interactions) may be biased due to the fact that the final model was not prespecified and decisions regarding what terms to include were based on statistical significance and/or effect size.

5.10 Predictions

To get a prediction (again, technically, an estimate of the mean outcome), you *could* plug the predictor values into the regression equation, multiplying each by its corresponding regression coefficient, summing the products, and adding the intercept. However, it is much easier to use `predict()` to do the calculation for you, with the added benefit of also getting a 95% confidence interval for the mean.

An optional second argument to `predict()` is a `data.frame` containing predictor values. The predictor values must be in the same format as those used to fit the model. For a numeric predictor, specify a number (not in quotes). For a categorical predictor, specify a level (in quotes). To make sure you enter a legitimate factor level, use `levels()` to check the spelling.

NOTE: If any categorical predictor level is misspelled, `predict()` will return an error. However, specifying one or more continuous predictor values beyond the range observed in the data will, unfortunately, not return an error or even a warning. Therefore, when making predictions, be careful to only predict at values within the range of the data used to fit the model. For example, if the model were fit using data from those age 18 years and older, a prediction for 10-year-olds would be invalid (see Section 5.26).

Example 5.1 (continued): Estimate the mean fasting glucose and its 95% confidence interval at the following values of waist circumference, smoking status, age, gender, race/ethnicity, and income:

- WC = 130 cm
- Smoker = Current
- Age = 50 years
- Gender = Male
- Race/Ethnicity = Non-Hispanic Black
- Income = $55,000+

First, check the spelling of the levels.

```
levels(nhanesf.complete$smoker)
levels(nhanesf.complete$RIAGENDR)
levels(nhanesf.complete$race_eth)
levels(nhanesf.complete$income)
# (results not shown)
```

Next, use `predict()` with the appropriate `data.frame` as the second argument, and `interval = "confidence"`.

```
# Use predict() with a data.frame with predictor levels
predict(fit.ex5.1, data.frame(
  BMXWAIST = 130,
  smoker   = "Current",
  RIDAGEYR = 50,
  RIAGENDR = "Male",
  race_eth = "Non-Hispanic Black",
  income   = "$55,000+"),
interval = "confidence")
```

```
##     fit   lwr   upr
## 1 7.168 6.711 7.625
```

Conclusion: The estimated mean fasting glucose among individuals with the specified predictor values is 7.17 mmol/L (95% CI = 6.71, 7.63).

5.11 Confidence intervals and prediction intervals

As with SLR (Section 4.7), there are three kinds of intervals we are interested in:

- Confidence intervals (CI) for the regression coefficients (CIs for βs)
- CI for the mean outcome (CI for $E(Y|X = x)$)
- Prediction interval (PI) for an individual observation (PI for $Y|X = x$)

The syntax is the same as for SLR, just with more predictors in the `data.frame`. Using our model from Example 5.1, and the predictor values we used for the prediction in the previous section we get the following.

```
# CIs for regression coefficients
confint(fit.ex5.1)
```

```
##                                  2.5 %   97.5 %
## (Intercept)                    2.23373  3.71228
## BMXWAIST                       0.01850  0.03052
## smokerPast                    -0.03443  0.45685
## smokerCurrent                 -0.19789  0.39342
## RIDAGEYR                       0.01856  0.03154
## RIAGENDRFemale                -0.53431 -0.12158
## race_ethNon-Hispanic White    -0.79371 -0.22356
## race_ethNon-Hispanic Black    -0.63860  0.14331
## race_ethNon-Hispanic Other    -0.40786  0.44075
## income$25,000 to <$55,000     -0.41978  0.21903
## income$55,000+                -0.38281  0.19519
```

```
# CI for the mean outcome
predict(fit.ex5.1, data.frame(
  BMXWAIST = 130,
  smoker   = "Current",
  RIDAGEYR = 50,
  RIAGENDR = "Male",
  race_eth = "Non-Hispanic Black",
  income   = "$55,000+"),
interval = "confidence")
```

```
##     fit   lwr   upr
## 1 7.168 6.711 7.625
```

```
# PI for an individual observation
predict(fit.ex5.1, data.frame(
  BMXWAIST = 130,
  smoker   = "Current",
  RIDAGEYR = 50,
  RIAGENDR = "Male",
  race_eth = "Non-Hispanic Black",
  income   = "$55,000+"),
interval = "prediction")
```

```
##     fit  lwr   upr
## 1 7.168 4.17 10.17
```

As discussed in Section 4.7, the estimate of the mean outcome and the prediction for an individual are the same, but the PI for an individual (the interval in which we expect 95% of observations to lie) will always be wider than the CI for the mean outcome (the interval resulting from a method that, if used on repeated samples, we expect to contain the true mean in 95% of samples). Single observations are more variable than the mean of many observations.

5.12 Which to use when? **car::Anova()**, **anova()**, **gmodels::estimable()**, or **predict()**

It is easy get `car::Anova()`, `anova()`, `gmodels::estimable()`, and `predict()` mixed up. This section summarizes when to use each, followed by some examples.

In general, `car::Anova()` and `anova()` are used to test whether one or more regression coefficients are equal to 0, whereas `gmodels::estimable()` and `predict()` are used to estimate a sum of regression coefficients each multiplied by some number.

Use **`car::Anova(, type = 3)`** to test whether all the adjusted regression coefficients associated with a single term in the model (e.g., a single predictor, a single interaction) are simultaneously zero. For binary categorical predictors, continuous predictors, and interactions between them, the `car::Anova()` output is redundant with the regression coefficient table output from `summary()` since each such term only has one corresponding regression coefficient. However, you must use `car::Anova()` to test the significance of a categorical predictor with more than two levels or of an interaction that involves a categorical predictor with more than two levels. `car::Anova()` carries out comparisons for each of a set of specific pairs of nested models – each comparison is between the full model and a reduced model with one predictor (or interaction) removed.

Use **`anova()`** to compare two *nested* models – a full model compared to a reduced model with some terms removed from the full model. The tests possible with `car::Anova()` are also possible with `anova()`, but the reverse is not true. For example, you must use `anova()` to test the significance of a term involved in an interaction by comparing the full model to the reduced model that lacks both that term's main effect and interaction. When you *can* use `car::Anova()`, however, it is often more convenient than `anova()` because it carries out more than one test at the same time, one for each term in the model (comparing the full model to the reduced model that lacks that term).

Use **`gmodels::estimable()`** to estimate and test the significance of a linear combination of coefficients. For example, use `gmodels::estimable()` to estimate the effect of one predictor in an interaction at a specific level of the other predictor (since this effect is the sum of a main effect and an interaction term).

Use **`predict()`** to estimate the mean outcome at specified levels of the predictors. Since a prediction is a linear combination of coefficients, you could use `gmodels::estimable()`, but for this purpose `predict()` is simpler to use since you do not have to remember the order of the terms or which levels are reference levels.

Both prediction (using `predict()`) and estimating an effect (using `gmodels::estimable()`) involve multiplying regression coefficients by numbers and then adding them up. However, there are a few differences. When using `predict()`, you supply values for every predictor in the model and you do not supply a value for the intercept. The intercept is always assumed to be included; otherwise, the result would not be a prediction. When using `gmodels::estimable()`; however, you supply numbers by which to multiply the regression coefficients. Typically, a zero is supplied for the intercept but not necessarily. Technically, you could use `gmodels::estimable()` to compute any prediction by supplying numbers corresponding to the values of predictors and 1 for the intercept. However, the reverse it not true – there are some effects which are not estimable using a single call to `predict()`.

Examples

1. Test the significance of smoking status (a factor with three levels) in the model for fasting glucose that includes waist circumference, smoking status, age, gender, race/ethnicity, and income (`fit.ex5.1`).

```
car::Anova(fit.ex5.1, type = 3)
```

```
## Anova Table (Type III tests)
##
## Response: LBDGLUSI
##              Sum Sq  Df F value            Pr(>F)
## (Intercept)     142   1   62.30 0.0000000000000091 ***
## BMXWAIST        146   1   64.02 0.0000000000000040 ***
## smoker            7   2    1.44            0.23638
## RIDAGEYR        131   1   57.42 0.0000000000000926 ***
## RIAGENDR         22   1    9.73            0.00188 **
## race_eth         39   3    5.68            0.00075 ***
## income            1   2    0.23            0.79082
## Residuals      1929 846
## ---
## Signif. codes:  0 '***' 0.001 '**' 0.01 '*' 0.05 '.' 0.1 ' ' 1
```

```
# Same test using anova()
# Fit model without smoker
fit0 <- lm(LBDGLUSI ~ BMXWAIST + RIDAGEYR +
             RIAGENDR + race_eth + income,
           data = nhanesf.complete)
# Compare reduced and full models
anova(fit0, fit.ex5.1)
```

```
## Analysis of Variance Table
##
## Model 1: LBDGLUSI ~ BMXWAIST + RIDAGEYR + RIAGENDR + race_eth + income
## Model 2: LBDGLUSI ~ BMXWAIST + smoker + RIDAGEYR + RIAGENDR + race_eth +
##     income
##   Res.Df  RSS Df Sum of Sq    F Pr(>F)
## 1    848 1936
## 2    846 1929  2      6.59 1.44   0.24
```

Although `anova()` gives the same test in this case, the `car::Anova()` code is more succinct and provides tests for all the other terms, as well.

2. Test the significance of the interaction (which has two levels) in the model for fasting glucose that includes gender, income, and their interaction (`fit.ex5.3.int`) (Section 5.9.10).

```
car::Anova(fit.ex5.3.int, type = 3)
```

```
## Anova Table (Type III tests)
##
## Response: LBDGLUSI
##                 Sum Sq  Df F value               Pr(>F)
## (Intercept)       2603   1  977.55 <0.0000000000000002 ***
## RIAGENDR             2   1    0.67                0.413
## income               1   2    0.11                0.898
## RIAGENDR:income     17   2    3.17                0.042 *
## Residuals         2266 851
## ---
## Signif. codes:  0 '***' 0.001 '**' 0.01 '*' 0.05 '.' 0.1 ' ' 1
```

```
# Same test using anova()
# Fit model without the interaction
fit0 <- lm(LBDGLUSI ~ RIAGENDR + income,
           data = nhanesf.complete)
# Compare reduced and full models
anova(fit0, fit.ex5.3.int)
```

```
## Analysis of Variance Table
##
## Model 1: LBDGLUSI ~ RIAGENDR + income
## Model 2: LBDGLUSI ~ RIAGENDR + income + RIAGENDR:income
##   Res.Df  RSS Df Sum of Sq    F Pr(>F)
## 1    853 2283
## 2    851 2266  2      16.9 3.17  0.042 *
## ---
## Signif. codes:  0 '***' 0.001 '**' 0.01 '*' 0.05 '.' 0.1 ' ' 1
```

3. Test the significance of income in `fit.ex5.3.int` by comparing the full model to the reduced model that lacks income and the interaction (Section 5.9.11). Is it possible to do this test using `car::Anova()`?

```
# Fit model without the main effect or interaction
fit0 <- lm(LBDGLUSI ~ RIAGENDR, data = nhanesf.complete)

# Compare reduced and full models
anova(fit0, fit.ex5.3.int)
```

```
## Analysis of Variance Table
##
## Model 1: LBDGLUSI ~ RIAGENDR
## Model 2: LBDGLUSI ~ RIAGENDR + income + RIAGENDR:income
##   Res.Df  RSS Df Sum of Sq    F Pr(>F)
## 1    855 2295
## 2    851 2266  4      29.6 2.78  0.026 *
## ---
## Signif. codes:  0 '***' 0.001 '**' 0.01 '*' 0.05 '.' 0.1 ' ' 1
```

The above test is not possible using `car::Anova()` since the reduced model has two terms removed, `income` and `RIAGENDR:income`, that are not simply all the levels of a single `factor` variable or all the interaction terms.

4. In `fit.ex5.3.int`, estimate the gender effect among those with income \$25,000 to <\$55,000 (Sections 5.9.7 and 5.9.9.2). Is it possible to use a single call to `predict()` to estimate this effect?

```
# Number of terms
length(coef(fit.ex5.3.int))
```

```
## [1] 6
```

```
# Check the naming of the slots
summary(fit.ex5.3.int)$coef[, 1, drop=F]
```

```
##                                          Estimate
## (Intercept)                                6.2328
## RIAGENDRFemale                             0.2158
## income$25,000 to <$55,000                  0.1007
## income$55,000+                             0.1021
## RIAGENDRFemale:income$25,000 to <$55,000  -0.7122
## RIAGENDRFemale:income$55,000+             -0.7418
```

```
# Estimate the effect
gmodels::estimable(fit.ex5.3.int,
                   c("RIAGENDRFemale"                            = 1,
                     "RIAGENDRFemale:income$25,000 to <$55,000" = 1),
                   conf.int = 0.95)
```

```
##                Estimate Std. Error t value  DF Pr(>|t|) Lower.CI Upper.CI
## (0 1 0 0 1 0)  -0.4964     0.2197  -2.259 851  0.02412  -0.9276 -0.06513
```

You cannot use a single call to `predict()` to estimate this effect because there is no way to exclude the intercept. To estimate the gender effect at the specified level of income, you would need two calls to `predict()`, as shown below. The first call estimates the mean for females at the middle level of income and the second for males. Subtracting these two provides the correct gender effect (females vs. males). However, the subtraction results in the wrong CI since the difference of CIs is not equal to the CI of the difference.

```
# Estimated mean outcome for females at middle income level
Y1 <- predict(fit.ex5.3.int,
        data.frame(RIAGENDR = "Female",
                   income   = "$25,000 to <$55,000"),
        interval = "confidence")

# Estimated mean outcome for males at middle income level
Y2 <- predict(fit.ex5.3.int,
        data.frame(RIAGENDR = "Male",
                   income   = "$25,000 to <$55,000"),
        interval = "confidence")

# Correct effect estimate but NOT the correct CI
Y1 - Y2
```

```
##       fit     lwr     upr
## 1 -0.4964 -0.4839 -0.5088
```

5. Estimate the mean fasting glucose among individuals with a waist circumference of 110 cm who are current smokers, age 50 years, male, non-Hispanic White, and have an income <$25,000. in `fit.ex5.1` (Section 5.10).

```
predict(fit.ex5.1,
        data.frame(BMXWAIST = 110,
                   smoker   = "Current",
                   RIDAGEYR = 50,
                   RIAGENDR = "Male",
                   race_eth = "Non-Hispanic White",
                   income   = "<$25,000"),
        interval = "confidence")
```

```
##      fit   lwr   upr
## 1 6.511 6.156 6.866
```

```
# Same result using gmodels::estimable()
gmodels::estimable(fit.ex5.1,
                   c("(Intercept)"                  = 1,
                     "BMXWAIST"                     = 110,
                     "smokerPast"                   = 0,
                     "smokerCurrent"                = 1,
                     "RIDAGEYR"                     = 50,
                     "RIAGENDRFemale"               = 0,
                     "race_ethNon-Hispanic White"   = 1,
                     "race_ethNon-Hispanic Black"   = 0,
                     "race_ethNon-Hispanic Other"   = 0,
                     "income$25,000 to <$55,000"    = 0,
                     "income$55,000+"               = 0),
                   conf.int = 0.95)
```

```
##                                Estimate Std. Error t value  DF Pr(>|t|) Lower.CI
## (1 110 0 1 50 0 1 0 0 0 0)        6.511     0.1809      36 846        0    6.156
##                                Upper.CI
## (1 110 0 1 50 0 1 0 0 0 0)        6.866
```

Although gmodels::estimable() can provide the correct answer, the code required is more cumbersome. Additionally, the output includes a test of the null hypothesis that the prediction is 0, a test which is not generally of interest.

5.13 Overview of regression diagnostics

As briefly mentioned at the close of the previous chapter, the validity of regression coefficient estimates, confidence intervals, and significance tests depends on a number of assumptions. A standard linear regression model makes strong assumptions about the source of the data. The model assumes the data are from a set of **independent** cases. Each case has an outcome and a set of predictor values, and the model assumes a specific relationship between them described by the regression equation 5.1, reproduced here.

$$Y = \beta_0 + \beta_1 X_1 + \beta_2 X_2 + \ldots + \beta_K X_K + \epsilon$$

Thus, the model assumes the outcome is approximately the sum of the predictors each multiplied by some number. This implies a **linear** relationship between the outcome and each predictor when holding other predictors constant. The relationship implied by the βX parts of the equation is not assumed to be exact, only true on average. That is where the error term comes into play – ϵ captures the influence of all the other predictors that we have not been able to include in our model, either because we do not have measures of those predictors or because we do not know they need to be in the model, as well as random variation in the outcome even among cases with all the same predictor values. The model assumes $\epsilon \sim N(0, \sigma^2)$, that is, that the error term has a **normal** distribution with a **variance that does not vary between cases**.

All the subsequent results – the estimates of regression coefficients, their confidence intervals, and their p-values – are derived based on formulas that make use of these assumptions:

- Independence
- Normality
- Linearity
- Constant variance

If the assumptions are not valid, the results might not be correct. To make matters worse, there may be a few individual observations that cause problems. There may be **outliers**, observations with outcomes that are very different from what the regression model predicts, that could lead to problems with the normality or constant variance assumptions. There may be **influential observations**, observations that if removed would result in large changes in the regression coefficient estimates. It is not encouraging to find that the results depend heavily on just a few observations!

Finally, **collinearity** is an issue that is not present in SLR that we have to consider in MLR. Collinearity has to do with how correlated the predictors are with each other. In the extreme case, if two predictors are completely redundant, the equation that is being solved when fitting the regression model has no unique solution. There is no unique coefficient for a predictor with the definition "the effect of this predictor while holding all other predictors fixed" if there is another predictor that is perfectly correlated with it; you cannot vary the predictor while holding the other fixed. Even when the redundancy is only partial, correlations between predictors can cause regression results to be unstable.

The following sections go into detail regarding each of the above. Each section describes the impact of an issue, its diagnosis, and potential solutions when the diagnosis is not favorable.

5.14 Checking the independence assumption

A linear regression model assumes that each observation is independent of the others. If the cases were drawn from a population using a simple random sample, then this assumption is met. The independence assumption would be violated if the observations are clustered, such as if there are repeated measures from the same individual, as in longitudinal data, or if households were first sampled followed by sampling individuals within households.

5.14.1 Impact of dependence

A violation of the assumption of independence results in incorrect confidence intervals and p-values, although in some cases regression coefficient estimates will still be unbiased (Liang and Zeger, 1993; Diggle et al., 2002; Fitzmaurice et al., 2011).

5.14.2 Diagnosis of dependence

Think about how the data were collected. Are there clusters? For example, are there data from individual patients clustered within hospitals? Or individuals clustered in families or neighborhoods? Are there repeated measures from the same individual? Answering "yes" to any of those questions implies the presence of dependent data.

5.14.3 Potential solutions for dependence

There are a number of methods of handling correlated data. If the correlation is the result of a complex sampling method, such as is utilized in NHANES (see NHANES Tutorials[4], accessed July 30, 2021), it is possible to adjust for this feature of the data using methods discussed in Chapter 8. If the clusters are a simple random sample, but there are multiple observations within clusters, then generalized least squares (Pinheiro and Bates, 2000), linear mixed models (Laird and Ware, 1982; Fitzmaurice et al., 2011), or generalized estimating equations (Liang and Zeger, 1986; Zeger and Liang, 1986; Diggle et al., 2002) can be used to account for the within-cluster correlations. A specific example of clustered data is longitudinal data, in which the clusters are individuals who are measured repeatedly over time. These methods are beyond the scope of this text.

5.15 Checking the normality assumption

Earlier in this chapter we stated that $\epsilon \sim N(0, \sigma^2)$ which is, in part, shorthand for "the errors are normally distributed around the mean outcome and the average of the errors is zero". This implies that at any given predictor value the distribution of Y given X is assumed to be normal. For example, in Figure 5.20, the SLR on the left meets the normality assumption, while the one on the right (from Example 4.1) does not.

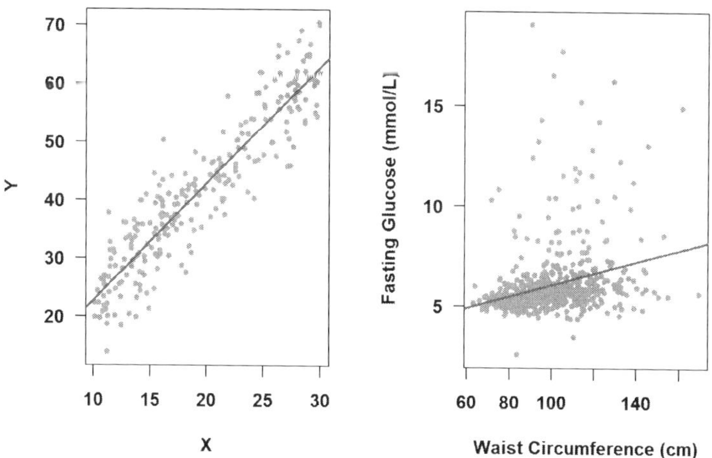

FIGURE 5.20 Normality assumption met vs. not met

[4]https://wwwn.cdc.gov/nchs/nhanes/tutorials/default.aspx

5.15.1 Impact of non-normality

A violation of the assumption of normality of errors may result in incorrect inferences in small samples. Both confidence intervals and p-values rely on the normality assumption, so if it is not valid then these may be inaccurate. However, in large samples (at least ten observations per predictor, although this is just a rule of thumb (Schmidt and Finan, 2018)) violation of the normality assumption does not have much of an impact on inferences (Weisberg, 2014).

5.15.2 Diagnosis of non-normality

The assumption is that the errors are normally distributed. We do not actually observe the errors, but we do observe the residuals and so we use those to diagnose normality (as well as other assumptions). There are formal tests of normality (e.g., the Shapiro-Wilk test); however, they are not needed in large samples since violation of the normality assumption will not impact inferences and they lack power in small samples. Instead, use a visual diagnosis.

While Figure 5.20 provided a general illustration, two diagnostic tools that allow us to directly compare the residuals to a normal distribution are a histogram of the residuals and a normal quantile-quantile plot (QQ plot), both of which are explained in the following example.

To illustrate the diagnosis of non-normality in a small sample size, we will use a random subset of 50 observations from our NHANES fasting subsample dataset (`nhanesf.complete.50_rmph.Rdata`) and re-fit the model from Example 5.1.

Example 5.1 (continued): Diagnose non-normality in the small sample size version of Example 5.1. Plot the residuals using a histogram and QQ plot.

```
load("Data/nhanesf.complete.50_rmph.Rdata")

# Re-fit the model using the small n dataset
fit.ex5.1.smalln <- lm(LBDGLUSI ~ BMXWAIST + smoker + RIDAGEYR +
                         RIAGENDR + race_eth + income,
                       data = nhanesf.complete.50)

# View results
round(cbind(summary(fit.ex5.1.smalln)$coef,
            confint(fit.ex5.1.smalln)),4)
```

```
##                              Estimate Std. Error t value Pr(>|t|)   2.5 %  97.5 %
## (Intercept)                    3.1771     1.5706  2.0228   0.0500  0.0002  6.3540
## BMXWAIST                       0.0258     0.0102  2.5390   0.0152  0.0052  0.0464
## smokerPast                    -0.4796     0.5278 -0.9088   0.3690 -1.5471  0.5879
## smokerCurrent                  0.5608     0.6197  0.9048   0.3711 -0.6928  1.8143
## RIDAGEYR                       0.0168     0.0139  1.2109   0.2332 -0.0113  0.0448
## RIAGENDRFemale                -1.0163     0.4812 -2.1120   0.0411 -1.9896 -0.0430
## race_ethNon-Hispanic White     0.0297     0.6071  0.0489   0.9612 -1.1983  1.2577
## race_ethNon-Hispanic Black     0.2479     0.8903  0.2785   0.7821 -1.5528  2.0487
## race_ethNon-Hispanic Other     2.2347     0.8800  2.5393   0.0152  0.4547  4.0147
## income$25,000 to <$55,000     -0.1413     0.7488 -0.1886   0.8514 -1.6559  1.3734
## income$55,000+                -0.2065     0.6999 -0.2951   0.7695 -1.6223  1.2092
```

```
# Compute residuals
RESID <- fit.ex5.1.smalln$residuals

# Sample size vs. # of predictors
length(RESID)
```

```
## [1] 50
```

```
length(coef(fit.ex5.1.smalln))-1
```

```
## [1] 10
```

There are fewer than 10 observations per predictor (50 / 10 = 5).

```
par(mfrow=c(1,2))
hist(RESID, xlab = "Residuals", probability = T,
     # Adjust ylim to be able to see more of the curves if needed
     ylim = c(0, 0.6))
# Superimpose empirical density (no normality assumption, for comparison)
lines(density(RESID, na.rm=T), lwd = 2, col = "red")
# Superimpose best fitting normal curve
curve(dnorm(x, mean = mean(RESID, na.rm=T), sd = sd(RESID, na.rm=T)),
      lty = 2, lwd = 2, add = TRUE, col = "blue")

# Normal quantile-quantile (QQ) plot
# (use standardized residuals)
qqnorm(rstandard(fit.ex5.1.smalln), col="red", pch=20)
abline(a=0, b=1, col="blue", lty=2, lwd=2)
```

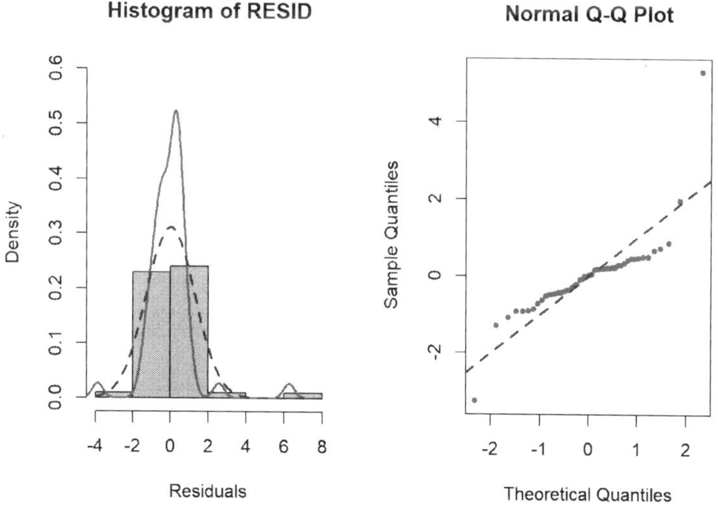

FIGURE 5.21 Diagnosis of normality of errors

In the **left panel** (histogram of residuals) of Figure 5.21, the solid line provides a smoothed estimate of the distribution of the observed residuals without making any assumption about the shape of the distribution. The dashed line, however, is the best fit normal curve. The closer these two curves are to each other, the better the normality assumption is met. In the **right panel** (a QQ plot), the ordered values of the n residuals are plotted against the corresponding n quantiles of the standard normal distribution. The closer the points are to the dashed line, the better the normality assumption is met. In this example, there is a lack of normality in the tails of the distribution. Since the sample size is small (n = 50, 10 predictors, fewer than 10 per predictor), non-normality may impact the results.

For convenience, instead of all the code above you can use the `check_normality()` function found in `Functions_rmph.R` (which you loaded at the beginning of this chapter).

```
# If you have not already loaded this file...
# ...do so now to make check_normality() available
source("Functions_rmph.R")

check_normality(fit.ex5.1.smalln)
```

5.15.3 Potential solutions for non-normality

If the sample size is large (at least ten observations per predictor, although this is just a rule of thumb (Schmidt and Finan, 2018)), non-normality does not need to be addressed (Weisberg, 2014).

If the sample size is not large: Some alternative methods of handling non-normality include the following.

- Non-normality may co-occur with non-linearity and/or non-constant variance. It is a good idea to check all the assumptions, and then resolve the most glaring assumption failure first, followed by re-checking all the assumptions to see what problems remain. Transforming the outcome is often a good place to start and can help with all three assumptions.
- **Normalizing transformation:** A transformation of the outcome Y used to correct non-normality of errors is called a "normalizing transformation". Common transformations are natural logarithm, square root, and inverse. These are all special cases of the Box-Cox family of outcome transformations which we will discuss in Section 5.18.
  - Unfortunately, these transformations only work for positive outcomes. An exception is the square-root transformation which can handle zeros, but still not negative values. For the other transformations, if the outcome is non-negative but includes some zero values, it is common to add 1 before transforming (e.g., $\log(X + 1)$). If the values of the raw variable are all very small or large, then instead of adding 1, add a number on a different scale (e.g., 0.01, 0.10, 10, 100). However, this approach is arbitrary, and the final results can depend on the number chosen. A sensitivity analysis can be useful to check if the results change when adding a different number (see Section 5.25).
  - If the outcome has many zero values, however, then no transformation will result in normality – there will always be a spike in the distribution at that (transformed) value. An alternative is to transform the variable into a binary or ordinal variable and use logistic regression (see Chapter 6), or use a regression method designed for non-negative data such as Poisson regression, zero-inflated Poisson regression, or negative binomial regression (Zeileis et al., 2008).

NOTE: A regression model does *not* assume that either the outcome or predictors are normally distributed. Rather, it assumes that the *errors* are normally distributed. It is often the case, however, that the normality or non-normality of the outcome (Y) does correspond to that of the errors.

Example 5.1 (continued): Continuing with our small sample size version of this example, examine if the natural logarithm, square-root, or inverse normalizing transformations correct the lack of normality.

Figures 5.22 to 5.25 display the histograms and QQ plots of residuals after these three transformations. Since the inverse transformation reverses the ordering of the variable, to preserve the original ordering also multiply by -1.

```
# Check for zero or negative values
summary(nhanesf.complete.50$LBDGLUSI)
```

```
##    Min. 1st Qu.  Median    Mean 3rd Qu.    Max.
##    4.55    5.27    5.61    5.97    6.11   16.20
```

```
# All positive so we can use log, sqrt, or inverse
fit.ex5.1.logY <- lm(log(LBDGLUSI) ~ BMXWAIST + smoker + RIDAGEYR +
      RIAGENDR + race_eth + income, data = nhanesf.complete.50)
fit.ex5.1.sqrtY <- lm(sqrt(LBDGLUSI) ~ BMXWAIST + smoker + RIDAGEYR +
      RIAGENDR + race_eth + income, data = nhanesf.complete.50)
fit.ex5.1.invY <- lm(-1/LBDGLUSI ~ BMXWAIST + smoker + RIDAGEYR +
      RIAGENDR + race_eth + income, data = nhanesf.complete.50)
```

```
# Histograms and QQ plots of residuals
# sample.size=F suppresses printing of the sample size
check_normality(fit.ex5.1.smalln, sample.size=F, main = "Y")
```

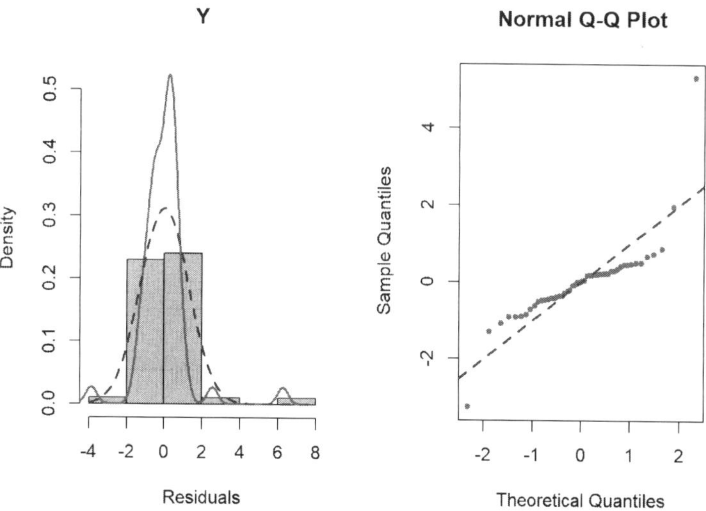

FIGURE 5.22 Checking the normality assumption (untransformed outcome)

```
check_normality(fit.ex5.1.logY,   sample.size=F, main = "log(Y)")
```

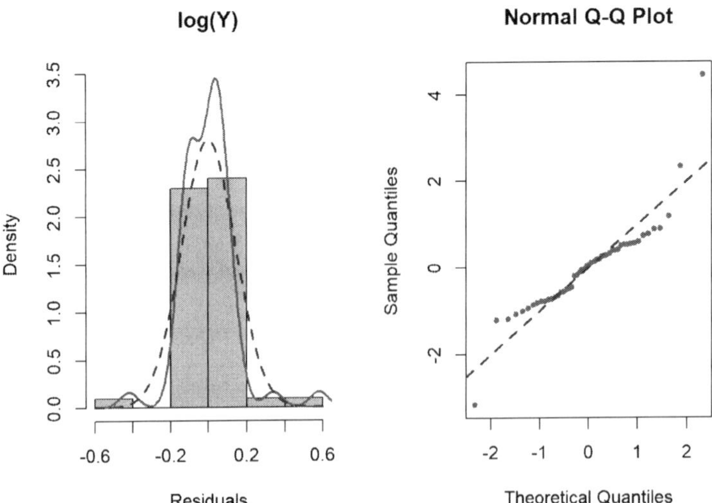

FIGURE 5.23 Checking the normality assumption (log-transformed outcome)

```
check_normality(fit.ex5.1.sqrtY,  sample.size=F, main = "Sqrt(Y)")
```

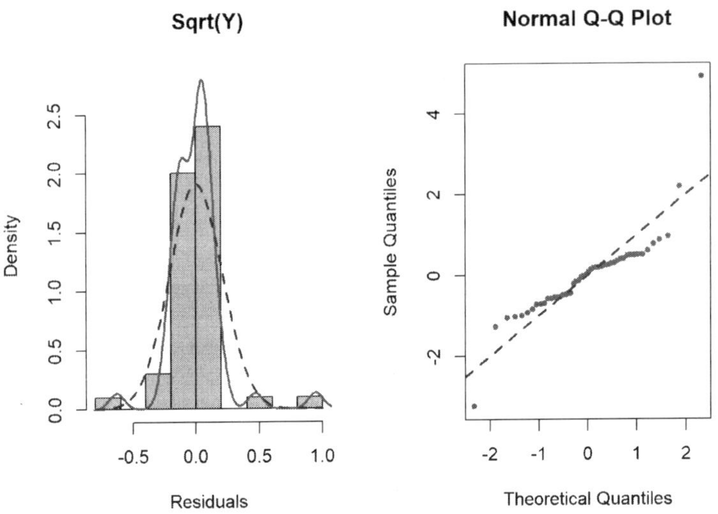

FIGURE 5.24 Checking the normality assumption (square root transformed outcome)

```
check_normality(fit.ex5.1.invY,   sample.size=F, main = "-1/Y")
```

None seem to resolve the non-normality completely, but of these the inverse transformation is the best. See Section 5.18 for how to implement the Box-Cox transformation which potentially can improve upon these. In this example, Box-Cox turns out not to help much – after reading Section 5.18, come back and try it as an exercise to confirm that conclusion.

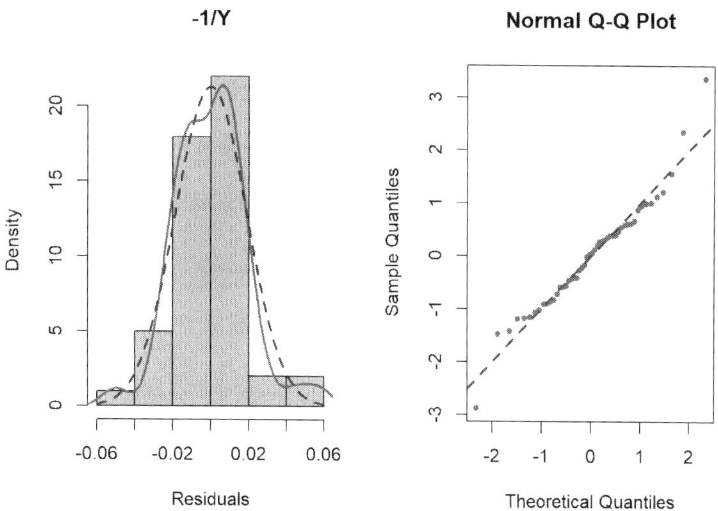

FIGURE 5.25 Checking the normality assumption (inverse outcome)

5.16 Checking the linearity assumption

A linear regression model assumes that the average outcome is linearly related to each term in the model when holding all others fixed. That is why plotted unadjusted (Section 5.4.6) or adjusted (Section 5.7) linear regression lines are straight lines – the model *assumes* straight lines. However, the underlying population relationship might not be linear – in that case, a straight line may not be a good fit to the data. For example, in Figure 5.26, the SLR on the left meets the linearity assumption, while the one on the right (from Example 4.3) does not (the dashed line is a smoother added to illustrate the shape of the trajectory when relaxing the linearity assumption).

There is an aspect of the linearity assumption that must be clarified before proceeding. The assumption is linearity in "each term in the model", not linearity in "each predictor". Why this distinction? Because sometimes a predictor is expressed in the model using multiple terms. Recall that for Example 5.1, the MLR model is the following.

$$\begin{aligned} \text{FG} &= \beta_0 \\ &+ \beta_1 \text{WC} \\ &+ \beta_2 I(\text{Smoker} = \text{Past}) + \beta_3 I(\text{Smoker} = \text{Current}) \\ &+ \beta_4 \text{Age} \\ &+ \beta_5 I(\text{Gender} = \text{Female}) \\ &+ \beta_6 I(\text{Race} = \text{Non-Hispanic White}) + \beta_7 I(\text{Race} = \text{Non-Hispanic Black}) \\ &+ \beta_8 I(\text{Race} = \text{Non-Hispanic Other}) \\ &+ \beta_9 I(\text{Income} = \$25{,}000 \text{ to } <\$55{,}000) + \beta_{10} I(\text{Income} = \$55{,}000+) + \epsilon \end{aligned}$$

For a continuous predictor (e.g., WC), the "term in the model" is the predictor itself. For a binary categorical predictor (e.g., gender), the "term in the model" is a single indicator function for one of the levels. In each of these cases, if you were to plot the mean outcome (holding all except this predictor fixed) vs. the predictor, you would see a straight line. For

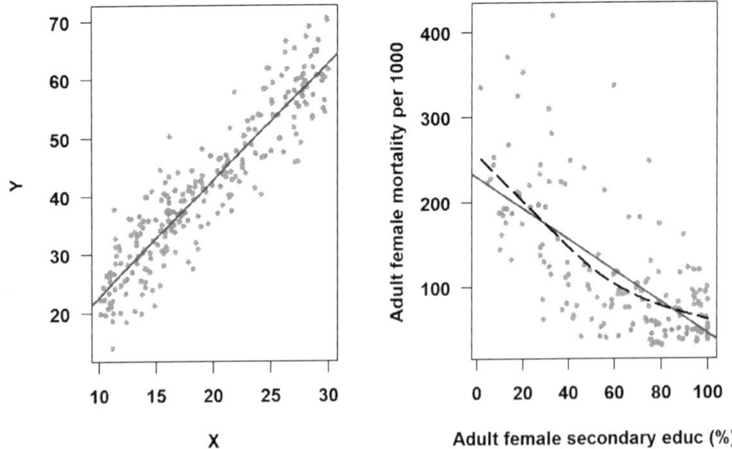

FIGURE 5.26 Linearity assumption met vs. not met

a continuous predictor, this is because the model assumes the relationship is linear, but in fact that assumption may be wrong and needs to be checked. For a binary predictor, the linearity assumption is always true – there are two means (the mean outcome at each level of the predictor) and a straight line always perfectly fits two points.

What about a categorical predictor with more than two levels (e.g., smoking status)? Figure 5.27 illustrates the unadjusted association between smoking status and fasting glucose in Example 5.1.

```
plotyx("LBDGLUSI", "smoker", nhanesf.complete,
       ylab = "FG (mmol/L)",
       xlab = "Smoking Status")
```

The line connecting the means at each level of smoking status is not straight. This is *not*, however, a violation of the linearity assumption as the "line" here was added after the fact just to connect the dots to provide a visual aid to discerning if any means are larger or smaller than the others – the line was not estimated by the regression model. If you were to plot the mean outcome vs. each individual indicator function, however, you would see just two means and, as mentioned above, a line always perfectly fits two points (Figure 5.28). Thus, even with categorical predictors with more than two levels, the underlying numeric indicator functions are each always linearly related to the outcome.

```
par(mfrow=c(1,2))
tmpdat <- nhanesf.complete %>%
  # Create an indicator function for each non-reference level
  mutate(Past    = factor(smoker == "Past"),
         Current = factor(smoker == "Current"))
plotyx("LBDGLUSI", "Past",    tmpdat,
       ylab = "FG (mmol/L)",
       xlab = "Smoking = Past")
plotyx("LBDGLUSI", "Current", tmpdat,
       ylab = "FG (mmol/L)",
       xlab = "Smoking = Current")
```

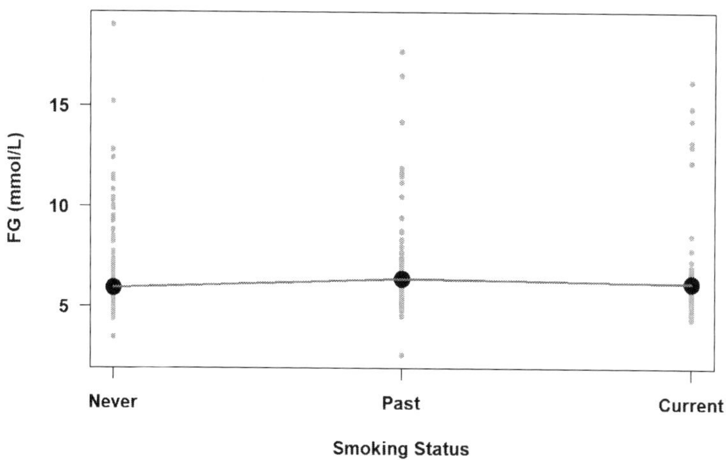

FIGURE 5.27 Categorical predictors do not violate the linearity assumption even if the plot appears non-linear

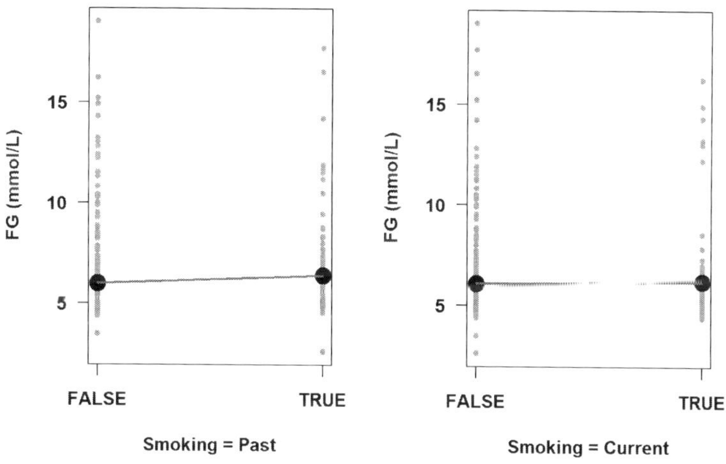

FIGURE 5.28 Linearity assumption met by default for indicator variables

Additionally, there are many applications of linear regression where you can fit something other than a straight line, as we did in Section 4.8 when we fit a polynomial, or if we transform variables in some other way (e.g., a log-transformation of the outcome or a predictor). Technically, the "linear" in "linear regression" refers to the outcome being linear in the parameters, the βs. The mathematics behind fitting a regression line involves taking the derivative of Y with respect to the parameters and when the relationship is linear the resulting equations can be solved. When the relationship is non-linear, however, estimating the parameters of the model requires more complex algorithms.

When we used a polynomial curve for a single predictor, we still could use "linear" regression because we could express the polynomial function as the sum of functions. You can use linear regression to fit any non-linear function of a predictor X that can be expressed as a sum of the product of coefficients and functions of X that do not have any coefficients inside the function. For example,

1. Linear regression (quadratic polynomial): $Y = \beta_0 + \beta_1 X + \beta_2 X^2 + \epsilon$
2. Linear regression (log-transformed predictor): $Y = \beta_0 + \beta_1 \log(X) + \epsilon$
3. Linear regression (log-transformed outcome): $\log(Y) = \beta_0 + \beta_1 X + \epsilon$
4. Non-linear regression: $Y = \beta_0 + \frac{\beta_1 X}{\beta_2 + \beta_3 (X+1)} + \epsilon$
5. Non-linear regression: $Y = \beta_0 e^{\beta_1 X} \epsilon$

In the first three examples above, even though the Y vs. X relationship is non-linear, the (possibly transformed) outcome is linearly related to each term in the model. That is not the case with the final two examples, which are non-linear regression models. Sometimes, however, you can convert a non-linear regression model into a linear one, but not always. To fit the fourth model above, you would have to use non-linear regression. But for the fifth, a natural log transformation of Y converts the model into the linear model $Y^* = \beta_0^* + \beta_1 X + \epsilon^*$ where $Y^* = \log(Y)$, $\beta_0^* = \log(\beta_0)$ and $\epsilon^* = \log(\epsilon)$.

In summary:

- You only need to check the linearity assumption for continuous predictors. For a categorical predictor, the linearity assumption is always met for each of the indicator functions since a straight line always fits two points exactly.
- The MLR model assumes linearity between the (possibly transformed) outcome and each "term in the model", each of which is either a (possibly transformed) continuous predictor or an indicator function for a level of a categorical predictor.
- In some cases, a transformation can convert a non-linear regression model into a linear regression model.

5.16.1 Impact of non-linearity

If the true model is not linear, then predictions based on a model that assumes linearity will be biased. For example, in the analysis of nation-level adult female mortality rate in Example 4.3 in Section 4.8, we found that if you predict mortality rate based on the linear fit, your prediction will be very different from the prediction based on the better fitting non-linear relationship, as shown in Figure 5.29 At some predictor values, the linear fit would result in an overestimate of the mean outcome, at others an underestimate.

Also, importantly, the test of significance of the regression coefficient for the predictor is testing the significance of a *linear* relationship. But if the true form of the relationship is not linear, then this is not the test you want.

5.16.2 Diagnosis of non-linearity

The linearity assumption requires that the Y vs. X relationship be linear *when holding all other predictors fixed.* A residual plot that takes that into account and is effective at diagnosing non-linearity is the **component-plus-residual plot** (CR plot), also known as a

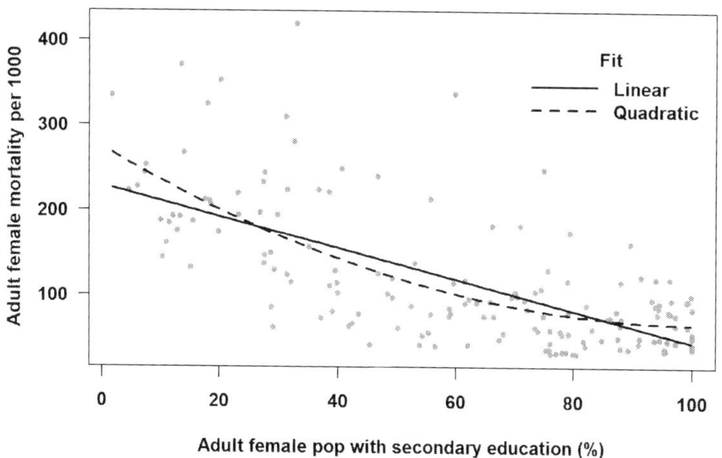

FIGURE 5.29 Not accounting for non-linearity can lead to biased predictions

partial-residual plot. Create a CR plot for each continuous predictor using the `car::crPlots()` function (Fox et al., 2023; Fox and Weisberg, 2019).

Example 5.1 (continued): Check the linearity assumption using a CR plot for each continuous predictor (waist circumference, age). The plots are shown in Figure 5.30.

NOTE: The `terms` argument is used below to avoid outputting unnecessary diagnostic plots for categorical predictors. If all predictors in the model are continuous, you can omit the `terms` argument below and you will get a plot for each.

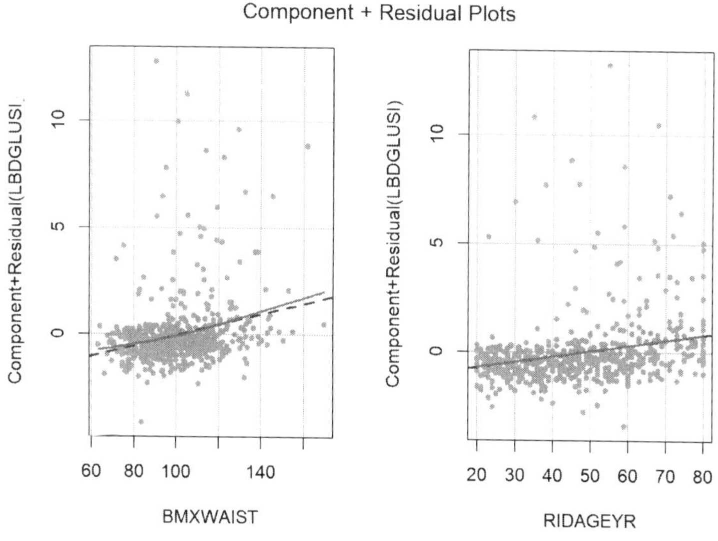

FIGURE 5.30 Residuals vs. predictors to check linearity assumption

```
# Use the "terms" option to specify the continuous predictors
# Override the default smoother -- gamLine seems to work better
car::crPlots(fit.ex5.1, terms = ~ BMXWAIST + RIDAGEYR,
             pch=20, col="gray",
             smooth = list(smoother=car::gamLine))
```

In each panel, the raw predictor values are plotted on the horizontal axis. The vertical axis plots the residuals plus the specific contribution of X, adjusted for all the other predictors in the model, $(\hat{\beta}_j x_{ij})$ added back in (the *component*). The *dashed line* is the linear regression of the component + residual vs. X. If there were no linear association between Y and X after adjusting for all other predictors in the model, the dashed line would be a horizontal line at 0. The *solid line* is a smoother that relaxes the linearity assumption. In each panel, if the solid line falls exactly on top of the dashed line, then the linearity assumption is perfectly met for that predictor. In this example, age is linear, but there seems to be a slight non-linearity for waist circumference.

5.16.2.1 Checking linearity in a model with an interaction

If the model includes an interaction, the linearity assumption becomes more complicated to check (and `car::crPlots()` will return an error). Although not ideal, an approximate solution is to diagnose linearity in the model without any interactions. Another option is to stratify the data by one of the predictors in the interaction, run `lm()` separately at each level of that predictor, and check the linearity assumption in each of those models (none of which have an interaction anymore) using `car::crPlots()`. If the stratifying predictor is continuous, then you can approximate this method by creating a categorical version of that predictor first (see, for example, `?cut`).

5.16.3 Potential solutions for non-linearity

- Fit a polynomial curve (see Section 4.8 and the example below).
- Transform X and/or Y to obtain a linear relationship. Common transformations are the natural logarithm, square root, and inverse. A Box-Cox transformation of the outcome may help, as well (Section 5.18). These transformations all assume X varies monotonically with Y (either X always increases with Y or always decreases with Y). A polynomial curve, however, can handle non-monotonic non-linearities (a curve that changes directions vertically).
- Use a spline function to fit a smoother (not discussed here, but see Harrell (2015) for more information).
- Transform X into a categorical variable (e.g., using tertiles or quartiles of X as cutoffs). It is generally better to not categorize a continuous predictor as you lose information and the cutoffs are arbitrary, but this method can be useful just to see what the trend looks like.
- Non-linearity may co-occur with non-normality and/or non-constant variance. As mentioned previously, check all the assumptions, resolve the most glaring assumption failure first, and then re-check all the assumptions.

Example 5.1 (continued): Above, we found that waist circumference violates the linearity assumption. Try adding a quadratic term and see if that helps.

```
fit.ex5.1.quad <- lm(LBDGLUSI ~ BMXWAIST + smoker + RIDAGEYR +
                     RIAGENDR + race_eth + income +
                     I(BMXWAIST^2), data = nhanesf.complete)
round(summary(fit.ex5.1.quad)$coef, 4)
```

```
##                                Estimate Std. Error t value Pr(>|t|)
## (Intercept)                      5.7468     1.4313  4.0152   0.0001
## BMXWAIST                        -0.0301     0.0274 -1.1004   0.2715
## smokerPast                       0.2089     0.1249  1.6720   0.0949
## smokerCurrent                    0.0730     0.1509  0.4835   0.6288
## RIDAGEYR                         0.0266     0.0034  7.8477   0.0000
## RIAGENDRFemale                  -0.3491     0.1055 -3.3098   0.0010
## race_ethNon-Hispanic White      -0.5407     0.1459 -3.7069   0.0002
## race_ethNon-Hispanic Black      -0.2770     0.1994 -1.3895   0.1650
## race_ethNon-Hispanic Other      -0.0125     0.2163 -0.0576   0.9541
## income$25,000 to <$55,000       -0.0818     0.1627 -0.5026   0.6154
## income$55,000+                  -0.0831     0.1471 -0.5650   0.5722
## I(BMXWAIST^2)                    0.0003     0.0001  2.0085   0.0449
```

The quadratic term is statistically significant; however, with a large sample size that might happen even if it does not change the model fit very much. Does it visually solve the failure of the linearity assumption? Enter each of the `BMXWAIST` terms, as well `RIDAGEYR`, into `car::crPlots()` and check the assumption again (Figure 5.31).

```
car::crPlots(fit.ex5.1.quad, terms = ~ BMXWAIST + I(BMXWAIST^2) + RIDAGEYR,
             pch=20, col="gray",
             smooth = list(smoother=car::gamLine))
```

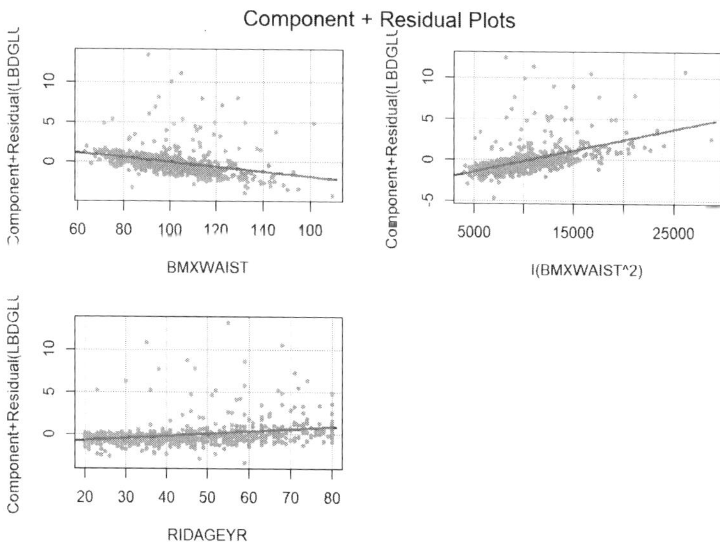

FIGURE 5.31 CR plots after adding a quadratic term

The linearity assumption is very well met now. It may seem strange that we have "linearity" when we have included waist circumference as a quadratic in the model! But remember that the linearity assumption has to do with each term in the model on its own, not with the entire quadratic polynomial for waist circumference. So as long as the CR plot shows linearity in each term individually, the linearity assumption is met.

5.16.4 Examples of non-linearities that are resolved by a transformation of the predictor

When the X values are highly skewed with a large range, which often happens with lab measurements or variables such as income or population size, then a **log transformation** often works well (Weisberg, 1985).

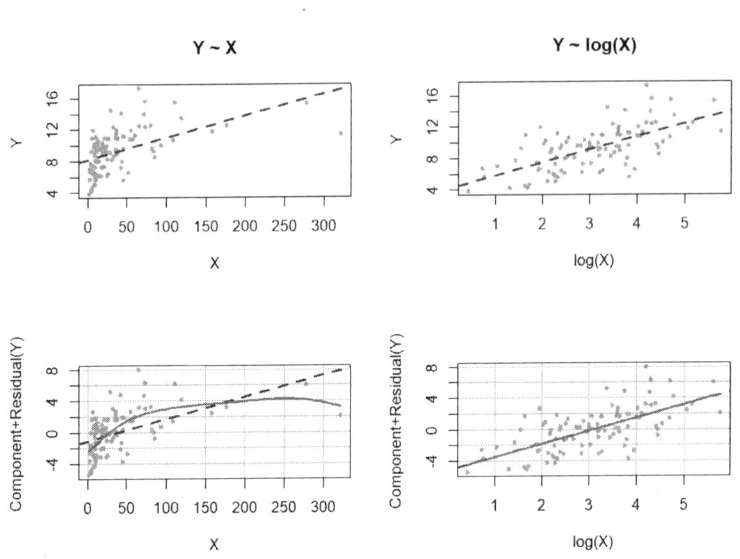

FIGURE 5.32 Log transformation of X to fix non-linearity

The top row of panels in Figure 5.32 are scatterplots of the outcome vs. the predictor, before (left panel) and after (right panel) the transformation. The panels on the bottom row are the corresponding CR plots. In this simulated example, the log transformation perfectly took care of the non-linearity. This illustrates the sort of Y vs. X relationship where a log transformation may help.

When the X values are skewed with a moderate range a **square-root transformation** often works well, as illustrated in Figure 5.33.

When there are many X values near zero but with some larger values, as well, then an **inverse transformation** often works well (Weisberg, 1985), as illustrated in Figure 5.34 (note that we multiply by –1 to preserve the ordering of the original variable).

5.16.5 Changing the amount of smoothing in a CR plot

The default options allow the smoother in `car::crPlots()` to choose the amount of smoothing. Sometimes the smoother is too unstable and we would like to change the amount of smoothing. The `k` argument (demonstrated below) has a default value of –1, which results in automatic smoothing (what you get if you leave out the `k` option altogether as we did in the previous

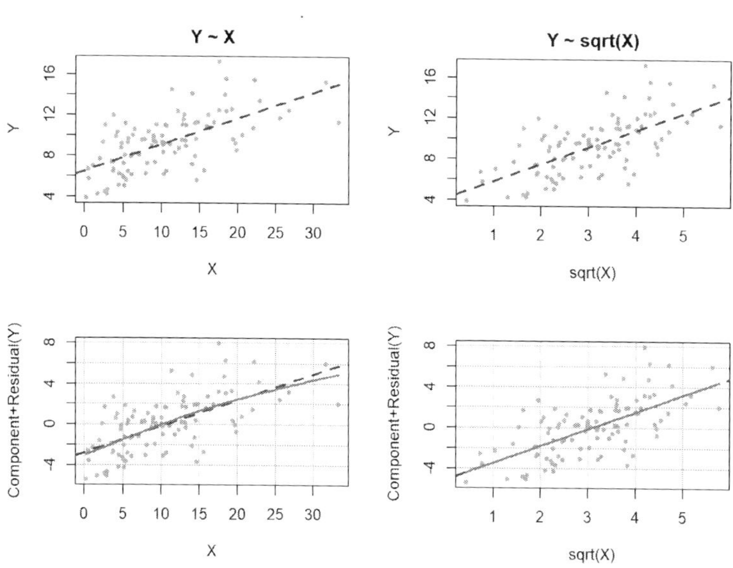

FIGURE 5.33 Square root transformation of X to fix non-linearity

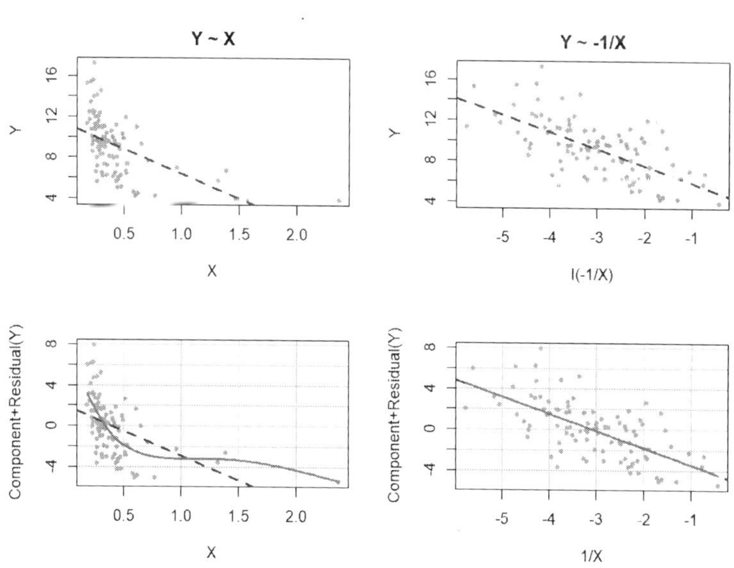

FIGURE 5.34 Inverse transformation of X to fix non-linearity

examples). If not the default, `k` must be at least 3 and larger values of `k` result in less smoothing.

Example 5.5: Use data from the United Nations 2020 Human Development Data (`unhdd2020.rmph.rData`, see Appendix A.2) (United Nations Development Programme, 2020) and regress the outcome Human Development Index (`hdi`) on the predictor tuberculosis incidence (2018, per 100,000 people) (`tb`), after using a square root transformation, and check the linearity assumption with smoothing argument values of –1 (automatic smoothing), 3 (maximum smoothing that is not a line), 5, and 10 (Figure 5.35).

```
load("Data/unhdd2020.rmph.rData")
fit.ex5.5 <- lm(hdi ~ sqrt(tb), data = unhdd)

par(mfrow=c(2,2))
car::crPlots(fit.ex5.5, terms = ~ sqrt(tb),
             pch=20, col="gray",
             smooth = list(smoother=car::gamLine, k = -1))
title("Default (automatic) smoothing")
car::crPlots(fit.ex5.5, terms = ~ sqrt(tb),
             pch=20, col="gray",
             smooth = list(smoother=car::gamLine, k = 3))
title("k = 3 (maximum smoothing)")
car::crPlots(fit.ex5.5, terms = ~ sqrt(tb),
             pch=20, col="gray",
             smooth = list(smoother=car::gamLine, k = 5))
title("k = 5")
car::crPlots(fit.ex5.5, terms = ~ sqrt(tb),
             pch=20, col="gray",
             smooth = list(smoother=car::gamLine, k = 10))
title("k = 10")
```

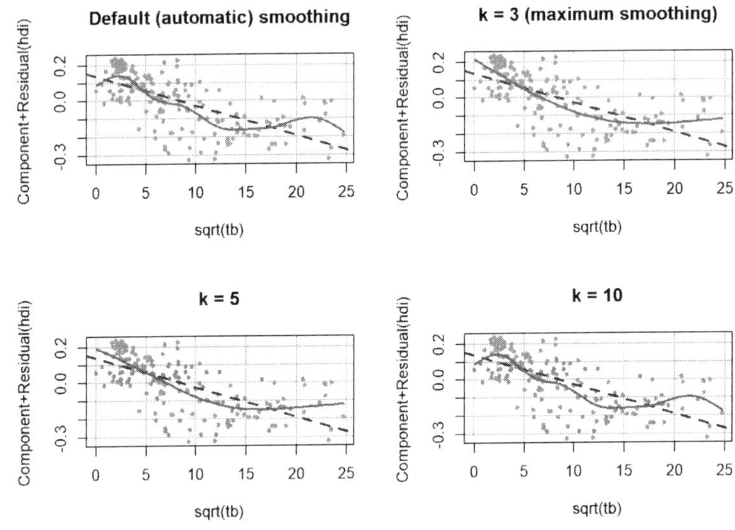

FIGURE 5.35 Changing the amount of smoothing in a CR plot

In this example, the default gives us a hint that there is non-linearity, but by smoothing more we see that it looks like a quadratic curve may be adequate. Note that adding a quadratic term in this example means adding the square of the square root, which is just `tb` itself, resulting in a model with both `sqrt(tb)` and `tb`. This will be explored in one of the exercises at the end of this chapter.

5.17 Checking the constant variance assumption

Not only does the assumption $\epsilon \sim N(0, \sigma^2)$ imply normality of errors, it also implies that the variance of the errors (σ^2) is the same for every observation. In particular, since observations have different predictor values, this implies that the variance does not depend on any of the predictors or on the fitted values (which are linear combinations of predictor values). This is the *constant variance* (or homoscedasticity) assumption. For example, in Figure 5.36, the SLR on the left meets the constant variance assumption, while the one on the right does not.

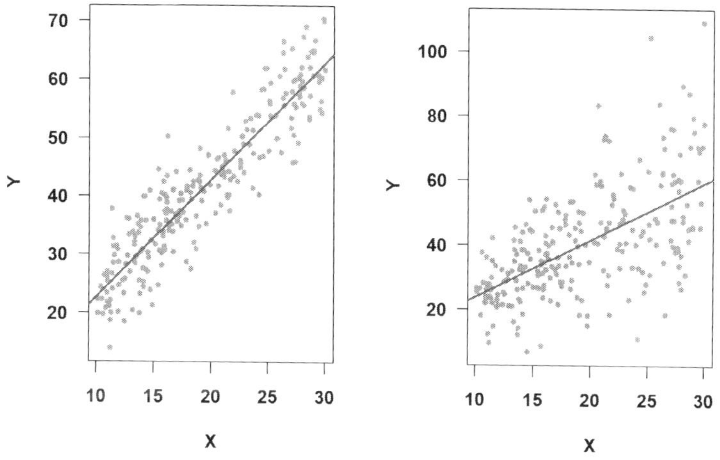

FIGURE 5.36 Constant variance assumption met vs. not met

5.17.1 Impact of non-constant variance

A violation of the constant variance assumption results in inaccurate confidence intervals and p-values, even in large samples, although regression coefficient estimates will still be unbiased (White, 1980; Hayes and Cai, 2007).

5.17.2 Diagnosis of non-constant variance

In SLR, the errors are assumed to have about the same amount of vertical variation from the regression line no matter where you are on the line, and this can be easily visualized, as in Figure 5.36, since there is just one predictor. In MLR, we visually diagnose the appropriateness of the constant variance assumption by examining a plot of residuals vs. fitted values (and, if we want to dig deeper, plots of residuals vs. each predictor) using `car::residualPlots()` (Fox et al., 2023; Fox and Weisberg, 2019). If the constant variance assumption is met, the spread of the points in the vertical direction should not vary much as you move in the horizontal direction.

Example 5.1 (continued): Evaluate the constant variance assumption for the MLR of fasting glucose on waist circumference, smoking status, age, gender, race, and income.

The residual vs. fitted value plot leads to the conclusion that we have non-constant variance since the variance is larger for larger fitted values (Figure 5.37).

```
# Residuals vs. fitted values
# fitted = T requests the residual vs. fitted plot
# terms = ~ 1 suppresses residual vs. predictor plots
# tests = F and quadratic = F suppresses hypothesis tests
# (see ?car::residualPlots for details)
car::residualPlots(fit.ex5.1,
                   pch=20, col="gray",
                   fitted = T, terms = ~ 1,
                   tests = F, quadratic = F)
```

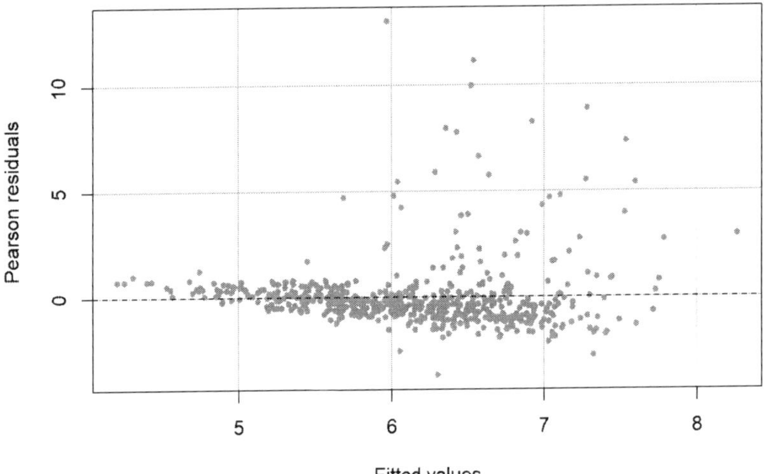

FIGURE 5.37 Residual vs. fitted plot to check the constant variance assumption

Examine residual vs. predictor plots for each predictor to see if the non-constant variance problem is related to any predictors more than others (Figure 5.38). In this example, these plots do not seem to add any information to the residual vs. fitted plot shown in Figure 5.37.

```
# Residuals vs. predictors
# fitted = F suppresses the residual vs. fitted plot
# ask = F prevents R from pausing and asking you to press
#     "Enter" to see additional plots
# layout = c(3,2) tells R to plot 6 panels per plot
#     in 3 rows and 2 columns
car::residualPlots(fit.ex5.1,
                   pch=20, col="gray",
                   fitted = F,
                   ask = F, layout = c(3,2),
                   tests = F, quadratic = F)
```

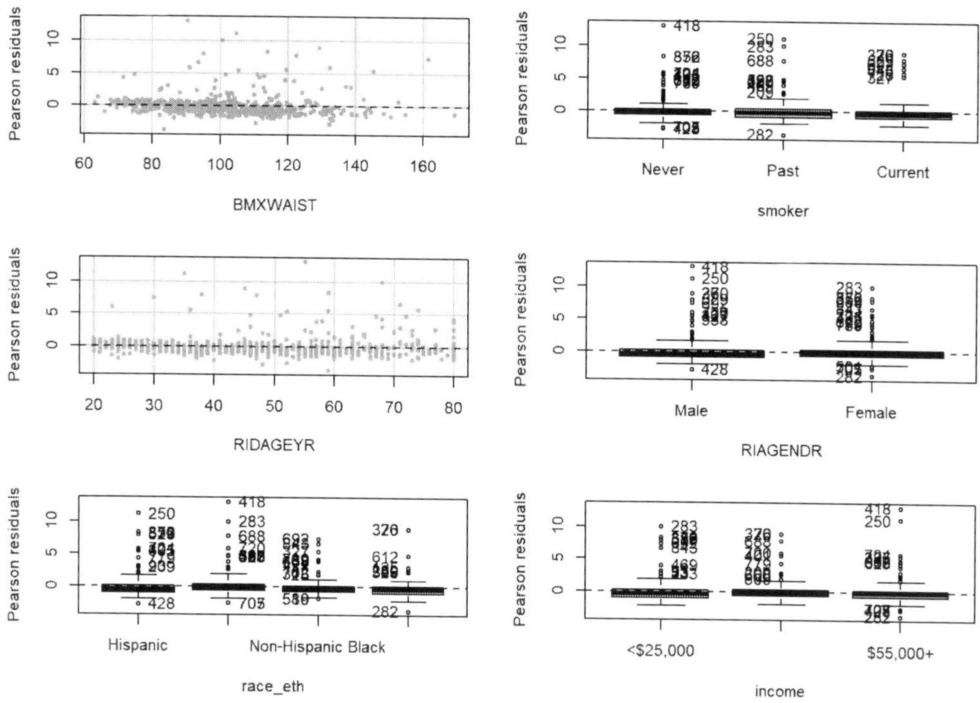

FIGURE 5.38 Residual vs. predictor plots to check the constant variance assumption

You can get all of these plots with one call to `car::residualPlots()` using the following syntax.

```
# Residuals vs. fitted, Residuals vs. predictors
car::residualPlots(fit.ex5.1,
                   pch=20, col="gray",
                   fitted = T,
                   ask = F, layout = c(2,2),
                   tests = F, quadratic = F)
```

5.17.3 Potential solutions for non-constant variance

- **Variance stabilizing transformation:** A transformation of the outcome Y used to correct non-constant variance is called a "variance stabilizing transformation". Again, common transformations are the natural logarithm, square root, inverse, and Box-Cox (Section 5.18).
- Advanced methods such as weighted or generalized least squares can be used to handle non-constant variance. These methods are beyond the scope of this text but see, for example, the `weights` option in `?lm`, (How to address heteroscedasticity in linear regression with R[5], accessed June 14, 2022), and `?nlme::gls` (Pinheiro et al., 2023).
- Non-constant variance may co-occur with non-linearity and/or non-normality. Again, check all the assumptions, resolve the most glaring assumption failure first, and then re-check all the assumptions.

[5]https://rpubs.com/mpfoley73/500818

5.18 Box-Cox outcome transformation

Continuous variables that are highly skewed are common in public health. For example, any variable that is derived as the sum of multiple indicators will have all non-negative values and this often leads to a skewed distribution. For example, in Rogers et al. (2021), researchers log-transformed the skewed outcome variable frailty index (FI), derived as the sum of 34 indicators across a number of health domains, prior to fitting linear regression studying the association between childhood socioeconomic status and frailty. Due to the presence of zero values, they used the transformation $\log(FI + 0.01)$. Other examples of skewed variables include population size, income, and costs. For example, in Sakai-Bizmark et al. (2021), researchers log-transformed cost of hospitalization in a linear regression investigating the association between homelessness and health outcomes in youth.

The previous sections included recommendations for handling assumption violations using variable transformations. The natural logarithm, square root, and inverse transformations are special cases of the more general **Box-Cox family of transformations** (Box and Cox, 1964). Note that, as presented here, the method only applies to transformation of the outcome, not predictors. Also, the method assumes that the outcome values are all positive. If there are a few zero or negative values, a modification of the Box-Cox that allows non-positive values (Box-Cox with negatives, BCN) can be used (Hawkins and Weisberg, 2017). If there are zero values, but no negative values, as mentioned previously an option is to add a small number before carrying out the transformation, although if there are a large number of zero values then it is possible no Box-Cox transformation will sufficiently normalize the errors or stabilize the variance of the errors.

Consider the family of transformations of a positive outcome Y defined by Y^λ. The square root transformation corresponds to $\lambda = 0.5$ and the inverse transformation to $\lambda = -1$. By convention, $\lambda = 0$ corresponds to the natural log transformation, not to Y^0 which would be a constant value of 1. When $\lambda < 0$, Y^λ reverses the ordering of the outcome values, so to preserve the original ordering, we will multiply by -1. We will use `MASS::boxcox()` (Ripley, 2023; Venables and Ripley, 2002) to find the value of λ that optimally transforms the outcome Y for a linear regression model.

In summary,

- if λ is close to 1, then no transformation is needed; otherwise,
- if λ is close to 0, replace Y with $\ln(Y)$; otherwise,
- if $\lambda > 0$, replace Y with Y^λ; otherwise,
- if $\lambda < 0$, replace Y with $-Y^\lambda$.

What is "close" to 1 or 0 is arbitrary. All else being equal, a model with no outcome transformation is most desirable as the results are easier to explain and all predictions are on the original scale. So unless λ is far enough from 1 to make a meaningful difference in the validity of the assumptions (e.g., constant variance is strongly violated), you might as well stick with no outcome transformation. Similarly, when a transformation is needed but λ is not far enough from 0 to make a meaningful difference, then use a log-transformation since it is commonly used and therefore understandable to a broader audience.

Example 5.1 (continued): In Sections 5.16 and 5.17 we found the following assumption violations:

- Possible non-linearity in waist circumference, and
- Non-constant variance of the errors – the variance was larger for observations with larger fitted values.

Additionally, the errors were not normally distributed, although this by itself is not an issue due to the large sample size.

The Box-Cox method searches over a range of possible transformations and finds the optimal one. `MASS::boxcox()` produces a plot of the profile log-likelihood values (the definition of this term is beyond the scope of this text) as a function of λ (Figure 5.39).

```
LAMBDA <- MASS::boxcox(fit.ex5.1)
```

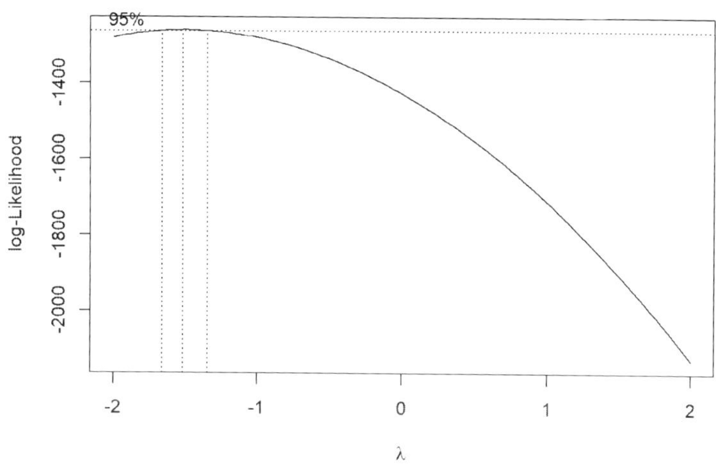

FIGURE 5.39 Box-Cox transformation: log-likelihood vs. lambda plot

The plot should have an upside-down U shape, and the optimal λ value is the one at which the log-likelihood is maximized. The object you assign the output to (below, `LAMBDA`) will contain elements x and y corresponding to the points being plotted to create the figure. These can be used to extract the optimal λ value using `which()`.

```
LAMBDA$x[which(LAMBDA$y == max(LAMBDA$y))]
```

```
## [1] -1.515
```

The optimal value for this example is -1.515, which we can round to -1.5. Thus, our Box-Cox transformation is $-Y^{-1.5}$. Interestingly, the optimal transformation turned out to be not that different from the inverse transformation ($-Y^{-1}$) that we earlier found in the small sample size version of this example to work better than a log or square root transformation.

Create the transformed version of the outcome and re-fit the model.

```
nhanesf.complete <- nhanesf.complete %>%
  mutate(LBDGLUSI_trans = -1*LBDGLUSI^(-1.5))

fit.ex5.1.trans <- lm(LBDGLUSI_trans ~ BMXWAIST + smoker +
    RIDAGEYR + RIAGENDR +  race_eth + income,
    data = nhanesf.complete)
```

NOTE: If the Box-Cox plot does not look like an upside down U, with a clear maximum, then increase the range over which to search λ. For example, the code below tells the function to search over the range from −3 to 3. Changing the range has the side effect of changing the resolution in the grid of possible values for λ, but this difference will typically not matter, especially if you round to the nearest 0.1 afterwards. To increase the resolution of the grid, decrease the range and/or increase the length of the `lambda` argument in `MASS::boxcox()`.

```
LAMBDA <- MASS::boxcox(fit.ex5.1, lambda = seq(-3, 3, length=100))
LAMBDA$x[which(LAMBDA$y == max(LAMBDA$y))]
```

After using the Box-Cox outcome transformation, re-check the regression assumptions.

Re-check normality

The normality assumption remains violated but, again, this is not an issued since the sample size is so large (Figure 5.40).

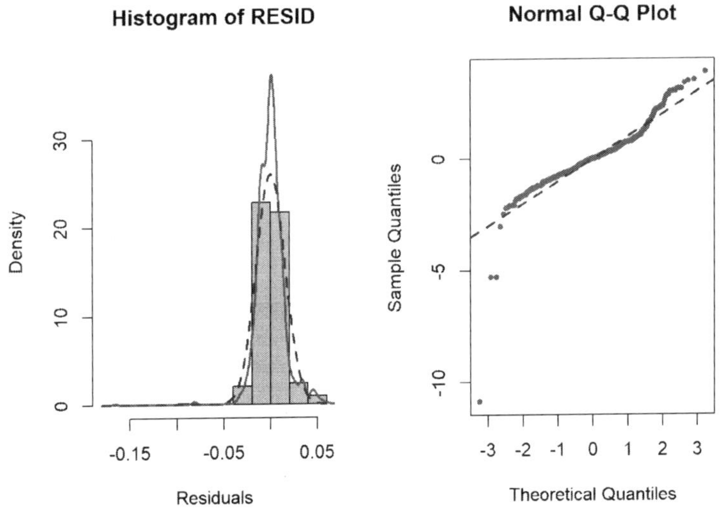

FIGURE 5.40 Re-check normality after Box-Cox transformation

Re-check linearity

Before the outcome transformation, we resolved non-linearity using a quadratic transformation of waist circumference. However, after the Box-Cox outcome transformation, there is no non-linearity so no need for the quadratic transformation. It is generally safe to ignore minor non-linearities in areas of the plot with few points, such as on the far left of the CR plot for waist circumference (Figure 5.41).

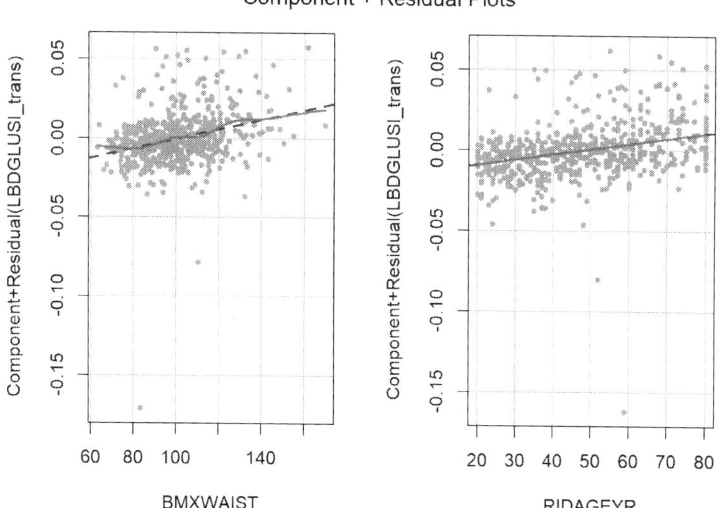

FIGURE 5.41 Re-check linearity after Box-Cox transformation

This illustrates that there may be multiple ways to resolve an assumption violation, and which you pick is more of an art than a science. In general, go with the solution that is simplest and easiest to explain. In this example, the outcome transformation is simpler than the quadratic predictor transformation in the sense that inference about the predictor's relationship to the outcome involves just one regression coefficient instead of the two you would have with a quadratic transformation. On the other hand, the outcome transformation is more complicated in the sense that it leads to predictions that are not on the original outcome scale.

Re-check constant variance

The variance still seems to increase with the fitted values, but not as much as before we used the Box-Cox transformation. Additionally, there are a few points with large negative residuals (Figure 5.42). We will examine these outliers in Section 5.21.

5.19 Interpreting a model with a transformed outcome

If the outcome has been transformed using a non-linear transformation such as a Box-Cox transformation or one of its special cases, then the relationships between the outcome and predictor based on the output of the regression model should be interpreted on the transformed scale, not the original scale. For example, in the final model we fit in Example 5.1, the regression coefficients are as follows:

```
round(summary(fit.ex5.1.trans)$coef, 4)
```

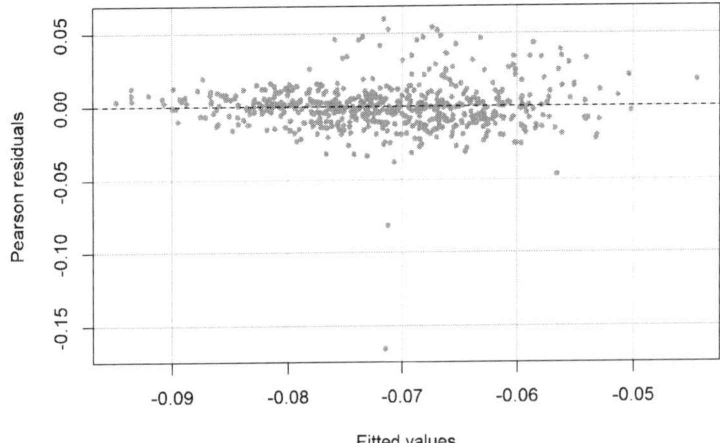

FIGURE 5.42 Re-check constant variance after Box-Cox transformation

```
##                             Estimate Std. Error  t value Pr(>|t|)
## (Intercept)                  -0.1127     0.0038 -29.2940   0.0000
## BMXWAIST                      0.0003     0.0000   9.7426   0.0000
## smokerPast                    0.0018     0.0013   1.4401   0.1502
## smokerCurrent                -0.0001     0.0015  -0.0732   0.9416
## RIDAGEYR                      0.0003     0.0000   9.7697   0.0000
## RIAGENDRFemale               -0.0047     0.0011  -4.4085   0.0000
## race_ethNon-Hispanic White   -0.0046     0.0015  -3.0799   0.0021
## race_ethNon-Hispanic Black   -0.0027     0.0020  -1.3160   0.1885
## race_ethNon-Hispanic Other   -0.0007     0.0022  -0.3116   0.7554
## income$25,000 to <$55,000     0.0006     0.0017   0.3760   0.7070
## income$55,000+               -0.0001     0.0015  -0.0689   0.9451
```

We conclude that waist circumference is positively associated with *transformed* fasting glucose and that a 1-unit difference in waist circumference is associated with a 0.0003 unit difference in mean $-FG^{-1.5}$, not with a 0.0003 difference in mean fasting glucose (mmol/L) on the original scale.

This relationship holds no matter the value of waist circumference – that is what it means to have a linear relationship. However, on the *original* scale of FG (mmol/L), the relationship is non-linear, as illustrated in the right-hand panel of Figure 5.43. On the original scale, the effect of waist circumference on fasting glucose depends on the value of waist circumference – at lower levels the effect is less and at greater levels the effect is greater.

Note: Back-transforming using the inverse of the outcome transformation does not lead to a valid estimate of the mean outcome on the original scale. However, if the assumptions of the linear regression model are met on the transformed scale, then the mean transformed outcome is equal to the median transformed outcome, and the back-transformed median is the estimated median on the original scale (Harrell, 2015, p. 391).

```
# Vector of WC values at which to predict
X     <- seq(min(nhanesf.complete$BMXWAIST),
             max(nhanesf.complete$BMXWAIST))

# Estimate mean outcome for these WC values
# Assume other predictors are at their mean
# or reference level
PRED <- predict(fit.ex5.1.trans,
                data.frame(BMXWAIST  = X,
                           smoker    = "Never",
                           RIDAGEYR  = mean(nhanesf.complete$RIDAGEYR),
                           RIAGENDR  = "Male",
                           race_eth  = "Hispanic",
                           income    = "<$25,000"))

# Plot on transformed scale
par(mfrow = c(1, 2))
plot(PRED ~ X, type = "l",
     ylab = "Mean Transformed Fasting Glucose",
     xlab = "Waist Circumference (cm)",
     main = "Transformed Outcome")

# LBDGLUSI_trans = -1*LBDGLUSI^(-1.5)
# Back-transform to the original scale
FG <- (-1*PRED)^(1/(-1.5))

# Plot on original scale
plot(FG ~ X, type = "l",
     ylab = "Median Fasting Glucose (mmol/L)",
     xlab = "Waist Circumference (cm)",
     main = "Original Scale")
```

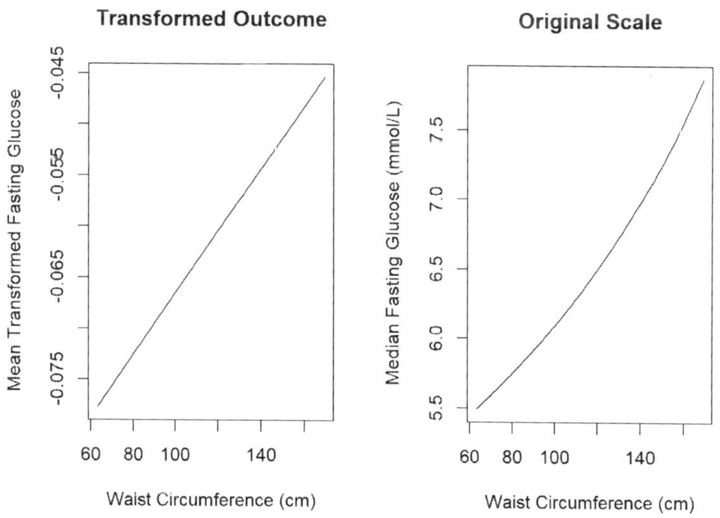

FIGURE 5.43 Interpreting a transformed outcome

This is similar to the effect of including a polynomial transformation of a predictor (see Section 4.8) – the relationship between the outcome and the predictor, on the original scale, becomes non-linear.

5.20 Collinearity

An important step in the process of fitting an MLR is checking for **collinearity** between predictors. This step was not needed in SLR because there was only one predictor.

Here's an extreme example which helps illustrate the concept. Suppose you want to know the relationship between fasting glucose and weight, and you fit a regression model that includes both weight in kilograms and weight in pounds. Clearly, these two predictors are redundant since they only differ in their units. With either one in the regression model, the other adds no new information. That is called **perfect collinearity** and it is mathematically impossible to include both in the model – you cannot estimate distinct regression coefficients for both predictors.

You also can have **approximate collinearity**. Suppose instead of weight in kilograms and weight in pounds, you have weight in kilograms and body mass index (BMI = weight/height^2). While weight and BMI are not *exactly* redundant, they are highly related. In this case, it *is* mathematically possible to include both in the model, but their regression coefficients are difficult to estimate accurately (they will have large variances) and will be difficult to interpret. The regression coefficient for weight is the difference in mean outcome associated with a 1-unit difference in weight while holding BMI constant. But in order to vary weight while holding BMI constant you have to vary height, as well. The "effect of greater weight adjusted for BMI" is therefore actually the "effect of greater weight and greater height", a combination of the effects of weight and height. In general, when two or more predictors are collinear, the interpretation of each while holding the other(s) fixed becomes more complicated.

Two predictors are **perfectly collinear** if their correlation is –1 or 1 (e.g., weight in kilograms and weight in pounds). The term "collinear" comes from the fact that one predictor (X_1) can be written as a linear combination of the other (X_2) as $X_1 = a + bX_2$. If you were to plot the pairs of points (x_{i1}, x_{i2}) for all the cases, they would fall exactly on a line.

Examples of perfect collinearity include:

- Two predictors that are exactly the same ($X_1 = X_2$);
- One predictor is twice the other ($X_1 = 2X_2$); and
- One predictor is 3 less than twice the other ($X_1 = -3 + 2X_2$).

If two predictors are perfectly collinear, then they are completely redundant for the purposes of linear regression. You could put either one in the model and the fit would be identical. Unless they are exactly the same ($X_1 = X_2$), you will get a different slope estimate using one or the other, but their p-values will be identical. If you put them both in the regression model at the same time, however, R will estimate a regression coefficient for just one of them, as shown in the following code.

```
# To illustrate...
# Generate random data from a standard normal distribution
set.seed(2937402)
y <- rnorm(100)
x <- rnorm(100)
# Create z perfectly collinear with x
z <- -3 + 2*x
lm(y ~ x + z)
```

```
##
## Call:
## lm(formula = y ~ x + z)
##
## Coefficients:
## (Intercept)            x            z
##     -0.0776       0.1418           NA
```

The problem is actually worse, however, for **approximate collinearity** because, if you put both in the model, R will estimate a regression coefficient for each but will have trouble estimating their unique contributions and you could get strange results, such as one large negative and one large positive estimated β and large standard errors for the estimates (implying that even with only a slightly different sample of individuals you may get very different estimates of the βs).

Two predictors that are approximately collinear are essentially measuring the same thing and so should have about the same association with the outcome. But when each is adjusted for the other, the results for each can be quite unstable and quite different from each other. In the example below, the second predictor is identical to the first except for having some random noise added. Compare their unadjusted and adjusted associations with the outcome.

```
# Generate random data but make the collinearity approximate
set.seed(2937402)
y <- rnorm(100)
x <- rnorm(100)
# Create z approximately collinear with x
z <- x + rnorm(100, sd=0.05)
# Each predictor alone
round(summary(lm(y ~ x))$coef, 4)
```

```
##             Estimate Std. Error t value Pr(>|t|)
## (Intercept)  -0.0776     0.1020 -0.7601   0.4490
## x             0.1418     0.1087  1.3045   0.1951
```

```
round(summary(lm(y ~ z))$coef, 4)
```

```
##             Estimate Std. Error t value Pr(>|t|)
## (Intercept)  -0.0783     0.1021  -0.767   0.4449
## z             0.1394     0.1079   1.292   0.1995
```

```
# Adjusted for each other
round(summary(lm(y ~ x + z))$coef, 4)
```

```
##                Estimate Std. Error t value Pr(>|t|)
## (Intercept)   -0.0742     0.1034 -0.7176   0.4747
## x              0.7256     2.3995  0.3024   0.7630
## z             -0.5799     2.3811 -0.2436   0.8081
```

Each predictor individually has a regression slope of about 0.14 with a standard error of about 0.11. However, when attempting to adjust for each other, their slopes are now in opposite directions and their standard errors are around 2.4! Again, a large standard error means that if you happened to have drawn a slightly different sample, you might end up with very different regression coefficient estimates in the model with both `x` and `z` – the results are not stable. If you want to verify this, re-run the above code multiple times starting after `set.seed()` (so it is not reset to the same value each time), and compare the results.

Going back to our example of weight and BMI, we can see that in our NHANES fasting subsample dataset these two predictors are highly collinear (Figure 5.44). The points do not fall *exactly* on a line, but they are generally close. Thus, weight and BMI seem to be approximately collinear. This makes sense because BMI is a measure of body mass computed as weight normalized by stature.

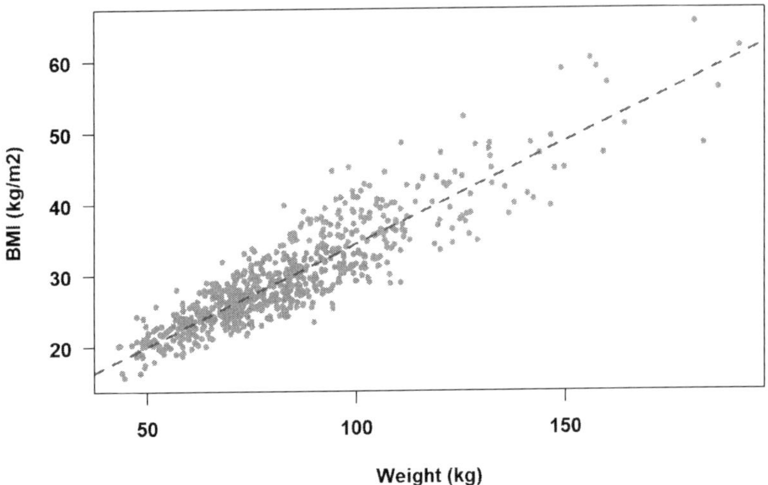

FIGURE 5.44 Body mass index and weight are highly collinear

All of the above examples were of pairwise collinearity. You can have collinearity between groups of predictors, as well, if some linear combination of one group of predictors is approximately equal to a linear combination of some other predictors. Such situations are harder to detect visually. However, in the following section, we will use a numeric diagnostic tool to help us detect collinearity even among groups of predictors.

5.20.1 Diagnosis of collinearity

The statistic we will use to diagnose collinearity is the **variance inflation factor (VIF)**. For each predictor, the VIF measures how much the variance of the regression coefficient estimate is inflated due to correlation between that predictor and the others. Compute VIFs using the `car::vif()` function (Fox et al., 2023; Fox and Weisberg, 2019).

5.20.1.1 VIFs when all predictors are continuous

Example 5.6: Use VIFs to evaluate the extent of collinearity in the regression of the outcome systolic blood pressure (`sbp`) on the predictors age (`RIDAGEYR`), weight (`BMXWT`), BMI (`BMXBMI`), and height (`BMXHT`) using NHANES 2017-2018 examination data from adults (`nhanes1718_adult_exam_sub_rmph.Rdata`). First, fit the regression model. Second, call `car::vif()` with the regression fit as the argument.

```
load("Data/nhanes1718_adult_exam_sub_rmph.Rdata")
nhanes <- nhanes_adult_exam_sub # Shorter name
rm(nhanes_adult_exam_sub)

fit.ex5.6 <- lm(sbp ~ RIDAGEYR + BMXWT + BMXBMI + BMXHT, data = nhanes)
car::vif(fit.ex5.6)
```

```
## RIDAGEYR    BMXWT   BMXBMI    BMXHT
##    1.014   88.835   69.934   18.764
```

One of the VIFs is near 1, while the other three are much larger. How do we interpret the magnitude of VIFs? How large is "too large"? Unfortunately, there is no clear-cut answer to that question. We do know that a value of 1 is ideal – that indicates that the variance of that predictor's regression coefficient estimate is not inflated at all by the presence of other predictors. Values above 2.5 may be of concern (Allison, 2012), and values above 5 or 10 are indicative of a more serious problem. In this example height, weight, and BMI clearly have serious collinearity issues; their variances are being inflated 19- to 89-fold by their collective redundancy.

5.20.1.2 Generalized VIFs when at least one predictor is categorical

Recall that a categorical predictor with L levels will be entered into a model as $L-1$ dummy variables ($L-1$ vectors of 1s and 0s). Why $L-1$? Because if you included all L of them the vectors would sum up to a vector of all 1s (since every observation falls in exactly one category) and that would be perfect collinearity. So one is always left out (the reference level).

Example 5.6 examined collinearity among predictors that were all continuous. What happens if there are any categorical predictors? It turns out that the usual VIFs computed on a model with a categorical predictor will differ depending on which level is designated as the reference level. In addition to getting inconsistent results, we are not interested in the VIF for each individual level of a categorical predictor, but rather of the predictor as a single entity. Fortunately, `car::vif()` automatically gives us a consistent VIF, called the generalized VIF (GVIF) (Fox and Monette, 1992), that is the same for each categorical predictor no matter the reference level, as long as the predictor is coded as a factor.

Example 5.7: Evaluate the extent of the collinearity in the regression of systolic blood pressure on the predictors age, BMI, height, and smoking status using our NHANES 2017-2018 examination subsample of data from adults (same dataset as in Example 5.6).

```
car::vif(lm(sbp ~ RIDAGEYR + BMXBMI + BMXHT + smoker, data = nhanes))
```

```
##              GVIF Df GVIF^(1/(2*Df))
## RIDAGEYR 1.052  1           1.026
## BMXBMI   1.006  1           1.003
## BMXHT    1.051  1           1.025
## smoker   1.081  2           1.020
```

The generalized VIF is found in the `GVIF` column. The `GVIF^(1/(2*Df))` column is the adjusted generalized standard error inflation factor (aGSIF) and is equal to the square-root of GVIF for continuous predictors and categorical predictors with just two levels (since, for those, `Df` $= 1$). Fox and Monette (1992) recommend using the aGSIF, however, since for categorical predictors with more than two levels it adjusts for the number of levels allowing comparability with the other predictors. A consequence is that when using aGSIF, we must take the square-root of our rules of thumb for what is a large value – aGSIF values above $\sqrt{2.5}$ (1.6) may be of concern, and values above $\sqrt{5}$ or $\sqrt{10}$ (2.2 or 3.2) are indicative of a more serious problem. Alternatively, you could square the aGSIF values and compare them to our original rule of thumb cutoffs.

Thus, the aGSIF for the three-level variable "smoker" is 1.02. In this example, the predictors do not exhibit strong collinearity (all the aGSIFs are near 1).

Run the code below if you would like to verify that the usual VIFs depend on which level is left out as the reference level but the GVIFs and aGSIFs remain unchanged.

```
# VIFs depend on the reference level
tmp <- nhanes %>%
  mutate(smoker1 = as.numeric(smoker == "Never"),
         smoker2 = as.numeric(smoker == "Past"),
         smoker3 = as.numeric(smoker == "Current"))
fit_ref1 <- lm(sbp ~ RIDAGEYR + BMXBMI + BMXHT + smoker2 + smoker3, data = tmp)
fit_ref2 <- lm(sbp ~ RIDAGEYR + BMXBMI + BMXHT + smoker1 + smoker3, data = tmp)
fit_ref3 <- lm(sbp ~ RIDAGEYR + BMXBMI + BMXHT + smoker1 + smoker2, data = tmp)
car::vif(fit_ref1)
car::vif(fit_ref2)
car::vif(fit_ref3)

# GVIF and aGSIF do not depend on the reference level
car::vif(lm(sbp ~ RIDAGEYR + BMXBMI + BMXHT + smoker, data = nhanes))
tmp <- nhanes %>%
  mutate(smoker = relevel(smoker, ref = "Past"))
car::vif(lm(sbp ~ RIDAGEYR + BMXBMI + BMXHT + smoker, data = tmp))
tmp <- nhanes %>%
  mutate(smoker = relevel(smoker, ref = "Current"))
car::vif(lm(sbp ~ RIDAGEYR + BMXBMI + BMXHT + smoker, data = tmp))
# (results not shown)
```

5.20.1.3 VIFs when there is an interaction or polynomial terms

Collinearity between terms involved in an interaction can be ignored (Allison, 2012). This applies also to collinearity between terms that together define a polynomial curve (e.g., X, X^2, etc.). In general, centering continuous predictors will reduce the collinearity. However, the sort of collinearity that is removed by centering actually has no effect on the fit of the model, only on the interpretation of the intercept and the main effects of predictors involved in the interaction. If you have interactions in the model, or multiple terms defining a polynomial, compute the VIFs or aGSIFs using a model without the interactions or higher order polynomial terms. Alternatively, instead of entering a polynomial using individual terms, use the `poly()` function. For example, for a quadratic in `BMXWT`, instead of entering `BMXWT + I(BMWXT^2)`, enter `poly(BMXWT, 2)` into `lm()`. `car::vif()` will then return one aGSIF value for the entire polynomial; note however, that the way polynomials are parameterized with `poly()` is different than in our examples where we entered terms individually.

5.20.1.4 VIF summary

- If all predictors in a model are continuous and/or binary (two levels) then `car::vif()` returns VIFs and our rule of thumb for what is large is that values above 2.5 may be of concern and values above 5 or 10 are indicative of a more serious problem.
- If any predictors in a model are categorical with more than two levels then `car::vif()` returns both GVIF and aGSIF values. Use the aGSIF values to evaluate collinearity since that allows predictors with different `Df` (different number of terms in the model) to be comparable. For aGSIFs, our rule of thumb for what is large is that values above 1.6 may be of concern and values above 2.2 or 3.2 are indicative of a more serious problem.
- Our rule of thumb regarding what is a large amount of inflation is arbitrary. It is useful as a guide, but there is no requirement to apply it strictly.
- If you have interactions in the model, or multiple terms defining a polynomial, compute the VIFs or aGSIFs using a model without the interactions or higher order polynomial terms. Alternatively, for polynomials, use `poly()` when fitting the model.

5.20.2 Impact of collinearity

When there is *perfect collinearity*, R will drop out predictors until that is no longer the case and so there is no impact on the final regression results. When there is *approximate collinearity*, however, there may be inflated standard errors and difficulty interpreting coefficients.

Example 5.7 (continued): It is reasonable to guess that the main driver of the collinearity is the strong correlation between weight and BMI. Compare the regression coefficients, their standard errors, and their p-values with and without weight in the model. First, create a complete-case dataset so differences between the models are not due to having different observations.

```
# Create a complete-case dataset so both models use the same sample
sbpdat <- nhanes %>%
  select(sbp, RIDAGEYR, BMXWT, BMXBMI, BMXHT) %>%
  drop_na()

# With weight in the model
fit_wt    <- lm(sbp ~ RIDAGEYR + BMXWT + BMXBMI + BMXHT, data = sbpdat)

# After dropping weight due to high collinearity
fit_nowt <- lm(sbp ~ RIDAGEYR         + BMXBMI + BMXHT, data = sbpdat)
```

Table 5.4 shows how the results differ after removing weight from the model. As expected, since its VIF was near 1, the "Age" row does not change much. Note in particular that, up to four decimal places, its standard error (SE) is exactly the same. This is what it means for a VIF to be 1 – that the variance (the square of the SE) is inflated by a factor of 1, in other words not inflated at all, by the presence of the other predictors.

TABLE 5.4 Demonstrating the impact of collinearity by comparing the models with and without weight

	With Weight				Without Weight			
Term	**B**	**SE**	**p**	**VIF**	**B**	**SE**	**P**	**VIF**
Intercept	137.34	36.41	<.001		73.52	8.96	<.001	
Age	0.4327	0.0292	<.001	1.01	0.4348	0.0292	<.001	1.01
Weight	0.3707	0.2050	.071	88.8	–	–	–	
BMI	-0.5093	0.5782	.379	69.9	0.5287	0.0692	<.001	1.00
Height	-0.3029	0.2169	.163	18.8	0.0787	0.0505	.119	1.01

Compare this, however, to the rows for "BMI" and "Height". After removing weight from the model, their estimates change drastically; both change direction, going from negative to positive. Also, their standard errors decrease dramatically, indicating that they are more precisely estimated. The squares of the ratios of their SEs with and without weight in the model are approximately the same as their VIFs in the model that includes weight.

```
(summary(fit_wt)$coef["BMXBMI", "Std. Error"] /
   summary(fit_nowt)$coef["BMXBMI", "Std. Error"])^2
```

```
## [1] 69.74
```

```
(summary(fit_wt)$coef["BMXHT",  "Std. Error"] /
   summary(fit_nowt)$coef["BMXHT",  "Std. Error"])^2
```

```
## [1] 18.49
```

The reason the variances decrease so much is that the definitions of "BMI holding other predictors fixed" and "height holding other predictors fixed" change after removing weight, with the old quantities (effects of BMI and height when holding age and weight constant) being much harder to estimate accurately (and harder to interpret) than the new (effects of BMI and height when holding age constant). Also, whereas "BMI when holding age, weight, and height constant" was not statistically significant ($p = .379$), "BMI when holding age and height constant" is ($p < .001$). Finally, after removing weight from the model, the VIFs for the remaining predictors are all near 1.

5.20.3 Potential solutions for collinearity

There are multiple ways to deal with collinearity:

- Remove one or more predictors from the model, or
- Combine predictors (e.g., average, sum, difference).

One could also try a biased regression technique such as LASSO or ridge regression. Those methods are beyond the scope of this text but see, for example, the R package `glmnet` (Friedman et al., 2010)).

5.20.3.1 Remove predictors to reduce collinearity

Ultimately, you want a set of predictors that are not redundant. One solution is to remove predictors suspected of being problematic, one at a time, and see if the VIFs become smaller.

Before removing any predictors, check the extent of missing data in the dataset. On one hand, when comparing VIFs between models with different predictors, each model should be fit to the same set of observations, which means starting with a complete case dataset. On the other hand, one or more predictors might have a much larger number of missing values than the others. In that case, it might be best to remove the predictors with a lot of missing values first, create a complete case dataset based on the remaining predictors, and then compare VIFs between models with different predictors.

Example 5.7 (continued): Remove one predictor at a time and see how the VIFs change. Use the complete case dataset created above so any differences are due to removing predictors rather than differing sets of observations with non-missing values.

```
# VIFs with all the predictors included
car::vif(fit_wt)
```

```
## RIDAGEYR    BMXWT   BMXBMI    BMXHT
##    1.014   88.835   69.934   18.764
```

```
# Drop weight
fit_nowt  <- lm(sbp ~ RIDAGEYR          + BMXBMI + BMXHT, data = sbpdat)
car::vif(fit_nowt)
```

```
## RIDAGEYR   BMXBMI    BMXHT
##    1.012    1.000    1.013
```

```
# Drop BMI
fit_nobmi <- lm(sbp ~ RIDAGEYR + BMXWT          + BMXHT, data = sbpdat)
car::vif(fit_nobmi)
```

```
## RIDAGEYR    BMXWT    BMXHT
##    1.012    1.271    1.284
```

```
# Drop height
fit_noht  <- lm(sbp ~ RIDAGEYR + BMXWT + BMXBMI         , data = sbpdat)
car::vif(fit_noht)
```

```
## RIDAGEYR    BMXWT   BMXBMI
##    1.010    4.794    4.786
```

With all four predictors in the model, BMI, weight, and height each exhibit high collinearity. After removing them one at a time we see that, in fact, weight and BMI are the problem variables (the VIFs are only large when both remain in the model). In this example, removing weight from the model best resolved the collinearity issue (the VIFs are smallest when weight was removed). In other examples, you may have to remove more than one predictor to obtain VIFs that are all small.

5.20.3.2 Combine predictors to reduce collinearity

Simply removing predictors is often the best solution. At other times, you really would like to keep all the predictors. In that case, combining them in some way may work. Common methods of combining predictors include taking the average, the sum, or the difference.

Example 5.8: Using the Natality dataset (see Appendix A.3), examine the extent of collinearity in the regression of the outcome birthweight (`DBWT`, g) on the ages of the mother (`MAGER`) and father (`FAGECOMB`) and resolve by combining predictors.

```
load("Data/natality2018_rmph.Rdata")
fit.ex5.8 <- lm(DBWT ~ MAGER + FAGECOMB, data = natality)
round(summary(fit.ex5.8)$coef, 4)
```

```
##              Estimate Std. Error t value Pr(>|t|)
## (Intercept) 3168.469     75.620 41.8999   0.0000
## MAGER          5.769      3.767  1.5317   0.1258
## FAGECOMB      -2.519      3.067 -0.8215   0.4115
```

```
car::vif(fit.ex5.8)
```

```
##    MAGER FAGECOMB
##     2.32     2.32
```

Ignoring statistical significance, this model implies that (1) for a given father's age, older mothers have children with greater birthweight and (2) for a given mother's age, older fathers have children with lower birthweight. In this particular dataset, the two predictors are not highly collinear, but their VIFs are close to the rule of thumb cutoff of 2.5. We will attempt to reduce their collinearity by combining them to form two predictors that are less collinear.

Replacing a pair of correlated predictors X_1 and X_2 with their average $Z_1 = (X_1 + X_2)/2$ and difference $Z_2 = X_1 - X_2$ may result in a pair of less correlated predictors. In this example, the transformed predictors would be interpreted as "average parental age" and "parental age difference".

```
natality <- natality %>%
  mutate(avg_parent_age          = (FAGECOMB + MAGER)/2,
         parent_age_difference =  FAGECOMB - MAGER)

fit.ex5.8_2 <- lm(DBWT ~ avg_parent_age + parent_age_difference, data = natality)
round(summary(fit.ex5.8_2)$coef, 4)
```

```
##                       Estimate Std. Error t value Pr(>|t|)
## (Intercept)           3168.469     75.620  41.900   0.0000
## avg_parent_age           3.250      2.483   1.309   0.1908
## parent_age_difference   -4.144      3.202  -1.294   0.1958
```

```
car::vif(fit.ex5.8_2)
```

```
##        avg_parent_age parent_age_difference
##                 1.099                 1.099
```

Ignoring statistical significance, these results imply that (1) for a given age difference, older parents have children with greater birthweight and (2) for a given average parental age, parents with a larger age difference have children with lower birthweight. By combining the predictors, we have reduced collinearity while retaining the ability to draw conclusions based on the combination of ages of both parents.

In other cases, you might combine the collinear predictors into a single predictor by replacing them with their sum or average. For example, if you have a set of items in a questionnaire that all are asking about the same underlying construct, each on a scale from, say, 1 to 5 (e.g., Likert-scale items), you may be able to average them together to produce a single summary variable. If taking this approach, be sure all the items are on the same scale and if they are not then standardize them before summing or averaging. Also, make sure the ordering of responses has the same meaning – sometimes one item is asking about agreement with the underlying construct and another is asking about disagreement, resulting in a low score on one item having the same meaning as a high score on another. If you have items with different response orderings, change some of them until all are in the same direction. For example, you could reverse the ordering of items asking about disagreement (e.g., replace a 1 to 5 scale with scores of 5 to 1) so they match items asking about agreement. While summing or averaging variables may work to reduce collinearity, it is important to also assess whether such a sum is valid using methods such as Cronbach's alpha and factor analysis (beyond the scope of this text but see, for example, `?psy::cronbach` (Falissard, 2022) and `?factanal`).

In general, take time to think about how your candidate predictors are related to each other and to the outcome. Are some completely or partially redundant (measuring the same underlying concept)? That will show up in the check for collinearity. Remove or combine predictors in a way that aligns with your analysis goals.

5.21 Outliers

While collinearity diagnostics look for problems in the regression model due to relationships between predictors, outlier (this section) and influence (next section) diagnostics look for problems due to individual observations (cases).

An **outlier** is an individual observation with a very large residual. "Large" here means large in magnitude – either very positive or very negative residuals could be outliers. Outliers are not *necessarily* extreme in either the outcome (Y) or any of the predictors (X). What makes an observation an outlier in regression is that the observed value is far from the predicted value. For example, in the *top left* panel in Figure 5.45, the filled-in point is extreme in both the X and Y directions but is close to the line; therefore, its residual is small and it is *not* a regression outlier. In the *top right* and *bottom left* panels, the filled-in point has a typical X or Y value, respectively, but is an outlier since it is far from the line. In the *bottom right* panel, the point is extreme in both the X and Y directions and is far from the line, but what makes it an outlier is being far from the line.

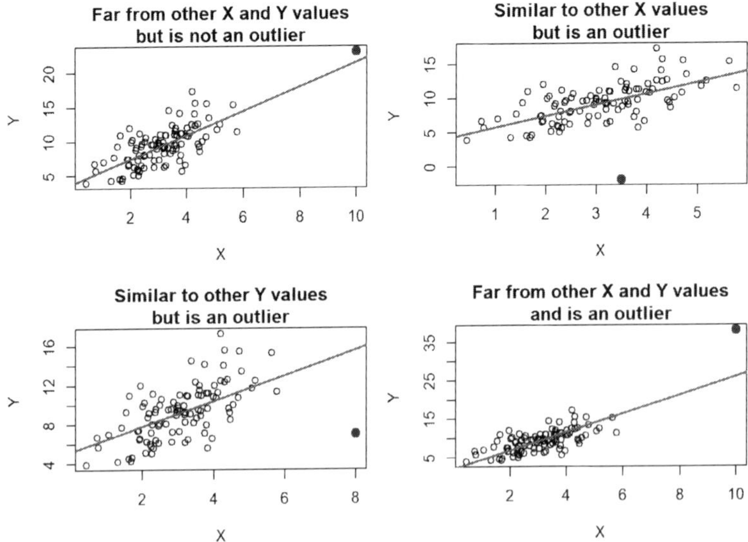

FIGURE 5.45 What is an outlier?

5.21.1 Impact of outliers

The presence of outliers can impact the validity of the normality and constant variance assumptions, resulting in invalid confidence intervals and p-values for regression coefficients. With a large sample size, these impacts will likely be small. Some outliers are also influential observations, a topic that will be discussed in Section 5.22. Finally, outliers are observations that are not well predicted by the model. Therefore, investigation of their characteristics may lead to new insights.

5.21.2 Diagnosis of outliers

We used Figure 5.45 to diagnose the presence of outliers in SLRs using outcome vs. predictor plots. In MLR, since there are multiple predictors, we instead detect outliers by looking at a plot of residuals vs. fitted values. Outliers are observations with large positive or negative residuals. How large is large enough to be considered an outlier? Recall that Studentized residuals have a t distribution (which approaches a standard normal distribution as the sample size increases). The cutoff for "large" is arbitrary, but we know that standard normal values larger in absolute value than 3 or 4 are very rare. They are less rare in larger samples, however, so we need a cutoff that changes with the sample size.

To diagnose and visualize outliers, we will (a) conduct a statistical test for outliers and (b) highlight the outliers in a plot of Studentized residuals vs. fitted values.

Example 5.1 (continued): Look for outliers in the model with the Box-Cox transformed outcome (`fit.ex5.1.trans`).

We start by carrying out a statistical test for outliers using `car::outlierTest()` (Fox et al., 2023; Fox and Weisberg, 2019). This tests each Studentized residual to see how likely we are to observe such an extreme value if the errors were truly t distributed. While large outliers are rare, they are less rare in larger samples. Therefore, a Bonferroni adjustment (see Section 5.24) is used to account for the increased chance of observing rare outcomes in larger samples. Observations are considered outliers if their `Bonferroni p` is <.05.

```
# Outlier test
# The default for n.max is 10. Using Inf leads to
# showing all the outliers if there are more than 10
car::outlierTest(fit.ex5.1.trans, n.max = Inf)
```

```
##        rstudent                    unadjusted p-value
## 1816   -11.708 0.000000000000000000000000000001908
## 66      -5.384 0.000000094625000000000000193942085
## 66.1    -5.384 0.000000094625000000000000193942085
##                            Bonferroni p
## 1816 0.00000000000000000000000001635
## 66   0.000081093000000000000438471481
## 66.1 0.000081093000000000000438471481
```

Three observations were flagged by this test as having unusually large negative residuals, indicating that their observed fasting glucose values are much lower than predicted by the model. After adjusting for multiple testing, these three had Bonferroni p-values < .05. We will discuss multiple testing in Section 5.24 – for now just know that the Bonferroni adjustment accounts for the fact that in a large sample size, we might expect a few really extreme outliers, so an adjustment is needed to make sure we only flag really, really extreme ones.

NOTE: A `car::outlierTest() Bonferroni p` value of `NA` corresponds to non-significance.

The row labels in the outlier test output can be used to identify the outlying observations. Make sure to put them in quotes when subsetting the data (these are row labels, not row numbers).

```
nhanesf.complete[c("1816", "66", "66.1"),
                 c("LBDGLUSI", "BMXWAIST", "smoker",
                   "RIDAGEYR", "RIAGENDR", "race_eth", "income")]
```

```
##      LBDGLUSI BMXWAIST smoker RIDAGEYR RIAGENDR               race_eth    income
## 1816     2.61     83.5   Past       59   Female   Non-Hispanic Other $55,000+
## 66       3.50    110.5  Never       52   Female Non-Hispanic White $55,000+
## 66.1     3.50    110.5  Never       52   Female Non-Hispanic White $55,000+
```

NOTE: Two of these rows are identical; this is an artifact of how the NHANES teaching datasets used in this book were created – by sampling *with replacement* from the full NHANES dataset (see Appendix A.1).

What makes these observations unusual? Looking at the regression coefficients below, we see that both age and waist circumference have positive associations with fasting glucose, and that past smokers have greater mean (transformed) fasting glucose. Looking at the overall distribution of the outcome and continuous predictors below, we see that these individuals have very low fasting glucose (`LBDGLUSI`), but some predictor values indicative of greater fasting glucose. For example, individual `1816` is a past smoker with above average age, and the other two individuals have above average waist circumference and above average age.

```
round(
  summary(fit.ex5.1.trans)$coef
  , 4)
```

```
##                             Estimate Std. Error  t value Pr(>|t|)
## (Intercept)                  -0.1127     0.0038 -29.2940   0.0000
## BMXWAIST                      0.0003     0.0000   9.7426   0.0000
## smokerPast                    0.0018     0.0013   1.4401   0.1502
## smokerCurrent                -0.0001     0.0015  -0.0732   0.9416
## RIDAGEYR                      0.0003     0.0000   9.7697   0.0000
## RIAGENDRFemale               -0.0047     0.0011  -4.4085   0.0000
## race_ethNon-Hispanic White   -0.0046     0.0015  -3.0799   0.0021
## race_ethNon-Hispanic Black   -0.0027     0.0020  -1.3160   0.1885
## race_ethNon-Hispanic Other   -0.0007     0.0022  -0.3116   0.7554
## income$25,000 to <$55,000     0.0006     0.0017   0.3760   0.7070
## income$55,000+               -0.0001     0.0015  -0.0689   0.9451
```

```
rbind(
  "Glucose" = summary(nhanesf.complete$LBDGLUSI),
  "Waist"   = summary(nhanesf.complete$BMXWAIST),
  "Age"     = summary(nhanesf.complete$RIDAGEYR)
)
```

```
##          Min. 1st Qu. Median   Mean 3rd Qu.  Max.
## Glucose  2.61    5.33   5.72   6.11    6.22  19.0
## Waist   63.20   88.30  98.30 100.82  112.20 169.5
## Age     20.00   34.00  47.00  47.79   61.00  80.0
```

You can visualize the outliers by highlighting them in a plot of Studentized residual vs. fitted values (Figure 5.46). To do this, we need a pair of cutoffs above and below which, respectively, are the residuals for the outliers identified by the outlier test (or just one cutoff if all the outliers are positive or all are negative). This can get tricky due to rounding, but just fiddle with the cutoffs until you get the right number of points highlighted in your plot. Again, due to the way this dataset was created, two of the outliers are identical so it will appear that only two points are highlighted since one is on top of the other.

```
# Compute Studentized residuals
RSTUDENT <- rstudent(fit.ex5.1.trans)

# Cutoff for flagging outliers based on the outlier test
SUB <- RSTUDENT < -5.38

# Check that you flagged the right number of outliers
sum(SUB)
```

```
## [1] 3
```

```
# Plot Studentized residuals vs. fitted values
car::residualPlots(fit.ex5.1.trans,
                   pch=20, col="gray",
                   fitted = T, terms = ~ 1,
                   tests = F, quadratic = F,
                   type = "rstudent")

# Highlight outliers
# NOTE: For points() the arguments are x, y not y ~ x
points(fitted(fit.ex5.1.trans)[SUB], RSTUDENT[SUB], pch=20, cex=2)
```

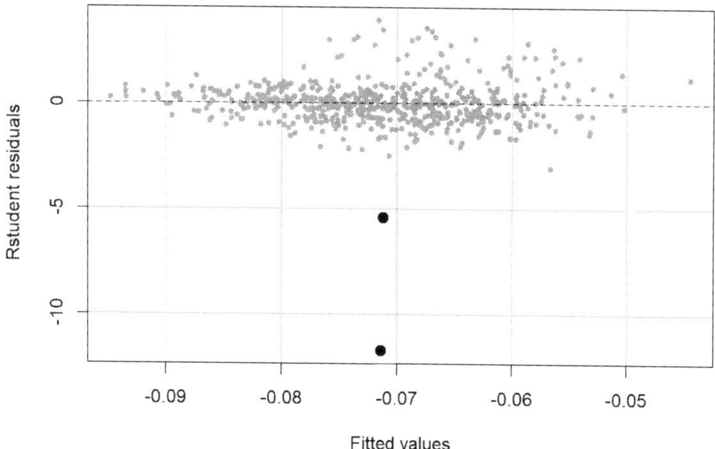

FIGURE 5.46 Highlighting outliers in a residual vs. fitted plot

NOTE: In this example, there were not any positive outliers. But, if there were, you would need to modify `SUB` above as in the examples below.

```
# Example 1: Suppose all the outliers have positive residuals with
#            the smallest being 4.62483
SUB <- RSTUDENT > 4.62

# Example 2: Suppose there are both positive and negative outliers
#            and, among the positive outliers, the smallest is 4.62483
#            and, among the negative outliers, the largest is -4.89398
SUB <- RSTUDENT > 4.62 | RSTUDENT < -4.89
```

5.21.3 Potential solutions for outliers

It is tempting to simply remove outliers. However, being an outlier does not alone justify removal of an observation. Removing outliers may make your model appear to fit better than it should, leading to overconfidence in your results. Given a set of observations with very large residuals, first check to see if the observed outcome and/or predictor values for those observations are data entry errors (if you have access to the raw data). If that is not possible, or if you have determined they are not data entry errors, then try one of the following options.

- **Outcome transformation:** If the Y distribution is very skewed, that can lead to large residuals. A transformation may solve this problem.
- **Perform a sensitivity analysis:** Fit the model with and without the outliers and see what changes (see Section 5.25). If the sample size is large, outliers will have little impact on your conclusions. But since there is no objective cutoff for "large" it is difficult to know if the outliers actually have little impact without carrying out a sensitivity analysis.

Finally, rather than simply being a problem, outliers may actually be some of the more interesting observations in the data. Observations that are not fit well by the model may warrant further investigation, possibly leading to new insights and hypotheses.

5.22 Influential observations

An **influential observation** is one which, when included in the dataset used to fit a model, alters the regression coefficients by a meaningful amount. A regression line attempts to provide a best fit to all the observations. Seen the other way around, each observation exerts some influence on the line, pulling the line toward itself. Observations that are extreme in the X direction are said to have high **leverage** – they have the *potential* to influence the line greatly. If you were to take a high leverage observation and change its Y value by a certain amount, the regression line would change more than if the same change were made to a low leverage observation (one near the center of the distribution of X).

However, not all high leverage observations are *actually* influential – those that are far from the regression line when the model is fit without them will exert more influence. Conversely, some less high leverage observations can be influential if their residuals are unusually large. The magnitude of influence is determined by the combination of leverage and magnitude of residual.

Ultimately, to assess whether an observation is actually influential, you must fit the model with and without the observation and see how the regression coefficients change. Figure 5.47 illustrates what sorts of single points are influential in an SLR with a continuous predictor. In each panel, the model is fit with and without the large solid point. The solid line is the regression line when the point is included, and the dashed line is the regression line when the point is excluded.

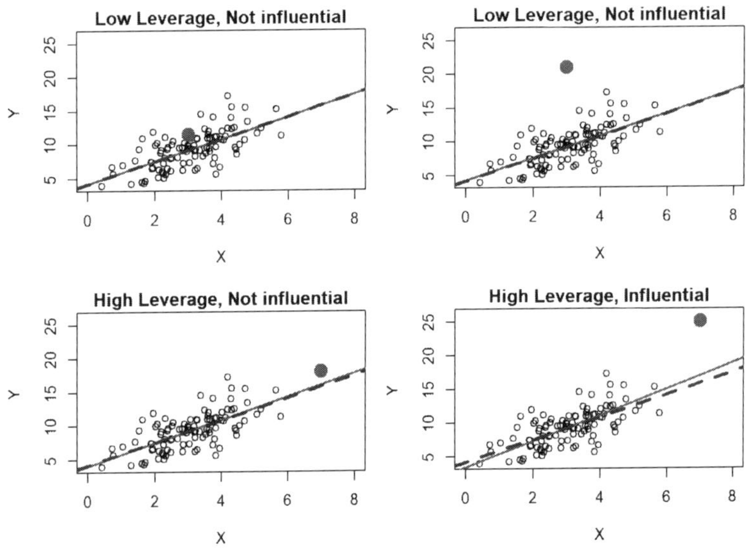

FIGURE 5.47 What are influential observations?

- **Low leverage, Close to the line fit without the point, Not influential:** In the top left panel, the solid point is close to the dashed regression line and not far from the other X values, both of which point to a lack of influence. Its lack of influence is confirmed by the fact that the regression lines are almost identical with or without this point included – the solid and dashed lines are one on top of the other.
- **Low leverage, Far from the line fit without the point, Not influential:** In the top right panel, despite the solid point now being very far from the dashed regression line (it is an outlier), it is not influential due to is low leverage. Again, the regression lines are almost identical with or without this point included.
- **High leverage, Close to the line fit without the point, Not influential:** In the bottom left panel, despite the solid point having high leverage, it is not influential since it is not far from the dashed regression line. With or without this point, the regression line remains about the same.
- **High leverage, Far from the line fit without the point, Influential:** In the bottom right panel, the solid point is both high leverage *and* far from the line fit without it, resulting in high influence. This point pulls the regression line quite a bit toward itself – when it is included, the slope increases noticeably.

5.22.1 Impact of influential observations

Influential observations pull the regression fit toward themselves. The results (predictions, parameter estimates, CIs, p-values) can be quite different with and without these cases included in the analysis. While influential observations do not necessarily violate any regression assumptions, they can cast doubt on the conclusions drawn from your sample. If a regression model is being used to inform real-life decisions, one would hope those decisions are not overly influenced by just one or a few observations.

5.22.2 Diagnosis of influential observations

If there is just one predictor, you could look at a scatterplot of Y vs. X, as in Figure 5.47. With multiple predictors, however, this will not work. Regarding leverage, with multiple predictors an observation can be typical for individual predictors yet be highly unusual jointly. What makes a point high leverage is how unusual it is when considering all the predictors together. Suppose, for example, you have a model that includes the predictors height and sex. An individual may have a height that is not extreme when just looking at height, but given sex it might be extreme (e.g., males are taller than females on average, so a very tall female or a very short male may be an unusual case). Fortunately, there are diagnostics that assess leverage and influence no matter how many predictors are in the model.

- The **hat value** measures how far an observation's predictors, taken together, are from those of other observations. Observations with large hat values have high leverage and are *potentially* (but not necessarily) influential. There is no objective cutoff above which a hat value is considered "large". Instead, look for observations with hat values that are much larger than those of the other observations.
- **Cook's distance** measures, for each observation, the difference in all the regression coefficient estimates taken together when fitting the model with and without that observation (Cook, R. D., 1977), providing a measure of global influence. There is no objective cutoff value above which a Cook's distance is considered "large". Instead, look

for observations with Cook's distances that are much larger than those of the other observations.

- **Standardized DFBetas** measure, for each observation, the standardized difference in *individual* regression coefficient estimates when fitting the model with and without that observation, providing a measure of influence on each coefficient. When we discuss sensitivity analyses in Section 5.25, we will see what happens to the model fit when we remove a group of observations, but DFBetas can tell us what happens when we remove each observation one-at-a-time. Standardized DFBetas can be compared to a cutoff value, with 0.2 suggested as a reasonable cutoff (Harrell, 2015, p504).

Example 5.1 (continued): Look for influential observations in the model with the transformed outcome. First, look at the Studentized residuals and hat values to see if any observations are both outliers and high leverage (as these are the most potentially influential). Then look at the Cook's distance and DFBetas to assess actual influence.

Use `car::influenceIndexPlot()` (Fox et al., 2023; Fox and Weisberg, 2019) to visualize the Studentized residuals, hat values, and Cook's distance, as shown in Figures 5.48, 5.49, and 5.50, respectively.

```
car::influenceIndexPlot(fit.ex5.1.trans, vars = "Studentized",
                        id=F, main = "Studentized Residuals")

car::influenceIndexPlot(fit.ex5.1.trans, vars = "hat",
                        id=F, main = "Leverage (hat values)")

car::influenceIndexPlot(fit.ex5.1.trans, vars = "Cook",
                        id=F, main = "Cook's distance")
```

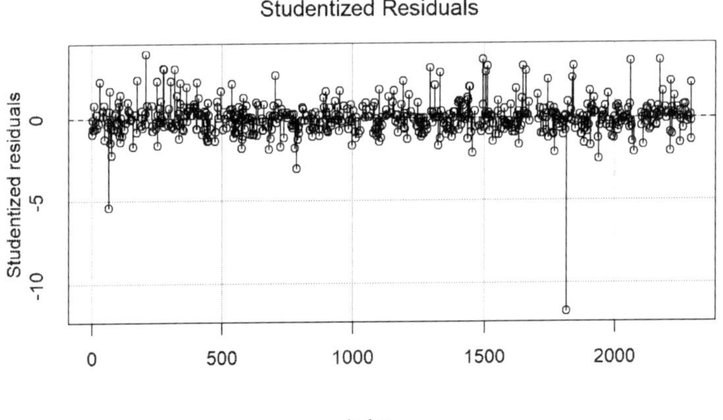

FIGURE 5.48 Studentized residuals

The points are plotted from left to right in the order they appear in the dataset, so their horizontal locations are not relevant. There appear to be a few hat values that stand out (Figure 5.49), indicating there are a few high leverage observations. However, those observations do not also have large residuals (Figure 5.48), so they do not actually exert

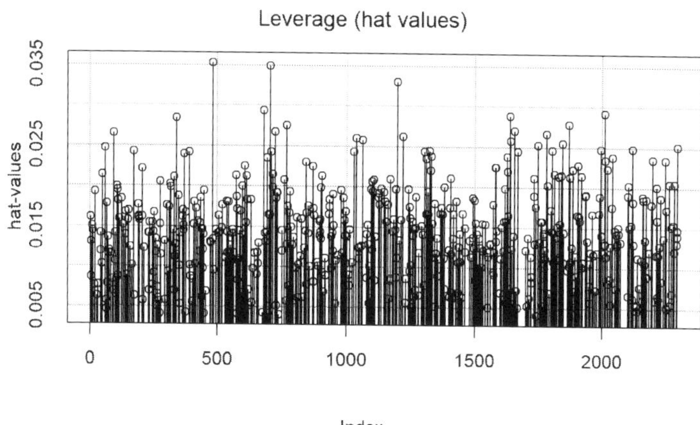

FIGURE 5.49 Hat values

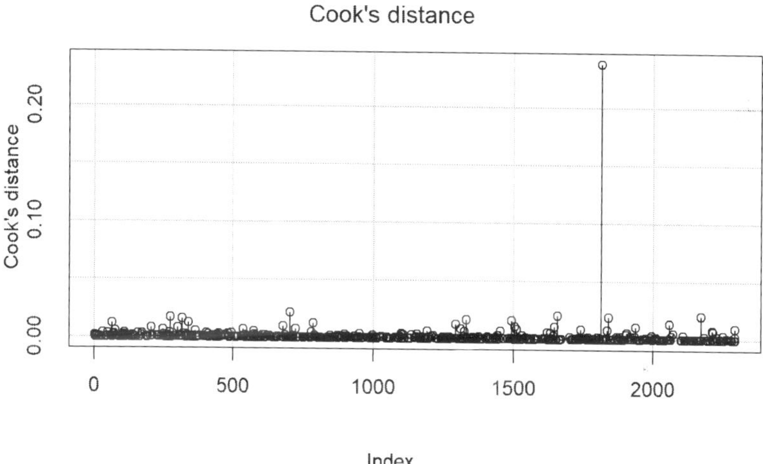

FIGURE 5.50 Cook's distance

much influence, as seen in their relatively small Cook's distances. The one observation with a very large negative residual, however, has enough leverage that it has a relatively large Cook's distance (Figure 5.50). It is the combination of leverage and being an outlier that results in high influence.

To compute these statistics directly, use the following functions.

```
hatvalues(fit.ex5.1.trans)
cooks.distance(fit.ex5.1.trans)
```

Cook's distance above identified one observation that highly influences the regression coefficients collectively. Next, use standardized DFBetas to assess influence on each regression coefficient individually, as shown in Figure 5.51 (only shown for a few of the predictors).

```
# Compute DFBETAS
DFBETAS <- dfbetas(fit.ex5.1.trans)
# Check the spelling of the terms and enter them accordingly
# in each plot() call below
# colnames(DFBETAS)
par(mfrow=c(2,3))
plot(DFBETAS[, "(Intercept)"], ylab="Intercept")
abline(h = c(-0.2, 0.2), lty = 2)
plot(DFBETAS[, "BMXWAIST"], ylab="WC")
abline(h = c(-0.2, 0.2), lty = 2)
plot(DFBETAS[, "smokerPast"], ylab="Past")
abline(h = c(-0.2, 0.2), lty = 2)
plot(DFBETAS[, "smokerCurrent"], ylab="Current")
abline(h = c(-0.2, 0.2), lty = 2)
plot(DFBETAS[, "RIDAGEYR"], ylab="Age")
abline(h = c(-0.2, 0.2), lty = 2)
plot(DFBETAS[, "RIAGENDRFemale"], ylab = "Female")
abline(h = c(-0.2, 0.2), lty = 2)
plot(DFBETAS[, "race_ethNon-Hispanic White"], ylab = "NHW")
abline(h = c(-0.2, 0.2), lty = 2)
plot(DFBETAS[, "race_ethNon-Hispanic Black"], ylab = "NHB")
abline(h = c(-0.2, 0.2), lty = 2)
plot(DFBETAS[, "race_ethNon-Hispanic Other"], ylab = "NHO")
abline(h = c(-0.2, 0.2), lty = 2)
plot(DFBETAS[, "income$25,000 to <$55,000"], ylab = "$25-$54")
abline(h = c(-0.2, 0.2), lty = 2)
plot(DFBETAS[, "income$55,000+"], ylab = "$55+")
abline(h = c(-0.2, 0.2), lty = 2)
```

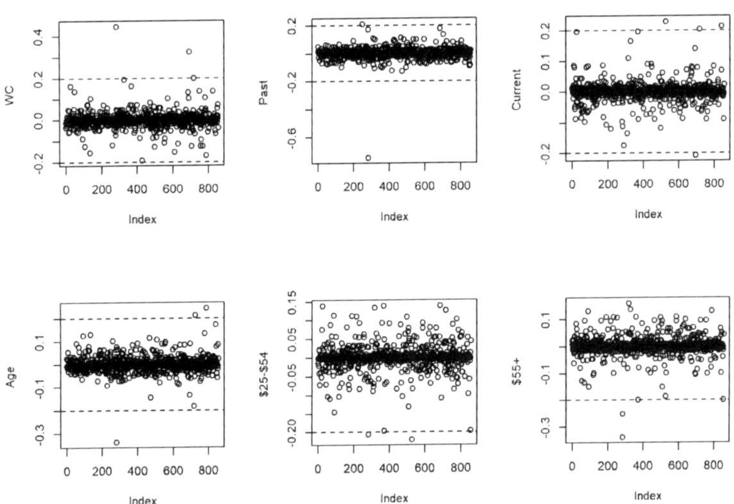

FIGURE 5.51 Standardized DFBetas

Each DFBeta plot shows, on the vertical axis, how much that predictor's regression coefficient changes when an observation is removed, on a standardized scale. A few points appear to have large standardized DFBetas (greater in absolute value than 0.2, indicating removal of that observation would change that β by more than 0.2 standard deviations).

5.22.3 Potential solutions for influential observations

- **Predictor transformation:** A highly skewed X distribution can lead to high leverage for points at the extreme. A transformation (e.g., logarithm, square root, inverse) may solve this problem.
- **Outcome transformation:** If the Y distribution is very skewed, that can lead to large residuals. A transformation may solve this problem.
- **Perform a sensitivity analysis:** Fit the model with and without the influential observations and see what changes (see Section 5.25). If a point is influential based on the diagnostics, that alone does not justify its removal. As with outliers, never simply remove observations from the data just because they might be problematic. Instead, do the analysis with and without them and state the differences in any discussion of the results.

5.23 Confirmatory vs. exploratory analysis

Model selection is the process of deciding which predictors should be included in a model and in what form each should be included. This decision depends on the goals of the analysis, and priority should be placed on subject-matter knowledge. How you go about this process depends on the purpose of your regression analysis.

If your goal is to confirm a pre-specified hypothesis, then you are doing a **confirmatory analysis**. Pre-specify a hypothesized model based on subject-matter knowledge – the outcome, predictors, and predictor interactions. In a confirmatory analysis, do not remove terms based on lack of statistical significance, although it is reasonable to alter the form of variables (e.g., a transformation, collapsing a sparse predictor) based on meeting regression assumptions and remove or combine predictors in order to reduce collinearity. Regression p-values are computed assuming you have a pre-specified model. To be able to make strong, confirmatory conclusions, you cannot arrive at the model after making decisions about the form of the model based on the relationships between the predictors and the outcome.

If, instead, your goal is to explore the data to try to find the best fitting model (the set of predictors that best explain the outcome) then you are doing an **exploratory analysis**, also known as hypothesis generating research. In an exploratory analysis, both which predictors to include and what form they take are flexible, and basing decisions on statistical significance is acceptable, as long as you do not later act as if significance tests were based on a pre-specified model. Inferences (e.g., confidence intervals, p-values) from an exploratory analysis that used significance of association with the outcome to determine which predictors to include in the model are not valid unless the decision-making process is considered (e.g., by using bootstrap resampling where the model selection process is systematically carried out within each resample) (Harrell, 2015). Stepwise regression methods, for example, are exploratory methods; see, for example, `help(package = "olsrr")` (requires installation of the `olsrr` library) (Hebbali, 2024).

5.24 Multiple testing

When carrying out a confirmatory analysis, you test the null hypothesis that the primary predictor regression coefficient is zero and, typically, you carry out this test at the $\alpha =$ .05 level of significance. This limits the probability of a Type I error to 5%. That is, if the null hypothesis is true (no association) then there is only a 5% chance that your sample will result (incorrectly) in a statistically significant result. But what happens if you want to confirm multiple hypotheses using **multiple tests**? Suppose you have V independent outcomes you want to test for association with an exposure. If you carry out V confirmatory regression analyses, each at the $\alpha = .05$ level of significance, then your actual Type I error rate over all V tests is larger than 5%. If there are no associations, the chance of at least one Type I error is expressed by the following.

$$P(\text{At least 1 Type I error among V tests}) = 1˘P(\text{no Type I errors}) = 1 - (1 - .05)^V$$

If you have two tests, this probability is $1˘(0.95)^2 = 0.0975$, almost twice the target Type I error rate. If you have five tests, it is $1˘(0.95)^5 = 0.2262$. Carrying out multiple statistical tests with no adjustment for the inflated Type I error results in a greater risk of spurious findings.

The same issue arises in a regression where you have a categorical predictor with more than 2 levels. Although you can use an F-test to get an overall p-value for such a predictor (using `car::Anova()`), you may also want to know which pairs of levels are significantly different. This is called making **post-hoc pairwise comparisons**, or **multiple comparisons**. If you have a categorical predictor with L levels, there are $L(L-1)/2$ possible pairwise comparisons.

How do we solve these problems? By using a **multiple testing adjustment.** The simplest adjustment is the Bonferroni correction which divides α by the number of tests. For example, if you have three tests, then you test each one at the $.05/3 = .0167$ level of significance. Equivalently, multiply each p-value by the number of tests to get an "adjusted p-value" which can be directly compared to .05. This results in an overall Type I error no greater than .05 and we express this by saying *the multiple testing adjustment preserves a familywise Type I error rate of* $\alpha = .05$. The term "familywise" means "over the set of tests under consideration" (in this case, three tests). The Bonferroni correction can be carried out easily in R using `p.adust(, method = "bonferroni")`.

The Bonferroni correction, however, is overly conservative. There are more powerful alternatives available in `p.adjust()`. The Hommel method (Hommel, 1988) is more powerful and is valid when the tests are independent or when they are positively associated (Sarkar, 1998; Sarkar and Chang, 1997) (see `?p.adjust`). While the Hommel method is effective with mildly correlated outcomes, in cases where tests are carried out on outcomes that are highly related to each other, it is recommended to use the more powerful (but more complicated to carry out) resampling-based stepdown MinP method (Blakesley et al., 2009; Westfall and Young, 1993) (see, for example `?NPC::FWE`).

NOTE: Simply having many tests does not always imply the need for multiple testing. See, for example, Rothman (1990), Greenland and Hofman (2019), and Rothman et al. (2008) (p237). If one has a set of associations of interest, each of interest on their own, then it is appropriate to test each association at the $\alpha = .05$ level with no adjustment for multiple testing. For example, suppose that, using data from a large national survey, an investigator decides to study ten different exposure-outcome relationships. These each can be tested on their own with no adjustment since each is of interest on its own. On the other hand, suppose an investigator is interested in the association between a particular exposure and a set of ten outcomes and wants to claim the exposure is harmful if *any* of the ten associations are statistically significant. In this case, it *is* necessary to adjust for multiple testing. Otherwise, if the exposure is actually benign then the chance of a spurious conclusion that the exposure is harmful is much larger than 5%.

Example 5.9: Carry out a confirmatory analysis to test the association between smoking status (exposure) and any of the following outcomes: body mass index (`BMXBMI`), systolic blood pressure (`sbp`), HDL cholesterol (`LBDHDDSI`), and triglycerides (`LBDTRSI`), adjusted for confounding due to age, gender, and race/ethnicity. Assume that we are interested in screening these outcomes for further study based on the significance of their associations with the exposure. In this setting, it is appropriate to adjust for multiple testing over these four tests since we will conclude smoking is harmful if it is associated with *any* of these outcomes. An adjustment is necessary to preserve a familywise .05 Type I error rate.

```
# Fit each model
fit1 <- lm(BMXBMI   ~ smoker + RIDAGEYR + RIAGENDR + RIDRETH3, data = nhanes)
fit2 <- lm(sbp      ~ smoker + RIDAGEYR + RIAGENDR + RIDRETH3, data = nhanes)
fit3 <- lm(LBDHDDSI ~ smoker + RIDAGEYR + RIAGENDR + RIDRETH3, data = nhanes)
fit4 <- lm(LBDTRSI  ~ smoker + RIDAGEYR + RIAGENDR + RIDRETH3, data = nhanes)

# Store each p-value
p1   <- car::Anova(fit1, type = 3)["smoker", "Pr(>F)"]
p2   <- car::Anova(fit2, type = 3)["smoker", "Pr(>F)"]
p3   <- car::Anova(fit3, type = 3)["smoker", "Pr(>F)"]
p4   <- car::Anova(fit4, type = 3)["smoker", "Pr(>F)"]

# Compute adjusted p-values
p.unadj  <- c(p1, p2, p3, p4)
p.bonf   <- p.adjust(p.unadj, method = "bonferroni")
p.hommel <- p.adjust(p.unadj, method = "hommel")

# Compare the results
DF    <- data.frame(Unadjusted = round(p.unadj,  3),
                    Bonferroni = round(p.bonf,   3),
                    Hommel     = round(p.hommel, 3))
rownames(DF) <- c("BMI", "SBP", "HDL", "Triglycerides")
DF
```

```
##               Unadjusted Bonferroni Hommel
## BMI                0.014      0.057  0.043
## SBP                0.853      1.000  0.853
## HDL                0.213      0.854  0.427
## Triglycerides      0.002      0.007  0.007
```

If we based our conclusions on the raw p-values, two of the four tests would be statistically significant (BMI and triglycerides). After adjusting for multiple testing, only one of the four is statistically significant when using the Bonferroni correction (which is overly conservative) but both remain significant after using the Hommel adjustment. The Hommel method is more powerful, resulting in smaller adjusted p-values. The only reason to use the Bonferroni adjustment is in a situation where you want to keep things very simple and easy to understand and if more powerful methods give the same result. Otherwise, use a more powerful method such as Hommel. In this example, our conclusions about statistical significance (that smoking status is significantly associated with BMI and triglycerides) were the same when using unadjusted p-values and after adjusting for multiple testing using the Hommel method, but this will not always be the case.

Example 5.10: Carry out a confirmatory analysis to test the association between smoking status (the exposure of interest) and triglycerides (`LBDTRSI`), adjusted for confounding due to age, gender, and race/ethnicity. Compare the mean outcome between the three levels of smoking status. There are three pairwise comparisons (Past vs. Never, Current vs. Never, Current vs. Past) and we will conclude there is a significant difference between levels if *any* of these three are different. Thus, it is appropriate to adjust for multiple comparisons.

```
fit.ex5.10 <- lm(LBDTRSI ~ smoker + RIDAGEYR + RIAGENDR + RIDRETH3, data = nhanes)
round(summary(fit.ex5.10)$coef, 4)
```

```
##                              Estimate Std. Error t value Pr(>|t|)
## (Intercept)                    1.0977     0.1587  6.9173   0.0000
## smokerPast                     0.1196     0.0907  1.3187   0.1880
## smokerCurrent                  0.3820     0.1068  3.5760   0.0004
## RIDAGEYR                       0.0029     0.0022  1.3004   0.1942
## RIAGENDRFemale                -0.1359     0.0778 -1.7461   0.0815
## RIDRETH3Other Hispanic         0.0030     0.1676  0.0179   0.9857
## RIDRETH3Non-Hispanic White    -0.0343     0.1209 -0.2838   0.7767
## RIDRETH3Non-Hispanic Black    -0.2585     0.1792 -1.4424   0.1499
## RIDRETH3Non-Hispanic Asian    -0.0034     0.2027 -0.0166   0.9867
## RIDRETH3Other/Multi           -0.0222     0.2013 -0.1103   0.9122
```

```
summary(fit.ex5.10)$coef[c("smokerPast",
                           "smokerCurrent"), "Pr(>|t|)"]
```

```
##    smokerPast smokerCurrent
##     0.1879633     0.0003881
```

```
# To get Current vs. Past, re-level smoker
tmpdat <- nhanes %>%
  mutate(smoker = relevel(smoker, ref = "Past"))
fit.ex5.10b <- lm(LBDTRSI ~ smoker + RIDAGEYR + RIAGENDR + RIDRETH3, data=tmpdat)
round(summary(fit.ex5.10b)$coef, 4)
```

```
##                              Estimate Std. Error t value Pr(>|t|)
## (Intercept)                    1.2173     0.1682  7.2365   0.0000
## smokerNever                   -0.1196     0.0907 -1.3187   0.1880
## smokerCurrent                  0.2624     0.1188  2.2088   0.0277
## RIDAGEYR                       0.0029     0.0022  1.3004   0.1942
## RIAGENDRFemale                -0.1359     0.0778 -1.7461   0.0815
## RIDRETH3Other Hispanic         0.0030     0.1676  0.0179   0.9857
## RIDRETH3Non-Hispanic White    -0.0343     0.1209 -0.2838   0.7767
## RIDRETH3Non-Hispanic Black    -0.2585     0.1792 -1.4424   0.1499
## RIDRETH3Non-Hispanic Asian    -0.0034     0.2027 -0.0166   0.9867
## RIDRETH3Other/Multi           -0.0222     0.2013 -0.1103   0.9122
```

```
summary(fit.ex5.10b)$coef["smokerCurrent", "Pr(>|t|)"]
```

```
## [1] 0.02771
```

Pulling out the three p-values of interest (comparing the mean outcome between levels of smoking status), we get the following adjustment for multiple comparisons.

```
# Multiple comparisons adjustment
p.unadj <- c(0.187963329, 0.000388096, 0.02770697)
p.hommel <- p.adjust(p.unadj, method = "hommel")
DF   <- data.frame(Unadjusted = round(p.unadj,  3),
                   Hommel     = round(p.hommel, 3))
rownames(DF) <- c("Past vs. Never",
                  "Current vs. Never",
                  "Current vs. Past")
DF
```

```
##                   Unadjusted Hommel
## Past vs. Never         0.188  0.188
## Current vs. Never      0.000  0.001
## Current vs. Past       0.028  0.055
```

Before adjusting for multiple comparisons, two comparisons (Current vs. Never, Current vs. Past) appear to be statistically significant. However, after adjusting for multiple comparisons, only Current vs. Never remains significant (adjusted $p < .05$).

5.24.1 Primary vs. secondary tests

When you have many tests or many comparisons, you lose power when using a multiple testing adjustment. That is, if there really is an association, you have less of a chance of finding it since with more tests you have a stricter criterion to meet to conclude statistical significance. However, if some tests are less important than others you can, when designing a study, split them up into two groups: primary tests (confirmatory) and secondary tests (exploratory, or "hypothesis generating"). Then, adjust for multiple testing over just the primary tests.

This step in research design requires careful thought – there is a tradeoff between maximizing the chance of confirming certain hypotheses of primary interest (by limiting the number of primary tests) and maximizing the number of hypotheses that you have the opportunity to confirm (by increasing the number of primary tests). For tests that are in the "secondary" group, if any have a low p-value, you cannot claim confirmation, only that they are interesting results that warrant further research.

NOTE: Decide which tests are primary *before* looking at the results. You cannot look at the p-values and decide how many tests are primary in order to confirm the most tests; your choice of primary tests or comparisons must be *a priori* – before you have seen the results.

5.25 Sensitivity analysis

A **sensitivity analysis** compares your conclusions between the analysis you carried out and another analysis in which you change some aspect of the approach. This method can be used to assess the sensitivity of your regression results (e.g., parameter estimates, 95% confidence intervals, p-values) to changes in your approach. Even in a confirmatory analysis, where you must pre-specify the approach, you can use a sensitivity analysis to assess what would have happened had you used a different approach.

When reporting the results of a sensitivity analysis, think about how your conclusions differ between approaches. Your results could differ **quantitatively** and/or **qualitatively**. A quantitative difference affects the strength of conclusions but may or may not affect the nature of the conclusions themselves. For example, suppose a regression coefficient estimate meaningfully differs in magnitude between two approaches, but is meaningfully large and in the same direction in both. This would be a quantitative difference, but not a qualitative difference.

A qualitative difference affects the nature of the conclusions. For example, when comparing two approaches, suppose an association changes in direction or changes from meaningfully large to close to no association or vice versa. These are qualitative differences. A change in statistical significance is also a qualitative difference, in that it affects conclusions based on a strict p-value cutoff, but since the typical .05 cutoff for statistical significance is arbitrary a change in significance really does not matter as much as changes in the parameter estimates themselves. For example, if two analyses yield a regression coefficient that is approximately the same magnitude, but in one case p = .049 and in the other p = .051, then really nothing has changed despite the fact that the former is "statistically significant" and the latter is not. Although some may insist on making much of this difference, there really is no meaningful difference.

In summary, report the nature of your sensitivity analysis (what you altered and why), summarize quantitative differences, comment on qualitative differences, and combine this information into a judgment of how sensitive your original analysis is to changes in the approach. Ideally, you will be able to report "we carried out a sensitivity analysis and our results did not meaningfully change and our conclusions remained the same". If, however, the results do differ meaningfully, then you may need to report both sets of results and note that it is not clear which better reflects reality.

To demonstrate, this section will assess sensitivity to:

- The choice of how to collapse a categorical predictor into fewer levels; and
- The presence of outliers and influential observations.

NOTES:

- When carrying out a sensitivity analysis, be careful to identify any changes that come along for the ride. For example, if you compare the inclusion of confounder X with, instead, the inclusion confounder Z, but X and Z have different amounts of missing data, then differences in results could be due to the change in confounder and/or the change in sample size. To isolate the effect of changing the confounder, use the same sample for both analyses, one that has no missing values for either X or Z.
- When removing outliers and/or influential observations, the sample size will always decrease. Thus, standard errors, width of confidence intervals, and p-values will always change just due to a reduction in sample size. However, typically we are only removing a few observations relative to the full sample size, so this will not make a large difference. Regardless, as always, pay more attention to changes in the magnitude of effects than to changes in p-values.
- When removing any observation, the characteristics of the remaining observations may change. For example, a non-omitted observation that was an outlier or influential in the original analysis may no longer be, or vice versa. A thorough analysis of sensitivity to outliers and/or influential observations would entail removing observations one at a time, assessing the effects on the model, and reassessing the remaining observations. DFBetas already tell us what happens to each regression coefficient when each observation is removed one at a time, but they do not tell us how the influence measures themselves (DFBetas, Cook's distance) change for the remaining observations.

- Be careful when assessing differences in coefficient magnitude between analyses that are on different scales.
  - If your sensitivity analysis involves changing the scale of a *predictor*, then the regression coefficient for that predictor will be on a different scale in the two analyses (for example, if you assess sensitivity to choice of predictor transformation).
  - If your sensitivity analysis involves changing the scale of the *outcome*, then *all* the regression coefficients in the two analyses will be on different scales (for example, if you assess sensitivity to choice of outcome transformation).

5.25.1 Example: Sensitivity to collapsing a categorical predictor

Example 5.1 (continued): Our final model (`fit.ex5.1.trans`) included `race_eth` (race/ethnicity) which was derived by collapsing `RIDRETH3` into fewer categories due to sparsity. Carry out a sensitivity analysis to assess how robust are the final conclusions about the primary predictors (waist circumference and smoking status) to this approach. To accomplish this, re-fit the model including `RIDRETH3` instead of `race_eth` and compare the results to the original model results.

```
# Original model
fit.ex5.1.trans <- lm(LBDGLUSI_trans ~ BMXWAIST + smoker +
    RIDAGEYR + RIAGENDR +  race_eth + income,
    data = nhanesf.complete)
# Alternative model
fit.ex5.1.trans_b <- lm(LBDGLUSI_trans ~ BMXWAIST + smoker +
    RIDAGEYR + RIAGENDR +  RIDRETH3 + income,
    data = nhanesf.complete)
```

`car::compareCoefs()` (Fox et al., 2023; Fox and Weisberg, 2019) provides a side-by-side comparison of the regression coefficients and their standard errors.

```
car::compareCoefs(fit.ex5.1.trans,
                  fit.ex5.1.trans_b,
                  pvals = T)
```

```
# Results only shown for waist circumference and smoking status

## Calls:
## 1: lm(formula = LBDGLUSI_trans ~ BMXWAIST + smoker + RIDAGEYR + RIAGENDR +
##   race_eth + income, data = nhanesf.complete)
## 2: lm(formula = LBDGLUSI_trans ~ BMXWAIST + smoker + RIDAGEYR + RIAGENDR +
##   RIDRETH3 + income, data = nhanesf.complete)
##
##                                 Model 1              Model 2
##
## BMXWAIST                      0.0003047            0.0003117
## SE                            0.0000313            0.0000315
## Pr(>|z|)          < 0.0000000000000002 < 0.0000000000000002
##
## smokerPast                      0.00184              0.00216
## SE                              0.00128              0.00129
## Pr(>|z|)                         0.1498               0.0922
##
## smokerCurrent                -0.0001127            0.0000915
## SE                            0.0015381            0.0015372
## Pr(>|z|)                         0.9416               0.9525
```

Looking at these results, we conclude that collapsing the race/ethnicity variable did not meaningfully change the magnitude (or precision) of the estimated regression coefficients for our primary predictors (waist circumference and smoking status). To see the impact on the overall multiple degree of freedom test of significance of `smoker`, use `car::Anova()` on each model to see that the conclusion is not sensitive to how we collapsed race/ethnicity compared to not collapsing (results not shown).

```
car::Anova(fit.ex5.1.trans,   type = 3)
car::Anova(fit.ex5.1.trans_b, type = 3)
```

5.25.2 Example: Sensitivity to outliers and influential observations

Example 5.1 (continued): For our final model (`fit.ex5.1.trans`), we identified three outliers (Section 5.21) and a number of potentially influential observations (Section 5.22). Carry out a sensitivity analysis to assess how robust are the final conclusions about the primary predictors (waist circumference and smoking status) to the presence of these observations. Re-fit the model after excluding these observations and compare the results to the original model.

As mentioned in the **NOTE** above, a more thorough sensitivity analysis would proceed by removing observations one at a time, assessing the effects on the model, and reassessing the remaining observations. In this example, we simply remove them all at once to illustrate the process of identifying and removing observations and assessing the results.

Identifying the outliers

Recall that when we carried out the outlier test in Section 5.21, we created a logical vector that identified these observations. This is reproduced here.

```
# Compute Studentized residuals
RSTUDENT <- rstudent(fit.ex5.1.trans)

# Use a numeric cutoff based on the
# outlier test results to identify outliers
SUB.OUTLIERS <- RSTUDENT < -5.38
RSTUDENT[SUB.OUTLIERS]
```

```
##    1816      66    66.1
## -11.708  -5.384  -5.384
```

Identifying the influential observations

Recall that when we used influence diagnostics in Section 5.22, we identified a few influential observations using Figures 5.50 and 5.51. The strategy for identifying these observations in the dataset is to compute the Cook's distances and DFBetas and create logical vectors using numeric cutoffs.

NOTES:

- For Cook's distance, use a cutoff based on looking at Figure 5.50 and choosing a value above which there are observations that stick out well above the rest (which is subjective). For DFBetas, use a cutoff of 0.2 for each regression coefficient.

- For Cook's distance, each observation has one value. For DFBetas, each observation has one value for each regression coefficient. Use `apply()` and `any()` below to identify rows for which *any* DFBeta is larger (in absolute value) than 0.2.

```
COOK    <- cooks.distance(fit.ex5.1.trans)
DFBETAS <- dfbetas(fit.ex5.1.trans)

# Subjective cutoff from figure
SUB.COOK <- COOK > 0.15

# Standard cutoff for DFBeta
SUB.DFBETAS <- apply(abs(DFBETAS) > 0.2, 1, any)

# View the extreme Cook's distance values and compare
# to plot to make sure you captured all you wanted to capture
COOK[SUB.COOK]
```

```
##   1816
## 0.2377
```

```
# View the extreme DFBetas - a large matrix so not shown
# DFBETAS[SUB.DFBETAS, ]
```

The following illustrates how to, instead, identify observations with large DFBetas for a single term in the regression (rather than for any term as was done above). Figure 5.51 illustrated that there was an observation with a DFBeta <-1 for the "Non-Hispanic Other" indicator variable of `race_eth`. The following code identifies that single observation.

```
SUB.DFBETA.NHO <- DFBETAS[, "race_ethNon-Hispanic Other"] < -1
DFBETAS[SUB.DFBETA.NHO, "race_ethNon-Hispanic Other"]
```

```
## [1] -1.057
```

Putting it all together

In this example, we are removing all the outliers and influential observations all at once. If you were removing them one at a time, then just set `SUB` below to be the logical vector that identifies the single observation you are removing at this step.

```
SUB <- SUB.OUTLIERS | SUB.COOK | SUB.DFBETAS
sum(SUB)
```

```
## [1] 21
```

Next, fit the model without these observations and compare the results before vs. after. Make sure to include the negation operator `!` before the logical vector `SUB` so as to include only observations that are *not* outliers or influential.

```
# Subsetted data
subdat <- subset(nhanesf.complete, !SUB)
nrow(nhanesf.complete)
```

```
## [1] 857
```

```
nrow(subdat)
```

```
## [1] 836
```

```
fit.ex5.1.trans_sens <- lm(LBDGLUSI_trans ~ BMXWAIST + smoker +
    RIDAGEYR + RIAGENDR +  race_eth + income,
    data = subdat)
```

```
car::compareCoefs(fit.ex5.1.trans,
                  fit.ex5.1.trans_sens,
                  pvals = T)
```

```
# Results only shown for waist circumference and smoking status

## Calls:
## 1: lm(formula = LBDGLUSI_trans ~ BMXWAIST + smoker + RIDAGEYR + RIAGENDR +
##   race_eth + income, data = nhanesf.complete)
## 2: lm(formula = LBDGLUSI_trans ~ BMXWAIST + smoker + RIDAGEYR + RIAGENDR +
##   race_eth + income, data = subdat)
##
##                                    Model 1              Model 2
##
## BMXWAIST                         0.0003047            0.0002929
## SE                               0.0000313            0.0000254
## Pr(>|z|)            < 0.0000000000000002 < 0.0000000000000002
##
## smokerPast                         0.00184              0.00249
## SE                                 0.00128              0.00102
## Pr(>|z|)                            0.1498               0.0146
##
## smokerCurrent                    -0.000113            -0.001259
## SE                                0.001538             0.001242
## Pr(>|z|)                            0.9416               0.3108
```

```
car::Anova(fit.ex5.1.trans,      type = 3)
car::Anova(fit.ex5.1.trans_sens, type = 3)
```

```
# Results only shown for smoking status
## Anova Table (Type III tests)
## Response: LBDGLUSI_trans
##                 Sum Sq  Df  F value                Pr(>F)
## smoker        0.000547   2   1.1509              0.316851

## Anova Table (Type III tests)
## Response: LBDGLUSI_trans
##                 Sum Sq  Df   F value                Pr(>F)
## smoker        0.001332   2    4.4632                0.0118 *
```

We find that the results for waist circumference have not changed meaningfully, but the results for smoking status have, both quantitatively and qualitatively. In particular, the size of the comparison between Past and Never smokers has increased from 0.0018 to 0.0025. Additionally, the p-value for this comparison dropped from non-significant ($p = .150$) to significant ($p = .015$), and the overall p-value for smoking status changed from not even close to significant ($p = .317$) to well below .05 ($p = .012$). While we still should focus on effect sizes rather than p-values, these changes are notably large.

5.26 Generalization / extrapolation / interpolation / overfitting

Generalization refers to using a model to predict outcome values for cases that were not in your dataset. Only generalize to cases that are similar to those that were used to fit the model.

- For example, if you fit a model to predict total cholesterol for individuals in the U.K., the model may not provide accurate predictions for individuals from another country.
- As another example, suppose you fit a model predicting lead in tap water using data only from cities. The model may not generalize to rural areas.

In these two examples, the observations included in the model fitting were restricted by some variable (country, urban/rural status) not included in the model. These variables were used to define **inclusion/exclusion criteria** and generalization is limited to cases that meet those criteria.

Additionally, generalization is limited to cases that, for predictors included in the model, have values similar to those of the cases in the dataset used to fit the model.

- **Extrapolation** refers to using a model to predict outcome values for cases with predictor values outside the range used to fit the model. For example, if you fit a model to predict total cholesterol from age for individuals age 40-65 years, do not make predictions for individuals younger than age 40 years or older than age 65 years.
- **Interpolation** refers to making a prediction for a case with a predictor value that is within the range of the observed values of that predictor in the dataset used to fit the model. Interpolation is valid IF the assumptions of linear regression are met and the model was appropriate. For example, if you fit a line when you should be fitting a curve, or left out important predictors, then even interpolations could be biased.

Another cause of lack of generalizability is **overfitting** – including so many predictors in a model that it fits the current data very well but does not predict future data well. A model with too many predictors is fitting not only the true relationships between the predictors and the outcome ("signal"), but also relationships that are specific to this sample and do not hold in general ("noise"). A rule of thumb is to limit the number of predictors in a linear regression model no more than $n/15$, where n is the sample size. Seen the other way around, if you are designing a study and plan to include K predictors, you need at least $15 \times K$ observations. Of course, you may need more observations to have sufficient power to test a specific hypothesis – this rule of thumb is only for generalizability. For binary logistic regression (Chapter 6) and Cox proportional hazards regression (Chapter 7), the n is replaced by the number of observations in the less prevalent outcome category and the number of events, respectively (Babyak, 2004; Harrell, 2015, pp.72-73). As with any rule of thumb, this is meant as guidance – there is no requirement that it be strictly applied.

5.27 Writing it up

This section demonstrates how to write up the regression methods and results for Example 5.1. Below are a few lines of code to extract the information we need for our write-up.

Outcome, predictors, and dataset name:

```
fit.ex5.1.trans$call
```

```
## lm(formula = LBDGLUSI_trans ~ BMXWAIST + smoker + RIDAGEYR +
##     RIAGENDR + race_eth + income, data = nhanesf.complete)
```

Number of observations: If we used a complete case analysis then the number of rows in the dataset (`nrow(nhanesf.complete)`) is the number of cases used in the analysis. But counting the number of residuals to get the sample size will work even when you have missing data.

```
length(fit.ex5.1.trans$residuals)
```

```
## [1] 857
```

Selected descriptives: Extract any descriptive summaries you might want to add to your write-up. For example, the age range of the individuals in the analysis.

```
summary(nhanesf.complete$RIDAGEYR)
```

```
##    Min. 1st Qu.  Median    Mean 3rd Qu.    Max.
##    20.0    34.0    47.0    47.8    61.0    80.0
```

Summary of the model: Regression coefficients, p-values, and R^2. The p-values shown here test the association with the outcome for continuous predictors and categorical predictors with exactly two levels; for categorical predictors, each p-value tests the difference in mean outcome between the level shown in that row and the reference level.

```
summary(fit.ex5.1.trans) # (results not shown)
```

Confidence intervals for regression coefficients:

```
confint(fit.ex5.1.trans) # (results not shown)
```

Predictor p-values: For continuous predictors and categorical predictors with exactly two levels, the Type III test p-values are the same as those in the regression coefficient table; for each categorical predictor with more than two levels, the Type III test provides the multiple degree of freedom p-value for the overall test of that predictor's association with the outcome.

```
car::Anova(fit.ex5.1.trans, type = 3) # (results not shown)
```

The above will also provide a test of significance of any interactions in the model. If there is an interaction in the model, however, the main effect p-value for each predictor involved in the interaction is *not* the overall p-value for that predictor. To get an overall test of a predictor involved in an interaction, see Section 5.9.11. To get tests of a predictor involved in an interaction at specific levels of the other predictor in the interaction, see Section 5.9.7.

Methods:

When writing up the Methods, describe the data, including the data source, sample size, and inclusion/exclusion criteria, if any. For continuous predictors, specify the units. For categorical predictors, list the levels and specify the reference level. Describe all the steps you took to process the data, fit the model, and diagnose the fit of the model, including any sensitivity analyses you carried out. Describe what steps you took to resolve any issues uncovered by regression diagnostics (e.g., a variable transformation).

We used data from a teaching dataset including 857 NHANES 2017-2018 participants aged 20 to 80 years (mean 47.8 years) to assess the association between fasting glucose (mmol/L) and each of waist circumference (cm) and smoking status (Never (reference), Past, Current) using multiple linear regression adjusted for confounding due to age (years), gender (Male (reference), Female), race/ethnicity (Hispanic (reference), non-Hispanic White, non-Hispanic Black, non-Hispanic Other), and annual household income (<\$25,000 (reference), \$25,000 to <\$55,000, \$55,000+). We checked the model assumptions of normality, linearity, and constant variance. Some issues were found but after fasting glucose was transformed as $-Y^{-1.5}$ (based on a Box-Cox transformation) these issues were sufficiently resolved. We checked for collinearity between the predictors and found no issues (all VIFs were near 1) *(you can verify this yourself using* `car::vif(fit.ex5.1.trans)`*)*. We also checked for outliers and influential observations and evaluated the robustness of our results to such observations using a sensitivity analysis.

Results: The final model explained 23.6% of the variation in (transformed) fasting glucose. Waist circumference was positively associated with (transformed) fasting glucose (B = 0.00030; 95% CI = 0.00024, 0.00037; p <.001). Smoking status was not significantly associated with (transformed) fasting glucose (p = .317), although this conclusion is sensitive to the presence of outliers and influential observations. After removing 21 potential outliers and influential observations, the magnitude of the smoking status effect comparing Past vs. Never smokers was much larger and statistically significant. Age, gender, and race/ethnicity were also significantly associated with the outcome.

See Table 5.5 for full regression results.

```
NOBS <- length(fit.ex5.1.trans$residuals)
library(gtsummary)
TABLE <- fit.ex5.1.trans %>%
  tbl_regression(intercept = T,
                 estimate_fun = function(x) style_sigfig(x, digits = 5),
                 pvalue_fun   = function(x) style_pvalue(x, digits = 3),
                 label  = list(BMXWAIST ~ "Waist Circumference (cm)",
                               smoker   ~ "Smoking Status",
                               RIDAGEYR ~ "Age (years)",
                               RIAGENDR ~ "Gender",
                               race_eth ~ "Race/Ethnicity",
                               income   ~ "Annual Income")) %>%
  add_global_p(keep = T) %>%
  modify_caption(paste("Linear regression results for (transformed) fasting
    glucose
  (-mmol/L^(-1.5)) vs. waist circumference (cm) (N = ", NOBS, ")", sep=""))
```

```
TABLE
```

TABLE 5.5 Linear regression results for (transformed) fasting glucose (-mmol/L^(-1.5)) vs. waist circumference (cm) (N = 857)

Characteristic	**Beta**	**95% CI**	**p-value**
(Intercept)	-0.11266	-0.12021, -0.10511	<0.001
Waist Circumference (cm)	0.00030	0.00024, 0.00037	<0.001
Smoking Status			0.317
Never	—	—	
Past	0.00184	-0.00067, 0.00435	0.150
Current	-0.00011	-0.00313, 0.00291	0.942
Age (years)	0.00033	0.00026, 0.00040	<0.001
Gender			<0.001
Male	—	—	
Female	-0.00473	-0.00684, -0.00263	<0.001
Race/Ethnicity			0.009
Hispanic	—	—	
Non-Hispanic White	-0.00457	-0.00748, -0.00166	0.002
Non-Hispanic Black	-0.00268	-0.00667, 0.00132	0.189
Non-Hispanic Other	-0.00069	-0.00502, 0.00364	0.755
Annual Income			0.846
<$25,000	—	—	
$25,000 to <$55,000	0.00062	-0.00264, 0.00389	0.707
$55,000+	-0.00010	-0.00305, 0.00285	0.945

[1] CI = Confidence Interval

One downside to the use of an outcome transformation is that the regression coefficients are no longer on the original scale. In Section 5.5.2 we interpreted the coefficients of the model before applying a transformation to the outcome. The interpretations are similar here, except now the outcome is on a different scale.

5.28 Summary of multiple linear regression

There was a lot of information in this chapter. This section attempts to put all the steps in the order in which you would carry them out in a real-life analysis, although in fact you may have to jump back and forth between steps as, invariably, things will not go as planned.

1. **Think about your research question**

- What is your research question? Based on that question, what is your outcome and what is your predictor or predictors of interest?
- For a given predictor of interest (see Section 5.8):
  - Are there possible confounders or mediators? Adjust for confounders, but not mediators.
  - Are there any moderators? If so, include corresponding interaction terms.

- For confirmatory analyses (see Section 5.23):
  - Do not base decisions about which predictors or interactions to include, or in what form to include them, on statistical significance or effect size. Base decisions on subject-matter knowledge, interpretability, meeting the regression assumptions, and avoiding overfitting (see Section 5.26).
  - Determine if you need to adjust for multiple tests of outcomes or multiple comparisons between levels of a categorical predictor. To maximize power for the tests or comparisons of most interest, designate some as primary and others as secondary. Adjust for multiple testing over the primary (confirmatory) tests (see Section 5.24) and consider the secondary tests to be exploratory.

2. **Assess and handle missing data**

- Summarize the data using `summary()` to assess the extent of missing data. Some variables may be unusable due to having a large number of cases with missing values. Some variables may need to be modified as, in many studies, missing values actually have meaning (e.g., a survey question was automatically skipped based on the answer to a previous question, in which case you may be able to logically infer the value).
- Multiple imputation of missing data is covered in Chapter 9.
- If not using multiple imputation, at this stage it is best to create a dataset with no missing values (complete-case analysis, see Section 5.3). This will ensure your "Table 1" of descriptive statistics will be based on the same sample size as your regression analysis. Also, if you decide to compare models with different sets of predictors, it is important that the number of observations with non-missing values be the same for each model. If you add or remove variables in any of the following steps, you may need to return and repeat this step as the number of observations with missing data may have changed or, if using multiple imputation the imputation model will need to be adjusted and re-fit.

3. **Examine the analysis variables and adjust as needed**

- Create a histogram and numerical summary of each continuous variable (see Section 5.4).
  - Any anomalous values? Check to see if they are data entry errors or should be set to missing based on being impossible values.
  - If the outcome is highly skewed, consider a transformation.
  - If a continuous predictor is highly skewed, consider a transformation to reduce the leverage of extreme values.
  - If in doubt, wait until you have looked at regression diagnostics before deciding on the need for a transformation of either the outcome or any continuous predictors.
- Create a frequency table for each categorical predictor (see Section 5.4).
  - Any anomalous values? Check to see if they are data entry errors or should be set to missing based on being impossible values.
  - Decide if you need to collapse any levels so there is sufficient sample size in each.
  - Make sure each categorical predictor is coded as a `factor` (see Section 4.4.1).
- Check for collinearity (see Section 5.20).
  - Compute VIFs or aGSIFs.
  - Remove and/or combine predictors to reduce redundancy.
- Visualize the unadjusted relationships (see Section 5.4.6).
- Create a "Table 1" of descriptive statistics (see Section 3.3).

4. **Fit the model**

- Fit the model, carry out Type III tests, and compute CIs for the regression coefficients (see Section 5.5).

- If the model included an interaction (see Section 5.9):
  - Visualize the interaction (see Section 5.9.5).
  - Test the significance of the interaction term (see Section 5.9.10).
  - Estimate and test the significance of each predictor in the interaction at levels of the other (see Sections 5.9.7, 5.9.9.2, and 5.9.9.3).
  - Test the overall significance of each predictor in the interaction (see Section 5.9.11).
- Multiple testing adjustments, if necessary (for multiple outcomes and/or *post-hoc* comparisons between levels of a categorical predictor) (see Section 5.24).

5. **Diagnostics**

- Check assumptions (independence, linearity, normality, constant variance) (see Sections 5.14 to 5.17)
- Check for outliers and influential observations (see Sections 5.21 and 5.22)
- Carry out a sensitivity analysis, if necessary (see Section 5.25).

6. **Adjustments**

- Adjust the model, if necessary, to resolve assumption violations or other issues.
- After re-fitting a model, re-do all diagnostic checks.
- If any adjustment results in a predictor being added or removed, go back to the missing data step. The new complete case dataset might have a different sample size or, if using multiple imputation, the imputation model will be different.

7. **Write up the methods and results** (see Section 5.27)

5.29 Exercises

1. Suppose the variable "age category" has levels <18 years, 18 to 25 years, 26 to 39 years, 40 to 55 years, 56 to 64 years, and 65 years and older. If you include this variable in a regression model, how many age category regression coefficients will be estimated?

2. True or false? The unstandardized residual for an individual case is the difference between the observed and predicted outcome for that case.

3. True or false? An assumption of the linear regression model is that the outcome is normally distributed.

4. Explain the difference between a confounder and a mediator. Which should be adjusted for in a regression model and which should not?

5. What are two ways of incorporating effect modification into a regression analysis?

6. Which is wider, a confidence interval for the mean outcome or a prediction interval for an individual observation? Why?

7. Name the assumptions of the linear regression model.

8. The natural log, inverse, and square root transformations are all special cases of what family of transformations?

9. What kind of additional analysis can be done to determine the impact of outliers and influential observations on the regression coefficients and the conclusions of the regression analysis?

10. Explain when a multiple testing adjustment would be necessary.

For Exercises 11 to 13, use the 2018 Natality subsample teaching dataset (`natality2018_rmph.Rdata`, see Appendix A.3).

11. Create a complete case analysis dataset that includes all cases with no missing data for the following variables: `DBWT` (birthweight (g)), `CIG_REC` (smoked during pregnancy), `risks` (risk factors reported), `MAGER` (mother's age), `MRACEHISP` (mother's race/ethnicity), `DMAR` (marital status), `MEDUC` (mother's education), `PRIORLIVE` (prior births now living), `PRIORDEAD` (prior births now dead), and `BMI` (mother's body mass index, kg/m^2). How many rows were in the original dataset? How many rows are in the complete case analysis dataset?

12. Examine the complete case analysis dataset you created in Exercise 11. Create a `summary()` and histogram for each continuous variable. For each categorical variable, check to see if it is coded as a factor and create a frequency table. If you are not sure whether a numeric (non-factor) variable should be treated as continuous or categorical, use `table()` to see how many unique values there are – if there are only a few, then consider it categorical.

13. Using the complete case analysis dataset you created in Exercise 11, treat `PRIORLIVE` and `PRIORDEAD` as factor variables and collapse them so they have fewer levels, since many of the levels are sparse. Create a new variable called `PRIORLIVE_cat` that is a factor with levels 0, 1, 2, 3, and 4+. Create a new variable called `PRIORDEAD_cat` that is a factor with levels 0 and 1+. Hint: Use `factor()` to turn them into factor variables first, then use `fct_collapse()` to collapse the levels. Finally, use `table()` to compare the variables before and after the derivation to make sure it was done correctly.

For Exercises 14 to 21, use the modified complete case analysis 2018 Natality subsample teaching dataset you created in Exercises 11 and 13. If you did not complete Exercises 11 and 13, you can still do these exercises using `PRIORLIVE` instead of `PRIORLIVE_cat` and `PRIORDEAD` instead of `PRIORDEAD_cat` (treat them as continuous variables). In these exercises, you will assess the association between birthweight (the outcome) and smoking during pregnancy (the primary predictor), adjusted for risk factors reported, mother's age, mother's race/ethnicity, marital status, mother's education, prior births now living (use the collapsed factor version if you completed Exercise 13), prior births now dead (use the collapsed factor version if you completed Exercise 13), and mother's body mass index.

14. Visualize the unadjusted relationship between the outcome and each predictor.

15. Fit both the unadjusted and adjusted regression models. Assess the unadjusted and adjusted association between birthweight and cigarette use during pregnancy. How do the effect estimate, its 95% confidence interval, and its statistical significance change after adjusting for other variables?

16. Create a regression table displaying the results of the adjusted model, including the regression coefficients, their 95% CIs and p-values, and multiple degree of freedom p-values for categorical variables with more than two levels. Include an appropriate brief label for each variable.

17. Interpret the regression coefficient(s) for each predictor in the adjusted model. For each, state the regression coefficient, 95% CI, whether or not the predictor is statistically

significant (along with its p-value), and provide an interpretation in terms of how differences in the predictor are associated with differences in the mean outcome.

18. Visualize the adjusted relationships. How are the solid lines in these plots related to the estimated regression coefficients?

19. Does the association between cigarette smoking during pregnancy and birthweight depend on mother's education? Modify the adjusted model appropriately to be able to answer this question. Regardless of statistical significance, create a plot illustrating the smoking effect at each level of mother's education and estimate the cigarette smoking effect at each level of mother's education.

20. Does the association between mother's BMI and birthweight differ between mothers of different race/ethnicity? Modify the adjusted model appropriately to be able to answer this question. Regardless of statistical significance, estimate the BMI effect at each level of mother's race/ethnicity and make a plot illustrating the BMI effect at each level of mother's race/ethnicity.

21. Does the association between mother's BMI and birthweight depend on mother's age? Modify the adjusted model appropriately to be able to answer this question. Regardless of statistical significance, estimate the BMI effect for mother's age 20, 30, and 40 years and make a plot illustrating the BMI effect at each of these ages.

For Exercises 22 and 23, use the CAMP teaching dataset `camp_0_48_rmph.rData` (see Appendix A.6).

22. In Exercise 26 in Chapter 4, we tested to see if post-bronchodilator Forced Expiratory Volume at 1 second (FEV1) (`POSFEV0`) differed significantly between treatment groups (Budesonide, Nedocromil, Placebo) (`TG`). Does the difference in `POSFEV0` between treatment groups significantly depend on whether or not the child has a dehumidifier in the home (`dehumid`)?

23. In Exercise 27 in Chapter 4, we estimated the two treatment effects (Budesonide vs. Placebo, Nedocromil vs. Placebo) along with their 95% CIs and p-values. Using the model you fit in the previous exercise, estimate these two treatment effects among those with a dehumidifier and among those without. Interpret these results.

For Exercises 24 and 25, the following code creates the NHANES 2017-2018 (Appendix A.1) teaching dataset needed:

```
load("Data/nhanes1718_adult_fast_sub_rmph.Rdata")
nhanesf <- nhanes_adult_fast_sub
rm(nhanes_adult_fast_sub)

nhanesf.complete <- nhanesf %>%
  select(LBDGLUSI, BMXWAIST, smoker, RIDAGEYR,
         RIAGENDR, RIDRETH3, income) %>%
  drop_na()
```

24. In Section 5.9.9.2, Example 5.3, we fit the following model

```
fit.ex5.3.int <- lm(LBDGLUSI ~ RIAGENDR + income + RIAGENDR:income,
                    data = nhanesf.complete)
```

and plotted fasting glucose vs. income by gender. In this exercise, plot fasting glucose vs. gender by income to visualize how the income effects differ between genders.

25. In Section 5.9.9.3, Example 5.4, we fit the following model

```
fit.ex5.4.int <- lm(LBDGLUSI ~ RIDAGEYR + BMXWAIST + RIDAGEYR:BMXWAIST,
                    data = nhanesf.complete)
```

and plotted fasting glucose vs. waist circumference by age. In this exercise, plot fasting glucose vs. age at waist circumference values of 80, 100, and 120 cm to visualize how the age effect differs between those with different levels of waist circumference.

In each of Exercises 26 to 28, draw a figure similar to those in Section 5.8 that illustrates the described hypothetical relationships between variables, and state whether this illustrates a confounder, mediator, or moderator.

26. Substance use affects risk of suicide in part through substance use disorder.

27. The effect of substance use on suicide differs between those with different gender identity.

28. The effect of substance use on suicide must be adjusted for gender identity because gender identity is associated with both variables but is not in the causal pathway.

For Exercises 29 to 35, use the UN Human Development Data (`unhdd2020.rmph.rData`, see Appendix A.1).

29. What is the association between the outcome life expectancy at birth (years) (`life`) and expected years of schooling (years) (`educ_expected`), adjusted for urban population (%) (`urban`), annual growth in per capita GDP (%) (`gdp_percap_growth`), and depth of food deficit (average dietary energy supply adequacy, 2017/2019, %) (`food_deficit`)? Estimate the adjusted regression coefficient, 95% CI, and p-value. Also, how does the adjusted association compare to the unadjusted and, based on this comparison, what would you conclude about the extent to which the three variables you adjusted for confound the association? A 10% difference in effects is often used as a cutoff for deciding if confounding is meaningful (of course, this cutoff is arbitrary). Make sure you fit the unadjusted model to the same sample as your adjusted model.

30. Using the adjusted model from the previous exercise, what is the estimated life expectancy (and its 95% CI) for a country with 10 years of expected schooling? What about for a country with 15 years of expected schooling? Assume 50% urban population (`urban` = 50), no change in annual per capita GDP (`gdp_percap_growth` = 0), and a depth of food deficit of 120% (`food_deficit` = 120).

31. For each of the two sets of predictor values in the previous exercise, within what bounds do we expect 95% of countries' life expectancy to fall?

32. Does the Gender Inequality Index (`gii`) differ between HDI groups (`hdi_group`), adjusted for urban population (%) (`urban`), median age (years) (`age_median`), and current health expenditure (2017, % of GDP) (`cur_health_expen`)? Estimate the adjusted differences between groups and the reference level ("Low"), their 95% CIs, and p-values. Also, how do the adjusted differences compare to the unadjusted and, based on these comparisons, what would you conclude about the extent to which the three variables you adjusted for confound the association? A 10% difference in effects is often used as a cutoff for deciding if confounding is meaningful (of course, this cutoff is arbitrary). Make sure you fit the unadjusted model to the same sample as your adjusted model.

33. Using the adjusted model from the previous exercise, what is the estimated GII (and its 95% CI) for a country in the "Low" HDI group? What about for a country in the "Very high" HDI group? For the other variables in the model (urban population, median age, and current

health expenditure), choose any plausible value. You can use `summary()` to summarize their distributions to help you decide what values are plausible.

34. For each of the two sets of predictor values in the previous exercise, within what bounds do we expect 95% of countries' GII to fall?

35. Is there an association between the predictor tuberculosis incidence (2018, per 100,000 people) (`tb`) and the outcome Human Development Index (`hdi`)? Plot the unadjusted association. Then fit the model and check the assumptions. Then transform the predictor (try each of log, inverse, and square root) and re-fit the model to answer the research question.

For Exercises 36-43, use our COVID-19 county-level data (`covid_20210908_rmph.rData`, see Appendix A.4).

36. Is the (predictor) county-level number of ICU beds per 100,000 population (`icu_beds_per_100k`) associated with the (outcome) number of deaths (`deaths.usafacts.20210908`)? Give a possible interpretation for the direction of association.

37. Is the predictor county-level number of ICU beds per 100,000 population (`icu_beds_per_100k`) associated with the outcome number of deaths (`deaths.usafacts.20210908`) after adjusting for population size (`PopulationEstimate2018`)?

38. Assess the regression model assumptions (normality, linearity, and constant variance) for the model you fit in the previous exercise.

39. In the model from the previous exercise, log-transform the outcome and both predictors, re-fit the model, and re-check the assumptions. Hint: Check for zeros for each variable, and if there are any then add a small number to the variable prior to the log-transformation.

40. Instead of a log-transformation of the outcome in the previous exercise, try a Box-Cox transformation, re-fit the model, and re-check the assumptions. If there were zeros and you added a small number prior to transformation in the previous exercise, do the same here prior to the Box-Cox transformation. Hint: Fit the model with log-transformed predictors but with the UN-transformed outcome (after adding a small number, if needed), then apply the `MASS::boxcox()` function to that model.

41. Fit a regression model for the outcome "estimated age-adjusted percentage of diagnosed diabetes in 2016 among adults age 20 and over" (`DiabetesPercentage`) that includes the following predictors:

- `X..Adults.with.Obesity`: percentage of the adult population (age 20 and older) that reports a body mass index (BMI) greater than or equal to 30 kg/m^2 (2016)
- `Rural.UrbanContinuumCode2013`: rural-urban continuum code
- `MedianAge2010`: median age of county in 2010
- `X..Fair.or.Poor.Health`: percentage of adults reporting fair or poor health (age-adjusted) (2017)
- `SVIPercentile`: the county's overall percentile ranking indicating the CDC's Social Vulnerability Index (SVI); higher ranking indicates greater social vulnerability
- `X..Uninsured`: percentage of population under age 65 without health insurance (2017)
- `X..Unemployed`: percentage of population ages 16 and older unemployed but seeking work (2018)
- `X..Children.in.Poverty`: percentage of people under age 18 in poverty (2018)

Check for collinearity. Resolve any collinearity issues you find to arrive at a model where all VIFs are below 2.5. Compare the regression coefficients and their standard errors in the full model and the final model. Did any change by a meaningful amount?

42. Using the final model from the previous exercise, determine if there are any outliers using an outlier test, and then visualize the outliers. Based on the results of the outlier test, examine the significant outliers by looking at their values of `X..Adults.with.Obesity`, `DiabetesPercentage`, and population (`pop.usafacts`). Compare their obesity % and diabetes % to the distribution of these variables in all counties in this dataset. What is unusual about these counties with respect to their obesity % and % of adults with diabetes (what makes them outliers in our model)?

43. Using the final model from the previous exercise, look for influential observations.

For Exercises 44 to 46, use our 2018 Natality teaching dataset (`natality2018_rmph.Rdata`, see Appendix A.3).

44. Using our natality teaching dataset, fit multiple models, each one testing the association between the outcome birthweight (g) (`DBWT`) and a single risk factor, and each adjusted for Mother's Age (`MAGER`), Mother's Education (`MEDUC`), Marital Status (`DMAR`), and Total Birth Order (`TBO_REC`). Fit an adjusted model for each of the following risk factors, and then adjust for multiple testing using the Bonferroni and Hommel methods to obtain adjusted p-values. Which were statistically significant before adjusting for multiple testing? After adjusting for multiple testing? Was the Bonferroni adjustment too conservative for any of the risk factors? NOTE: The need for multiple testing here assumes that the researcher is screening for "any risk factor" and will conclude they have found an association if any of them are significant.

- RF_GDIAB: Gestational Diabetes
- RF_PHYPE: Pre-pregnancy Hypertension
- RF_GHYPE: Gestational Hypertension
- RF_PPTERM: Previous Preterm Birth
- RF_INFTR: Infertility Treatment Used
- RF_CESAR: Previous Cesarean

45. Regress the outcome birthweight (g) (`DBWT`) on the following predictors: `MRACEHISP` (mother's race/ethnicity), `risks` (risk factors reported), `MAGER` (mother's age), `PRIORLIVE` (prior births now living), `PRIORDEAD` (prior births now dead), `DMAR` (marital status), `MEDUC` (mother's education), and `BMI` (mother's body mass index, kg/m^2). Compute the p-values for each of the six pairwise comparisons of estimated birthweight between the four levels of `MRACEHISP` and use the Hommel method to adjust for multiple comparisons. Which comparisons were significant before adjusting for multiple comparisons? After? Hint: You will have to re-level the predictor in two different ways to get all the comparisons.

46. Referring back to the model in the previous exercise, if mother's race/ethnicity is the primary predictor of interest, does it make sense to adjust for all the other variables in the model? Think about each other predictor and evaluate whether it is a confounder or a mediator of racial disparities in birthweight. Remove the mediators and re-fit the model. Compare the magnitude of the estimated racial disparities before and after removing the mediators. Explain the difference.

Exercises 47 and 48 refer back to Exercises 41-43. The following code creates the final model that was fit in Exercise 41.

```
fit1 <- lm(DiabetesPercentage ~ X..Adults.with.Obesity +
             Rural.UrbanContinuumCode2013 +
           MedianAge2010 + X..Fair.or.Poor.Health + X..Uninsured +
           X..Unemployed, data = covid)
```

47. Do a sensitivity analysis for `fit1` to assess the robustness of model conclusions to the presence of outliers (which you looked for in Exercise 42).

48. Do a sensitivity analysis for `fit1` to assess the robustness of model conclusions to the presence of influential observations (which you looked for in Exercise 43).

49. Suppose you have a dataset with $n = 200$ observations. In order to avoid overfitting, what is the maximum number of predictors you should include in a regression model?

50. You are designing a study for which you wish to fit a regression model with eight predictors. What is the minimum sample size you need to avoid overfitting?

51. Using the NHANES fasting subset teaching dataset (`nhanes1718_adult_fast_sub_rmph.Rdata`, see Appendix A.1), regress log systolic blood pressure (`log(sbp)`) on log body mass index (`log(BMXBMI)`) and age (`RIDAGEYR`).

- Suppose you want to predict the log SBP for an individual with a BMI of 70 kg/m^2. Would that be a valid prediction? Why or why not?
- Suppose you want to predict the log SBP for an individual who is age 25 years. Would that be a valid prediction? Why or why not?
- Suppose you want to predict the log SBP for an individual with a BMI of 55 kg/m^2 and an age of 70 years. Would that be a valid prediction? Why or why not?

6

Binary Logistic Regression

In this chapter, you will learn how to:

- Interpret the equation for a binary logistic regression model;
- Compute and interpret odds and odds ratios (OR);
- Estimate unadjusted and adjusted ORs using binary logistic regression;
- Interpret the estimated regression coefficients;
- Create a forest plot to visualize estimated ORs and their confidence intervals;
- Compute predicted probabilities from the model;
- Test interactions between predictors;
- Diagnose the fit of the model;
- Appropriately summarize the methods and results for a binary logistic regression analysis;
- Fit an ordinal logistic regression model;
- Fit a conditional logistic regression model for matched case-control data; and
- Fit a log-binomial regression model to estimate a risk ratio (RR) or prevalence ratio (PR).

To use the code in this chapter, first load `tidyverse` and `Functions_rmph.R` (downloadable from RMPH Resources[1]).

```
library(tidyverse)
source("Functions_rmph.R")
```

6.1 Introduction

Binary logistic regression is similar in principle to linear regression – it models an outcome as a function of one or more predictors. The main distinction, however, is the form of the outcome. In linear regression, the outcome is a continuous variable, taking on values over a range of possible values. In binary logistic regression, the outcome is binary: a categorical variable with exactly two possible values. Hereafter, unless a distinction is necessary, this method is simply referred to as "logistic regression".

Given a binary outcome Y (e.g., "disease") that takes on values 0 ("no") and 1 ("yes") and K predictors, the logistic regression model

$$\ln\left(\frac{p}{1-p}\right) = \beta_0 + \beta_1 X_1 + \beta_2 X_2 + \ldots + \beta_K X_K \tag{6.1}$$

[1] https://github.com/rwnahhas/RMPH_Resources

models p, the probability that $Y = 1$ (e.g., "has the disease"), as a function of the predictors. Unlike linear regression, there is no normally distributed error term in Equation 6.1. Instead, the model assumes that, for a group of n individuals with the same predictors, the number of individuals with $Y = 1$ has a binomial distribution with sample size n and probability p. In practice, the coded values of Y can be any two values, not just 0 and 1.

Logistic regression could be used to model, for example, the relationship between the binary outcome coronary heart disease status (yes or no) and age, adjusted for confounding due to other factors. As with linear regression, the model can be used to test the null hypothesis of no association, estimate the magnitude of the association and its 95% confidence interval, and predict the outcome at a given age. For logistic regression, the raw prediction from the model is on the log-odds scale but can be transformed to produce a predicted probability. All of these steps will be demonstrated in the following sections.

6.2 Interpretation of the logistic regression coefficients

How do we interpret the logistic regression coefficients? To answer this question, we need to dive into some mathematical details, although, in the end, we will use R to do all the computations for us.

If p is the probability of an event, then $p/(1-p)$ is the "odds" of the event, the ratio of how likely the event is to occur and how likely it is to not occur. The left-hand side of the logistic regression equation $\ln(p/(1-p))$ is the natural logarithm of the odds, also known as the "log-odds" or "logit". To convert log-odds to odds, apply the inverse of the natural logarithm which is the exponential function e^x. To convert log-odds to a probability, apply the inverse logit function $e^x/(1+e^x)$.

Intercept

Plugging $X = 0$ into Equation 6.1, we find that the intercept β_0 is $\ln(p/(1-p))$, the log-odds when all predictors are 0 or at their reference level. Applying the exponential function demonstrates that e^{β_0} is the corresponding *odds* of the outcome. Applying the inverse logit function demonstrates that $e^{\beta_0}/(1+e^{\beta_0})$ is the corresponding *probability* of the outcome.

Predictor coefficients

In linear regression, β_k was the difference in the outcome associated with a 1-unit difference in X_k (or between a level and the reference level). Similarly, in logistic regression, it is the difference in the *log-odds* of the outcome associated with a 1-unit difference in X_k. It turns out that e^{β_k} is the **odds ratio (OR)** comparing individuals who differ by 1-unit in X_k. To see why, start by exponentiating both sides of the logistic regression equation to get the odds as a function of the predictors.

$$\frac{p}{1-p} = e^{\beta_0+\beta_1 X_1+\beta_2 X_2+\ldots+\beta_K X_K}$$

For a continuous predictor X_1, the ratio of the odds at $X_1 = x_1 + 1$ to the odds at $X_1 = x_1$ (a one-unit difference) can be expressed as the following ratio, for which all the terms cancel except e^{β_1}.

$$e^{\beta_0+\beta_1(X_1+1)+\beta_2 X_2+\ldots+\beta_K X_K}/e^{\beta_0+\beta_1 X_1+\beta_2 X_2+\ldots+\beta_K X_K} = e^{\beta_1}$$

If the first predictor is instead categorical and we want the OR comparing the first non-reference level to the reference level, then we want the ratio of the odds at $X_1 = 1$ to the odds at $X_1 = 0$. This is a 1-unit difference, so the derivation above also applies to categorical predictors.

Summary of interpretation of regression coefficients

- The intercept is the log-odds of the outcome when all predictors are at 0 or their reference level. Use the exponential function (e^{β_0}) to convert the intercept to odds and the inverse logit function $\left(e^{\beta_0}/(1+e^{\beta_0})\right)$ to convert the intercept to a probability.
- For a continuous predictor, the regression coefficient is the log of the odds ratio comparing individuals who differ in that predictor by one unit, holding the other predictors fixed.
- For a categorical predictor, each regression coefficient is the log of the odds ratio comparing individuals at a given level of the predictor to those at the reference level, holding the other predictors fixed.
- To compute an OR for X_k, exponentiate the corresponding regression coefficient, e^{β_k}, thus converting the log of the odds ratio to an OR.
- When there are multiple predictors, ORs are called adjusted ORs (AORs).

6.3 Why not use linear regression for a binary outcome?

Example 6.1: Figure 6.1 displays the linear regression fit for simulated data from 60 individuals aged 20 to 80 years where Y = coronary heart disease (CHD) status and X = age. The possible values for Y are "yes" and "no" coded as $Y = 1$ and 0, respectively. The gray dots at $Y = 1$ (along the top of the figure) are plotted at the ages of individuals with CHD, while the dots at $Y = 0$ (along the bottom of the figure) are for those without CHD. The solid line is the linear regression line, which attempts to estimate the proportion of CHD = Yes (the mean outcome) as a function of age using a straight line. The dashed line is a smoother which tracks the observed proportion without any constraint on its shape.

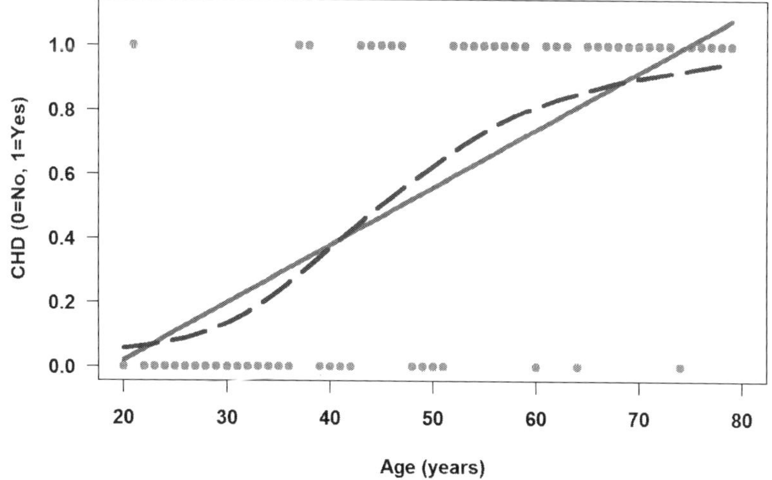

FIGURE 6.1 Linear regression does not fit well for a binary outcome

A straight line is not a good fit to this data in the sense that most of the points are very far from the line. However, it turns out to not be a terrible fit when comparing it to the relationship between the *proportion* of CHD = Yes and age (the dashed line). However, in general, logistic regression does even better at estimating the proportion.

Logistic regression models the mean (the probability of $Y = 1$) using the complicated looking logit function $\ln(p/(1-p))$ on the left-hand side of the equation. Why not have Y or p on the left-hand side? One reason is that, ideally, the range of possible values on the left-hand side should match that of the right-hand side. The right-hand side of the model equation is $\beta_0 + \beta_1 X_1 + \beta_2 X_2 + \ldots + \beta_K X_K$ and can take on any value from $-\infty$ to ∞, whereas Y can only be 0 or 1 and p must be in the range 0 to 1. However, the logit of p, $\ln(p/(1-p))$, *can* take on any value, matching the range of the right-hand side.

When transformed back to the probability scale, logistic regression fits an S-shaped curve that estimates the proportion of 1s at a given value of the predictor. Figure 6.2 demonstrates how the logistic regression fit (solid line) closely tracks the smoother (dashed line) and the predicted probabilities are all in [0, 1]. At the highest observed ages, however, the height of the estimated *linear* regression line (Figure 6.1) is greater than 1, resulting in predictions outside of [0, 1].

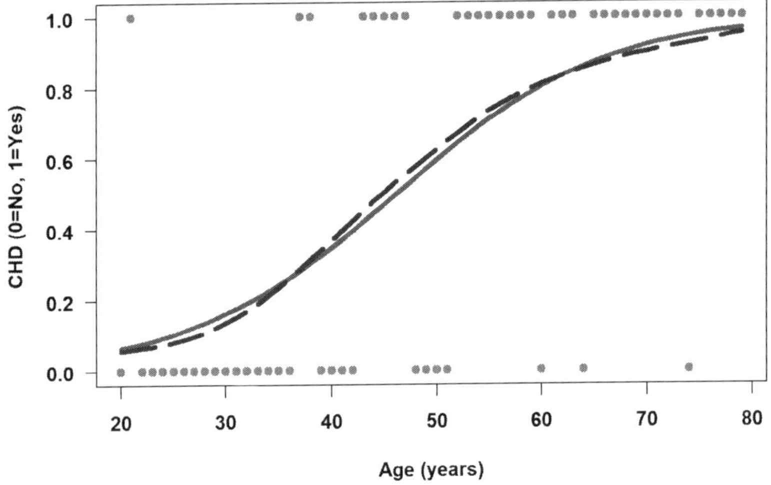

FIGURE 6.2 Logistic regression fits an S-shaped curve modeling P(Y = 1)

6.4 Odds and odds ratios

What are odds?

The odds of an event are the ratio of how likely the event is to occur and how likely it is to not occur. If p (probability) tells use how likely an event is to occur then $1 - p$ tells us how likely it is not to occur and the odds are the ratio of these two, $p/(1-p)$. For example, suppose an event has a probability of 75%. The odds of that event are $75\%/25\% = 3:1$, read as "3 to 1". As another example, suppose you attend a meeting of five people, including yourself. You each write your name on a piece of paper for a randomly drawn door prize.

Your chance, or probability, of winning the prize is 1/5 (0.20). Your odds of winning, however, are 1 to 4 (1:4) (0.25). There is one piece of paper with your name and four without, so you have one chance of winning and four chances to lose.

There is a one-to-one relationship between odds and probability. They are two ways of expressing the same thing, but on different scales. The odds are greater (less) than 1 if and only if the probability is greater (less) than 0.5, and the odds are exactly 1 if and only if the probability is 0.5.

What is an odds ratio (OR)?

An odds ratio (OR) compares the odds of an event between two different groups. For example, suppose the probability of disease is 0.35 for men and 0.25 for women. The OR is, therefore, $[0.35/(1-0.35)] \div [0.25/(1-0.25)] = 0.54 \div 0.33 = 1.62$.

Interpreting an OR

An OR > 1 (< 1) implies a positive (negative) association between a continuous predictor and the outcome. An OR of 1 implies no association. For an OR > 1, 100% $\times$ (OR $-$ 1) represents the % greater odds for $X = x + 1$ compared to $X = x$ for a continuous predictor, or for a level compared to the reference level for a categorical predictor. For an OR < 1, 100% $\times$ (1 $-$ OR) represents the % lower odds. Also, if you reverse the order of the groups being compared, the OR will be inverted.

For example:

- OR = 1.45 implies that the first group has 45% greater odds of the outcome than the second group, or 1.45 times the odds of the second group.
- OR = 0.81 implies that the first group has 19% lower odds of the outcome than the second group, or 0.81 times the odds of the second group.
- If the OR comparing group A to group B is 0.667, then the OR comparing group B to group A is 1/0.667 = 1.50.

The "times the odds" style of interpreting an OR may be more clear than the "% greater odds" style when OR ≥ 2. For example, OR = 2.50 could be interpreted as the first group having "150% greater odds than" or "2.5 times the odds of" the second group. Both interpretations are correct, but the latter may be more clear.

6.5 Estimating an OR using a 2 × 2 table

For the special case of a binary predictor, the OR can be computed from a 2 × 2 table. In Table 6.1, the OR comparing the odds of "positive" between risk factor levels 2 and 1 is the cross-product $(d \times a) \div (b \times c)$. We started the cross-product at d since we wanted the odds of "positive" comparing risk factor level 2 to level 1. If, instead, we wanted the odds of "positive" comparing level 1 to level 2, we would start at b and compute the OR as $(b \times c) \div (d \times a)$ (as previously mentioned, if you change the order of comparison then you invert the OR).

Example 6.2: Using the 2019 National Survey of Drug Use and Health (NSDUH) teaching dataset (Section A.5), compute the OR comparing the odds of lifetime marijuana use (`mj_lifetime`) between males and females (`demog_sex`).

TABLE 6.1 Computing an odds ratio from a two-way table

	Condition negative	Condition positive
Risk factor level 1	a	b
Risk factor level 2	c	d

```
load("Data/nsduh2019_adult_sub_rmph.RData")
# Shorter name
nsduh <- nsduh_adult_sub
TAB <- table(nsduh$demog_sex, nsduh$mj_lifetime)
TAB
```

```
##
##          No Yes
##   Male   206 260
##   Female 285 249
```

The way the table is set up, the cross-product $(d \times a) \div (b \times c)$ will give us the odds of "Yes" comparing females to males.

```
# (d*a) / (b*c)
(TAB[2,2]*TAB[1,1])/(TAB[1,2]*TAB[2,1])
```

```
## [1] 0.6922
```

However, the question asked for the OR comparing males to females, so we need to use the formula $(b \times c) \div (d \times a)$.

```
# (b*c) / (d*a)
(TAB[1,2]*TAB[2,1]) / (TAB[2,2]*TAB[1,1])
```

```
## [1] 1.445
```

Thus, the estimated OR for lifetime marijuana use comparing males to females is 1.445; males have 44.5% greater odds of lifetime marijuana use than females. We could have also obtained this OR by inverting the OR comparing females to males (1 / 0.692 = 1.445).

To review, here is a reminder of where the value of 1.445 comes from:

- P(lifetime marijuana use among males) = 260 ÷ (206 + 260) = 0.558
- P(lifetime marijuana use among females) = 249 ÷ (285 + 249) = 0.466
- Odds for males = 0.558 ÷ (1 – 0.558) = 1.262 (odds > 1 because probability > 0.5)
- Odds for females = 0.466 ÷ (1 – 0.466) = 0.874 (odds < 1 because probability < 0.5)
- Odds ratio = 1.262 ÷ 0.874 = 1.445 (OR > 1 because odds for males > odds for females)

Below are some useful functions for converting probabilities to odds and ORs, along with the logit function and its inverse which we will use in the next section. These functions are all in `Functions_rmph.R` which you loaded at the beginning of this chapter.

```
odds       <- function(p)      p/(1-p)
odds.ratio <- function(p1, p2) odds(p1)/odds(p2)
logit      <- function(p)      log(p/(1-p))
ilogit     <- function(x)      exp(x)/(1+exp(x))
# exp() is the exponential function

# Example 6.2
PM <- TAB[1,2]/(TAB[1,2]+TAB[1,1]) # P(Yes | Male)
PF <- TAB[2,2]/(TAB[2,2]+TAB[2,1]) # P(Yes | Female)
OM <- odds(PM)                     # PM / (1 - PM)
OF <- odds(PF)                     # PF / (1 - PF)
OR.MvsF <- odds.ratio(PM, PF)      # Odds ratio
round(c(PM, PF, OM, OF, OR.MvsF), 3)
```

```
## [1] 0.558 0.466 1.262 0.874 1.445
```

6.6 Estimating an OR using logistic regression

Unadjusted, categorical predictor

We can also estimate an OR using logistic regression via the `glm()` function ("generalized linear model").

Example 6.2 (continued): Use logistic regression to estimate the OR comparing the odds of lifetime marijuana use (`mj_lifetime`) between males and females (`demog_sex`).

First, re-level `demog_sex` to make "Female" the reference level in order to estimate the OR comparing males to females.

```
nsduh <- nsduh %>%
  mutate(demog_sex = relevel(demog_sex, ref = "Female"))
```

Next, use `glm()` with `family = binomial` to fit a logistic regression and `summary()` to view the output.

```
fit.ex6.2 <- glm(mj_lifetime ~ demog_sex,
                 family = binomial, data = nsduh)
summary(fit.ex6.2)
```

```
##
## Call:
## glm(formula = mj_lifetime ~ demog_sex, family = binomial, data = nsduh)
##
## Coefficients:
##               Estimate Std. Error z value Pr(>|z|)
## (Intercept)    -0.1350     0.0867   -1.56   0.1195
## demog_sexMale   0.3678     0.1274    2.89   0.0039 **
## ---
## Signif. codes:  0 '***' 0.001 '**' 0.01 '*' 0.05 '.' 0.1 ' ' 1
##
## (Dispersion parameter for binomial family taken to be 1)
##
```

```
##     Null deviance: 1386.0  on 999  degrees of freedom
## Residual deviance: 1377.6  on 998  degrees of freedom
## AIC: 1382
##
## Number of Fisher Scoring iterations: 3
```

We showed previously that the intercept (-0.1350) is the log-odds of lifetime marijuana use when all the predictors are zero or at their reference level. Use `ilogit()` to convert a log-odds to a probability.

```
ilogit(coef(fit.ex6.2)["(Intercept)"])
```

```
## (Intercept)
##      0.4663
```

Thus, we estimate that the probability of marijuana use among females (the reference level) is $p = e^{-0.1350}/(1 + e^{-0.1350}) = 0.466$ (the same value we computed using the 2×2 table).

The regression coefficient for `demog_sexMale` (0.3678) represents the log of the OR for lifetime marijuana use comparing males to females. To get the OR itself, exponentiate the coefficient using `exp()`. Thus, we estimate the OR to be $e^{0.3678} = 1.445$ (again, the same value we computed using the 2×2 table).

```
# [-1] drops the first row which corresponds to the intercept)
exp(coef(fit.ex6.2)[-1])
```

```
## demog_sexMale
##         1.445
```

Use `confint()` to compute 95% CIs for the regression coefficients, and exponentiate these to get 95% CIs for the ORs (dropping the intercept since e^{β_0} is an odds not an OR).

NOTE: `confint()` uses a profile likelihood method rather than the Wald method used by `summary()` (see Section 6.18). Thus, the 95% CIs may not correspond exactly to the p-values. That is, you may have $p < .05$ but a 95% CI that contains 1, or $p > .05$ but a 95% CI that does not contain 1.

```
# CIs for regression coefficients
confint(fit.ex6.2)
```

```
##                   2.5 %  97.5 %
## (Intercept)     -0.3055 0.03476
## demog_sexMale    0.1186 0.61808
```

```
# CI for OR (use [-1,] to drop the first row which corresponds to the intercept)
# and drop=F to retain the rownames since there is only one predictor term
exp(confint(fit.ex6.2))[-1, , drop=F]
```

```
##               2.5 % 97.5 %
## demog_sexMale 1.126  1.855
```

In this example, the categorical predictor had just two levels, so get the p-value testing its association with the outcome from the regression coefficient table (0.0039). In general, use `car::Anova()` to obtain the p-value for any predictor, continuous or categorical with any number of levels. Unlike with linear regression, there are two different options here: a likelihood ratio test or a Wald test. We will use the Wald test here to be consistent with the p-values in the regression coefficient table (which are equivalent to Wald tests). However, see Section 6.18 for how to obtain likelihood ratio tests throughout.

```
car::Anova(fit.ex6.2, type = 3, test.statistic = "Wald")
```

```
## Analysis of Deviance Table (Type III tests)
##
## Response: mj_lifetime
##              Df Chisq Pr(>Chisq)
## (Intercept)  1  2.42     0.1195
## demog_sex    1  8.34     0.0039 **
## ---
## Signif. codes:  0 '***' 0.001 '**' 0.01 '*' 0.05 '.' 0.1 ' ' 1
```

Conclusion: Males have significantly greater odds of lifetime marijuana use than females (OR = 1.445; 95% CI = 1.126, 1.855; p = .0039). Males have 44% greater odds of lifetime marijuana use than females.

For this simple case with a binary outcome and a binary predictor, the OR is exactly the same as that obtained via the cross-product from a 2 × 2 table. An advantage of using logistic regression, however, is that, it provides the ability to adjust for other predictors to obtain an adjusted OR (AOR), as we will see in Section 6.6.3.

Unadjusted, continuous predictor

Next, let's fit a logistic regression with a continuous predictor and interpret the results.

Example 6.3: Using the 2019 National Survey of Drug Use and Health (NSDUH) teaching dataset (Section A.5), what is the association between lifetime marijuana use (`mj_lifetime`) and age at first use of alcohol (`alc_agefirst`)?

Fit the model and interpret the coefficients.

```
fit.ex6.3 <- glm(mj_lifetime ~ alc_agefirst, family = binomial, data = nsduh)
round(summary(fit.ex6.3)$coef, 4)
```

```
##                Estimate Std. Error z value Pr(>|z|)
## (Intercept)      5.3407     0.4747   11.25        0
## alc_agefirst    -0.2835     0.0267  -10.62        0
```

```
ilogit(coef(fit.ex6.3)["(Intercept)"])
```

```
## (Intercept)
##      0.9952
```

The inverse logit of the intercept (5.3407) is the probability of the outcome when all predictors are zero or at their reference level. Thus, the estimated probability of lifetime marijuana use among those who started drinking alcohol at age 0 years is 0.9952.

Examining a summary of the predictor, we see that the range of age of first use of alcohol is 3 to 45 years, and the median value is 17 years.

```
# Examine the predictor
summary(nsduh$alc_agefirst)
```

```
##    Min. 1st Qu.  Median    Mean 3rd Qu.    Max.    NA's
##     3.0    15.0    17.0    17.5    19.0    45.0     157
```

Thus, the probability of marijuana use for those who first drank alcohol at age 0 years is an extrapolation. Had we centered the predictor, the intercept would be more interpretable. For example, had we centered it at the median, 17 years, the intercept would be interpreted as the log-odds at the median age of first alcohol use, with a corresponding estimated prevalence of lifetime marijuana use among 17-year-olds of 62.73%.

```
nsduh <- nsduh %>%
  mutate(calc_agefirst = alc_agefirst - 17)
fit.ex6.3.centered <- glm(mj_lifetime ~ calc_agefirst,
                          family = binomial, data = nsduh)
ilogit(coef(fit.ex6.3.centered)["(Intercept)"])
```

```
## (Intercept)
##      0.6273
```

The regression coefficient for `alc_agefirst` (-0.2835) is the difference in log-odds associated with a 1-year difference in age at first use of alcohol. Converting this to an OR by exponentiating, we get $e^{-0.2835} = 0.7531$.

```
exp(coef(fit.ex6.3)[-1])
```

```
## alc_agefirst
##       0.7531
```

Thus, starting drinking alcohol 1 year later is associated with 24.7% lower odds of lifetime marijuana use (since $1 - 0.753 = 0.247$). Finally, compute a 95% CI for the OR using `confint()` and exponentiating.

```
exp(confint(fit.ex6.3)[-1,])
```

```
##  2.5 % 97.5 %
## 0.7135 0.7923
```

Conclusion: Age at first alcohol use is significantly negatively associated with lifetime marijuana use (OR = 0.753; 95% CI = 0.713, 0.792; p <.001). Individuals who first used alcohol at, say, age 19 years have 24.7% lower odds of having ever used marijuana than those who first used alcohol at age 18 years.

6.6.1 OR associated with other than a 1-unit difference

If, after fitting the model, you want to know the increase or reduction in odds associated with other than a 1-unit difference in a continuous predictor, multiply the regression coefficient by that factor *before* exponentiating. Similarly, to compute the corresponding confidence interval, multiply the confidence limits for the regression coefficient by that factor prior to exponentiating.

If you want to have the logistic regression model itself estimate an OR corresponding to other than a 1-unit difference then, prior to fitting the model, *divide* the continuous predictor by the desired factor. In this way a 1-unit difference on the new scale will correspond to the desired difference on the original scale.

Example 6.3 (continued): To estimate the reduction in odds associated with starting drinking alcohol 3 years later, do *not* multiply 24.7% by 3 (which would be 74.1%). Instead, first compute the OR as $e^{-0.2835 \times 3} = 0.427$ and then compute the percent change in odds. In other words, those who started using alcohol 3 years later have 57.3% lower odds of lifetime marijuana use (not 74.1%).

```
# OR for a 3-unit difference
exp(3*coef(fit.ex6.3))[-1]
```

```
## alc_agefirst
##       0.4272
```

```
# 95% CI for the OR for a 3-unit difference
exp(3*confint(fit.ex6.3))[-1,]
```

```
##  2.5 % 97.5 %
## 0.3632 0.4973
```

```
# Same result accomplished by transforming the
# predictor prior to fitting the model
tmpdat <- nsduh %>%
  mutate(alc_agefirst3 = alc_agefirst/3)

tmpfit <- glm(mj_lifetime ~ alc_agefirst3,
              family = binomial, data = tmpdat)

exp(coef(tmpfit))[-1]
```

```
## alc_agefirst3
##        0.4272
```

```
exp(confint(tmpfit))[-1,]
```

```
##  2.5 % 97.5 %
## 0.3632 0.4973
```

6.6.2 Make sure you know what probability `glm()` is modeling

The outcome Y must be binary (exactly two unique values, not including `NA`) and have numeric values 0 and 1 or be coded as a factor. If Y is numeric, then `glm()` will model $p = P(Y = 1)$. If Y is a factor, then `glm()` will model $p = P(Y = \text{the non-reference level})$. If the outcome is not coded to these specifications, or if you want to model the probability of the other level of the outcome, then recode it (see Section 4.4.1).

It is essential to know which level of the outcome corresponds to the probability p when using logistic regression. Use `is.factor()` to check if Y is a factor, and `levels()` to examine the levels (if it is a factor). The first level listed by `levels()` is the reference level, so the second is the level whose probability is being modeled. If the outcome is not a factor, use `unique()` to check the unique values – they should be 0 and 1.

Example 6.3 (continued): For the outcome Y = `mj_lifetime`, what probability is `glm()` modeling?

```
is.factor(nsduh$mj_lifetime)
```

```
## [1] TRUE
```

```
# If it is a factor, check the levels
levels(nsduh$mj_lifetime)
```

```
## [1] "No"  "Yes"
```

The possible values are No (reference level, since it is listed first) and Yes. Thus, `glm()` models $p = P(Y = \text{Yes})$.

```
# If it is not a factor, check
# the unique values.
# Should be 0 and 1
# unique(x)
```

6.6.3 Adjusted OR

Just as you can move from simple to multiple linear regression by adding more predictors, so you can move from simple to multiple logistic regression. When there are multiple predictors in a logistic regression model, the resulting odds ratios are called adjusted odds ratios (AORs).

Example 6.3 (continued): What is the association between lifetime marijuana use (`mj_lifetime`) and age at first use of alcohol (`alc_agefirst`), adjusted for age (`demog_age_cat6`), sex (`demog_sex`), and income (`demog_income`)? Fit the model and compute the AOR, its 95% CI, and the p-value that tests the significance of age at first use of alcohol. While we are not primarily interested in the confounders, it is common to also report their AORs, 95% CIs, and p-values. Since there are some categorical confounders with more than two levels, use `car::Anova()` to compute multiple df p-values.

```
fit.ex6.3.adj <- glm(mj_lifetime ~ alc_agefirst + demog_age_cat6 + demog_sex +
                     demog_income, family = binomial, data = nsduh)
# Regression coefficient table
round(summary(fit.ex6.3.adj)$coef, 4)
```

```
##                        Estimate Std. Error z value Pr(>|z|)
## (Intercept)              6.2542     0.5914 10.5759   0.0000
## alc_agefirst            -0.2754     0.0276 -9.9922   0.0000
## demog_age_cat626-34     -0.2962     0.3286 -0.9012   0.3675
## demog_age_cat635-49     -0.8043     0.2966 -2.7120   0.0067
## demog_age_cat650-64     -0.6899     0.2985 -2.3109   0.0208
## demog_age_cat665+       -1.2748     0.3043 -4.1893   0.0000
```

```
## demog_sexMale                    -0.0609     0.1618 -0.3763     0.7067
## demog_income$20,000 - $49,999 -0.5309     0.2664 -1.9927     0.0463
## demog_income$50,000 - $74,999 -0.0793     0.3049 -0.2601     0.7948
## demog_income$75,000 or more     -0.3612     0.2532 -1.4264     0.1538
```

```
# AORs and 95% CIs
# Use cbind to view the AORs and CIs side-by-side
# Add [-1,] to drop the row for the intercept
# since exp(intercept) is not an odds ratio
OR.CI <- cbind("AOR" = exp(coef(fit.ex6.3.adj)),
               exp(confint(fit.ex6.3.adj)))[-1,]
round(OR.CI, 3)
```

```
##                                  AOR 2.5 % 97.5 %
## alc_agefirst                   0.759 0.718  0.800
## demog_age_cat626-34            0.744 0.387  1.409
## demog_age_cat635-49            0.447 0.247  0.791
## demog_age_cat650-64            0.502 0.275  0.891
## demog_age_cat665+              0.279 0.152  0.502
## demog_sexMale                  0.941 0.684  1.291
## demog_income$20,000 - $49,999 0.588 0.347  0.987
## demog_income$50,000 - $74,999 0.924 0.507  1.680
## demog_income$75,000 or more    0.697 0.421  1.139
```

```
# P-values
car::Anova(fit.ex6.3.adj, type = 3, test.statistic = "Wald")
```

```
## Analysis of Deviance Table (Type III tests)
##
## Response: mj_lifetime
##              Df  Chisq           Pr(>Chisq)
## (Intercept)   1 111.85 < 0.0000000000000002 ***
## alc_agefirst  1  99.84 < 0.0000000000000002 ***
## demog_age_cat6 4  23.01             0.00013 ***
## demog_sex     1   0.14             0.70669
## demog_income  3   5.44             0.14197
## ---
## Signif. codes:  0 '***' 0.001 '**' 0.01 '*' 0.05 '.' 0.1 ' ' 1
```

Interpretation of the output:

- The AOR for our primary predictor `alc_agefirst` is 0.759. This AOR corresponds to the odds ratio comparing the odds of lifetime marijuana use between those who differ by one year in age at first use of alcohol but have the same age, sex, and income.
- The remaining AORs compare levels of categorical predictors to their reference level, adjusted for the other predictors in the model. For example, comparing individuals with the same age of first alcohol use, sex, and income, 35- to 49-year-olds have 55.3% lower odds of lifetime marijuana use than 18- to 25-year-olds (OR = 0.447; 95% CI = 0.247, 0.791; p = .007). The p-value for this specific comparison of ages comes from the `Coefficients` table. An overall, 4 df p-value for age, can be read from the `Type III Test` table (0.00013).
- The `Type III tests` output contains the multiple df Wald tests for categorical predictors with more than two levels. For continuous predictors, or for categorical predictors with exactly two levels, the Type III Wald test p-values are identical to those in the `Coefficients` table.

Conclusion: After adjusting for age, sex, and income, age at first alcohol use is significantly negatively associated with lifetime marijuana use (AOR = 0.759; 95% CI = 0.718, 0.800; p <.001). Individuals who first used alcohol at a given age have 24.1% lower odds of having ever used marijuana than those who first used alcohol one year earlier.

NOTE: The steps of extracting the coefficients and CIs and exponentiating them are demonstrated here to make it clear where the numbers are coming from. However, the `car::S()` function (Fox et al., 2023; Fox and Weisberg, 2019) can be used instead to conveniently produce output showing the `summary()`, as well as the AORs and their 95% CIs (but you still need `car::Anova()` for the Type III tests).

```
car::S(fit.ex6.3.adj)
# (output not shown)
```

6.7 Visualizing ORs

In addition to reporting the numeric results of a logistic regression, it is helpful to create a **forest plot** to visualize the AORs and their 95% CIs. You can easily create a forest plot using `plot_model()` from the `sjPlot` library (Lüdecke, 2023). See `?plot_model` for customization options.

Example 6.3 (continued): Create a forest plot displaying the AORs for the model. To put the AOR for age of first alcohol use on a similar scale as the categorical predictors, use the AOR and 95% CI for a 4-year difference in age (the interquartile range).

For the purpose of creating a forest plot, changing the age scale must be made via the method that transforms age prior to fitting the model (see Section 6.6.1). Using the interquartile range results in an AOR that compares the adjusted odds of the outcome between individuals at the 75th and 25th percentiles of the predictor distribution.

```
IQR(nsduh$alc_agefirst, na.rm=T)
```

```
## [1] 4
```

```
fpdat <- nsduh %>%
  mutate(alc_agefirst4 = alc_agefirst/4)

fit.fp <- glm(mj_lifetime ~ alc_agefirst4 + demog_age_cat6 + demog_sex +
                  demog_income, family = binomial, data = fpdat)

# Use [-1,] to exclude the intercept since exp(intercept) is not an OR
CI <- exp(confint(fit.fp))[-1,]
```

The forest plot is shown in Figure 6.3.

```
# Forest plot
library(sjPlot)
plot_model(fit.fp,
           axis.lim = c(min(CI), max(CI)),
           auto.label = F)
```

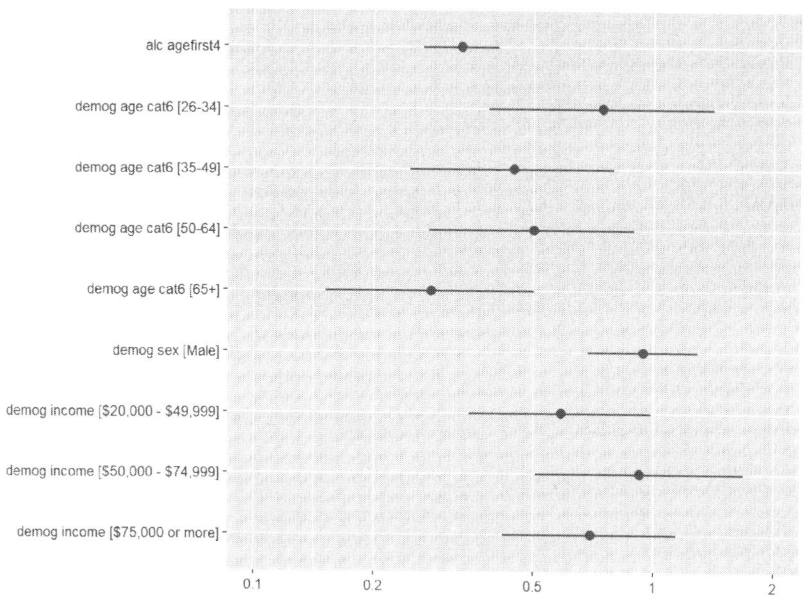

FIGURE 6.3 Forest plot of AORs and their 95% CIs for the logistic regression analysis of lifetime marijuana use

6.8 Prediction

As with MLR, `predict()` computes predicted values from the model. The default output will be on the log-odds scale, but adding `type = "response"` will instead give a predicted probability. Unfortunately, `predict()` does not have a CI option for `glm()` models. Instead, use `gmodels::estimable()` to obtain the prediction along with its 95% CI.

Example 6.3 (continued): What is the predicted probability (and 95% CI) of lifetime marijuana use for someone who first used alcohol at age 13 years, is currently age 30 years, male, and has an income of <$20,000? What if they first used alcohol at age 21 years?

The following demonstrates the use of `predict()` to compute these predicted probabilities.

```
predict(fit.ex6.3.adj,
        newdata = data.frame(alc_agefirst   = c(13, 21),
                             demog_age_cat6 = "26-34",
                             demog_sex      = "Male",
                             demog_income   = "Less than $20,000"),
        type = "response")
```

```
##      1      2
## 0.9103 0.5283
```

However, if you also want a CI, then use `gmodels::estimable()`. The code is slightly more complicated than using `predict()`. First, `predict()` automatically includes the intercept in its calculation, but since `gmodels::estimable()` is also used for estimates that are not predictions you have to explicitly tell it you want to include the intercept. Second, you must enter the names of the terms as spelled in the coefficient output rather than the variable names in the model (these are different for factor predictors). Third, you must use the inverse logit function to convert the result to a probability. Lastly, unlike predict, where we included two values for `alc_agefirst` to get two predictions, if you want multiple predictions you must run `gmodels::estimable()` multiple times.

First, examine the `rownames()` of the coefficient output to make sure you spell each term correctly.

```
rownames(summary(fit.ex6.3.adj)$coef)
```

```
##  [1] "(Intercept)"                    "alc_agefirst"
##  [3] "demog_age_cat626-34"            "demog_age_cat635-49"
##  [5] "demog_age_cat650-64"            "demog_age_cat665+"
##  [7] "demog_sexMale"                  "demog_income$20,000 - $49,999"
##  [9] "demog_income$50,000 - $74,999" "demog_income$75,000 or more"
```

Then enter the needed terms and their multiplicative factors in `gmodels::estimable()`, surrounded by `ilogit()`. Note that we only extract the estimate and the CI, not the other parts of the `gmodels::estimable()` output as the other parts would not make sense after applying `ilogit()`.

```
# Always include the intercept for prediction
# Specify a 1 for the intercept, a # for each continuous predictor
# and a 1 for each non-reference level of a categorical variable.
# If a predictor is at its reference level, it should be left out.
ilogit(gmodels::estimable(fit.ex6.3.adj,
  c("(Intercept)"         = 1,
    "alc_agefirst"        = 13,
    "demog_age_cat626-34" = 1,
    "demog_sexMale"       = 1),
  conf.int = 0.95))[c("Estimate", "Lower.CI", "Upper.CI")]
```

```
##                          Estimate Lower.CI Upper.CI
## (1 13 1 0 0 0 1 0 0 0)    0.9103   0.8387   0.9519
```

```
ilogit(gmodels::estimable(fit.ex6.3.adj,
  c("(Intercept)"         = 1,
    "alc_agefirst"        = 21,
    "demog_age_cat626-34" = 1,
    "demog_sexMale"       = 1),
  conf.int = 0.95))[c("Estimate", "Lower.CI", "Upper.CI")]
```

```
##                          Estimate Lower.CI Upper.CI
## (1 21 1 0 0 0 1 0 0 0)    0.5283   0.3724   0.6789
```

Conclusion: We predict that among individuals who first used alcohol at age 13 years, are age 30 years, male, and have an income <\$20,000 per year, 91.0% have used marijuana (95% CI = 83.9%, 95.2%). Among those who instead waited until age 21 years to first use alcohol, the proportion drops to 52.8% (95% CI = 37.2%, 67.9%).

6.9 Interactions

Just like with linear regression, you can include an interaction in a logistic regression model to evaluate if a predictor is an effect modifier, that is, if the OR for one predictor depends on another (the effect modifier).

Example 6.3 (continued): After adjusting for age and income, does the association between lifetime marijuana use and age at first use of alcohol differ by sex?

Adding the interaction term `alc_agefirst:demog_sex` to the model, we get the following results.

```
fit.ex6.3.int <- glm(mj_lifetime ~ alc_agefirst + demog_age_cat6 + demog_sex +
                     demog_income + alc_agefirst:demog_sex,
          family = binomial, data = nsduh)
round(summary(fit.ex6.3.int)$coef, 4)
```

```
##                                   Estimate Std. Error z value Pr(>|z|)
## (Intercept)                         7.3735     0.8292  8.8923   0.0000
## alc_agefirst                       -0.3378     0.0423 -7.9876   0.0000
## demog_age_cat626-34                -0.2951     0.3314 -0.8902   0.3733
## demog_age_cat635-49                -0.8160     0.2987 -2.7320   0.0063
## demog_age_cat650-64                -0.6866     0.3007 -2.2832   0.0224
## demog_age_cat665+                  -1.2600     0.3064 -4.1130   0.0000
## demog_sexMale                      -2.1381     0.9847 -2.1713   0.0299
## demog_income$20,000 - $49,999      -0.5304     0.2682 -1.9773   0.0480
## demog_income$50,000 - $74,999      -0.0931     0.3073 -0.3029   0.7620
## demog_income$75,000 or more        -0.3629     0.2547 -1.4250   0.1542
## alc_agefirst:demog_sexMale          0.1189     0.0555  2.1420   0.0322
```

To answer the question of significance, look at the row in the regression coefficients table corresponding to the interaction (`alc_agefirst:demog_sex`). The p-value is 0.0322, so the interaction is statistically significant. Had there been more than one row (e.g., if the interaction involved a categorical predictor with more than two levels), then use a Type III test to get the test of interaction. Since, in this case, there is just one term corresponding to the interaction, the Type III Wald test p-value from `car::Anova()` is exactly the same as the p-value in the coefficients table.

```
car::Anova(fit.ex6.3.int, type = 3, test.statistic = "Wald")
```

```
## Analysis of Deviance Table (Type III tests)
##
## Response: mj_lifetime
##                        Df Chisq              Pr(>Chisq)
## (Intercept)             1 79.07 < 0.0000000000000002 ***
## alc_agefirst            1 63.80   0.0000000000000014 ***
## demog_age_cat6          4 22.33              0.00017 ***
## demog_sex               1  4.71              0.02991 *
## demog_income            3  5.23              0.15555
## alc_agefirst:demog_sex  1  4.59              0.03219 *
## ---
## Signif. codes:  0 '***' 0.001 '**' 0.01 '*' 0.05 '.' 0.1 ' ' 1
```

Conclusion: The association between age of first alcohol use and lifetime marijuana use differs significantly between males and females (p = .032).

6.9.1 Overall test of a predictor involved in an interaction

As shown in Section 5.9.11 for MLR, to carry out an overall test of a predictor involved in an interaction in a regression model, compare the full model to a reduced model with both the main effect and interaction removed. As with MLR, use `anova()` to compare the models, but for a logistic regression you must also specify `test = "Chisq"` to get a p-value. If you did not remove missing data prior to fitting the full model, you must do so before comparing models if the main effect predictor being removed to create the reduced model has missing values. You can either create a new dataset with missing values removed or use the `subset` option.

Example 6.3 (continued): In the model that includes an interaction, is age at first alcohol use significantly associated with the outcome?

```
# Fit the reduced model using the subset option
# to remove missing values for the predictor
# being removed to create the reduced model
fit0 <- glm(mj_lifetime ~ demog_age_cat6 + demog_sex + demog_income,
            family = binomial, data = nsduh,
            subset = complete.cases(alc_agefirst))
# Compare to full model
anova(fit0, fit.ex6.3.int, test = "Chisq")
```

```
## Analysis of Deviance Table
##
## Model 1: mj_lifetime ~ demog_age_cat6 + demog_sex + demog_income
## Model 2: mj_lifetime ~ alc_agefirst + demog_age_cat6 + demog_sex + demog_income
   +
##     alc_agefirst:demog_sex
##   Resid. Df Resid. Dev Df Deviance               Pr(>Chi)
## 1       834       1085
## 2       832        931  2      154 <0.0000000000000002 ***
## ---
## Signif. codes:  0 '***' 0.001 '**' 0.01 '*' 0.05 '.' 0.1 ' ' 1
```

The 2 df test (2 df because the models differ in two terms – the main effect and the interaction) results in a very small p-value.

Conclusion: Age at first alcohol use is significantly associated with the outcome (p <.001).

6.9.2 Estimate the OR at each level of the other predictor

Use `gmodels::estimable()` to compute separate ORs for a predictor involved in an interaction at each level of the other predictor in the interaction. When there is no interaction, or when there is one but the other predictor in the interaction is zero or at its reference level, the OR for a predictor is its exponentiated regression coefficient. When there is an interaction and the other predictor is non-zero or at a non-reference level, add the appropriate interaction term multiplied by the value of the other predictor to the main effect before exponentiating. Do *not* add the intercept or any other terms in the model since when computing the odds *ratio* all other terms drop out of the equation. In particular, do not add the main effect coefficient for the other predictor in the interaction.

Example 6.3 (continued): In the model that includes an interaction, what are the AORs for age at first alcohol use for females and males?

Before using `gmodels::estimable()`, look at the coefficient names in the `glm` object to make sure you spell them correctly.

```
rownames(summary(fit.ex6.3.int)$coef)
```

```
##  [1] "(Intercept)"                   "alc_agefirst"
##  [3] "demog_age_cat626-34"           "demog_age_cat635-49"
##  [5] "demog_age_cat650-64"           "demog_age_cat665+"
##  [7] "demog_sexMale"                 "demog_income$20,000 - $49,999"
##  [9] "demog_income$50,000 - $74,999" "demog_income$75,000 or more"
## [11] "alc_agefirst:demog_sexMale"
```

```
# ORs for alc_agefirst
# Other predictor in the interaction is sex

# Females (reference level of other predictor)
EST.F <- gmodels::estimable(fit.ex6.3.int,
         c("alc_agefirst"                = 1),
         conf.int = 0.95)

# Males (non-reference level of other predictor)
EST.M <- gmodels::estimable(fit.ex6.3.int,
         c("alc_agefirst"                = 1,
           "alc_agefirst:demog_sexMale" = 1),
         conf.int = 0.95)

rbind(EST.F, EST.M)
```

```
##                           Estimate Std. Error X^2 value DF           Pr(>|X^2|)
## (0 1 0 0 0 0 0 0 0 0 0)  -0.3378    0.04229     63.80  1 0.000000000000001332
## (0 1 0 0 0 0 0 0 0 0 1)  -0.2189    0.03626     36.43  1 0.000000001581411446
##                          Lower.CI Upper.CI
## (0 1 0 0 0 0 0 0 0 0 0)  -0.4217  -0.2539
## (0 1 0 0 0 0 0 0 0 0 1)  -0.2908  -0.1469
```

These estimates are on the log-odds-ratio scale. We need to exponentiate to get AORs. But we have to be careful because the `gmodels::estimable()` output includes a p-value, and we do not want to exponentiate that. The code below exponentiates just the estimate and 95% CI.

```
# AORs and 95% CIs (do use exp here)
rbind(exp(EST.M[c("Estimate", "Lower.CI", "Upper.CI")]),
      exp(EST.F[c("Estimate", "Lower.CI", "Upper.CI")]))
```

```
##                          Estimate Lower.CI Upper.CI
## (0 1 0 0 0 0 0 0 0 0 1)   0.8034   0.7476   0.8633
## (0 1 0 0 0 0 0 0 0 0 0)   0.7133   0.6559   0.7758
```

```
# P-values (do NOT use exp here)
rbind(EST.M["Pr(>|X^2|)"],
      EST.F["Pr(>|X^2|)"])
```

```
##                                          Pr(>|X^2|)
## (0 1 0 0 0 0 0 0 0 0 1) 0.0000000001581411446
## (0 1 0 0 0 0 0 0 0 0 0) 0.0000000000000001332
```

Conclusion: After adjusting for age and income, age at first alcohol use is significantly associated with lifetime marijuana use (p <.001 – from the test comparing the model with an interaction to the model with no `alc_agefirst` main effect or interaction). The association between age at first alcohol use and lifetime marijuana use differs significantly between males and females (interaction p = .032). An age of first alcohol use 1 year greater is associated with 19.7% lower odds of lifetime marijuana use for males (AOR = 0.803; 95% CI = 0.748, 0.863; p <.001) and 28.7% lower odds for females (AOR = 0.713; 95% CI = 0.656, 0.776; p <.001).

NOTE: The above example was for a continuous × categorical interaction. For a categorical × categorical or continuous × continuous interaction, imitate the code in Section 5.9.9.2 or 5.9.9.3, respectively, and then exponentiate the resulting estimates and CIs (but do not exponentiate the p-values).

6.9.3 Visualize an interaction

It is always a good idea to visualize the magnitude of an interaction effect. Remember, in a large sample an effect can be statistically significant without being meaningfully large (practically significant) and, in a small sample, an effect can lack statistical significance but still be meaningfully large. Visualizing the effect can help with assessing practical significance.

Example 6.3 (continued): Visualize the interaction between age of first alcohol use and sex. The code below produces Figure 6.4.

We use `plot_model()` as we did with multiple linear regression. Note the addition of `[all]` after `alc_agefirst` which specifies plotting at all the possible values of the variable. Typically, `[all]` is not needed, but leaving it out in this example results in a curve that does not look smooth.

```
library(sjPlot)

# Effect of alc_agefirst at each level of demog_sex
plot_model(fit.ex6.3.int,
           type = "eff",
           terms = c("alc_agefirst [all]", "demog_sex"),
           title = "",
           axis.title = c("Age of First Alcohol Use (years)",
                          "P(Lifetime Marijuana Use)"),
           legend.title = "Sex")
```

Conclusion: We previously concluded that the interaction was statistically significant. In linear regression, this is equivalent to concluding that the fitted lines are significantly different from parallel. The same is true for logistic regression, but only on the log-odds scale.

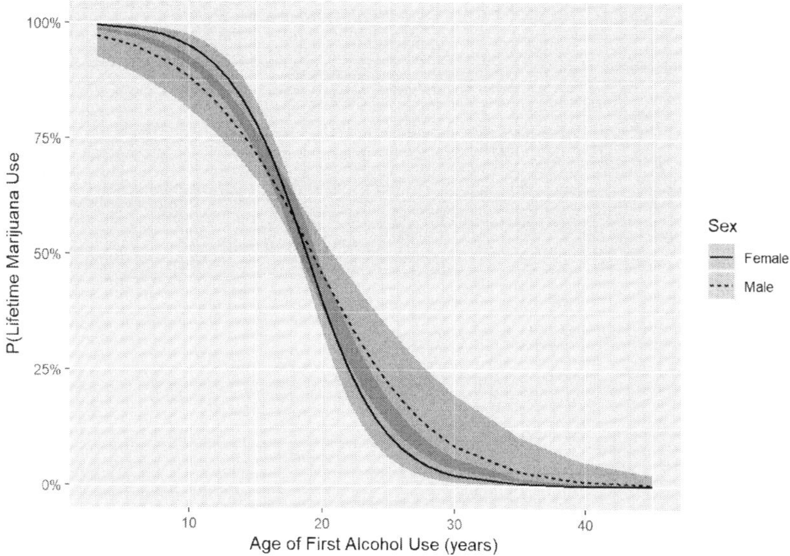

FIGURE 6.4 Logistic regression interaction plot

On the probability scale, as shown in Figure 6.4, we have fitted curves, not lines. However, the shape of the curves do appear to be meaningfully different from each other, indicating that the magnitude of the interaction is meaningfully large.

NOTE: The above example demonstrated the code for a continuous × categorical interaction. If you have a categorical × categorical or continuous × continuous interaction, modify the code accordingly by imitating the code in Section 5.9.9.2 or 5.9.9.3, respectively.

6.10 Separation

6.10.1 Quasi-complete separation

Consider the following two-way table describing how the presence of a condition is related to a risk factor with two levels.

	Condition negative	Condition positive
Risk factor level 1	100	0
Risk factor level 2	50	25

Using the cross-product method (Section 6.5), the OR comparing the odds of "positive" between risk factor levels 2 and 1 is $(25 \times 100)/(0 \times 50) = 2500/0 = \infty$. There are no positive individuals in risk factor level 1, so the observed odds of being positive at that level are 0. When compared to 0, any number of positives in the other risk factor level leads to an infinite OR. In general, if there is a 0 in the table there will be a 0 in the numerator (leading to an OR of zero) or in the denominator (leading to an infinite OR).

If, within one or more levels of a predictor, you can perfectly predict the outcome in the sample, then you cannot compute a confidence interval for the usual estimate of the OR, nor can you test its significance. Prediction may be perfect in the sample, but the usual methods cannot estimate the accuracy of an estimated OR of 0 or ∞. The situation where you can perfectly predict the outcome within some but not all levels of the risk factor is called **quasi-complete separation** and, depending on the software, when using logistic regression will lead to an error, warning, and/or output that is nonsense. Even without an error or warning, you can diagnose the problem by examining the output.

```
mydat <- data.frame(condition  = c(rep(1, 100), rep(1, 50), rep(2, 25)),
                    riskfactor = c(rep(1, 100), rep(2, 50), rep(2, 25))) %>%
  mutate(condition = factor(condition,
                            levels = 1:2,
                            labels = c("Negative", "Positive")),
         riskfactor = factor(riskfactor,
                             levels = 1:2,
                             labels = c("1", "2")))
table(mydat$riskfactor, mydat$condition)
```

```
##
##     Negative Positive
##   1      100        0
##   2       50       25
```

```
fit <- glm(condition ~ riskfactor, family = binomial, data = mydat)

# Very large estimates and SEs
summary(fit)$coef
```

```
##             Estimate Std. Error  z value Pr(>|z|)
## (Intercept)   -20.57       1773 -0.01160   0.9907
## riskfactor2    19.87       1773  0.01121   0.9911
```

```
# OR approaching infinite
exp(coef(fit)["riskfactor2"])
```

```
## riskfactor2
##   427267715
```

The sample OR is actually ∞, but R attempts to estimate it anyway and the results are nonsense – when you see very large standard errors, large positive or negative estimates, or ORs approaching ∞ or 0 then you likely have a problem with separation.

6.10.2 Complete separation

Consider the following two-way table:

	Condition negative	**Condition positive**
Risk factor level 1	100	0
Risk factor level 2	0	25

In the previous example (quasi-complete separation), we could perfectly predict the outcome at some but not all levels of the risk factor. In this example, we can perfectly predict the outcome at *every* level of the risk factor. This may seem like a good thing, but numerically it causes the same problem as quasi-complete separation: logistic regression fails when the sample OR is 0 or ∞. This is called **complete separation** of the data (also known as perfect prediction). When there is complete separation, R gives a lack-of-convergence warning (with quasi-complete separation, sometimes there is no warning).

```
mydat <- data.frame(condition  = c(rep(1, 100), rep(2, 25)),
                    riskfactor = c(rep(1, 100), rep(2, 25))) %>%
  mutate(condition = factor(condition,
                            levels = 1:2,
                            labels = c("Negative", "Positive")),
         riskfactor = factor(riskfactor,
                             levels = 1:2,
                             labels = c("1", "2")))
table(mydat$riskfactor, mydat$condition)
```

```
##
##     Negative Positive
##   1      100        0
##   2        0       25
```

```
# Lack of convergence warning
fit <- glm(condition ~ riskfactor, family = binomial, data = mydat)
```

```
## Warning: glm.fit: algorithm did not converge
```

```
# Very large estimates and SEs
summary(fit)$coef
```

```
##             Estimate Std. Error    z value Pr(>|z|)
## (Intercept)   -26.57      35612 -0.0007460   0.9994
## riskfactor2    53.13      79632  0.0006672   0.9995
```

```
# Undefined OR
exp(coef(fit)["predictor2"])
```

```
## <NA>
##   NA
```

6.10.3 Diagnosing separation

Separation is more likely to occur with an outcome that is rare or one that almost always occurs (sample proportion near 0 or 1). However, it can also occur with a less rare outcome if you have a categorical predictor with at least one level with a small sample size.

Before running a logistic regression, always check for separation. For each categorical predictor, create a two-way table of the predictor vs. the outcome. **Make sure to use a complete-case dataset so that the sample used in each table is the same sample**

that would be used in a regression that includes all of these variables. If there is an interaction between categorical predictors in the model, create a three-way table for the predictors involved in the interaction (predictor1 × predictor2 × outcome). Look for zeros in the cells of the tables.

If you forget to check for separation and instead run the regression first, hopefully you will notice any separation problem that exists by noticing standard errors that are unusually large or ORs that are very large or near zero.

Example 6.4: Heroin use is more rare than marijuana use. Using our NSDUH dataset, check for separation in the regression of lifetime heroin use (`her_lifetime`) on age at first use of alcohol (`alc_agefirst`), age (`demog_age_cat6`), and sex (`demog_sex`).

What happens if we fit the logistic regression model without examining the data first?

```
# Complete-case dataset
nsduh_complete <- nsduh %>%
  select(her_lifetime, alc_agefirst, demog_age_cat6, demog_sex) %>%
  drop_na()

fit.ex6.4 <- glm(her_lifetime ~ alc_agefirst + demog_age_cat6 +
  demog_sex, family = binomial, data = nsduh_complete)
# Regression coefficients
round(summary(fit.ex6.4)$coef, 4)
```

```
##                       Estimate Std. Error z value Pr(>|z|)
## (Intercept)           -16.2988  1024.2091 -0.0159   0.9873
## alc_agefirst           -0.2438     0.0638 -3.8215   0.0001
## demog_age_cat626-34    15.5392  1024.2087  0.0152   0.9879
## demog_age_cat635-49    15.4474  1024.2086  0.0151   0.9880
## demog_age_cat650-64    15.6021  1024.2086  0.0152   0.9878
## demog_age_cat665+      15.3629  1024.2087  0.0150   0.9880
## demog_sexMale           1.2376     0.6526  1.8965   0.0579
```

```
# AORs and CIs
OR.CI <- cbind("AOR" = exp(coef(fit.ex6.4)),
               exp(confint(fit.ex6.4)))[-1,]
round(OR.CI, 3)
```

```
##                             AOR 2.5 % 97.5 %
## alc_agefirst              0.784 0.689  0.886
## demog_age_cat626-34 5605018.279 0.000     NA
## demog_age_cat635-49 5113370.493 0.000     NA
## demog_age_cat650-64 5969190.982 0.000     NA
## demog_age_cat665+   4699268.444 0.000     NA
## demog_sexMale             3.447 1.079 15.290
```

```
## Warning: glm.fit: fitted probabilities numerically 0 or 1 occurred
```

The warning indicates there is an issue with separation, as do the extremely large estimates and standard errors and the ORs approaching ∞. There may be a separation problem with `demog_age_cat`. Examine each categorical predictor vs. the outcome to look for separation.

```
table(nsduh_complete$demog_age_cat6, nsduh_complete$her_lifetime)
```

```
##
##          No Yes
##   18-25 102   0
##   26-34 121   3
##   35-49 225   5
##   50-64 198   5
##   65+   180   4
```

```
table(nsduh_complete$demog_sex, nsduh_complete$her_lifetime)
```

```
##
##            No Yes
##   Female 431   3
##   Male   395  14
```

There is a zero in one of the cells for age; in this dataset, no one in the 18- to 25-year-old group reported having ever used heroin. This is causing quasi-complete separation – we can perfectly predict the outcome among 18- to 25-year-olds.

6.10.4 Resolving separation

For each categorical predictor that has a zero in any cell of its two-way table comparing it to the outcome, do one of the following:

- **Filter:** Remove predictor levels which have zero observations at either level of the outcome (this only works if the predictor has more than two levels).
- **Collapse:** Collapse levels together until all cells have some observations (this only works if the predictor has more than two levels).
- **Remove:** Remove the predictor from the model.

Collapsing is often the best solution when it is possible, but sometimes removing the predictor is the only option available among these three. Although zeros in cells are definitely a problem, cells with only a few observations can be problematic, as well (this would be **approximate separation**). Such cases may not cause as serious a problem, but they may result in unstable estimates. If in doubt, compare the regression coefficients and standard errors from the original fit with a fit that deals with such cases in one of the ways mentioned above to see if approximate separation is impacting your conclusions.

NOTE: Separation is one example of the more general **sparse data bias** problem which can lead to inflated odds ratios, and the problem becomes more severe after adjusting for confounding (Greenland et al., 2016a; Mansournia et al., 2018). **Penalized logistic regression** (Heinze and Schemper, 2002; Greenland et al., 2016a), not covered in this text, addresses this more general problem and can be implemented with the `logistf` package (Heinze et al., 2023b), for the Firth adjustment, or with the more general `plogit()` function (Greenland et al., 2016a, web appendix).

After resolving separation and re-fitting the model, look at the regression coefficients, standard errors, and ORs. If any still seem unusually large, you may have missed a problem with separation. In that case, go back and re-examine the data to see if there is a predictor that needs to be dealt with. If your outcome is very rare, then you may simply not have enough data to obtain reliable estimates for *any* predictors.

Example 6.4 (continued): Resolve the separation issue we just found in the regression of lifetime heroin use (`her_lifetime`) on age at first use of alcohol, age, and sex.

We could try any of the following:

- **Filter** the data to remove individuals age 18-25 years.
- **Collapse** individuals age 18-25 and 26-34 years into one group.
- **Remove** `demog_age_cat6` from the model.

Which choice you make depends on your goals. What age range do you want to be able to draw conclusions about? How important is age as a potential confounder? All three options are described below, along with their relative merits.

6.10.4.1 Filter

Filtering the data to remove predictor levels for which cells have zero observations may not be a good option if this results in a large drop in sample size or removes a subpopulation you are interested in. If you filter out the level that has a cell with zero observations, you remove individuals in another cell. For example, if there are no "Yes" outcomes at $X = x$ and you filter out everyone with $X = x$, then you end up removing individuals with $X = x$ and a "No" outcome.

The following points address additional issues with filtering.

- If filtering would result in just one level for a predictor then, instead, remove it from the model (see Section 6.10.4.3).
- Since filtering reduces the sample size, it may lead to a problem with separation for a different predictor that previously did not have a problem. If you filter, make sure to re-check the other predictors for separation.
- Filtering changes the scope of the analysis. *All* individuals at the filtered out levels are excluded, both those with a "Yes" outcome and those with a "No" outcome, even if only one level of the outcome has zero observations. Therefore, results do not generalize to the population at the filtered out levels.

Example 6.4 (continued): Resolve separation by filtering the data to remove individuals age 18-25 years.

```
# Filter
nsduh_complete_filtered <- nsduh_complete %>%
  filter(demog_age_cat6 != "18-25")

# Re-check separation
table(nsduh_complete_filtered$demog_age_cat6,
      nsduh_complete_filtered$her_lifetime)
```

```
##
##          No Yes
##   18-25   0   0
##   26-34 121   3
##   35-49 225   5
##   50-64 198   5
##   65+   180   4
```

Although there are zeros in this table, they all occur in the 18- to 25-year-old group which was filtered out. When a row in the table has *all* zeros (i.e., there are no individuals in that level in the dataset) that level will not cause a problem in the model fit.

Filtering reduced the sample size. This could have led to a separation problem for demog_sex, but fortunately it did not. In hindsight, this is obvious for this example since we filtered out a group of individuals with no lifetime heroin use so the cells in the sex vs. heroin use table with small numbers of cases were unaffected.

```
table(nsduh_complete_filtered$demog_sex,
      nsduh_complete_filtered$her_lifetime)
```

```
##
##          No Yes
##   Female 372   3
##   Male   352  14
```

Finally, re-fit the model using the filtered dataset.

```
fit.ex6.4.filter <- glm(her_lifetime ~ alc_agefirst + demog_age_cat6 + demog_sex,
  family = binomial, data = nsduh_complete_filtered)

round(summary(fit.ex6.4.filter)$coef, 4)
```

```
##                      Estimate Std. Error z value Pr(>|z|)
## (Intercept)           -0.7597     1.2292 -0.6180   0.5366
## alc_agefirst          -0.2438     0.0638 -3.8218   0.0001
## demog_age_cat635-49   -0.0918     0.7571 -0.1213   0.9035
## demog_age_cat650-64    0.0629     0.7597  0.0829   0.9340
## demog_age_cat665+     -0.1763     0.8232 -0.2141   0.8305
## demog_sexMale          1.2376     0.6524  1.8969   0.0578
```

```
OR.CI <- cbind("AOR" = exp(coef(fit.ex6.4.filter)),
                        exp(confint(fit.ex6.4.filter)))[-1,]
round(OR.CI, 3)
```

```
##                       AOR 2.5 % 97.5 %
## alc_agefirst        0.784 0.689  0.886
## demog_age_cat635-49 0.912 0.213  4.651
## demog_age_cat650-64 1.065 0.248  5.461
## demog_age_cat665+   0.838 0.161  4.639
## demog_sexMale       3.447 1.079 15.290
```

Conclusion: All the estimates, standard errors, AORs, and 95% CIs look reasonable. Filtering took care of the separation problem, but at the cost of reducing the sample size and limiting the inference to individuals age 26 years and older.

6.10.4.2 Collapse

Another option is to collapse levels together until all cells have some observations. This only works if the predictor has more than two levels. The downside to collapsing levels of a predictor together is that we must assume the odds of the outcome are the same among those in each of the collapsed levels.

Example 6.4 (continued): Resolve separation by collapsing individuals age 18-25 and 26-34 years into one group.

```
# Collapse
nsduh_collapsed <- nsduh_complete %>%
  mutate(demog_age_cat_new = fct_collapse(demog_age_cat6,
                                           "18-34" = c("18-25", "26-34")))

# Re-check separation (use the new predictor name)
table(nsduh_collapsed$demog_age_cat_new,
      nsduh_collapsed$her_lifetime)
```

```
##
##          No Yes
##   18-34 223   3
##   35-49 225   5
##   50-64 198   5
##   65+   180   4
```

There are now no zeros in the age table.

Finally, re-fit the model using the collapsed dataset.

```
fit.ex6.4.collapse <- glm(her_lifetime ~ alc_agefirst +
  demog_age_cat_new + demog_sex,
  family = binomial, data = nsduh_collapsed)

round(summary(fit.ex6.4.collapse)$coef, 4)
```

```
##                          Estimate Std. Error z value Pr(>|z|)
## (Intercept)               -1.1906     1.2444 -0.9567   0.3387
## alc_agefirst              -0.2520     0.0650 -3.8779   0.0001
## demog_age_cat_new35-49     0.4231     0.7484  0.5653   0.5718
## demog_age_cat_new50-64     0.5822     0.7498  0.7764   0.4375
## demog_age_cat_new65+       0.3161     0.8259  0.3828   0.7019
## demog_sexMale              1.2761     0.6519  1.9576   0.0503
```

```
OR.CI <- cbind("AOR" = exp(coef(fit.ex6.4.collapse)),
                       exp(confint(fit.ex6.4.collapse)))[-1,]
round(OR.CI, 3)
```

```
##                          AOR 2.5 % 97.5 %
## alc_agefirst           0.777 0.682  0.881
## demog_age_cat_new35-49 1.527 0.362  7.659
## demog_age_cat_new50-64 1.790 0.423  9.010
## demog_age_cat_new65+   1.372 0.260  7.591
## demog_sexMale          3.583 1.124 15.876
```

Conclusion: Collapsing took care of the separation problem, with the advantages of no loss in sample size and retaining the ability to make inferences about all adults, but at the cost of assuming the odds of heroin use are the same among those age 18-25 and 26-34 years.

6.10.4.3 Remove

Yet another option for resolving separation is to remove the predictor from the model. The downsides to removing a predictor are that the AORs for the remaining predictors are no longer adjusted for the removed predictor, and there is no estimate of effect for the removed predictor. However, of the three discussed here, removing the predictor is the only viable option if filtering and collapsing result in just one level for the predictor.

Example 6.4 (continued): Resolve separation by removing `demog_age_cat6` from the model.

```
# Re-fit the model without age
fit.ex6.4.remove <- glm(her_lifetime ~ alc_agefirst + demog_sex,
  family = binomial, data = nsduh_complete)

round(summary(fit.ex6.4.remove)$coef, 4)
```

```
##                 Estimate Std. Error z value Pr(>|z|)
## (Intercept)      -0.9017     1.0656 -0.8462   0.3975
## alc_agefirst     -0.2491     0.0614 -4.0559   0.0000
## demog_sexMale     1.3054     0.6504  2.0070   0.0447
```

```
OR.CI <- cbind("AOR" = exp(coef(fit.ex6.4.remove)),
               exp(confint(fit.ex6.4.remove)))[-1,]
round(OR.CI, 3)
```

```
##                 AOR 2.5 % 97.5 %
## alc_agefirst  0.780 0.690   0.88
## demog_sexMale 3.689 1.161  16.31
```

Conclusion: Removing `demog_age_cat6` from the model solved the separation problem, but at the cost of the other effects no longer being adjusted for age and no estimate of the age effect.

6.10.4.4 Summary

When the outcome can be perfectly predicted at one or more levels of a categorical predictor, logistic regression runs into numerical difficulties and returns an error, a warning, and/or nonsense values. Always examine two-way tables (or three-way if there are interactions) of categorical predictors vs. the outcome and look for zeros. If there are zeros and the predictor has more than two levels, try **collapsing** groups so there are no zeros. Alternatively, **filter** the data to remove zeros or **remove** the predictor from the model, although these options come with stronger side effects that you may not want. If the predictor has only two levels then, of these three, the only option that works is to remove it from the model. Another alternative, beyond the scope of this text, is to use penalized logistic regression via the `logistf` package.

6.11 Collinearity

Collinearity (see Section 5.20) is a characteristic of the predictors, not the outcome, and so can be a problem not only in linear regression, but also in all other types of regression models. For logistic regression, diagnose collinearity in the same way as for MLR, using VIFs or aGSIFs.

Example 6.3 (continued): Evaluate collinearity in the adjusted model without an interaction. The aGSIFs are all small, so we conclude there is no problem with collinearity.

```
car::vif(fit.ex6.3.adj)
```

```
##                 GVIF Df GVIF^(1/(2*Df))
## alc_agefirst    1.044  1           1.022
## demog_age_cat6 1.094  4           1.011
## demog_sex       1.038  1           1.019
## demog_income    1.086  3           1.014
```

6.12 Assumptions

As with MLR, logistic regression assumes the observations are **independent** and this is checked in the same way as described in Section 5.14.

Unlike MLR, the logistic regression model does not include an error term (ϵ). While it is possible to compute various kinds of residuals in a logistic regression, there is no assumption that they be normally distributed or have constant variance. Like MLR, however, logistic regression assumes that continuous predictors have a **linear** relationship with the outcome (in this case, with the log-odds of the probability of the outcome). Use a CR plot to diagnose non-linearity (see Section 5.16.2).

Example 6.3 (continued): Assess the linearity assumption in the adjusted model without an interaction.

```
# Enter just the continuous predictors in terms,
# separated by + signs. In this case, there is only one.
car::crPlots(fit.ex6.3.adj, terms = ~alc_agefirst,
             pch=20, col="gray",
             smooth = list(smoother=car::gamLine))
```

```
# # If you want to try more smoothing change k
# car::crPlots(fit.ex6.3.adj, terms = ~alc_agefirst,
#               pch=20, col="gray",
#               smooth = list(smoother=car::gamLine, k = 5))
```

Figure 6.5 displays the CR plot for `alc_agefirst`. In a logistic regression, since the outcome can only take on two values, there will typically be a strong pattern in the points, but for diagnosis of non-linearity just focus on the dashed and solid lines. The dashed line illustrates the relationship between the continuous predictor and the outcome assuming linearity, while the solid line relaxes the linearity assumption. If these two are very different, then consider modeling the predictor using a non-linear function (e.g., adding a quadratic term, using a log or other non-linear transformation). In a logistic regression, make sure to only transform predictors, not the outcome.

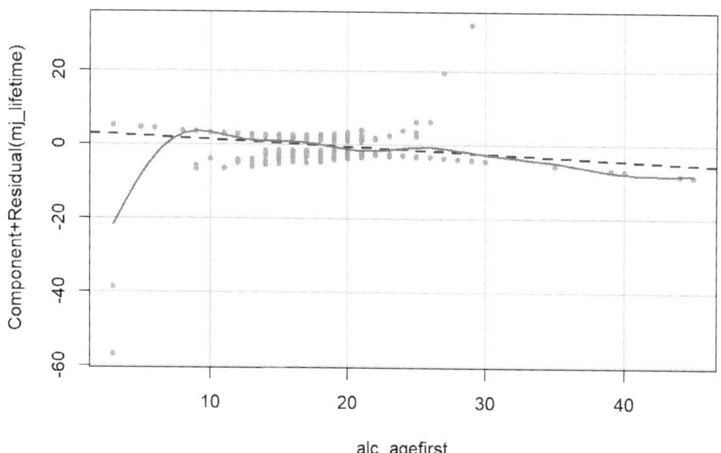

FIGURE 6.5 Component + residual plot for a continuous predictor in a logistic regression

Conclusion: There is no strong non-linearity. The dip in the solid line at lower ages of first alcohol use is in an area with few points so can be ignored – smoothers are highly influenced by local fluctuations.

6.13 Outliers

Outliers are observations with predictions that are very far from their observed values. In linear regression, outliers signaled a possible problem with the normality and/or constant variance assumptions, particularly in small samples. Logistic regression does not have these assumptions; however, it is still useful to examine outliers to find observations that are not predicted well by the model.

In a logistic regression, the observed values are always each 0 or 1 (even if the response is coded as a factor, numerically `glm()` uses 0s and 1s) and the predicted probabilities are always between 0 and 1, so the differences between these are always between −1 and 1. The residuals in a logistic regression, called **deviance residuals**, are more complex, however, than simple differences between observed outcomes and predicted probabilities. For a `glm` object, the usual residual functions `resid()`, `rstandard()`, and `rstudent()` compute deviance residuals and their standardized and Studentized counterparts, respectively.

Examine outliers by highlighting points on a residual plot and examining their predictor and outcome values. Again, here we are simply examining observations that are fit poorly by the model. After identifying them, we do not remove them from the model, although you could do a sensitivity analysis to evaluate their impact on your conclusions.

Example 6.3 (continued): Look for outliers in the model that does not include an interaction. The cutoff you choose here is arbitrary. In this example, a cutoff of 2.5 highlighted a few observations and you can see from Figure 6.6 that the two with the most negative residuals especially seem to stand out as being highly unusual.

```
RSTUDENT <- rstudent(fit.ex6.3.adj)
SUB <- abs(RSTUDENT) > 2.5
sum(SUB, na.rm = T)
```

```
## [1] 4
```

```
car::residualPlots(fit.ex6.3.adj, terms = ~ 1,
                   tests = F, quadratic = F, fitted = T,
                   type = "rstudent", pch = 20, col = "gray")
# NOTE: For points() the arguments are x, y not y ~ x
points(logit(fitted(fit.ex6.3.adj))[SUB],
       RSTUDENT[SUB],
       pch=20, cex=1.5)
abline(h = c(-2.5, 2.5), lty = 2)
```

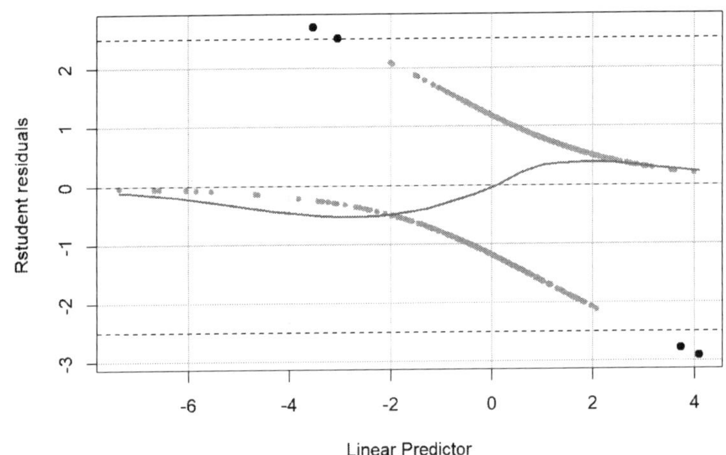

FIGURE 6.6 Identifying outliers in a logistic regression

Outliers in a logistic regression are observations with either (a) $Y = 1$ but predictors at levels at which the predicted probability is very low (resulting in a large positive residual) or (b) $Y = 0$ but predictors at levels at which the predicted probability is very high (resulting in a large negative residual). By way of reminder, here are the AORs for the terms in this model.

```
OR.CI <- cbind("AOR" = exp(coef(fit.ex6.3.adj)),
                       exp(confint(fit.ex6.3.adj)))[-1,]
round(OR.CI, 3)
```

```
##                                     AOR 2.5 % 97.5 %
## alc_agefirst                      0.759 0.718  0.800
## demog_age_cat626-34               0.744 0.387  1.409
## demog_age_cat635-49               0.447 0.247  0.791
## demog_age_cat650-64               0.502 0.275  0.891
## demog_age_cat665+                 0.279 0.152  0.502
## demog_sexMale                     0.941 0.684  1.291
## demog_income$20,000 - $49,999 0.588 0.347  0.987
## demog_income$50,000 - $74,999 0.924 0.507  1.680
## demog_income$75,000 or more       0.697 0.421  1.139
```

If we examine the observation with the largest negative residual more closely, we see that it falls into the latter category. This individual did not use marijuana ($Y = 0$) but is in the income group with the highest odds of lifetime marijuana use and has an extremely young age of first alcohol use, also corresponding to greater odds of the outcome, and these discrepancies result in a large negative residual. If one had access to the raw data, it would be worth checking to see if the value of 3 for `alc_agefirst` is a data entry error.

```
# Which row has the largest negative residual?
RSTUDENT[SUB]
```

```
##  10412   3043  48455  40288
##  2.510  2.695 -2.909 -2.776
```

```
# Examine that row
nsduh[c("48455"),
      c("mj_lifetime",
        "alc_agefirst",
        "demog_age_cat6",
        "demog_sex",
        "demog_income")]
```

```
##       mj_lifetime alc_agefirst demog_age_cat6 demog_sex      demog_income
## 48455          No            3            65+      Male Less than $20,000
```

6.14 Influential observations

Checking for influential observations in logistic regression is the same as for MLR (Section 5.22). Fit the model, plot the Cook's distances and DFBetas, and, if there are observations with extreme values, conduct a sensitivity analysis to see if their removal impacts your conclusions (Section 5.25).

Example 6.3 (continued): Look for influential observations in the model that includes an interaction (Figures 6.7 and 6.8).

```
car::influenceIndexPlot(fit.ex6.3.int, vars = "Cook",
                        id=F, main = "Cook's distance")
```

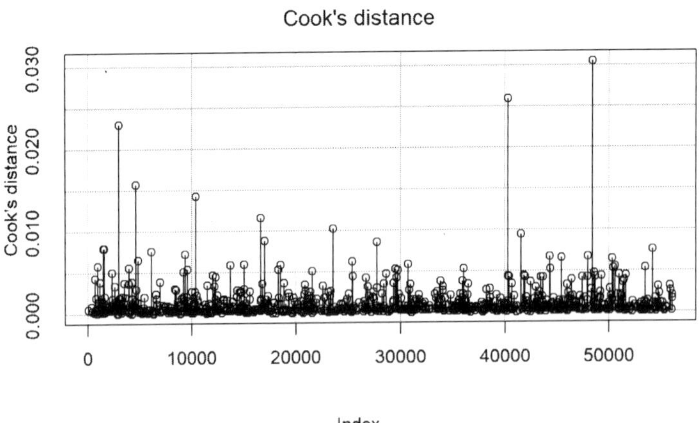

FIGURE 6.7 Cook's distance plot for a logistic regression

```
# Compute DFBETAS
DFBETAS <- dfbetas(fit.ex6.3.int)

# To see the spelling of the terms
# colnames(DFBETAS)
# Index plot for each predictor
# (results only shown for one plot)

plot(DFBETAS[, "alc_agefirst"], ylab="AlcAge")
abline(h = c(-0.2, 0.2), lty = 2)
```

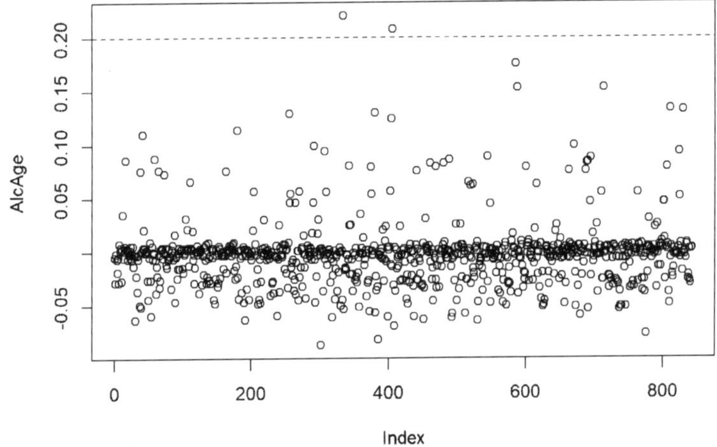

FIGURE 6.8 Plot of DFBETAS for a logistic regression

Conclusion: There appear to be a few potentially influential observations based on Cook's Distance, and a few observations that are influential for the main effect of age of first alcohol use.

6.15 Generalization / overfitting

As discussed in Section 5.26, it is important to limit the number of predictors in a model to avoid overfitting and ensure generalizability. In a logistic regression with n observations, the rule of thumb is to have no more than $n_{min}/15$ predictors, where n_{min} is the number of observations in the less common outcome level (Babyak, 2004; Harrell, 2015, pp.72-73). This can also be expressed in terms of the sample proportion – have no more than $n \times p_{min}/15$ predictors, where p_{min} is the proportion of observations with the less common outcome level. Seen the other way around, if you are designing a study and plan to include K predictors, you need at least $15 \times K/p_{min}$ observations, where p_{min} is your best guess for the smaller of the two population prevalences. As with any rule of thumb, this is meant as guidance – there is no requirement that it be strictly applied.

For example, suppose your outcome is "occurrence of disease within one-year post-exposure". The outcome levels are "disease" and "no disease". If you have a sample of size $n = 300$ and 23% of the sample developed the disease (so 77% did not and the less prevalent outcome level is "disease"), you should include no more than $n \times p_{min}/15 = 300 \times 0.23/15 = 4.6$ predictors (you can round up to 5) in a logistic regression model.

Suppose instead you were designing a study in which you expect 65% of the individuals to experience the outcome and you would like to include five predictors. In this case, the prevalences are 65% and 35%, so the lower prevalence is that of not experiencing the outcome. To ensure generalizability, you need at least $15 \times K/p_{min} = 15 \times 5/0.35 = 214.29$ observations in your sample (round up to 215).

NOTE: A sample size sufficient to ensure generalizability may or may not be sufficient for the purpose of having enough power to test a hypothesis.

6.16 Goodness-of-fit

Since a binary outcome can only take on two possible values, numerically represented as 1 for the condition of interest and 0 otherwise, how close individual predicted values are to observed values is not a great indication of goodness-of-fit. The logistic regression curve does not actually go through most of the 1s and 0s (see Figure 6.2). However, within a *group* of individuals, we can expect a good-fitting model to produce predicted probabilities that are similar to the observed proportion of $Y = 1$ values within that group.

6.16.1 Hosmer-Lemeshow test

The **Hosmer-Lemeshow test** (Hosmer and Lemeshow, 2000) (HL test) quantifies this expectation by splitting the data into ten groups based on their predicted probabilities and then, within groups, comparing the observed and expected proportions. The first group is the 10% of observations with the lowest predicted probabilities, the second is the 10% with the next lowest, and so on. If the model fits well, we expect the observed proportions of $Y = 1$ in these groups to be similar to their within-group average predicted probabilities.

The HL test compares the observed to expected within each group and sums over groups to obtain a statistic that has a chi-square distribution. If the overall deviation is large, then the statistic will be large and the p-value will be small, resulting in rejection of the null hypothesis that the data arise from the specified model. In other words, a large p-value indicates a good fit and a small p-value indicates a poor fit.

You can carry out the HL test in R using `ResourceSelection::hoslem.test()` (Lele et al., 2023). The function requires you to input the outcome values (as 0s and 1s) and fitted probabilities. Extract the outcome and fitted values from a model fit using `fit$y` and `fit$fitted.values`. Using `fit$y` is safer than extracting the outcome from the dataset (e.g., `nsduh$mj_lifetime`) – if there are missing values, the outcome extracted from the dataset will have a different length than the fitted values.

Example 6.3 (continued): Assess goodness-of-fit for the model with an interaction.

```
HL <- ResourceSelection::hoslem.test(fit.ex6.3.int$y,
                                     fit.ex6.3.int$fitted.values)

# View the test
HL
```

```
##
## Hosmer and Lemeshow goodness of fit (GOF) test
##
## data:  fit.ex6.3.int$y, fit.ex6.3.int$fitted.values
## X-squared = 2.9, df = 8, p-value = 0.9
```

```
# Observed number of individuals with Y = 0 (y0) and Y = 1 (y1)
HL$observed
```

```
##
## cutyhat              y0 y1
##   [0.000118,0.24] 76 11
##   (0.24,0.379]    56 28
##   (0.379,0.495]   50 34
##   (0.495,0.563]   37 45
##   (0.563,0.635]   40 53
##   (0.635,0.695]   24 53
##   (0.695,0.74]    25 59
##   (0.74,0.795]    15 68
##   (0.795,0.855]   15 70
##   (0.855,0.99]     8 76
```

```
# Expected number of individuals with Y = 0 (yhat0) and Y = 1 (yhat1)
HL$expected
```

```
##
## cutyhat             yhat0  yhat1
##    [0.000118,0.24] 76.771 10.229
##    (0.24,0.379]    57.228 26.772
##    (0.379,0.495]   46.307 37.693
##    (0.495,0.563]   38.208 43.792
##    (0.563,0.635]   36.395 56.605
##    (0.635,0.695]   25.522 51.478
##    (0.695,0.74]    23.549 60.451
##    (0.74,0.795]    19.160 63.840
##    (0.795,0.855]   14.828 70.172
##    (0.855,0.99]     8.032 75.968
```

In the output above, we see that the test split the observations into ten groups. Although the sample sizes in the groups are not shown, they are approximately equal (to verify this, sum each row in the observed table). Within each group, the HL test compares the observed frequencies to the predicted, combines this information into a chi-square test statistic, and computes a p-value as the probability of observing a chi-square value at least as large under the null hypothesis of a perfect fit.

Conclusion: In this example, $p > .05$ so we do not reject the null hypothesis of a perfect fit, and we conclude the fit is adequate.

NOTE: The HL test with the default of ten bins may fail (resulting in a p-value of `NA`) in a model with few predictors, as there will not be enough unique fitted values to create ten bins. You can reduce the number of bins using the g option.

```
ResourceSelection::hoslem.test(fit.ex6.3.int$y,
                               fit.ex6.3.int$fitted.values,
                               g = 9)
```

```
##
##  Hosmer and Lemeshow goodness of fit (GOF) test
##
## data:  fit.ex6.3.int$y, fit.ex6.3.int$fitted.values
## X-squared = 3.1, df = 7, p-value = 0.9
```

6.16.2 Calibration plot

P-values depend heavily on the sample size. Therefore, we must be careful since the HL test may lack power to detect a poor fit in a small sample (and we would therefore incorrectly conclude the fit is good) and may detect even minor deviations in a large sample (and we would therefore incorrectly conclude the fit is poor). It is helpful, along with looking at the HL test p-value, to create a **calibration plot** (Harrell, 2015; Steyerberg, 2009) to visualize the relationship between the observed and expected values. The `calibration.plot()` function (in `Functions_rmph.R` which you loaded at the beginning of this chapter) creates a calibration plot corresponding to the HL test, and displays the HL test p-value on the plot.

Example 6.3 (continued): Use a calibration plot to visualize goodness-of-fit for the model with an interaction.

```
calibration.plot(fit.ex6.3.int)
```

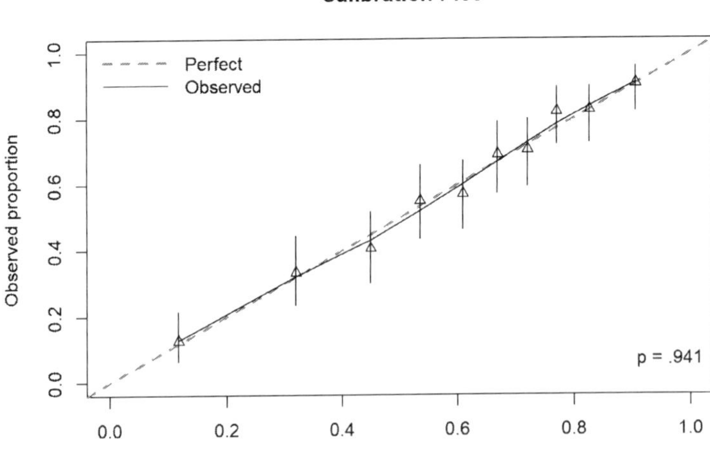

FIGURE 6.9 Calibration plot for a logistic regression, including Hosmer-Lemeshow test p-value

In Figure 6.9, the first triangle is plotted at (x, y) where x = the average predicted probability among those in the first bin and y = the observed proportion of outcomes in the first bin (11 / (11 + 76) = 0.126). The first triangle is very close to the 45° line indicating the model is predicting probabilities in the first bin that are, on average, close to the observed proportion of outcomes. The remaining triangles are plotted similarly, one for each bin. Perfect calibration would be indicated by all the triangles falling exactly on the 45° line.

The vertical lines going through each triangle represent 95% CIs for the population proportions. The solid "Observed" curve is a smoother and since, in this example, it pretty much tracks along the 45° line, the calibration of this model is very good, on average.

Plotting the same plot along with the bin boundaries and the observed values makes it easier to understand where this plot is coming from (Figure 6.10).

```
# To see the bins and points
calibration.plot(fit.ex6.3.int,
                 show.bins = T,
                 show.points = T)
```

The gray points (individual observations) are plotted left to right in order from lowest to highest predicted probability, those with $Y = 1$ at the top and those with $Y = 0$ at the bottom (jittered to make it easier to see how many there are). The vertical dashed lines are the bin boundaries and, by design, each bin has about the same number of observations.

Conclusion: The calibration plot confirms the conclusion of the HL test, that the model fits this data well.

Rare outcome

If you are modeling a rare outcome, some adjustments can be made to the calibration plot to zoom in on a smaller range of probabilities.

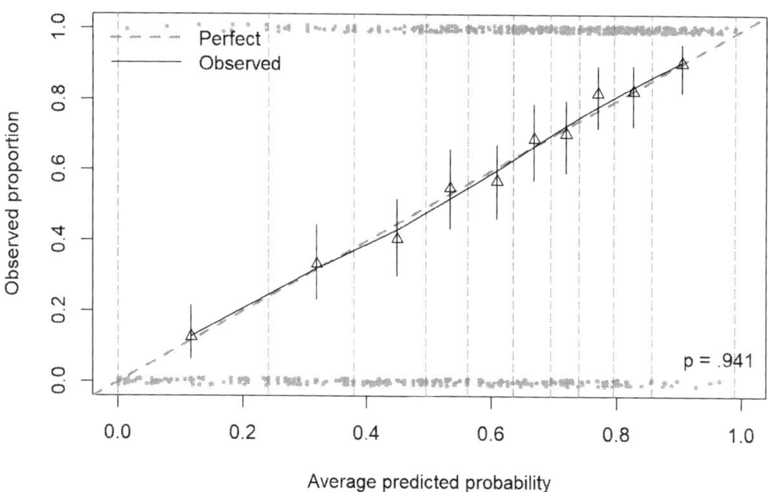

FIGURE 6.10 Calibration plot including bins and points

Example 6.4 (continued): Visualize goodness-of-fit for the model for lifetime heroin use that we obtained after resolving separation by collapsing age categories.

```
calibration.plot(fit.ex6.4.collapse,
                 show.bins = T,
                 show.points = T)
```

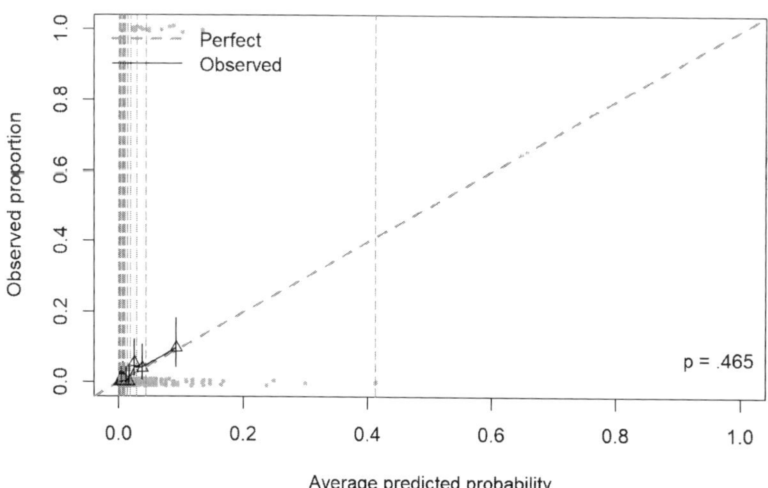

FIGURE 6.11 Calibration plot for a logistic regression with a rare outcome, before zooming

The outcome lifetime heroin use is more rare than lifetime marijuana use. Thus, there are very few large predicted probabilities. The result is a plot that is hard to read due to having one large bin for the largest probabilities and all the other bins being squished together (Figure 6.11). The `calibration.plot()` function has options that can solve this problem by zooming in on the points: `zoom`, `zoom.x`, or `zoom.y`, each of which can be `FALSE`, the default, `TRUE` to automatically zoom, or a vector of limits to customize the amount of zooming.

Figure 6.12 displays the calibration plot for the prediction of lifetime heroin use after zooming in on the x and y axes (so much so that the bin furthest to the right is cut out altogether). The model does not appear to be well calibrated since many triangles seem "far" from the diagonal. However, be careful when zooming in, as even small differences between observed proportions and predicted probabilities can appear large. This does illustrate, however, the difficulty in estimating rare probabilities – a very large sample size is needed to obtain reliable estimates of very small probabilities.

Whether this model fits well or not comes down to how important it is to estimate the smallest probabilities well. If knowing that a probability is "small" is good enough, then this model may be considered to fit well. The first seven bins all have CIs that are bounded above by 0.06. So, while some bins have triangles "far" from the diagonal, if any distinction beyond "this probability is small" is not important, then the model is doing its job well. If, on the other hand, you really want to make distinctions between small probabilities, then perhaps it does not fit well.

```
calibration.plot(fit.ex6.4.collapse,
                 zoom.x = c(0, 0.12),
                 zoom.y = c(0, 0.10))
```

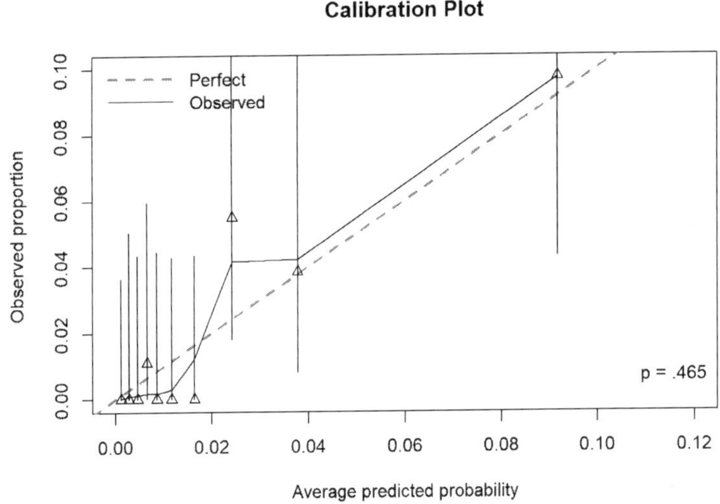

FIGURE 6.12 Calibration plot for a logistic regression with a rare outcome, after zooming

NOTES:

- The calibration plot with the default of ten bins may return an error in a model with few predictors, as there will not be enough unique fitted values to create ten bins. You can reduce the number of bins using the g option (this will also then use this value of g for the HL test p-value that is printed on the plot).

- There are some cases where `g=10` works for `ResourceSelection::hoslem.test()` but fails for `calibration.plot()`. Despite not being able to create enough bins, the former ends up with one bin that is much larger than the others. If you need to use a smaller `g` for `calibration.plot()`, then also use that same `g` for `ResourceSelection::hoslem.test()`.

```
calibration.plot(fit.ex6.3.int, g=9)
```

6.16.3 Relationship between the HL test p-value and the calibration plot

The null hypothesis for the HL test is that the triangles fall *exactly* on the diagonal in the population. If the triangles in the sample fall "close enough" to the diagonal, then the model is well calibrated. However, when using a statistical hypothesis test, what counts as "close enough" changes with the sample size.

In general, the p-value of the HL test is smaller (indicating poorer fit) when the CIs do not overlap the diagonal. Triangles far from the diagonal, an indication of a poor fit, are more likely to have CIs that do not overlap the diagonal, so this seems reasonable. However, small samples result in less precise estimates and wider CIs. Therefore, in a small sample, the triangles could be far from the diagonal but have CIs that are so wide they cross it, leading to the incorrect conclusion of a good fit. In general, goodness-of-fit tests lack power to detect poor fit in small samples.

Conversely, the p-value of the HL test is larger (indicating better fit) when the CIs do cross the diagonal. Triangles close to the diagonal, an indication of a good fit, are more likely to have CIs that overlap the diagonal, so this seems reasonable. However, large samples result in more precise estimates and narrower CIs. Therefore, in a large sample, the triangles could be close to the diagonal but have CIs that are so narrow they fail to cross it, leading to the incorrect conclusion of a poor fit. In general, goodness-of-fit tests are overly sensitive to minor deviations from perfect fit in large samples.

There are two other possibilities: triangles far from the diagonal with CIs that do not cross it (poor fit) or triangles close to the diagonal with CIs that do cross it (good fit). In each of these scenarios, the HL test and the calibration plot are more likely to be consistent. However, in general, when assessing goodness-of-fit, it is wise to also view a visualization rather than relying solely on a statistical test.

6.17 Writing it up

This section demonstrates how to write up the methods and results sections corresponding to a logistic regression analysis.

Example 6.3 (continued): Using the 2019 National Survey of Drug Use and Health (NSDUH) teaching dataset (Section A.5), what is the association between lifetime marijuana use (`mj_lifetime`) and age at first use of alcohol (`alc_agefirst`), adjusted for age (`demog_age_cat6`), sex (`demog_sex`), and income (`demog_income`)? Write up the Methods and Results for the analyses that answered this research question using models without (`fit.ex6.3.adj`) and with an interaction (`fit.ex6.3.int`).

6.17.1 Writing up logistic regression results (no interaction)

Below are a few lines of code to extract the information we need for our write-up.

Outcome, predictors, and dataset name: The `call` component of the `glm` object provides the call to `glm`, including the variable and dataset names.

```
fit.ex6.3.adj$call
```

```
## glm(formula = mj_lifetime ~ alc_agefirst + demog_age_cat6 + demog_sex +
##     demog_income, family = binomial, data = nsduh)
```

Number of observations: If we used a complete case analysis, then the number of rows in the dataset (`nrow(nsduh)`) is the number of cases used in the analysis. But counting the number of residuals to get the sample size will work even when you have missing data.

```
length(fit.ex6.3.adj$residuals)
```

```
## [1] 843
```

Selected descriptive statistics: Extract any descriptive summaries you might want to add to your write-up. For example, the proportion of individuals in each age group. In this example, we did not remove missing data first before fitting the model, so we need to do that here before creating our summary. We want our summary to be based on the same sample as our regression model, and a regression will always carry out a complete case analysis whether you removed the cases with missing values first or not.

```
tmp <- nsduh %>%
  drop_na(mj_lifetime, alc_agefirst, demog_age_cat6,
          demog_sex, demog_income)
round(100*prop.table(table(tmp$demog_age_cat6)), 1)
```

```
##
## 18-25 26-34 35-49 50-64   65+
##  12.1  14.7  27.3  24.1  21.8
```

Regression coefficients, p-values, AORs, and 95% confidence intervals: The p-values in the regression coefficient table test the association with the outcome for continuous predictors and categorical predictors with exactly two levels. For categorical predictors, each p-value tests the difference in log-odds of the outcome between the level shown in that row and the reference level.

```
# Regression coefficients, p-values, AORs, 95% CIs for AORs
car::S(fit.ex6.3.adj)
# 95% CIs for the regression coefficients
confint(fit.ex6.3.adj)
# (results not shown)
```

Type III Wald test p-values: For each categorical predictor with more than two levels, the Type III Wald test provides the multiple degree of freedom p-value for the overall test of that predictor's association with the outcome.

```
car::Anova(fit.ex6.3.adj, type = 3, test.statistic = "Wald")
# (results not shown)
```

Methods: We used data from a teaching dataset including 843 participants in the 2019 National Survey of Drug Use and Health aged 18 years and older (18-25 = 12.1%, 26-34 = 14.7%, 35-49 = 27.3%, 50-64 = 24.1%, 65+ = 21.8%) to assess the association between lifetime marijuana use and age at first use of alcohol, adjusted for age group (ref = 18-25), sex (Female (ref), Male), and income (Less than \$20,000 (ref), \$20,000 to \$49,999, \$50,000 to \$74,999, \$75,000 or more) using binary logistic regression. We assessed the linearity assumption and checked for separation, collinearity, and outliers (no issues were found). We evaluated the robustness of our results to the presence of a few potentially influential observations using a sensitivity analysis (we did not carry this out in this chapter, but this demonstrates where you would report such an analysis). A forest plot is provided to illustrate the AORs and their 95% CIs.

Results: After adjusting for age, sex, and income, age at first alcohol use was significantly negatively associated with lifetime marijuana use (AOR = 0.759; 95% CI = 0.718, 0.800; p <.001). Individuals who first used alcohol at a given age have 24.1% lower odds of having ever used marijuana than those who first used alcohol one year earlier. Current age was also significantly associated with the outcome, with older individuals having lower odds of lifetime marijuana use.

See Table 6.2 for full regression results and Figure 6.3 for a forest plot of the AORs and their 95% CIs.

The code below illustrates the use of `gtsummary::tbl_merge()` (Sjoberg et al., 2021, 2023) to merge together two regression tables. In this case, we are merging one table that includes the

TABLE 6.2 Logistic regression results for lifetime marijuana use vs. age at first alcohol use (years) (N = 843)

	Adjusted Coefficient		Adjusted Odds Ratio		
Characteristic	**log(AOR)**	**95% CI**	**AOR**	**95% CI**	**p-value**
Intercept	0.470	-0.150, 1.120	1.601	0.860, 3.064	0.145
Age First Alc - 21y (y)	-0.275	-0.331, -0.223	0.759	0.718, 0.800	<0.001
Age (y)					<0.001
18-25	—	—	—	—	
26-34	-0.296	-0.949, 0.343	0.744	0.387, 1.409	0.367
35-49	-0.804	-1.400, -0.234	0.447	0.247, 0.791	0.007
50-64	-0.690	-1.289, -0.116	0.502	0.275, 0.891	0.021
65+	-1.275	-1.885, -0.689	0.279	0.152, 0.502	<0.001
Sex					0.707
Female	—	—	—	—	
Male	-0.061	-0.379, 0.256	0.941	0.684, 1.291	0.707
Income					0.142
Less than \$20,000	—	—	—	—	
\$20,000 - \$49,999	-0.531	-1.059, -0.013	0.588	0.347, 0.987	0.046
\$50,000 - \$74,999	-0.079	-0.679, 0.519	0.924	0.507, 1.680	0.795
\$75,000 or more	-0.361	-0.864, 0.130	0.697	0.421, 1.139	0.154

regression coefficients (which are on the logit scale) and another table that includes the AORs (the exponentiated regression coefficients). The intercept, however, is not a log-odds ratio, like the other regression coefficients; it is the log-odds of the outcome when all predictors are 0 or at their reference level. This distinction is added in a footnote to the table. To make the intercept more interpretable, the one continuous variable in this model was centered (age of first alcohol use, centered at age 21 years).

Notice how some features are included in one table but not the other to avoid redundancies (e.g., p-values only in the second table). The labels, however, must be identical in both tables in order for the merge to work correctly. Notice also (in the code comments) the difference between using `style_sigfig()`, which keeps a specified number of significant digits, and `style_number()`, which rounds to a specified number of decimal places. For example, the number 1.532 with two significant digits would be 1.5, while rounded to two decimal places would be 1.53. In this particular table, rounding makes everything line up nicely. In other tables, especially when different regression terms are on different scales, specifying the number of significant digits may be preferred.

```
# Re-fit the model with a centered version of alc_agefirst
tmpdat <- nsduh %>%
  mutate(calc_agefirst = alc_agefirst - 21)

fit.ex6.3.adj.b <- glm(mj_lifetime ~ calc_agefirst + demog_age_cat6 + demog_sex +
                       demog_income, family = binomial, data = tmpdat)

NOBS <- length(fit.ex6.3.adj.b$residuals)

library(gtsummary)
t1 <- fit.ex6.3.adj.b %>%
  tbl_regression(intercept = T,
               # Use style_number to round to 3 digits (e.g., 1.27)
               estimate_fun = function(x) style_number(x, digits = 3),
               # Use style_sigfig to keep 3 significant digits (e.g., 1.3)
               # estimate_fun = function(x) style_sigfig(x, digits = 3),
               label  = list('(Intercept)'  ~ "Intercept",
                             calc_agefirst  ~ "Age First Alc - 21y (y)",
                             demog_age_cat6 ~ "Age (y)",
                             demog_sex      ~ "Sex",
                             demog_income   ~ "Income")) %>%
  modify_column_hide(p.value) %>%
  modify_caption(paste("Logistic regression results for lifetime marijuana use vs.
                 age at first alcohol use (years) (N = ", NOBS, ")", sep=""))

t2 <- fit.ex6.3.adj.b %>%
  tbl_regression(intercept = T,
               exponentiate = T,  # OR = exp(B)
               # Use style_number to round to 3 digits (e.g., 1.27)
               estimate_fun = function(x) style_number(x, digits = 3),
               # Use style_sigfig to keep 3 significant digits (e.g., 1.3)
               # estimate_fun = function(x) style_sigfig(x, digits = 3),
               pvalue_fun   = function(x) style_pvalue(x, digits = 3),
               label  = list('(Intercept)'  ~ "Intercept",
                             calc_agefirst  ~ "Age First Alc - 21y (y)",
                             demog_age_cat6 ~ "Age (y)",
                             demog_sex      ~ "Sex",
                             demog_income   ~ "Income")) %>%
  # Add test.statistic = "Wald" here to get Wald Type III tests
  add_global_p(keep = T, test.statistic = "Wald")
```

```
TABLE <- tbl_merge(
  tbls = list(t1, t2),
  tab_spanner = c("**Adjusted Coefficient**", "**Adjusted Odds Ratio**")
  ) %>%
  modify_footnote(everything() ~ NA, abbreviation = TRUE) %>%
  modify_header(update = list(
    estimate_1 = "**log(AOR)**",
    estimate_2 = "**AOR**"
))
```

```
TABLE
```

6.17.2 Writing up logistic regression results (with an interaction)

For the model with an interaction (`fit.ex6.3.int`), more information must be added to the Methods and Results. Specifically, mention the inclusion of an interaction, the test of interaction, the overall test of the primary predictor of interest that tested the main effect and the interaction at the same time (Section 6.9.1), and the separate AORs for the primary predictor of interest at the levels of the other predictor in the interaction (Section 6.9.2).

The following code displays the raw regression coefficients, 95% CIs, and p-values.

```
# Regression coefficients, p-values, AORs, 95% CIs for AORs
car::S(fit.ex6.3.int)
# 95% CIs for the regression coefficients
confint(fit.ex6.3.int)
# Pvalues
car::Anova(fit.ex6.3.int, type = 3, test.statistic = "Wald")
# (Results not shown)
```

The following code displays the overall test of the predictor.

```
# Overall test of alc_agefirst
fit0 <- glm(mj_lifetime ~ demog_age_cat6 + demog_sex + demog_income,
            family = binomial, data = nsduh,
            subset = complete.cases(alc_agefirst))
anova(fit0, fit.ex6.3.int, test = "Chisq")
# (Results not shown)
```

The following code displays the AORs at each level of the other term in the interaction. We store these as the objects `EST.M` and `EST.F` because we will be re-using them below when we create a table illustrating the interaction effect.

```
# AOR for Males
EST.M <- gmodels::estimable(fit.ex6.3.int,
       c("alc_agefirst"                   = 1,
         "alc_agefirst:demog_sexMale" = 1),
       conf.int = 0.95)
# AOR for Females
EST.F <- gmodels::estimable(fit.ex6.3.int,
       c("alc_agefirst"                   = 1),
       conf.int = 0.95)
# (Results not shown)
```

TABLE 6.3 Logistic regression results for lifetime marijuana use vs. age at first alcohol use (years) including an interaction with sex (N = 843)

Characteristic	Adjusted Coefficient		Adjusted Odds Ratio		
	log(AOR)	95% CI	AOR	95% CI	p-value
Intercept	0.279	-0.380, 0.958	1.322	0.684, 2.607	0.411
Age First Alc - 21y (y)	-0.338	-0.425, -0.259	0.713	0.654, 0.772	<0.001
Age (y)					<0.001
18-25	—	—	—	—	
26-34	-0.295	-0.954, 0.350	0.744	0.385, 1.419	0.373
35-49	-0.816	-1.417, -0.242	0.442	0.243, 0.785	0.006
50-64	-0.687	-1.291, -0.108	0.503	0.275, 0.897	0.022
65+	-1.260	-1.875, -0.671	0.284	0.153, 0.511	<0.001
Sex					0.156
Female	—	—	—	—	
Male	0.359	-0.137, 0.858	1.432	0.872, 2.359	0.156
Income					0.156
Less than $20,000	—	—	—	—	
$20,000 - $49,999	-0.530	-1.062, -0.009	0.588	0.346, 0.991	0.048
$50,000 - $74,999	-0.093	-0.697, 0.510	0.911	0.498, 1.665	0.762
$75,000 or more	-0.363	-0.869, 0.131	0.696	0.419, 1.140	0.154
Age First Alc - 21y (y) * Sex					0.032
Age First Alc - 21y (y) * Male	0.119	0.011, 0.229	1.126	1.011, 1.257	0.032

Methods: *(Same as for the model with no interaction with the following additions.)* An interaction between age at first alcohol use and sex was included in the model to assess if the association with the outcome differed between females and males. The overall significance of age at first alcohol use was assessed by comparing the full model to a reduced model with both its main effect and the interaction removed. Additionally, we estimated the AOR for age at first alcohol use separately for females and for males.

Results: The interaction effect was statistically significant (B = 0.119; 95% CI = 0.011, 0.229; p = .032). Overall, after adjusting for age, sex, and income, age at first alcohol use was significantly associated with lifetime marijuana use (p <.001). The association was significant for both females (OR = 0.713; 95% CI = 0.656, 0.776; p <.001) and males (OR = 0.803; 95% CI = 0.748, 0.863; p <.001). Current age was also significantly associated with the outcome.

See Table 6.3 for full regression results.

NOTE: Be careful interpreting the "AOR" for the interaction term (1.126). It is not an AOR for a specific predictor. Rather, it is a ratio of odds ratios and can be interpreted in two ways. First, it is the ratio of the AOR for a 1-unit difference in age of first alcohol use for males divided by that AOR for females (0.803 / 0.713). Second, it is the ratio of the AOR comparing males to females for those of a given age of first alcohol use divided by that AOR for those who started one year earlier. For example, for age of first use 17 years, this ratio can be derived as exp(0.359 + (17-21) × 0.119) / exp(0.359 + (16-21) × 0.119) = 10.817 / 9.605.

Since we have an interaction, it is helpful to also table the AORs for each term in the interaction at specific levels of the other term. We have already done this for `calc_agefirst` at each level of demog_sex (0.803 and 0.713 in Results above), but we compute them again here. We now also do this for demog_sex at specific values of calc_agefirst. You can choose any values for calc_agefirst within the range of the data, whatever you feel best illustrates how the AOR varies between meaningfully different predictor values; here we use 15, 18, and 21 years (and subtract 21 from each since we centered this variable at 21 years).

```
# calc_agefirst at demog_sex = "Male"
EST.M <- gmodels::estimable(fit.ex6.3.int.b,
        c("calc_agefirst"                   = 1,
          "calc_agefirst:demog_sexMale" = 1),
        conf.int = 0.95)

# calc_agefirst at demog_sex = "Female"
EST.F <- gmodels::estimable(fit.ex6.3.int.b,
        c("calc_agefirst"                   = 1),
        conf.int = 0.95)

# demog_sex effect at calc_agefirst = 15
EST.15 <- gmodels::estimable(fit.ex6.3.int.b,
          c("demog_sexMale"                  = 1,
            "calc_agefirst:demog_sexMale" = 15 - 21),
          conf.int = 0.95)

# demog_sex effect at calc_agefirst = 18
EST.18 <- gmodels::estimable(fit.ex6.3.int.b,
          c("demog_sexMale"                  = 1,
            "calc_agefirst:demog_sexMale" = 18 - 21),
          conf.int = 0.95)

# demog_sex effect at calc_agefirst = 21
EST.21 <- gmodels::estimable(fit.ex6.3.int.b,
          c("demog_sexMale"                  = 1,
            "calc_agefirst:demog_sexMale" = 21 - 21),
          conf.int = 0.95)

# Exponentiate the AORs and 95% CIs but not the p-values
DF <- cbind(
        rbind(exp(EST.F[ c("Estimate", "Lower.CI", "Upper.CI")]),
              exp(EST.M[ c("Estimate", "Lower.CI", "Upper.CI")]),
              exp(EST.15[c("Estimate", "Lower.CI", "Upper.CI")]),
              exp(EST.18[c("Estimate", "Lower.CI", "Upper.CI")]),
              exp(EST.21[c("Estimate", "Lower.CI", "Upper.CI")])),
        # P-values
        rbind(EST.F["Pr(>|X^2|)"],
              EST.M["Pr(>|X^2|)"],
              EST.15["Pr(>|X^2|)"],
              EST.18["Pr(>|X^2|)"],
              EST.21["Pr(>|X^2|)"])
    )

names(DF)    <- c("AOR", "Lower", "Upper", "p-value")

rownames(DF) <- c("Age of first alcohol use effect @ Sex = Female",
                  "Age of first alcohol use effect @ Sex = Male",
                  "Male vs. Female @ Age of first alcohol use = 15y",
                  "Male vs. Female @ Age of first alcohol use = 18y",
                  "Male vs. Female @ Age of first alcohol use = 21y")

round(DF, 3)
```

```
##                                                       AOR Lower Upper p-value
## Age of first alcohol use effect @ Sex = Female      0.713 0.656 0.776   0.000
## Age of first alcohol use effect @ Sex = Male        0.803 0.748 0.863   0.000
## Male vs. Female @ Age of first alcohol use = 15y 0.702 0.460 1.072   0.097
## Male vs. Female @ Age of first alcohol use = 18y 1.003 0.723 1.390   0.987
## Male vs. Female @ Age of first alcohol use = 21y 1.432 0.866 2.369   0.156
```

6.18 Likelihood ratio test vs. Wald test

For binary logistic regression models fit with `glm()`, the p-values obtained from `summary()` are, by default, what are referred to as "Wald" tests. Above, we used the `test.statistic = "Wald"` option in `car::Anova()` to obtain p-values that would be consistent with those from `summary()`. An alternative is to use a "likelihood ratio" (LR) test, which is typically more powerful (more likely to lead to rejection of the null hypothesis when it is, in fact, false). In fact, `confint()` uses a profile likelihood method to obtain confidence intervals for a model fit with `glm()` and these correspond to LR hypothesis tests.

The default for `car::Anova()` is to compute LR test p-values for binary logistic regression models. This works for continuous predictors, binary predictors, and for multiple df tests for categorical predictors. For tests of individual levels of a categorical predictor vs. the reference level, an additional step is needed. After fitting the model using categorical predictors with more than two levels coded as factors to get the multiple df p-values, re-fit the model after replacing these factor variables with the corresponding sets of binary indicator variables. Using `car::Anova()`, one can then get the corresponding LR tests for the comparisons between levels.

Example 6.3 (continued): What is the association between lifetime marijuana use (`mj_lifetime`) and age at first use of alcohol (`alc_agefirst`), adjusted for age (`demog_age_cat6`), sex (`demog_sex`), and income (`demog_income`)? Use LR tests for all the p-values.

```
# Same model as fit previously
fit.ex6.3.adj <- glm(mj_lifetime ~ alc_agefirst + demog_age_cat6 + demog_sex +
                     demog_income, family = binomial, data = nsduh)
```

The p-values displayed by `summary()` and the p-values computed by `car::Anova(, type = 3, test.statistic = "Wald")` are all Wald tests.

```
round(summary(fit.ex6.3.adj)$coef, 4)
```

```
##                             Estimate Std. Error z value Pr(>|z|)
## (Intercept)                   6.2542     0.5914 10.5759   0.0000
## alc_agefirst                 -0.2754     0.0276 -9.9922   0.0000
## demog_age_cat626-34          -0.2962     0.3286 -0.9012   0.3675
## demog_age_cat635-49          -0.8043     0.2966 -2.7120   0.0067
## demog_age_cat650-64          -0.6899     0.2985 -2.3109   0.0208
## demog_age_cat665+            -1.2748     0.3043 -4.1893   0.0000
## demog_sexMale                -0.0609     0.1618 -0.3763   0.7067
## demog_income$20,000 - $49,999 -0.5309     0.2664 -1.9927   0.0463
## demog_income$50,000 - $74,999 -0.0793     0.3049 -0.2601   0.7948
## demog_income$75,000 or more   -0.3612     0.2532 -1.4264   0.1538
```

```
car::Anova(fit.ex6.3.adj, type = 3, test.statistic = "Wald")
```

```
## Analysis of Deviance Table (Type III tests)
##
## Response: mj_lifetime
##              Df  Chisq            Pr(>Chisq)
## (Intercept)   1 111.85 < 0.0000000000000002 ***
## alc_agefirst  1  99.84 < 0.0000000000000002 ***
## demog_age_cat6 4 23.01              0.00013 ***
## demog_sex     1   0.14              0.70669
## demog_income  3   5.44              0.14197
## ---
## Signif. codes:  0 '***' 0.001 '**' 0.01 '*' 0.05 '.' 0.1 ' ' 1
```

The following computes a LR test p-value for each predictor.

```
car::Anova(fit.ex6.3.adj, type = 3, test.statistic = "LR")
```

```
## Analysis of Deviance Table (Type III tests)
##
## Response: mj_lifetime
##                LR Chisq Df            Pr(>Chisq)
## alc_agefirst      149.6  1 < 0.0000000000000002 ***
## demog_age_cat6     24.0  4             0.000081 ***
## demog_sex           0.1  1                 0.71
## demog_income        5.5  3                 0.14
## ---
## Signif. codes:  0 '***' 0.001 '**' 0.01 '*' 0.05 '.' 0.1 ' ' 1
```

Use the multiple df LR p-values provided here in place of Wald p-values we obtained earlier from `car::Anova(, type = 3, test.statistic = "Wald")`, and the 1 df LR p-values provided here in place of the Wald p-values we obtained earlier from `summary()` for continuous and binary predictors. However, we need an additional step to get LR p-values to replace the remaining Wald p-values in `summary()`, those for comparisons between a level of a categorical predictor with more than two levels and a reference level.

```
# Create indicator variables
nsduh <- nsduh %>%
  mutate(age26_34 = as.numeric(demog_age_cat6 == "26-34"),
         age35_49 = as.numeric(demog_age_cat6 == "35-49"),
         age50_64 = as.numeric(demog_age_cat6 == "50-64"),
         age65    = as.numeric(demog_age_cat6 == "65+"),
         income1  = as.numeric(demog_income   == "$20,000 - $49,999"),
         income2  = as.numeric(demog_income   == "$50,000 - $74,999"),
         income3  = as.numeric(demog_income   == "$75,000 or more"))

# Re-fit model
fit.ex6.3.adj.2 <- glm(mj_lifetime ~ alc_agefirst +
                         age26_34 + age35_49 + age50_64 + age65 +
                         demog_sex + income1 + income2 + income3,
                       family = binomial, data = nsduh)

# LR p-values
car::Anova(fit.ex6.3.adj.2, type = 3, test.statistic = "LR")
```

```
## Analysis of Deviance Table (Type III tests)
##
## Response: mj_lifetime
##             LR Chisq Df           Pr(>Chisq)
## alc_agefirst   149.6  1 < 0.0000000000000002 ***
## age26_34         0.8  1               0.3655
## age35_49         7.8  1               0.0054 **
## age50_64         5.6  1               0.0182 *
## age65           18.8  1             0.000014 ***
## demog_sex        0.1  1               0.7066
## income1          4.0  1               0.0445 *
## income2          0.1  1               0.7947
## income3          2.1  1               0.1506
## ---
## Signif. codes:  0 '***' 0.001 '**' 0.01 '*' 0.05 '.' 0.1 ' ' 1
```

In summary, other than in this section, we have used Wald tests for simplicity when introducing binary logistic regression. However, LR tests are more powerful. Now that you are familiar with all the steps needed to carry out a binary logistic regression, you should be able to add in this last extra step and use LR tests.

6.19 Summary of binary logistic regression

The steps of carrying out a logistic regression are similar to MLR (Section 5.28), but with the following differences.

- The outcome in a logistic regression is categorical, not continuous, and should not be transformed.
- Make sure the outcome is coded as a factor and you know which probability `glm()` is modeling (see Section 6.6.2).
- Check for separation and resolve any issues (see Section 6.10).
- If any step leads to a change in sample size or in the included categorical predictors, then re-evaluate separation.
- Evaluate linearity (Section 6.12), outliers (Section 6.13), and influential observations (Section 6.14). In logistic regression, there are no normality or constant variance assumptions
- Test goodness-of-fit using the Hosmer-Lemeshow test and assess goodness-of-fit visually using a calibration plot (see Section 6.16).

6.20 Conditional logistic regression for matched case-control data

Some case-control studies employ matching in an attempt to ensure the controls are comparable to the cases on confounding variables. Matching, however, must be taken into account in the analysis method. Matched case-control data can be validly analyzed using **conditional logistic regression** which stratifies the analysis by groups defined by the unique combinations of the matching variables. To carry out a conditional logistic regression in R, use the `clogit()` function (Gail et al., 1981; Logan, 1983) in the `survival` library (Therneau, 2023) with the matching variables listed as `strata` in the model. `clogit()` expects the outcome to be numeric with possible values 0 and 1, and the output does not contain an intercept. The conditional logit model does not estimate associations between the `strata` variables and the outcome. However, since their purpose was to control for confounding, this is not typically an issue.

Example 6.6: A matched case-control dataset of births was created from a subset of the 2018 U.S. Natality teaching dataset, containing 195 births that were followed by admission to the newborn intensive care unit (`AB_NICU`) and 1375 births that were not, matched on maternal education (`MEDUC`) and age (`MAGER`). Assess the association between admission to the newborn intensive care unit and previous preterm birth (`RF_PPTERM`), accounting for the matching in the analysis.

After loading the data, check the formatting of the outcome and convert to 0/1, if needed.

```
load("Data/natality_CC_rmph.Rdata")

table(natality_CC$AB_NICU)
```

```
##
##   No  Yes
## 1375  195
```

```
# clogit() expects a 0/1 outcome
# Convert from No/Yes to 0/1
natality_CC <- natality_CC %>%
  mutate(NICU = as.numeric(AB_NICU == "Yes"))

# Check derivation
table(natality_CC$AB_NICU, natality_CC$NICU, useNA = "ifany")
```

```
##
##          0    1
##   No  1375    0
##   Yes    0  195
```

Next, verify that the distributions of the matching variables are the same for cases and controls.

```
# MEDUC is categorical
addmargins(
  prop.table(
    table(natality_CC$MEDUC, natality_CC$NICU), 2), 1)
```

```
##
##                      0       1
##   <HS          0.10036 0.12308
##   HS           0.28945 0.33333
##   Some college 0.27709 0.26154
##   Bachelor     0.23636 0.17436
##   Adv Degree   0.09673 0.10769
##   Sum          1.00000 1.00000
```

```
# MAGER is continuous
boxplot(MAGER ~ NICU,
        data = natality_CC)
```

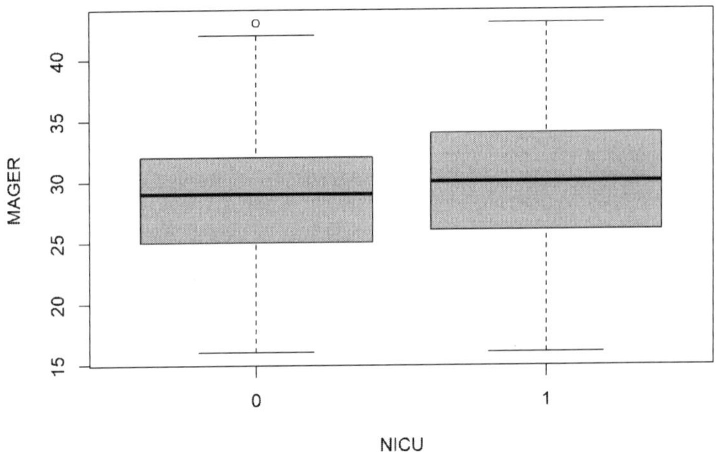

Matching is not always perfect. The distributions are not identical between cases and controls but are pretty close.

Finally, fit the conditional logit model and examine the output.

```
library(survival)
fit.clr <- clogit(NICU ~ RF_PPTERM + strata(MEDUC, MAGER),
                  data = natality_CC)

# Regression coefficient
round(summary(fit.clr)$coef, 4)
```

```
##                coef exp(coef) se(coef)     z Pr(>|z|)
## RF_PPTERMYes 1.123     3.074    0.316 3.554   0.0004
```

```
# OR and 95% CI
OR.CI <- cbind("OR" = exp(coef(fit.clr)),
               exp(confint(fit.clr)))
round(OR.CI, 3)
```

```
##                 OR 2.5 % 97.5 %
## RF_PPTERMYes 3.074 1.655   5.71
```

```
# Type III test
car::Anova(fit.clr, type = 3, test.statistic = "Wald")
```

```
## Analysis of Deviance Table (Type III tests)
##
## Response: Surv(rep(1, 1570L), NICU)
##              Df Chisq Pr(>Chisq)
## RF_PPTERM   1  12.6    0.00038 ***
## ---
## Signif. codes:  0 '***' 0.001 '**' 0.01 '*' 0.05 '.' 0.1 ' ' 1
```

Conclusion: Previous preterm birth is significantly associated with admission to the NICU (p <.001). Infants born to mothers with a previous preterm birth have 3.1 times the odds of admission to the NICU (OR = 3.07; 95% CI = 1.65, 5.71; p <.001).

6.21 Log-binomial regression to estimate a risk ratio or prevalence ratio

Logistic regression is a special case of a family of models known as **generalized linear models**. Each member of this family has an assumed distribution for the outcome and a **link function** that connects the mean outcome to a linear combination of predictors $\beta_0 + \beta_1 X_1 + \beta_2 X_2 + \ldots + \beta_K X_K$ (the **linear predictor**). In logistic regression, the outcome is assumed to have a binomial distribution and the link function is the logit function $\ln(p/(1-p))$. Linear regression is also a special case, with a normal distribution and an identity link function (the mean is assumed to be equal to the linear predictor).

Another special case of a generalized linear model is the **log-binomial regression** model which, like logistic regression, assumes a binomial distribution for a binary outcome but, unlike logistic regression, uses a log link function as shown in Equation 6.2.

$$\ln p = \beta_0 + \beta_1 X_1 + \beta_2 X_2 + \ldots + \beta_K X_K \tag{6.2}$$

With logistic regression, the left-hand side is the log of the odds, whereas in log binomial regression it is the log of the probability. Exponentiating a regression coefficient in logistic regression results in an odds ratio. Similarly, exponentiating a regression coefficient in log-binomial regression results in a risk ratio (RR) or prevalence ratio (PR). The model described by Equation 6.2 can be used to estimate an RR from incidence data or a PR from prevalence data. Thus, for a predictor X_k, the RR or PR is e^{β_k}.

A disadvantage of log-binomial regression is that the left-hand side ($\ln p$) is constrained to be positive, while the right-hand side can be anything from $-\infty$ to ∞. This leads to convergence issues at times (Williamson et al., 2013). One method for fitting a log-binomial model is to use `glm()` with `family = binomial(link="log")`. Alternatively, use the `logbin()` function in the `logbin` package (Donoghoe and Marschner, 2018) which may converge even in cases where `glm()` fails.

Example 6.2 (continued): Logistic regression estimated an OR comparing lifetime marijuana use between males and females of 1.44. Use log-binomial regression to compute the corresponding prevalence ratio.

```
library(logbin)
fit.ex6.2.logbin <- logbin(mj_lifetime ~ demog_sex,
                           data = nsduh,
                           method = "em")

# Summary of model
round(summary(fit.ex6.2.logbin)$coef, 4)
```

```
##                Estimate Std. Error z value Pr(>|z|)
## (Intercept)     -0.7629     0.0463 -16.479   0.0000
## demog_sexMale    0.1794     0.0620   2.894   0.0038
```

```
# PR, and 95% CI for PR
PR.CI <- cbind("PR" = exp(coef(fit.ex6.2.logbin)),
               exp(confint(fit.ex6.2.logbin)))[-1,]
round(PR.CI, 3)
```

```
##    PR  2.5 % 97.5 %
## 1.197  1.060  1.351
```

Although not needed for this example, if the predictor were categorical with more than two levels, then you can obtain a Type III multiple df test as usual.

```
# Type III test
car::Anova(fit.ex6.2.logbin, type = 3, test.statistic = "Wald")
```

Conclusion: Males are 1.20 times as likely to have ever used marijuana than females (PR = 1.20; 95% CI = 1.06, 1.35; p = .004).

In the interpretation, we used the phrase "times as likely" rather than "times the odds" because log-binomial regression models the log of the probability, not the log-odds. We could also say that the prevalence of marijuana use is 20% greater among males. If this were incidence data, we could say that males have 20% greater risk. To compute an adjusted RR or PR, simply add the confounding variables to the model formula.

NOTES:

- If you use `predict()` or `gmodels::estimable()` to estimate a probability from a log-binomial model, use `exp()` rather than `ilogit()` when transforming the prediction to the probability scale.
- `logbin()` does not allow interaction terms using the : notation. If `glm()` with `family(link = "log")` converges, then that is the simplest way to include an interaction since it does allow the : notation. To include an interaction with `logbin`, you must create variables corresponding to the interaction terms outside the model and then include those variables in the model (see Section 9.6.4.2 for an example, from a different context, of how to do this).

6.22 Ordinal logistic regression

An **ordinal** variable is a categorical variable in which the levels have a natural ordering (e.g., depression categorized as Minimal, Mild, Moderate, Moderately Severe, and Severe).

Ordinal logistic regression can be used to assess the association between predictors and an ordinal outcome. You can fit an ordinal logistic regression model in R with `MASS::polr()` (proportional odds logistic regression) (Ripley, 2023).

WARNING: Use the syntax `MASS::polr()` rather than loading the `MASS` library, as loading it masks the `select()` function in both `gtsummary()` and `tidyverse()` leading to errors when using those packages.

NOTE: The proportional odds model is only one possible model for ordinal data. See Greenland (1994) for a discussion of an alternative model which may be a better choice if, for example, the underlying distribution is truly ordinal (as opposed to having an ordinal variable that is the result of collapsing an underlying continuous variable into groups), or the proportional odds assumption is not met.

6.22.1 Ordinal model

In binary logistic regression, the outcome Y has two levels. If the two levels are referred to as 0 and 1, we model the probability that $Y = 1$. What if you have an outcome with more than two levels, and the levels are ordered? The **ordinal logistic regression model**, as parameterized by `MASS::polr()`, for an outcome Y with levels $\ell = 1, 2, ..., L$ is

$$\ln\left(\frac{P(Y \leq \ell)}{P(Y > \ell)}\right) = \zeta_\ell - \eta_1 X_1 - \eta_2 X_2 - \ldots - \eta_K X_K \tag{6.3}$$

for each level $\ell = 1, 2, ..., L-1$. Equation 6.3 only works for levels up to $L-1$ because if we went up to L we would have $P(Y > L) = 0$ in the denominator. The letter ζ is pronounced "zeta" and the letter η is pronounced "eta". Just like binary logistic regression, the left-hand side is the log-odds of a probability, but now instead of a probability of an outcome being at one level, it is the **cumulative probability** of an outcome being at any level up to and including a specified level.

How do we interpret the regression terms?

- ζ_ℓ is an intercept and represents the log-odds of $Y \leq \ell$ when all the predictors are at 0 or their reference level. Thus, $P(Y \leq \ell)$ is the inverse-logit of ζ_ℓ. Unlike other forms of regression we have worked with, an ordinal logistic regression model has multiple intercepts, one for each level of Y from 1 to $L-1$.
- Each $-\eta_k$ for $k = 1, 2, ..., K$ is the log of the odds ratio comparing the odds of $Y \leq \ell$ between individuals who differ by 1-unit in X_k (or comparing individuals at a level of X_k and the reference level).
- $e^{-\eta_k}$ is the odds ratio comparing the odds of $Y \leq \ell$ between those differing by 1-unit in X_k.
- e^{η_k} is the odds ratio comparing the odds of $Y > \ell$ between those differing by 1-unit in X_k.

Since `MASS::polr()` returns estimates of η, not $-\eta$, an OR $>$ 1 corresponds to greater odds of the outcome being at *higher* levels.

For each individual level ℓ, this is similar to a binary logistic regression for a dichotomized version of Y where the two levels are $Y > \ell$ and $Y \leq \ell$. However, the ordinal logistic regression model goes further. It fits the model for all the levels of Y at the same time, estimates an intercept for each level of the outcome (other than the highest level), and *assumes the regression coefficient for each predictor X_k is the same for all ℓ*. This assumption is called the **proportional odds assumption**, because it implies that the odds ratio for a predictor X is the same at all levels of Y – no matter where you split Y into $Y > \ell$ and $Y \leq \ell$, the resulting odds at any two values of X are proportional (have the same ratio).

6.22.2 Transforming a continuous outcome into an ordinal outcome

In addition to outcomes that are inherently ordinal, we often encounter outcomes that are continuous, highly skewed, and bounded below by 0. When fitting a linear regression to such an outcome, the regression assumptions can fail and no transformation is effective at resolving the issues. One solution is to collapse the outcome into an ordinal outcome.

NOTE: This section is necessary for Example 6.5, but if you are referring to this chapter in order to carry out an analysis of an outcome that is already coded as ordinal, skip ahead to Section 6.22.3.

Example 6.5: The PHQ-9 is the depression module of the Patient Health Questionnaire (Kroenke et al., 2001; Kroenke and Spitzer, 2002). It consists of nine questions, each scored from 0 to three and summed for a total score ranging from 0 to 27. Scores in the ranges 0-4, 5-9, 10-14, 15-19, and 20-27 are labeled as Minimal, Mild, Moderate, Moderately Severe, and Severe depression, respectively. A summary and histogram of PHQ-9 scores (`phq9`) from the NHANES examination teaching dataset (`nhanes1718_adult_exam_sub_rmph.Rdata`) reveals that it has many zeros and is highly skewed. Create a three-level ordinal variable corresponding to Minimal, Mild, and Moderate to Severe depression.

```
load("Data/nhanes1718_adult_exam_sub_rmph.Rdata")

sum(nhanes_adult_exam_sub$phq9 == 0, na.rm=T)
```

```
## [1] 278
```

```
hist(nhanes_adult_exam_sub$phq9)
```

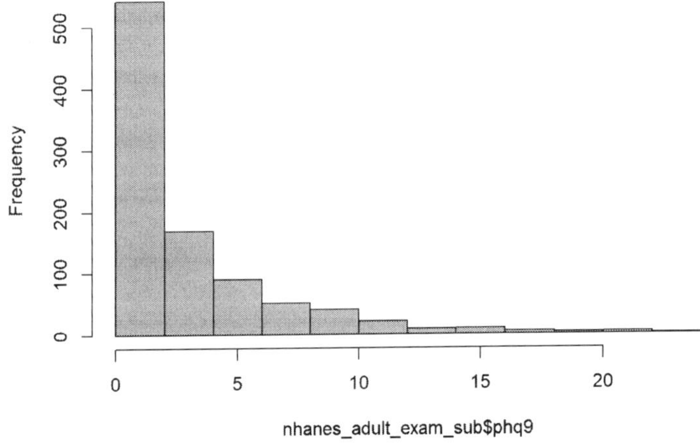

We demonstrate here two methods for turning the raw PHQ-9 variable, which ranges from 0 to 27, into a three-level variable. The `cut()` function is very useful but a little tricky to get right due to the way it requires you to specify whether intervals are closed on the left and/or right. A nice feature of `cut()` is that the result is formatted as a `factor` with levels labeled with the interval definitions. After trying out the different combinations of `right` and `include.lowest`, we find we need `right = F` and `include.lowest = T` to get the intervals we want.

```
# Options for cut()
# right = T leads to intervals closed on the right
# right = F leads to intervals open on the right (unless include.lowest=T)
# For right = T, include.lowest = T leads to the FIRST interval closed on the left
# For right = F, include.lowest = T leads to the LAST interval closed on the right
# (so for right = F, "include.lowest" is better thought of as "include.highest")
# right include.lowest  first_interval  middle_intervals last_interval
# T     T               [0,5]           (5, 10]          (10, 27]
# T     F               (0,5]           (5, 10]          (10, 27]
# F     T               [0,5)           [5, 10)          [10, 27]
# F     F               [0,5)           [5, 10)          [10, 27)

# Create three-level PHQ-9 total using cut
nhanes <- nhanes_adult_exam_sub %>%
  mutate(depression = cut(phq9,
                          c(0, 5, 10, 27),
                          right = F,
                          include.lowest = T))

# Is it a factor?
is.factor(nhanes$depression)
```

```
## [1] TRUE
```

```
# Did all values go to the right category, and no new NAs?
table(nhanes_adult_exam_sub$phq9, nhanes$depression, useNA = "ifany")
```

```
##
##        [0,5) [5,10) [10,27] <NA>
##   0      278      0       0    0
##   1      153      0       0    0
##   2      110      0       0    0
##   3       86      0       0    0
##   4       83      0       0    0
##   5        0     57       0    0
##   6        0     33       0    0
##   7        0     27       0    0
##   8        0     24       0    0
##   9        0     21       0    0
##   10       0      0      20    0
##   11       0      0      14    0
##   12       0      0       7    0
##   13       0      0       4    0
##   14       0      0       5    0
##   15       0      0       7    0
##   16       0      0       3    0
##   18       0      0       5    0
##   19       0      0       1    0
##   20       0      0       2    0
##   21       0      0       3    0
```

```
##    22          0         0          1    0
##    24          0         0          1    0
##    <NA>        0         0          0   55
```

```
# Check the range of values in each category
tapply(nhanes_adult_exam_sub$phq9, nhanes$depression, range)
```

```
## $`[0,5)`
## [1] 0 4
##
## $`[5,10)`
## [1] 5 9
##
## $`[10,27]`
## [1] 10 24
```

```
# Final distribution
table(nhanes$depression, useNA = "ifany")
```

```
##
##    [0,5)  [5,10) [10,27]    <NA>
##      710     162      73      55
```

The derivation was correct, although in this dataset the maximum value was 24 rather than the maximum possible of 27.

Another approach is to use `case_when()` in which we specify a series of logical statements to define the intervals, and then convert the result to a factor. The new variable takes on the value `1` (the value after the `~` in the first row) if the first condition is met, the value `2` if the first condition is *not* met but the second is, and so on. This leads to the exact same result as `cut()`. Regardless of which you use, always check the derivation afterward to make sure your new variable is as intended.

```
# Create three-level PHQ-9 total using case_when
nhanes <- nhanes_adult_exam_sub %>%
  mutate(depression = case_when(phq9 <   5 ~ 1,
                                phq9 <  10 ~ 2,
                                phq9 >= 10 ~ 3),
         depression = factor(depression,
                             levels = 1:3,
                             labels = c("Minimal",
                                        "Mild",
                                        "Moderate to Severe")))
```

```
# Check derivation (results identical to before)
# Is it a factor?
is.factor(nhanes$depression)
# Did all values go to the right category, and no new NAs?
table(nhanes_adult_exam_sub$phq9, nhanes$depression, useNA = "ifany")
# Check the range of values in each category
tapply(nhanes_adult_exam_sub$phq9, nhanes$depression, range)
```

```
# Final distribution
table(nhanes$depression, useNA = "ifany")
```

```
##
##               Minimal                        Mild Moderate to Severe               <NA>
##                   710                         162                73                  55
```

6.22.3 Make sure you know what probabilities `polr()` is modeling

Before fitting an ordinal logistic regression, check the ordering of the outcome variable. An OR > 1 corresponds to a risk factor that is associated with greater probability of higher levels of the outcome variable. Therefore, typically, the ordering desired for an outcome variable is from less severe to more severe. Whatever the ordering desired, check using `levels()` and, if needed, reorder using `factor()`.

Example 6.5 (continued): What is the ordering for the outcome `depression`? Is that the ordering desired?

```
levels(nhanes$depression)
```

```
## [1] "Minimal"                "Mild"                   "Moderate to Severe"
```

The ordering is from less severe to more severe. We would like an OR > 1 to correspond to a risk factor associated with greater probability of a more severe outcome, so this is the ordering desired.

The code below demonstrates how to reorder the levels of an outcome variable, if needed. For this demonstration, we create an artificial dataset with a factor `x` that is ordered from high to low severity (the opposite of what we would like).

```
# Just for this example (skip if not needed)
set.seed(5)
dat <- data.frame(x = factor(sample(1:3, 100, replace=T),
                             levels = 1:3,
                             labels = c("High", "Med", "Low")))

# Current level ordering
levels(dat$x)
```

```
## [1] "High" "Med"  "Low"
```

Currently, the levels are in order from high to low. Let's reverse that ordering.

```
# Reorder
dat <- dat %>%
  mutate(x_new = factor(x,
                        levels = c("Low", "Med", "High")))

# Check derivation
table(dat$x, dat$x_new, useNA = "ifany")
```

```
##
##          Low Med High
##   High     0   0   34
##   Med      0  32    0
##   Low     34   0    0
```

```
# Check new level ordering
levels(dat$x_new)
```

```
## [1] "Low"  "Med"  "High"
```

The new variable `x_new` is exactly the same as the old variable `x` (see the two-way "Check derivation" table) except that now the levels are ordered from low to high.

6.22.4 Separation

As with binary logistic regression, an ordinal logistic regression model could suffer from separation if there are levels of a categorical predictor at which the outcome can be perfectly predicted. But, since an ordinal model works with *cumulative* probabilities, there are some situations when a zero in the two-way table of a categorical predictor vs. the ordinal outcome does not cause a problem. In particular, if there is a zero in just one cell then the model may still fit and provide reasonable results. For example, consider the following two-way table of a Yes/No predictor vs. a three-level ordinal outcome.

```
table(dat$predictor, dat$outcome)
#         1   2   3
# Yes 215  38  17
# No  495   0  56
```

No individuals with "No" for the predictor were at level 2 of the outcome. However, despite this zero, an ordinal logistic regression model will converge and produce reasonable results. If the single zero is in the first or last level of the outcome, however, the model may converge but violate the proportional odds assumption (see Section 6.22.10) because there will be separation in one of the binary logistic regressions used to check the assumption.

Being able to perfectly predict an ordinal outcome at a level of a predictor requires zeros in the table *at all but one level of the outcome*, such as the following two-way table.

```
table(dat$predictor, dat$outcome)
#         1   2   3
# Yes 215  38  17
# No    0   0  56
```

An ordinal logistic regression fit to the data from the above table will suffer from quasi-complete separation. In this particular example, the model seems to converge but produces a huge odds ratio and an error when attempting to compute confidence intervals.

To diagnose separation for an ordinal logistic regression, fit the model and examine the output for errors, warnings, and nonsense output (large standard errors, inverse logit of intercepts near 0 or 1, or odds ratios approaching 0 or infinite). If there seems to be an issue, check for zeros using two-way tables of categorical predictors vs. the outcome to determine which predictor is the problem, or if the problem is a sparse outcome. If the problem is a predictor, then filter, collapse, or remove the predictor (see Section 6.10.4). If the problem is the outcome, then filter or collapse the outcome. If doing so results in a two-level outcome, then the model is reduced to a binary logistic regression.

6.22.5 Fitting the model

This section demonstrates how to use `MASS::polr()` to fit the ordinal logistic regression model. Include the `Hess=T` option to speed up subsequent calls to `summary()`.

Example 6.5 (continued): Use ordinal logistic regression to test the association between depression (`depression`) and engaging in vigorous recreational activities (Yes/No, `PAQ650`).

```
fit.olr <- MASS::polr(depression ~ PAQ650,
                      data = nhanes,
                      Hess= T)
summary(fit.olr)
```

```
## Call:
## MASS::polr(formula = depression ~ PAQ650, data = nhanes, Hess = T)
##
## Coefficients:
##          Value Std. Error t value
## PAQ650No 0.347      0.173       2
##
## Intercepts:
##                           Value  Std. Error t value
## Minimal|Mild              1.360  0.151       9.037
## Mild|Moderate to Severe   2.739  0.180      15.220
##
## Residual Deviance: 1347.12
## AIC: 1353.12
## (55 observations deleted due to missingness)
```

The output does not include any p-values; we will compute them in Section 6.22.8.

6.22.6 Interpreting the coefficients

Intercepts

Since the outcome in our example has three levels, there are two intercepts, stored in a component of the model object called `zeta`. These are log-odds of cumulative probabilities, which can be seen by using `ilogit()` (in `Functions_rmph.R`) and comparing the result to the raw cumulative probabilities at the reference level of the predictor (`PAQ650` = "Yes").

```
ilogit(fit.olr$zeta)
```

```
##            Minimal|Mild Mild|Moderate to Severe
##                  0.7958                  0.9393
```

```
cumsum(
  prop.table(
    table(nhanes$depression[nhanes$PAQ650 == "Yes"])))
```

```
##            Minimal                    Mild Moderate to Severe
##             0.7963                  0.9370             1.0000
```

In other words, for the predictor at its reference level, the inverse logit of the first intercept is $P(Y \leq 1)$ and the inverse logit of the second intercept is $P(Y \leq 2)$. There is no third intercept because Y only has three levels, so $P(Y \leq 3)$ is always 1 and there is no need to estimate it.

The first intercept is the log odds of Minimal depression. The label "Minimal|Mild" is expressing the fact that the odds correspond to $P(\text{Depression} \leq \text{Minimal})/P(\text{Depression} > \text{Minimal})$ which is equivalent to $P(\text{Depression} \leq \text{Minimal})/P(\text{Depression} \geq \text{Mild})$, hence the label "Minimal|Mild". Similarly, the second intercept is the log odds of "up to Mild depression" (which includes Minimal and Mild). The label "Mild|Moderate to Severe" is expressing the fact that the odds correspond to $P(\text{Depression} \leq \text{Mild})/P(\text{Depression} \geq \text{Moderate to Severe})$.

Conclusion: The estimated proportion of individuals with Minimal depression is 0.796 and the estimated proportion of individuals with Minimal or Mild depression is 0.939.

NOTE: Above, the estimated cumulative probabilities are similar but not exactly equal to the corresponding raw cumulative probabilities. Had there been no predictor in the model, these would be identical, as shown below.

```
fit0 <- MASS::polr(depression ~ 1,
                   data = nhanes,
                   Hess= T)
ilogit(fit0$zeta)
```

```
##           Minimal|Mild Mild|Moderate to Severe
##                 0.7513                  0.9228
```

```
cumsum(
  prop.table(
    table(nhanes$depression)))
```

```
##          Minimal                    Mild Moderate to Severe
##           0.7513                  0.9228             1.0000
```

Predictor coefficients

Each predictor's coefficient is the log odds ratio for that predictor. Exponentiate to get OR $= e^{\eta}$ and use `confint()` and exponentiate to compute a 95% confidence interval for the OR.

```
# OR and 95% CI
CI <-  confint(fit.olr)
data.frame(
  OR    = exp(fit.olr$coefficients),
  lower = exp(CI[1]),
  upper = exp(CI[2])
)
```

```
##              OR lower upper
## PAQ650No 1.415 1.013 2.001
```

```
# Use the following if there is more than one predictor row
# e.g., CI[,1] instead of CI[1] since CI is now a matrix
data.frame(
  OR    = exp(fit.olr$coefficients),
  lower = exp(CI[,1]),
  upper = exp(CI[,2])
)
```

Conclusion: Individuals who do not engage in vigorous recreational activities have 41.5% greater odds of more severe depression (OR = 1.415, 95% CI = 1.013, 2.001).

NOTE: The interpretation of the OR > 1 for vigorous recreational activity is expressed as corresponding to "greater odds of *more severe* depression" because the probability is shifting toward higher values of the outcome (illustrated in the next section). In a binary logistic regression of the outcome "depression" (yes vs. no), we would just say "greater odds of depression".

6.22.7 What does an OR > 1 mean?

In the end, practically, an ordinal logistic regression results in exactly the same number of ORs as a binary logistic regression – one for each continuous predictor and one less than the number of levels for each categorical predictor. When interpreting the effect of a predictor on the outcome, the interpretation will be very similar between binary and ordinal logistic regression.

For example, if you fit a binary logistic regression and estimate the OR for a continuous variable X to be 1.50, you conclude that those with 1-unit greater X have 50% greater odds of the outcome. Implied in this statement is that they have 50% greater odds of having $Y = 1$ compared to $Y = 0$ because the odds are the ratio $P(Y = 1)/P(Y = 0)$. The interpretation of an OR of 1.50 in an ordinal logistic regression is that those with a 1-unit greater X have 50% greater odds of having a *greater* outcome – 50% greater odds of $Y > 1$ compared to $Y \leq 1$, 50% greater odds of $Y > 2$ compared to $Y \leq 2$, ..., and 50% greater odds of $Y > L - 1$ compared to $Y \leq L - 1$.

Again, in an ordinal logistic regression, an OR > 1 means that one level of a predictor is associated with greater probability of more severe levels of the outcome. This section goes a bit deeper to illustrate what that statement means, as well as how the aforementioned proportional odds assumption constrains the shift in probability toward more severe levels.

Example 6.5 (continued): To further clarify the interpretation of an OR estimated from an ordinal model, consider the distribution of depression at each level of vigorous recreational activity. The fact that the OR is larger than 1 corresponds to the probability distribution shifting to more severe levels of depression when moving from the reference level to the non-reference level.

```
TAB.Yes <- table(nhanes$depression[nhanes$PAQ650 == "Yes"])
TAB.No  <- table(nhanes$depression[nhanes$PAQ650 == "No"])

OUT <- rbind(prop.table(TAB.Yes),
             prop.table(TAB.No))
rownames(OUT) <- c("P(Depression) @ Physical activity = Yes (ref)",
                   "P(Depression) @ Physical activity = No")

round(OUT,      3)
```

```
##                                               Minimal  Mild Moderate to Severe
## P(Depression) @ Physical activity = Yes (ref)   0.796 0.141              0.063
## P(Depression) @ Physical activity = No          0.733 0.184              0.083
```

Individuals in the non-reference level (in the second row of the table above) have a greater chance of having more severe depression. This "shift toward higher probabilities" happens in a manner that is constrained by the proportional odds assumption. To help clarify this assumption, consider the following set of empirical ORs (based on the raw data, not estimated from a model), each comparing the odds of being in higher levels vs. lower levels at different outcome cutoff values.

```
# Empirical ORs, computed based on raw probabilities

# Odds of > Minimal vs. <= Minimal at PA = Yes
ODDS.Y <- (0.141 + 0.063) / (1 - 0.141 - 0.063)

# Odds of > Minimal vs. <= Minimal at PA = No
ODDS.N <- (0.184 + 0.083) / (1 - 0.184 - 0.083)

OR1.RAW <- ODDS.N / ODDS.Y

# Odds of > Mild vs. <= Mild at PA = Yes
ODDS.Y <- 0.063 / (1 - 0.063)

# Odds of > Mild vs. <= Mild at PA = No
ODDS.N <- 0.083 / (1 - 0.083)

OR2.RAW <- ODDS.N / ODDS.Y

c(OR1.RAW, OR2.RAW)
```

```
## [1] 1.421 1.346
```

Next, consider the corresponding ORs computed using probabilities estimated from the model, and compare them to the empirical ORs above, as well as to the OR computed by exponentiating the estimated regression coefficient from the model. This code introduces the use of `predict()` with a `polr` object, a topic which will be discussed in Section 6.22.9.

```
# Distributions estimated from the model
predict(fit.olr,
        data.frame(PAQ650 = c("Yes", "No")),
        type="p")
```

```
##    Minimal   Mild Moderate to Severe
## 1   0.7958 0.1435            0.06073
## 2   0.7336 0.1826            0.08384
```

```
# ORs computed from probabilities estimated by the model

# Odds of > Minimal vs. <= Minimal at PA = Yes
ODDS.Y <- (0.1434646 + 0.06072555) / (1 - 0.1434646 - 0.06072555)

# Odds of > Minimal vs. <= Minimal at PA = No
ODDS.N <- (0.1825769 + 0.08383594) / (1 - 0.1825769 - 0.08383594)

OR1.EST <- ODDS.N / ODDS.Y

# Odds of > Mild vs. <= Mild at PA = Yes
ODDS.Y <- 0.06072555 / (1 - 0.06072555)

# Odds of > Mild vs. <= Mild at PA = No
```

```
ODDS.N <- 0.08383594 / (1 - 0.08383594)

OR2.EST <- ODDS.N / ODDS.Y

list("Empirical ORs"                            = c(OR1.RAW, OR2.RAW),
     "ORs computed from estimated probabilities" = c(OR1.EST, OR2.EST),
     "OR = exp(beta) from model" = exp(fit.olr$coefficients["PAQ650No"]))
```

```
## $`Empirical ORs`
## [1] 1.421 1.346
##
## $`ORs computed from estimated probabilities`
## [1] 1.415 1.415
##
## $`OR = exp(beta) from model`
## PAQ650No
##    1.415
```

This illustrates how the ordinal logistic regression model *assumes* the ORs comparing outcome groups based on different cutoffs are the same at all cutoffs (the proportional odds assumption). The OR computed from the model by exponentiating the estimated regression coefficient is equal to each of the ORs computed using the estimated probabilities. How close these are to the empirical ORs is a measure of the validity of the proportional odds assumption (to be discussed further in Section 6.22.10).

6.22.8 Adjusted model

As with all regression models, control for confounding by including additional predictors in the model.

Example 6.5 (continued): Use ordinal logistic regression to test the association between depression and engaging in vigorous recreational activities (Yes/No, `PAQ650`), adjusted for age (`RIDAGEYR`) and income (`income`).

In this example, we will also compute p-values. The p-values for the AORs are based on Wald tests. However, `car::Anova()` only does likelihood ratio (LR) tests for `polr` objects, so the p-values for 1 df predictors may not match between the two. In general, LR tests are preferred (see Section 6.18).

```
fit.olr.adj <- MASS::polr(depression ~ PAQ650 + RIDAGEYR + income,
                          data = nhanes,
                          Hess= T)

# AORs, 95% CIs, and normal approximation p-values
CI    <- confint(fit.olr.adj)
TSTAT <- summary(fit.olr.adj)$coef[1:nrow(CI), "t value"]
data.frame(
  AOR   = exp(fit.olr.adj$coefficients),
  lower = exp(CI[,1]),
  upper = exp(CI[,2]),
  p     = 2*pnorm(abs(TSTAT), lower.tail = F)
)
```

```
##                                AOR  lower  upper             p
## PAQ650No                    1.5078 1.0282 2.2416 0.03853833290
## RIDAGEYR                    1.0005 0.9911 1.0099 0.91957309935
## income$25,000 to <$55,000 0.5230 0.3418 0.7995 0.00275511997
## income$55,000+              0.3208 0.2151 0.4790 0.00000002487
```

```
# Type 3 test p-values (LR test)
car::Anova(fit.olr.adj, type = 3)
```

```
## Analysis of Deviance Table (Type III tests)
##
## Response: depression
##           LR Chisq Df Pr(>Chisq)
## PAQ650        4.43  1      0.035 *
## RIDAGEYR      0.01  1      0.920
## income       30.73  2 0.00000021 ***
## ---
## Signif. codes:  0 '***' 0.001 '**' 0.01 '*' 0.05 '.' 0.1 ' ' 1
```

```
# Use the following if there is only one predictor
# (the code above will produce an error in that case
#  since CI is a vector instead of a matrix)
CI    <- confint(fit.olr.adj)
TSTAT <- summary(fit.olr.adj)$coef[1, "t value"]
data.frame(
  OR    = exp(fit.olr.adj$coefficients),
  lower = exp(CI[1]),
  upper = exp(CI[2]),
  p     = 2*pnorm(abs(TSTAT), lower.tail = F)
)
```

Conclusion: After adjusting for age and income, vigorous recreational activity is significantly associated with depression (p = .035). Those who do not engage in vigorous recreational activities have 50.8% greater odds of having more severe depression than those who do (OR = 1.51, 95% CI = 1.03, 2.24). Age was not significantly associated with depression (p = .920), but income was (p <.001), with those with greater income having lower odds of more severe depression.

6.22.9 Prediction

Use `predict()` with `type = "p"` to estimate the probability of an ordinal outcome at each of its levels. If you leave out `type = "p"`, the default will be to return the level with the greatest estimated probability. As usual, pass a `data.frame()` to `predict()` containing values for all the predictors in the model.

Example 6.5 (continued): What is the estimated probability of each depression category for individuals who are age 45 years, who have an annual household income of at least $55,000, and who do and do not engage in vigorous recreational activities.

```
NEWDAT <- data.frame(PAQ650   = c("Yes", "No"),
                     RIDAGEYR = 45,
                     income   = "$55,000+")
cbind(NEWDAT,
      predict(fit.olr.adj, NEWDAT, type="p"))
```

```
##   PAQ650 RIDAGEYR    income Minimal   Mild Moderate to Severe
## 1    Yes       45 $55,000+  0.8593 0.1055             0.03516
## 2     No       45 $55,000+  0.8021 0.1459             0.05208
```

Conclusion: Among 45-year-olds making at least $55,000 per year who do not engage in vigorous recreational activities, an estimated 5.2% have moderate to severe depression, compared to 3.5% among those who do.

The following confirms that these predictions are constrained by the proportional odds assumption. The function `odds.ratio()` (in `Functions_rmph.R`) computes an OR from two probabilities.

```
# OR comparing odds of Depression > Minimal
# to the odds of Depression <= Minimal
# between PAQ650 = No and Yes
odds.ratio(0.1458677 + 0.05207582,
           0.1055049 + 0.03515519)
```

```
## [1] 1.508
```

```
# OR comparing odds of Depression > Mild
# to the odds of Depression <= Mild
# between PAQ650 = No and Yes
odds.ratio(0.05207582,
           0.03515519)
```

```
## [1] 1.508
```

```
# OR comparing PAQ650 = No vs. Yes
# estimated by ordinal logistic regression
exp(fit.olr.adj$coefficients)["PAQ650No"]
```

```
## PAQ650No
##    1.508
```

6.22.10 Checking the proportional odds assumption

See Fagerland and Hosmer (2013) for a formal statistical test of the proportional odds assumption. In general, however, such goodness-of-fit tests can lack power in small sample sizes and in large sample sizes can detect practically non-meaningful deviations from the assumption. Therefore, whether one chooses to conduct a formal test or not, it is recommended to examine the assumption numerically and/or visually. We present here a numeric examination of the proportional odds assumption that helps both in understanding what the assumption means and in assessing its validity.

Suppose that instead of an ordinal logistic regression you fit a series of binary logistic regressions with the outcomes $Y > 1$, $Y > 2$, ... $Y > L - 1$. Each of these regressions estimates a different OR. The proportional odds (PO) assumption is equivalent to assuming that all of these ORs are equal. That is, the effect of a predictor with OR > 1 on the outcome is to shift the probability toward larger values of Y and the magnitude of the shift (on the logit scale) is the same between all levels of the ordinal variable.

This implies that one way to assess the adequacy of the proportional odds assumption is to fit this series of binary logistic regressions and, for each predictor, compare the estimated ORs from the series to the single estimated OR from the corresponding ordinal logistic regression.

NOTE: If a continuous predictor has a very large range, then its OR may be near 1 and may appear to be similar between the ordinal and binary models when, in fact, it is not. Rescale such continuous predictors so their ORs will be easier to judge. Dividing by the interquartile range is a reasonable choice.

Example 6.5 (continued): Assess the proportional odds assumption in the adjusted model by comparing the estimated ordinal logistic regression AORs to those from a series of binary logistic regression models. Before fitting the models, rescale age so a 1-unit difference corresponds to a difference in age equal to the interquartile range.

Since the outcome `depression` has three levels (Minimal, Mild, Moderate to Severe), create two binary outcome variables, one corresponding to ($>$ Minimal) vs. ($\leq$ Minimal) and one corresponding to ($>$ Mild) vs. ($\leq$ Mild). Also, to be sure the correct level is the reference level, use `relevel`. We want the less severe category to be the reference level so the binary logistic regression AORs will be in the same direction as the ordinal.

```
levels(nhanes$depression)
```

```
## [1] "Minimal"            "Mild"               "Moderate to Severe"
```

```
# Create a series of binary variables
nhanes <- nhanes %>%
  mutate(# Y1 = Depression > Minimal
         Y1 = fct_collapse(depression,
                           ">Minimal"  = c("Mild", "Moderate to Severe"),
                           "<=Minimal" = "Minimal"),
         Y1 = relevel(Y1, ref = "<=Minimal"),
         # Y2 = Depression > Mild
         Y2 = fct_collapse(depression,
                           ">Mild"   = "Moderate to Severe",
                           "<=Mild" = c("Minimal", "Mild")),
         Y2 = relevel(Y2, ref = "<=Mild"))

# Check derivations
table(nhanes$depression, nhanes$Y1, exclude = F)
```

```
##
##                      <=Minimal >Minimal <NA>
##   Minimal                  710        0    0
##   Mild                       0      162    0
##   Moderate to Severe         0       73    0
##   <NA>                       0        0   55
```

```
table(nhanes$depression, nhanes$Y2, exclude = F)
```

```
##
##                          <=Mild >Mild <NA>
##     Minimal                 710     0    0
##     Mild                    162     0    0
##     Moderate to Severe        0    73    0
##     <NA>                      0     0   55
```

Next, rescale age by dividing by its interquartile range.

```
summary(nhanes$RIDAGEYR)
```

```
##     Min. 1st Qu.  Median    Mean 3rd Qu.    Max.
##     20.0    32.0    47.0    47.7    61.0    80.0
```

```
IQR(nhanes$RIDAGEYR)
```

```
## [1] 29
```

```
nhanes <- nhanes %>%
  mutate(age29 = RIDAGEYR/29)
```

Re-fit the ordinal model and fit the series of binary logistic regression models. Make sure to include the rescaled continuous variable (`age29`) in all the models.

```
fit.ordinal <- MASS::polr(depression ~ PAQ650 + age29 + income,
                          data = nhanes,
                          Hess= T)

fit.binary1 <- glm(Y1 ~ PAQ650 + age29 + income,
                   family = binomial, data = nhanes)

fit.binary2 <- glm(Y2 ~ PAQ650 + age29 + income,
                   family = binomial, data = nhanes)
```

Finally, compare the AORs. For the binary models, include `[-1]` so the intercept is not included (for the ordinal model `$coefficients` already does not include any intercepts).

```
exp(
  cbind(
    "ordinal"  = fit.ordinal$coefficients,
    "binary 1" = fit.binary1$coefficients[-1],
    "binary 2" = fit.binary2$coefficients[-1]
  )
)
```

```
##                            ordinal binary 1 binary 2
## PAQ650No                    1.5078   1.5149   1.5154
## age29                       1.0141   0.9907   1.1333
## income$25,000 to <$55,000   0.5230   0.4976   0.6919
## income$55,000+              0.3208   0.3010   0.4923
```

Conclusion: The proportional odds assumption is questionable for age and income. However, the assumption is well met for our primary predictor, physical activity. Thus, the violation of the proportional odds assumption for the other predictors does not impact the primary conclusions. Also, for income, the conclusions regarding direction (all AORs are < 1 for each level of income) and ordering (odds of the outcome decrease with greater income) are the same for all three models.

NOTE: If, for any predictor, the binary ORs are wildly different from the ordinal OR, in particular if they look like they are the inverse of the ordinal OR, then check the code in which you derived the binary outcomes. Make sure each binary outcome is of the form $Y > \ell$ vs. $Y \leq \ell$ where the latter is the reference level.

6.23 Exercises

1. True or false? Binary logistic regression is used for outcomes with three or more possible levels.

2. True or false? Binary logistic regression can be used to estimate an odds ratio adjusted for confounding due to other variables.

3. Write a sentence interpreting an odds of 2:1 for an event.

4. Write a sentence interpreting an odds ratio of 1.90.

5. Write a sentence interpreting an odds ratio of 2.10.

6. Write a sentence interpreting an odds ratio of 0.42.

7. What is the interpretation of the intercept (β_0) in a binary logistic regression model?

8. How can you convert the intercept in a binary logistic regression model to a probability?

9. What is the interpretation of the regression coefficient (β) for a continuous predictor X in a binary logistic regression model?

10. What is the interpretation of the regression coefficients for a categorical predictor in a binary logistic regression model?

11. How can you convert a regression coefficient in a binary logistic regression model to an odds ratio?

12. What assumption(s) of a linear regression model are not assumption(s) of a binary logistic regression model? What assumption(s) do they have in common?

13. What statistical method can be used to estimate a risk ratio (or prevalence ratio)?

For Exercises 14 to 25, use the NHANES 2017-2018 examination subsample teaching dataset (`nhanes1718_adult_exam_sub_rmph.Rdata`, see Appendix A.1). Create a binary version of PHQ-9 representing mild depression (PHQ-9 $\geq$ 5) using the following code prior to answering the questions.

```
load("Data/nhanes1718_adult_exam_sub_rmph.Rdata")

# Create dichotomized PHQ-9

# "PHQ-9 scores of 5, 10, 15, and 20 represented mild, moderate,
# moderately severe, and severe depression, respectively"
# (https://www.ncbi.nlm.nih.gov/pmc/articles/PMC1495268/)

nhanes <- nhanes_adult_exam_sub %>%
  mutate(depression_mild = factor(phq9 >= 5,
                                  levels = c(F, T),
                                  labels = c("No", "Yes")))
# Check
table(nhanes$phq9, nhanes$depression_mild, useNA = "ifany")
tapply(nhanes$phq9, nhanes$depression_mild, range, na.rm=T)
```

14. Create a 2×2 table of mild depression (`depression_mild`) vs. walk or bicycle (`PAQ635`, "In a typical week do you walk or use a bicycle for at least 10 minutes continuously to get to and from places?") and, using the table, compute the odds ratio comparing the odds of mild depression between those who do not and those who do answer "Yes" to the walk or bicycle question.

15. Create a 2×2 table of mild depression (`depression_mild`) vs. trouble sleeping (`SLQ050`, "Have you ever told a doctor or other health professional that you have trouble sleeping?") and, using the table, compute the odds ratio comparing the odds of mild depression between those who do and those who do not answer "Yes" to the sleeping question.

16. Suppose you are going to fit a logistic regression where the outcome is mild depression. What probability is `glm()` modeling? If it is not already, modify the outcome variable so `glm()` will model P(mild depression = Yes).

17. Compute the odds ratio comparing the odds of mild depression (`depression_mild`) between those who do and those who do not answer "Yes" to the sleeping question (`SLQ050`) using logistic regression, as well as its 95% confidence interval. Assume `depression_mild` is the outcome.

18. Do the odds of mild depression (`depression_mild`) differ between individuals of different income levels (`income`)? Test the global significance of income using a Type III Wald test and compute the OR, 95% CI, and p-value comparing each possible pair of levels.

19. Is mild depression (`depression_mild`) associated with the number of days someone engages in vigorous recreational activities (`PAQ655`, "In a typical week, on how many days do you do vigorous-intensity sports, fitness or recreational activities?")? What is the OR comparing mild depression between individuals who differ by 1 day in days of vigorous recreation? What about between those who differ by 5 days?

20. Is mild depression (`depression_mild`) significantly associated with trouble sleeping (`SLQ050`) after adjusting for age (`RIDAGEYR`), gender (`RIAGENDR`), income (`income`), and days someone engages in vigorous recreational activities (`PAQ655`)? Answer the question and report the AORs, 95% confidence intervals, and p-values. Also, interpret the AOR for trouble sleeping.

21. Create a forest plot to illustrate the AORs and their 95% CIs for the model from the previous exercise. For each continuous predictor, plot the AOR corresponding to a difference in the predictor equal to its interquartile range (IQR).

22. Using the model from Exercise 20, what is the predicted prevalence (and 95% CI) of mild depression among those with and without trouble sleeping who are age 40 years, male, earn $25,000 to <$55,000 per year, and do not engage in any days of vigorous recreational activities?

23. After adjusting for age (`RIDAGEYR`), gender (`RIAGENDR`), income (`income`), and days someone engages in vigorous recreational activities (`PAQ655`), does the association between mild depression (`depression_mild`) and trouble sleeping (`SLQ050`) depend on gender?

24. Using the model from the previous exercise, test the overall significance of trouble sleeping.

25. Using the model from Exercise 23, estimate the AOR for mild depression comparing those without and with trouble sleeping separately for males and females (along with their 95% CIs and p-values). Based on the answer to Exercise 23, are these two AORs significantly different from each other?

For Exercises 26 to 27, use the NHANES 2017-2018 fasting subsample teaching dataset (`nhanes1718_adult_fast_sub_rmph.Rdata`, see Appendix A.1).

26. Fit a logistic regression model to test the association between the outcome "Ever told had congestive heart failure?" (`MCQ160B`) and "How often do you snort or stop breathing?" (`SLQ040`), adjusted for age (`RIDAGEYR`), gender (`RIAGENDR`), and income (`income`). Look at the table of regression coefficients. Do you see any indicators of a problem with quasi- or complete separation?

27. Check for quasi- or complete separation in the model from the previous exercise, resolve any issues you find, and re-fit the model.

For Exercises 28 to 30, use the 2019 National Survey of Drug Use and Health (NSDUH) teaching dataset (`nsduh2019_adult_sub_rmph.RData`, see Appendix A.5).

28. Fit a logistic regression for the outcome substance use treatment (`tx_substance_lifetime`) vs. the predictor age of first cigarette use (`cig_agefirst`) and check the linearity assumption. Is the relationship linear?

29. For the model you fit in the previous exercise, relax the linearity assumption in each of the following three ways (not all at once): (1) log transformation of the predictor, (2) square-root transformation of the predictor, and (3) add a quadratic term for the predictor. Re-check the linearity assumption for each. Which transformation would you choose and why?

30. Compare the models with log, square-root, and quadratic transformed predictors from the previous exercise based on influential observations and goodness-of-fit (Hosmer-Lemeshow test, calibration plot). Is one preferable to the other?

31. Suppose you have a dataset with $n = 1000$ observations. In order to avoid overfitting, what is the maximum number of predictors you should include in a logistic regression model if the proportion of observations with the outcome is 0.15? What if the proportion is 0.70? What can you say about the relationship between the sample proportion and the number of predictors for a given sample size?

32. You are designing a study for which you wish to fit a logistic regression model with seven predictors. Assuming the population prevalence is 0.80, what is the minimum sample size you need to avoid overfitting? What if the population prevalence is 0.35? What can you say about the relationship between the population prevalence and the minimum sample size for a given number of predictors?

33. The matched case-control teaching dataset `nhanes_CC_rmph.Rdata` was created from a subset of adults from the 2017-2018 NHANES teaching dataset (Appendix A.1) containing 91 individuals who were ever told they had cancer or malignancy and 799 individuals who were not, matched on gender (`RIAGENDR`) and income (`income`). Assess the association between having been told one has cancer or malignancy and smoking status (`smoker` = Never, Past, or Current), accounting for the matching in the analysis. Compare all three levels of smoking status and give a plausible explanation of the results. NOTE: As in the example in the text, you must first convert the outcome to a 0/1 variable.

34. Repeat Exercise 20, but this time instead of computing the AOR, compute the adjusted prevalence ratio (APR).

For Exercises 35 and 36, use the 2018 Natality subsample teaching dataset (`natality2018_rmph.Rdata`, see Appendix A.3).

35. Is cigarette smoking during pregnancy (`CIG_REC`) associated with the outcome birthweight category (`BWTR4` = Normal, Low birthweight (LBW), or Very low birthweight (VLBW)), adjusted for maternal age (`MAGER`), maternal education (`MEDUC`), and the presence of risk factors (`risks`)?

36. Check the proportional odds (PO) assumption for the model you fit in the previous exercise.

7

Survival Analysis

In this chapter, you will learn how to:

- Identify a time-to-event outcome;
- Identify types and mechanisms of censoring;
- Interpret the survival and hazard functions;
- Use the Kaplan-Meier method to estimate the survival function;
- Use the log-rank test to compare survival between groups;
- Visualize the Kaplan-Meier estimate of the survival function;
- Visualize the estimated hazard function;
- Fit a Cox proportional hazards regression model, including the following:
  - Write and interpret the Cox regression equation;
  - Estimate unadjusted and adjusted hazard ratios;
  - Estimate the probability that an event has not yet occurred as of a given time;
  - Estimate the hazard of an event relative to a reference group;
  - Visualize the Cox regression estimate of the survival function;
  - Test interactions between predictors;
  - Incorporate time-varying predictors;
  - Check the proportional hazards assumption;
  - Allow non-proportional hazards using a time interaction or stratification;
  - Check the linearity assumption, examine outliers, and identify influential observations; and
  - Appropriately summarize the methods and results.

To use the code in this chapter, first load the `tidyverse` and `survival` (Grambsch and Therneau, 2000; Therneau, 2023) packages.

```
library(tidyverse)
library(survival)
```

7.1 Introduction

Survival analysis is a commonly used term for the analysis of an outcome that is a time to an event. The name refers to its use to study time to death, but the event can be anything. For example, consider the following examples:

- **Time to preterm birth:** In a prospective study of the effect of maternal periodontal disease on preterm birth, researchers randomized pregnant women to either receive a treatment intervention or be in a control group and then recorded gestational age at the end of the pregnancy (Michalowicz et al., 2006). The event is preterm birth and the time is gestational age.

- **Time to myocardial infarction (MI):** The Framingham Heart Study[1] is a longitudinal study that began in 1948 in Framingham, Massachusetts, with the goal of elucidating the etiology of cardiovascular disease (Dawber et al., 1951; Mahmood et al., 2014). Among the numerous outcomes recorded were hospitalization for MI ("heart attack"). For this outcome, the event is hospitalized MI, and the time is years from the baseline study examination to hospitalized MI.
- **Transition to heroin:** In a prospective natural history study of users of non-prescribed pharmaceutical opioids ("pain pills") who had never used heroin, researchers followed individuals every 6 months for 3 years after baseline to determine what factors influenced the time from first illicit pain pill use to first use of heroin (Carlson et al., 2016). The event is first use of heroin, and the time is years from initiation of illicit use of pain pills to first use of heroin.
- **Mental health / substance use referrals and juvenile recidivism:** Researchers used retrospective data to compare time from release to re-offense (recidivism) among first-time juvenile offenders between those who did and did not receive a mental health or substance use referral (Zeola et al., 2017). The event is recidivism, and the time is days from release to re-offense.

In each of these examples, the outcome is the **time to an event**. A complication is that not all individuals experience the event before they are lost to follow-up, before they experience a competing event that removes the possibility of experiencing the event of interest, or before the study ends. We consider such an individual's event time to be **censored**; if they were to ever experience the event, their time-to-event would be larger than the time at which we last observed them. In other words, we do not know their hypothetical event time, only that it is larger than some number. Survival analysis is designed to handle censored data.

Table 7.1 describes the time-to-event outcome and censoring for the studies listed above:

In the time to MI example, an individual who dies prior to experiencing MI has an event time that is censored at the time of death, but they *could have* experienced the event later had they not died. Censoring in the preterm birth example, however, is conceptually different. In that example, a pregnancy that reaches full term (37 weeks) has an event time that is censored at 37 weeks, but it *could not have* resulted in a preterm birth later because a birth at 37 weeks or later is not preterm, by definition. So, in some sense, their time to preterm birth is infinite. For the methods we will be learning in this chapter (Kaplan-Meier estimate of survival, Cox regression), the results will be identical whether we consider these event times censored at 37 weeks, 40 weeks, or 400 weeks. Despite the awkwardness in the definition of censoring, survival analysis allows us to compare not only the occurrence of preterm birth (as would be the case if we used logistic regression) but also the timing.

NOTE: For three of the four examples above, individuals were under observation from their time origin on. For the transition to heroin example, however, the time origin was prior to the start of observation. The study included only those who had begun using pain pills but not yet transitioned to heroin. Therefore, individuals who otherwise would have been sampled and included in the study but had already transitioned to heroin were excluded. This is an example of **left-truncation** – the distribution of observed event times is truncated on the left, excluding those whose time origin is prior to the beginning of observation. Those individuals have, on average, shorter times to event. Fortunately, the method used for handling time-varying predictors in Section 7.14 also handles left-truncation.

[1] https://www.nhlbi.nih.gov/science/framingham-heart-study-fhs

TABLE 7.1 Time origin, event, time-to-event, and censoring for four studies

Study	Time Origin	Event	Time-to-Event	Censoring
Time to preterm birth	First day of last menstrual cycle	Preterm birth (live or non-live) (gestational age < 37 weeks)	T = Weeks from first day of last menstrual cycle to preterm birth	Termination of pregnancy or loss to follow-up prior to 37 weeks: T > gestational age at that time; Gestational age at least 37 weeks: T > 37
Time to myocardial infarction (MI)	Baseline examination	MI	T = Years from baseline to MI	Loss to follow-up, end of follow-up, or death prior to MI: T > time at last follow-up or death
Transition to heroin	First illicit use of prescribed pain pills	First use of heroin	T = Years to transition to heroin	Loss to follow-up prior to heroin use or no heroin use at final (36m) interview: T > time at last follow-up
Juvenile recidivism	Release from first offense	Next offense	T = Days to recidivism	No repeat offense before aging out of system, loss to follow-up, or study conclusion: T > time at last follow-up

Comparison to linear and logistic regression

Why not use linear regression? Event times are numeric variables, after all. However, even if the assumptions of a linear regression model were met, standard linear regression cannot handle censoring.

Why not use logistic regression? Events are binary variables, after all – the event either happened or not within the follow-up period. While this approach would be able to handle censoring due to follow-up ending, it would not correctly handle censoring due to earlier loss to follow-up. Also, logistic regression fails to account for *when* the event occurred. For example, in our preterm birth example, treating the outcome as binary (preterm birth vs. not preterm birth) would allow us to compare the odds of preterm birth between groups (with those lost to follow-up treated as missing data). But suppose that in one group not only were there more preterm births, but they also tended to occur earlier. Logistic regression, which ignores event timing, would underestimate the risk associated with being in this group. Survival analysis, however, would estimate the risk based on both the occurrence and timing of preterm births.

In summary, unlike either logistic or linear regression, survival analysis handles both the "if" and "when" of events. It also handles the fact that the "if" may only be known to be "not yet", resulting in a censored "when", because individuals may exit a study for a reason other than the event prior to the end of the study, and because observation time is finite.

7.2 Censoring

An event time T is **censored** if it is not known exactly but is known to be in some range. Three kinds of censored times are:

- **Right-censored:** $T > t$. The event has not yet occurred as of time t and may never occur.
- **Left-censored:** $T < t$. The event occurred before time t.
- **Interval-censored:** $t_1 < T < t_2$. The event occurred between times t_1 and t_2.

In this text, we will only consider right-censored event times.

Example 7.1: Figure 7.1 illustrates censoring for 6 of 2000 individuals from the 2019 US Natality data teaching dataset. As described in Appendix A.3, gestational ages >37 weeks were censored at 37 weeks (as in Michalowicz et al. (2006)) and a random subset of gestational ages were censored at times <37 weeks. Preterm birth times are denoted by solid circles, and censored times are denoted by open circles. Thirty-seven weeks is denoted by the vertical dashed line. Individuals 1, 3, and 5 experienced a preterm birth. Individual 2 was lost to follow-up at gestational age 20 weeks, so their event time is censored at 20 – all we know is that if this pregnancy ended in a preterm birth, it did so after 20 weeks. Individuals 4 and 6 had not yet given birth as of 37 weeks, so they, too, have censored event times (at 37 weeks).

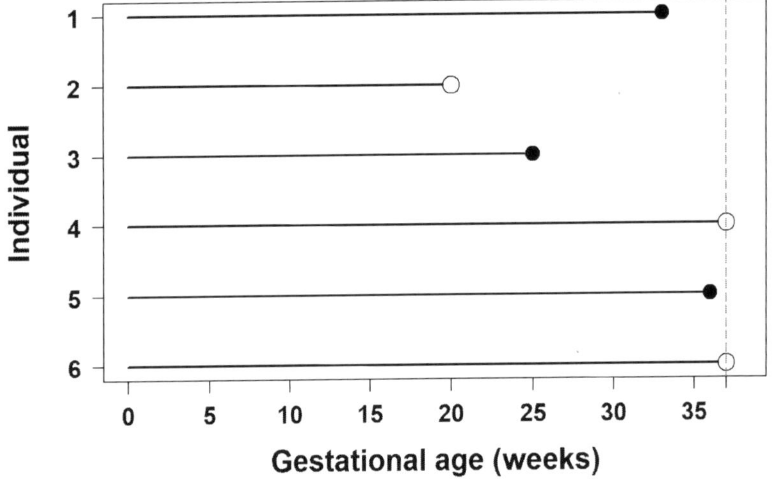

FIGURE 7.1 Censored and non-censored event times

There are multiple mechanisms that generate censored data:

- **Type I censoring:** The end of study time is pre-specified, making the censoring time(s) known before any data are collected.
- **Type II censoring:** The number of events is pre-specified. Thus, the end of study time and censoring time(s) are not known until after the data have been collected.
- **Random censoring:** The censoring times are not under the researcher's control.

In the studies listed above, all of which have right-censored event times, the censoring is a mix of Type I and random. Each had a specific end time (e.g., 37 weeks for the preterm birth study; 36-month interview for the transition to heroin study), but some times were censored for other, random, reasons (e.g., loss to follow-up).

NOTE: Even with a known end of study time resulting in Type I censoring, the end-of-study censoring time may vary between individuals. For example, in the transition to heroin study, once the first interview was completed, everyone's time from origin to end of study was known but, since not everyone began using pain pills at the same time, those times varied between participants.

7.2.1 Non-informative censoring assumption

The methods we will discuss assume that censoring is **non-informative.** Consider two individuals with the same risk factors, neither of whom have experienced the event of interest as of time t. One is lost to follow-up at that time (and so their event time is right-censored at t) and the other continues in the study. Assuming non-informative censoring means that we assume these two individuals have the same subsequent risk of experiencing the event – knowing that one of them has a censored time does not add any information.

For example, consider a study of an intervention (vs. control) intended to increase survival after cancer surgery. The study runs for 1 year, and patients who survive to the end of the study have a survival time that is right-censored at 1 year. The fact that these times are censored does not give you any information about these patients' subsequent likelihood of survival – the censoring had nothing to do with their condition other than the fact they survived to 1 year. Therefore, the censoring is non-informative. Suppose, instead, that an individual drops out of the study earlier than 1 year because they are not doing well. This individual may be less likely to survive than an individual who continued in the study beyond that time. Therefore, their censoring is informative. In this case, assuming non-informative censoring would lead to a biased assessment of the intervention. Censoring of individuals who drop out of the study early because they moved to another city where follow-up was not possible, however, would be non-informative (assuming the move had nothing to do with how well they were doing).

7.3 Survival function

The distribution of any random variable X can be described using its cumulative distribution function (CDF) $F(x) = P(X \leq x)$. This represents the probability that X takes on a value up to and including x. For a random event time T, the CDF is $F(t) = P(T \leq t)$, the probability that the event occurs prior to or at time t, but not later than t. In survival analysis, however, we typically work with the **survival function** $S(t)$ which is the probability that an individual has survived past t (not yet experienced the event as of time t). This is the complement of the CDF – instead of the cumulative probability up to and including t, it is the cumulative probability *after* t. So $S(t) = 1 - F(t) = P(T > t)$.

Example 7.1 (continued): The estimated survival function for the outcome preterm birth in the Natality teaching dataset is shown in Figure 7.2. The "event" is preterm birth, so "survival" past time t here means not yet having given birth as of time t.

For example, the function value at $t = 32$ weeks, $S(32)$, is 0.976, indicating that the probability of "survival" (not yet having a preterm birth) past 32 weeks is 97.6%. Equivalently, only

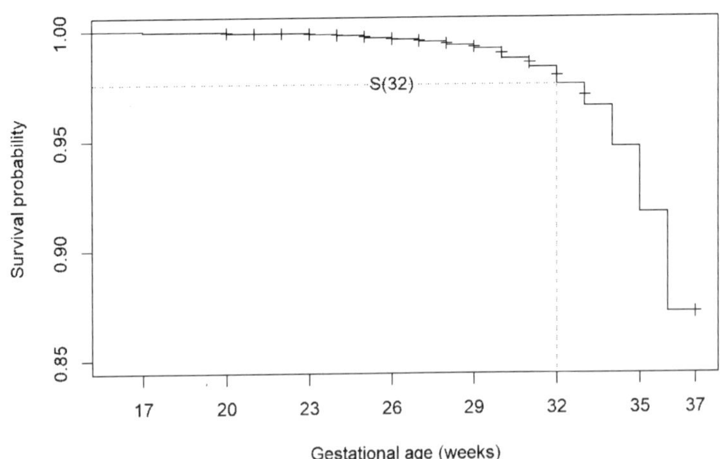

FIGURE 7.2 Survival function for time to preterm birth

2.4% of pregnancies resulted in a preterm birth prior to or at 32 weeks. The estimated survival function has a stair-step pattern. When reading a survival function at an event time where the function is drawn as a vertical line, the value of $S(t)$ is at the bottom of the step.

Out of $n = 2000$ pregnant women in this dataset, none experienced a preterm birth prior to gestational age 17 weeks, so $S(t) = 1$ for all times up to 17 weeks, at which time there was one preterm birth. The function starts at 1 (100% no event) and drops by $1/n$ at 17 weeks to 1999/2000, the proportion of individuals under observation who had not experienced a preterm birth as of that time.

The survival function drops at each time when an event occurred. Plus signs in the plot indicate censored times. The survival function does not drop at censored times (unless there is an observed event at that time, as well), but censored times do influence the size of the drop at the next observed event time since they are no longer in the denominator (the "risk set", discussed in Section 7.6). As time increases, moving left to right, the survival probability either stays the same or drops. At the end of the observation time, in this case at week 37 when the risk of preterm birth ends by definition, the remaining individuals have censored times and the estimated survival function beyond that day is not defined.

We will discuss in more detail how the survival function is estimated when we discuss the Kaplan-Meier method in Section 7.6.

7.4 Hazard function

The **cumulative risk** of an event is the probability that the event occurs within a specified time period. The **instantaneous risk** of an event is known as the **hazard** of the event. In survival analysis, we are interested in how the hazard changes over time, how it differs between groups, and how it depends on risk factors. The **hazard function** $h()$ at time t is defined mathematically as follows.

$$h(t) = \lim_{\Delta t \to 0} \frac{P(t \leq T < t + \Delta t | T \geq t)}{\Delta t} \tag{7.1}$$

Equation 7.1 describes the following process: given that a person has not yet experienced the event prior to time t (the $|T \geq t$ part of the numerator), compute the probability of the event occurring in a subsequent time window starting at t (the $t \leq T < t + \Delta t$ part of the numerator) per unit time (the Δt in the denominator) and then shrink the length of the time window to zero ($\lim_{\Delta t \to 0}$). Thus, the hazard function at time t is the instantaneous risk of experiencing the event among those who are still at risk. The hazard function is sometimes referred to as the **hazard rate**.

Example 7.1 (continued): The hazard is related to the rate of change in survival. For the Natality teaching dataset, the hazard function (solid line) for preterm birth and the corresponding survival function (dashed line) are shown in Figure 7.3. The hazard function is near 0 at times when the survival function is relatively flat (few events) and is greater when the survival function is steeper (events occurring more frequently). In this teaching dataset, the hazard of preterm birth is near 0 through around gestational age 21 weeks and there is a marked increase in hazard after around 28 weeks.

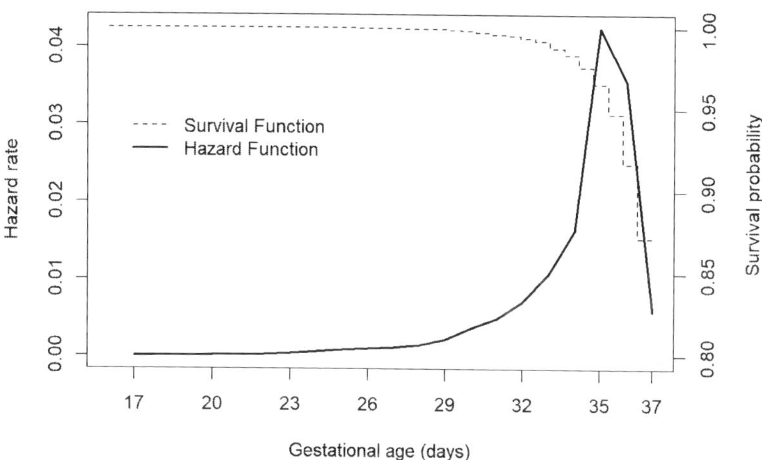

FIGURE 7.3 Hazard function for time to preterm birth data

7.5 Survival analysis dataset structure

The R functions we will use for survival analysis require a dataset with a specific structure. There must be a numeric **event time** variable and a binary **event indicator** variable, coded as numeric, with values of 1 for events, 0 for censored event times, and no other non-missing values. There can also be additional variables representing predictors. In the basic set-up, the predictors do not vary over time and so there is one row per individual. Later, we will discuss time-varying predictors (Section 7.14) which require a dataset with multiple rows per individual.

Example 7.1 (continued): The first five rows of the Natality teaching dataset, shown below, include the event time (`gestage37`), the indicator of preterm birth (`preterm01`), and a few time-invariant demographic variables and other risk factors – mother's age (`MAGER`), mother's race/Hispanic origin (`MRACEHISP`), previous preterm birth (`RF_PPTERM`), and previous Cesarean birth (`RF_CESAR`). Four of the five births have a gestational age censored at 37 weeks (`preterm01` = 0), and one was preterm at gestational age 31 weeks (`preterm01` = 1).

```
load("Data/natality2018_rmph.Rdata")
natality %>%
  select(gestage37, preterm01, MAGER, MRACEHISP, RF_PPTERM, RF_CESAR) %>%
  head(5)
```

```
## # A tibble: 5 x 6
##   gestage37 preterm01 MAGER MRACEHISP RF_PPTERM RF_CESAR
##       <dbl>     <dbl> <dbl> <fct>     <fct>     <fct>
## 1        37         0    35 Hispanic  No        Yes
## 2        31         1    28 NH White  Yes       No
## 3        37         0    22 NH Black  No        No
## 4        37         0    35 NH White  No        No
## 5        37         0    30 NH White  No        No
```

Verify the event time variable (`gestage37`) is numeric using `is.numeric()` and summarize the event times using `summary()`.

```
is.numeric(natality$gestage37)
```

```
## [1] TRUE
```

```
summary(natality$gestage37)
```

```
##    Min. 1st Qu.  Median    Mean 3rd Qu.    Max.
##    17.0    37.0    37.0    36.4    37.0    37.0
```

Similarly, verify the event indicator (`preterm01`) is numeric and use `table()` to verify its only non-missing values are 0 and 1.

```
is.numeric(natality$preterm01)
```

```
## [1] TRUE
```

```
table(natality$preterm01, useNA = "ifany")
```

```
##
##    0    1
## 1748  252
```

In Example 7.1, the event indicator variable was already in the correct format. What if it is not?

Example 7.2: The Digitalis Investigation Group (DIG) teaching dataset (`dig_rmph.rData`, see Appendix A.6) contains data from a clinical trial investigating the safety and efficacy of

Digoxin for treating congestive heart failure. One of the endpoints measured was toxicity (`DIG`). Examine if this event indicator variable is numeric with values 0 and 1 and, if it is not, then convert it to that form.

```
load("Data/dig_rmph.Rdata")
is.numeric(dig$DIG)
```

```
## [1] FALSE
```

```
table(dig$DIG, useNA = "ifany")
```

```
##
##     No Event First Event
##         6702          98
```

The variable is not numeric, but it does have just two values. Create a numeric event indicator variable that is 1 when the original variable is "First Event", and use `table()` to check the derivation. The syntax `dig$DIG == "First Event"` creates a logical vector of `TRUE` and `FALSE` values, and `as.numeric()` converts that logical vector to numeric, converting `TRUE` to 1 and `FALSE` to 0.

```
dig$DIG_event <- as.numeric(dig$DIG == "First Event")
table(dig$DIG, dig$DIG_event, useNA = "ifany")
```

```
##
##                  0    1
##   No Event    6702    0
##   First Event    0   98
```

The datasets used in this text all include an event time variable. However, in your future work you may encounter datasets for which you have to compute the event time. For example, you may be given the dates the individuals started being observed and dates that events occurred (or were censored). Computing the event times, the times between those dates, is facilitated in R by using date-formatted variables and functions specifically designed to count time units between date-formatted variables. See, for example, the chapter "Dates and Times" in **R for Data Science**[2] (Wickham et al., 2023).

7.6 Kaplan-Meier estimate of the survival function

The **Kaplan-Meier** (KM) estimate of the survival function is a non-parametric, or empirical, estimate. These terms just mean that the KM estimate makes no assumption about the shape of the survival function. By analogy, a probability histogram is an empirical estimate of the distribution function of a random variable, whereas a normal curve assumes the distribution has a specific form. Figure 7.2 is a KM estimate of the survival function. Rather than smoothing over the data, it has a stair-step pattern, dropping exactly at times when there are events.

[2] https://r4ds.hadley.nz/datetimes

Example 7.1 (continued): Use R to compute the KM estimate of the survival function for the Natality teaching dataset. We will then examine in more detail how it is computed.

```
surv.ex7.1 <- survfit(Surv(gestage37, preterm01) ~ 1, data = natality)
```

The syntax for the left-hand side of the model formula is `Surv(TIME, EVENT)` where `TIME` = the time to event variable and `EVENT` = the event indicator variable (taking on the value 1 for events and 0 for censored times). The `~ 1` on the right-hand side tells the function to estimate the survival function for all the individuals pooled together, not stratified by any characteristics.

Typing the name of the `survfit` object (`surv.ex7.1`) displays the following basic results:

```
surv.ex7.1
```

```
## Call: survfit(formula = Surv(gestage37, preterm01) ~ 1, data = natality)
##
##         n events median 0.95LCL 0.95UCL
## [1,] 2000    252     NA      NA      NA
```

- the number of individuals (2000),
- the number of events (252 preterm births), and
- the median survival time and its 95% confidence interval. For this data, the median is missing (`NA`) because the survival function never reaches 0.50. We will discuss the median survival time more in Section 7.6.4, including an example where the median is not missing.

Use `summary()` to see more information. Use `print()` to see the results to more than the default of three significant digits.

```
print(summary(surv.ex7.1), digits = 4)
```

```
## Call: survfit(formula = Surv(gestage37, preterm01) ~ 1, data = natality)
##
##  time n.risk n.event survival   std.err lower 95% CI upper 95% CI
##    17   2000       1   0.9995 0.0004999       0.9985       1.0000
##    20   1999       1   0.9990 0.0007068       0.9976       1.0000
##    23   1988       1   0.9985 0.0008668       0.9968       1.0000
##    24   1983       1   0.9980 0.0010020       0.9960       1.0000
##    25   1978       2   0.9970 0.0012291       0.9946       0.9994
##    26   1969       2   0.9960 0.0014212       0.9932       0.9988
##    27   1966       2   0.9950 0.0015901       0.9918       0.9981
##    28   1962       3   0.9934 0.0018141       0.9899       0.9970
##    29   1958       3   0.9919 0.0020130       0.9880       0.9959
##    30   1952       9   0.9873 0.0025156       0.9824       0.9923
##    31   1942       8   0.9833 0.0028871       0.9776       0.9889
##    32   1931      15   0.9756 0.0034735       0.9689       0.9825
##    33   1915      20   0.9654 0.0041173       0.9574       0.9736
##    34   1892      36   0.9471 0.0050506       0.9372       0.9570
##    35   1856      59   0.9170 0.0062279       0.9048       0.9293
##    36   1797      89   0.8716 0.0075542       0.8569       0.8865
```

This extended output contains one row for every unique time at which an event occurred (non-censored event times). The `n.risk` column provides the number of individuals who are still in the risk set at each time. At a given time, the **risk set** is the group of individuals who *could* be observed to experience the event at that time, specifically those who have not yet experienced the event and are still under observation (not previously censored). The `n.event` column displays the number of non-censored events that occurred at each time. The `survival` column shows the KM estimate of $S(t)$, and the remaining columns provide the standard error and 95% CI for the estimate of $S(t)$.

There were no preterm births before 17 weeks so at 17 weeks the risk set is all 2000 births (`n.risk` = 2000). There was 1 preterm birth (`n.event` = 1) at 17 weeks.

At 20 weeks, there are 1999 individuals in the risk set because the preterm birth at 17 weeks is no longer at risk. There was 1 preterm birth at 20 weeks, so we would expect there to be 1999 − 1 = 1998 individuals in the risk set at the next event time, 23 weeks. But at 23 weeks, there are only 1988 individuals in the risk set. Why? Because, in addition to the one preterm birth, there are 10 individuals with censored times since the last event time (inclusive of the last event time). So, in total, the size of the risk set decreased by 11 between 20 and 23 weeks.

```
natality %>%
  filter(gestage37 >= 20 & gestage37 < 23 & preterm01 == 0) %>%
  select(gestage37, preterm01)
```

```
## # A tibble: 10 x 2
##      gestage37 preterm01
##          <dbl>     <dbl>
##  1          21         0
##  2          22         0
##  3          20         0
##  4          22         0
##  5          21         0
##  6          21         0
##  7          21         0
##  8          22         0
##  9          21         0
## 10          22         0
```

How is the survival probability (`survival`) column computed? At a non-censored event time, the estimate of survival is made up of a product of probabilities. The conditional probability rule states that we can decompose the probability of an event A into the product of two probabilities involving another event B as $P(A) = P(A|B)P(B)$. Also, $P(\text{not } A) = 1 - P(A)$. Therefore,

$$
\begin{aligned}
S(t) &= P(T > t) \\
&= P(T > t|T \geq t)P(T \geq t) \\
&= [1 - P(T \leq t|T \geq t)]\, P(T \geq t) \\
&= [1 - P(T = t|T \geq t)]\, P(T \geq t) \\
&= [1 - P(T = t|T \geq t)]\, P(T > t_{\text{prev}}) \\
&= [1 - P(T = t|T \geq t)]\, S(t_{\text{prev}})
\end{aligned} \tag{7.2}
$$

The $T \leq t|T \geq t$ turns into $T = t|T \geq t$ in the fourth line because if both $T \leq t$ and $T \geq t$ then T must be t. Also, since the times are discrete, $P(T \geq t)$ turns into $P(T > t_{\text{prev}})$ in the fifth line. That is, the chance of survival from t on is the chance of surviving past the last time at which there was an event (t_{prev}). Finally, since by definition $P(T > t) = S(t)$, $P(T > t_{\text{prev}}) = S(t_{\text{prev}})$.

What does all this math tell us? That to compute the KM estimate of survival at time t, take one minus the proportion of events at time t among those at risk, $[1 - P(T = t|T \geq t)]$, and multiply it by the probability of survival past the previous event time, $S(t_{\text{prev}})$. Putting all this together using the column names from the `summary()` of the `survfit` object, we conclude that $S(t) = (1 - \text{n.event}/\text{n.risk})S(t_{\text{prev}})$. Since there were no events prior to the first event time, $S(t_{\text{prev}}) = 1$ at all t_{prev} earlier than the first event time.

Verify this using the following code, which uses the `lag()` function (in the `dplyr` library which is automatically loaded when you load the `tidyverse` library) to get $S(t_{\text{prev}})$. Since there is no lagged first value, this results in a missing value at the first time which we replace with a 1. As shown below, the estimate of $S(t)$ computed by `survfit()` is identical to the estimate computed using Equation 7.2.

```
time        <- summary(surv.ex7.1)$time
n.event     <- summary(surv.ex7.1)$n.event
n.risk      <- summary(surv.ex7.1)$n.risk
S_t         <- summary(surv.ex7.1)$surv
S_tprev     <- lag(S_t)
S_tprev[1] <- 1
cbind(time, n.event, n.risk, S_tprev,
      "(1 - n.event/n.risk)*S_tprev" = (1 - n.event/n.risk)*S_tprev,
      "S(t) from survfit"            = S_t)
```

```
##         time n.event n.risk S_tprev (1 - n.event/n.risk)*S_tprev
##   [1,]    17       1   2000  1.0000                       0.9995
##   [2,]    20       1   1999  0.9995                       0.9990
##   [3,]    23       1   1988  0.9990                       0.9985
##   [4,]    24       1   1983  0.9985                       0.9980
##   [5,]    25       2   1978  0.9980                       0.9970
##   [6,]    26       2   1969  0.9970                       0.9960
##   [7,]    27       2   1966  0.9960                       0.9950
##   [8,]    28       3   1962  0.9950                       0.9934
##   [9,]    29       3   1958  0.9934                       0.9919
##  [10,]    30       9   1952  0.9919                       0.9873
##  [11,]    31       8   1942  0.9873                       0.9833
##  [12,]    32      15   1931  0.9833                       0.9756
##  [13,]    33      20   1915  0.9756                       0.9654
##  [14,]    34      36   1892  0.9654                       0.9471
##  [15,]    35      59   1856  0.9471                       0.9170
##  [16,]    36      89   1797  0.9170                       0.8716
##         S(t) from survfit
##   [1,]             0.9995
##   [2,]             0.9990
##   [3,]             0.9985
##   [4,]             0.9980
##   [5,]             0.9970
##   [6,]             0.9960
##   [7,]             0.9950
##   [8,]             0.9934
##   [9,]             0.9919
##  [10,]             0.9873
##  [11,]             0.9833
##  [12,]             0.9756
##  [13,]             0.9654
##  [14,]             0.9471
##  [15,]             0.9170
##  [16,]             0.8716
```

For example, the KM estimates of the survival probabilities at weeks 17, 23, and 31 are as follows:

- $S(t)$ at 17 weeks is $(1 - 1/2000) \times 1 = 0.9995$.
- $S(t)$ at $t = 23$ weeks is $(1 - 1/1988) \times 0.9990 = 0.9985$.
- $S(t)$ at $t = 31$ weeks is $(1 - 8/1942) \times 0.9873 = 0.9833$.

Looking at the final row in the table, the estimated probability of "survival" past 36 weeks (the probability of not having a preterm birth) is 0.8716.

NOTES:

- The estimated survival probability only changes at non-censored event times, but the number in the risk set changes after both censored and non-censored event times. Thus, $S(t)$ decreases only at times with actual events, but the amount by which it decreases depends both on the number of actual events at that time and the number of individuals censored since the last event time.
- The way the KM estimator handles censored times is a compromise between treating them as non-events and treating them as non-censored events. If we treat individuals with censored times as if they *never* experienced the event, then they would always remain in the risk set and $S(t)$ would be too large at the next event time. If we treat them as if they experienced the event at the censored time, they would prematurely add an event and $S(t)$ would be too small at that time.

7.6.1 Plotting the survival function

Use `plot()` on the `survfit` object to plot the survival function. The default plot, however, might not be easy to read due to the scale (Figure 7.4).

```
plot(surv.ex7.1)
```

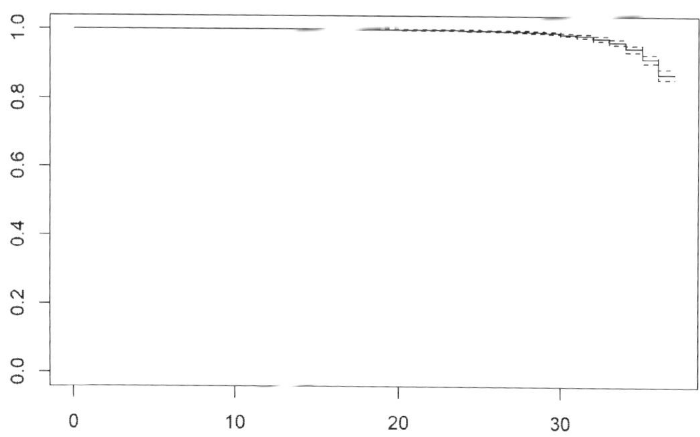

FIGURE 7.4 Default survival function plot

Figure 7.5 zooms in on the x- and y-axes using `xlim` and `ylim`, adds some labels, customizes the x-axis tick mark locations, and adds markings to identify censored times.

```
plot(surv.ex7.1, xlab = "Gestational age (weeks)", ylab = "Survival probability",
     xlim = c(16, 37), ylim = c(0.85, 1),
     mark.time = T, # Identify censored times
     xaxt = "n")    # Suppress the x-axis so we can customize it
axis(1, at = c(seq(17, 36, 3), 37)) # Customize location of tick marks
```

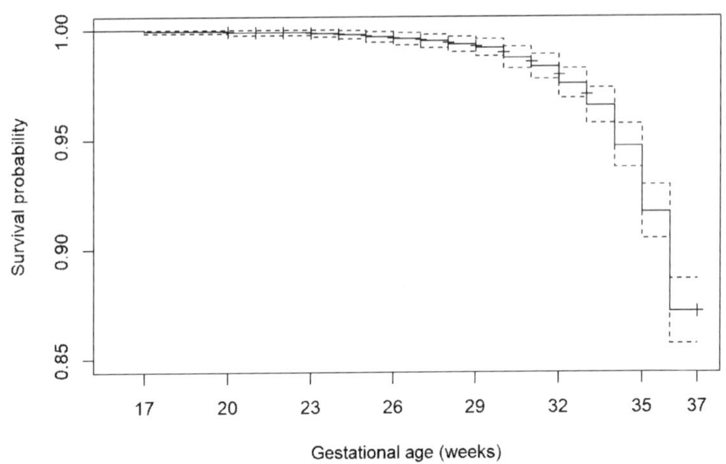

FIGURE 7.5 Customized survival function plot

While it is nice to have confidence bands, if they obscure the survival function they can be removed using `conf.int = F` (Figure 7.6).

```
plot(surv.ex7.1, xlab = "Gestational age (weeks)", ylab = "Survival probability",
     xlim = c(16, 37), ylim = c(0.85, 1),
     mark.time = T, # Identify censored times
     xaxt = "n",    # Suppress the x-axis so we can customize it
     conf.int = F)
axis(1, at = c(seq(17, 36, 3), 37)) # Customize location of tick marks
```

7.6.2 Computing and plotting the hazard function

Use the `bshazard()` function in the `bshazard` library to compute the hazard function (Rebora et al., 2018). When using `bshazard`, it is possible to have non-convergence leading to an error. The method uses an algorithm to smooth the estimated curve, and the `nk` option controls the amount of smoothing. In this case, the default value of 31 leads to an error. If there is an error, or if you just want to see what the function looks like with a different amount of smoothing, enter a different value for `nk` (Figure 7.7).

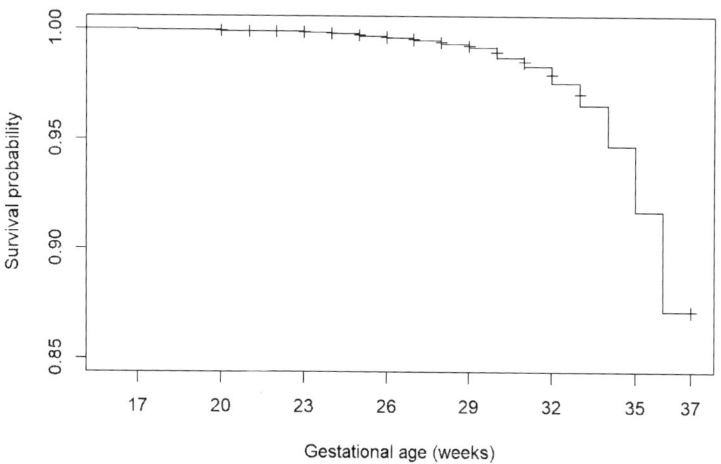

FIGURE 7.6 Survival function plot without confidence bands

```
haz.ex7.1 <- bshazard::bshazard(Surv(gestage37, preterm01) ~ 1,
  data = natality,
  verbose = F, # Suppress display of the iteration steps
  nk=15)
plot(haz.ex7.1, conf.int = F, xlab = "Gestational age (days)",
     xlim = c(16, 37), lwd = 2, xaxt = "n")
axis(1, at = c(seq(17, 36, 3), 37))
```

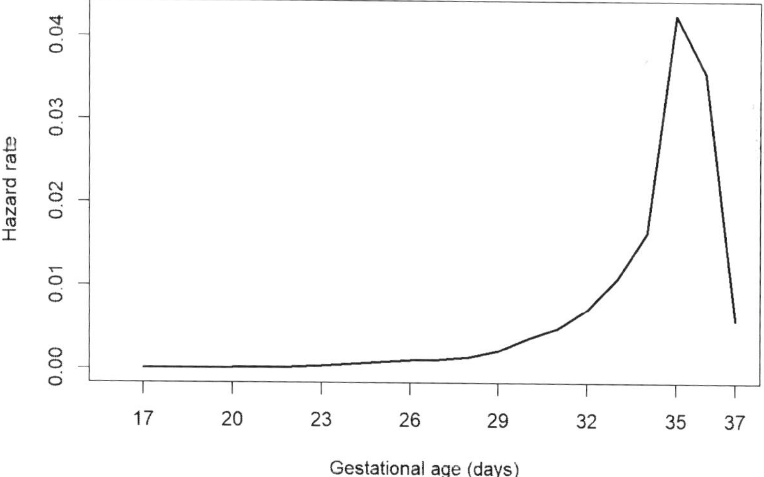

FIGURE 7.7 Hazard function plot

7.6.3 Estimating the event probability within a time interval

Use the KM estimate of $S(t)$ to estimate the probability an event occurs within a specific time interval. For example, in the Natality teaching dataset, what is the estimated probability that a woman will experience a preterm birth from 33 to 35 weeks, inclusive? Be careful when figuring out how to compute this probability – the correct answer depends on whether or not we include or exclude each endpoint of the interval.

We start with the answer and then explain how it was derived.

$$P(33 \leq T \leq 35) = S(32) - S(35)$$

Why $S(32)$ on right right-hand side instead of $S(33)$? Why $S(32) - S(35)$ instead of $S(35) - S(32)$? $S(35) = P(T > 35)$ is the probability of survival *past* 35 weeks and so is the sum of the probabilities of preterm birth occurring at 36 weeks and beyond (because the times are discrete, $P(T > 35) = P(T \geq 36)$). Similarly, $S(32)$ is the sum for 33 weeks and beyond. So if we subtract these two, we are left with $P(T = 33) + P(T = 34) + P(T = 35)$, as shown in Figure 7.8. The distance between the horizontal gray dashed lines is the sum of the drops in $S(t)$ at weeks 33, 34, and 35.

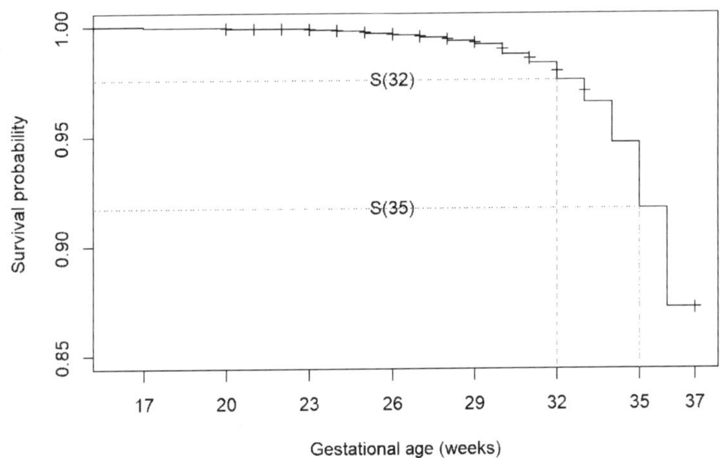

FIGURE 7.8 Probabiliy of an event between two times

Extract these probabilities from the `summary()` of the `survfit` object using the `times` option and subtract to compute $P(33 \leq T \leq 35) = 0.0587$.

```
S32 <- summary(surv.ex7.1, times=32)$surv
S35 <- summary(surv.ex7.1, times=35)$surv
c(S32, S35, S32 - S35)
```

```
## [1] 0.97564 0.91697 0.05867
```

Extracting information from a `survfit` object at a subset of times

The syntax `summary(surv.ex7.1)` produces a table showing the number at risk, number of events, and survival probability at each unique event time. We used the `times` option above to extract just the survival probability at a single time. More generally, you can use `times` to extract all the summary information at any subset of times. For example, here is the information for the first five event times.

```
print(summary(surv.ex7.1, times=c(17, 20, 23, 24, 25)), digits = 4)
```

```
## Call: survfit(formula = Surv(gestage37, preterm01) ~ 1, data = natality)
##
##  time n.risk n.event survival   std.err lower 95% CI upper 95% CI
##    17   2000       1   0.9995 0.0004999       0.9985       1.0000
##    20   1999       1   0.9990 0.0007068       0.9976       1.0000
##    23   1988       1   0.9985 0.0008668       0.9968       1.0000
##    24   1983       1   0.9980 0.0010020       0.9960       1.0000
##    25   1978       2   0.9970 0.0012291       0.9946       0.9994
```

You can also include times without events to get even more information than was obtained from `summary()`. For example, enter a range of times that includes times with and without events and including censored times to see how `n.risk` and $S(t)$ change depending on the timing of events and censored times.

```
print(summary(surv.ex7.1, times=17:25), digits = 4)
```

```
## Call: survfit(formula = Surv(gestage37, preterm01) ~ 1, data = natality)
##
##  time n.risk n.event survival   std.err lower 95% CI upper 95% CI
##    17   2000       1   0.9995 0.0004999       0.9985       1.0000
##    18   1999       0   0.9995 0.0004999       0.9985       1.0000
##    19   1999       0   0.9995 0.0004999       0.9985       1.0000
##    20   1999       1   0.9990 0.0007068       0.9976       1.0000
##    21   1997       0   0.9990 0.0007068       0.9976       1.0000
##    22   1992       0   0.9990 0.0007068       0.9976       1.0000
##    23   1988       1   0.9985 0.0008668       0.9968       1.0000
##    24   1983       1   0.9980 0.0010020       0.9960       1.0000
##    25   1978       2   0.9970 0.0012291       0.9946       0.9994
```

`survival` is constant over some of the rows, because it only changes at times with an event. `n.risk` is also constant over some rows, because it only changes at times for which there was an event or censored time at the *previous* time. The table does not show the number of censored times, but they can be inferred by looking to see where `n.risk` drops by more than the previous `n.event`. For example, at $t = 21$ weeks, `n.risk` drops by 2 despite there being only one event at the previous time. Thus, we infer that there was one individual censored at $t = 20$ weeks.

When using the `times` option, `n.event` at the first time requested is the *cumulative* number of events up to and including that time. The table below is identical to the last three rows of the table above, except for `n.event` in the first row.

```
print(summary(surv.ex7.1, times=23:25), digits = 4)
```

```
## Call: survfit(formula = Surv(gestage37, preterm01) ~ 1, data = natality)
##
##  time n.risk n.event survival   std.err lower 95% CI upper 95% CI
##    23   1988       3   0.9985 0.0008668       0.9968       1.0000
##    24   1983       1   0.9980 0.0010020       0.9960       1.0000
##    25   1978       2   0.9970 0.0012291       0.9946       0.9994
```

Similarly, if you request a single time, `n.event` is the cumulative number of events up to and including that time.

```
print(summary(surv.ex7.1, times=25), digits = 4)
```

```
## Call: survfit(formula = Surv(gestage37, preterm01) ~ 1, data = natality)
##
##  time n.risk n.event survival  std.err lower 95% CI upper 95% CI
##    25   1978       6    0.997 0.001229       0.9946       0.9994
```

If you want to extract the number of events at a given time, you can do so by subtraction of the cumulative `n.event` values at the current and previous times.

```
summary(surv.ex7.1, times=25)$n.event -
  summary(surv.ex7.1, times=24)$n.event
```

```
## [1] 2
```

Estimating the event probability within other time intervals

What if, instead of the probability in a two-sided interval, you want the probability of preterm birth at a specific time, prior to a certain time, or after a certain time? All the possibilities are listed below. For each, the correct computation formula depends on whether inequalities are strict or not strict.

The actual observed (non-censored) times in a dataset are finite, and you can sort them and list them in order from least to greatest. Let t_{prev} refer to the observed event time in the data that precedes t. For example, if the data includes observed events at 5, 8, 12, and 13 days, then, for $t = 8$, $t_{\text{prev}} = 5$. If t is the earliest observed event time, then t_{prev} can be any number smaller than t.

- $P(t_1 \leq T \leq t_2) = S(t_{1\text{prev}}) - S(t_2)$
- $P(t_1 < T \leq t_2) = S(t_1) - S(t_2)$
- $P(t_1 \leq T < t_2) = S(t_{1\text{prev}}) - S(t_{2\text{prev}})$
- $P(t_1 < T < t_2) = S(t_1) - S(t_{2\text{prev}})$
- $P(T > t) = S(t)$
- $P(T \geq t) = S(t_{\text{prev}})$
- $P(T \leq t) = 1 - S(t)$
- $P(T < t) = 1 - S(t_{\text{prev}})$
- $P(T = t) = S(t_{\text{prev}}) - S(t)$

In the Natality teaching dataset, the observed event times are

```
SUB <- natality$preterm01 == 1
sort(unique(natality$gestage37[SUB]))
```

```
##  [1] 17 20 23 24 25 26 27 28 29 30 31 32 33 34 35 36
```

Suppose, for example, that $t_1 = 23$ and $t_2 = 30$. Then $t_{1\text{prev}} = 20$ and $t_{2\text{prev}} = 29$. Therefore:

- $P(23 \leq T \leq 30) = S(20) - S(30)$
- $P(23 < T \leq 30) = S(23) - S(30)$
- $P(23 \leq T < 30) = S(20) - S(29)$
- $P(23 < T < 30) = S(23) - S(29)$
- $P(T > 23) = S(23)$
- $P(T \geq 23) = S(20)$
- $P(T \leq 23) = 1 - S(23)$
- $P(T < 23) = 1 - S(20)$
- $P(T = 23) = S(20) - S(23)$

After computing each of the four $S(t)$ values needed,

```
S20 <- summary(surv.ex7.1, times=20)$surv
S23 <- summary(surv.ex7.1, times=23)$surv
S29 <- summary(surv.ex7.1, times=29)$surv
S30 <- summary(surv.ex7.1, times=30)$surv

survprobs <- data.frame(c(20, 23, 29, 30),
                        c(S20, S23, S29, S30))
names(survprobs) <- c("t", "S(t)")
survprobs
```

```
##    t   S(t)
## 1 20 0.9990
## 2 23 0.9985
## 3 29 0.9919
## 4 30 0.9873
```

Use the formulas above to derive the following:

```
data.frame(Interval = c("P(23 <= T <= 30)",
                        "P(23 < T <= 30)",
                        "P(23 <= T < 30)",
                        "P(23 < T < 30)",
                        "P(T > 23)",
                        "P(T >= 23)",
                        "P(T <= 23)",
                        "P(T < 23)",
                        "P(T = 23)"),
           Probability = c(S20 - S30,
                           S23 - S30,
                           S20 - S29,
                           S23 - S29,
                           S23,
                           S20,
                           1 - S23,
                           1 - S20,
                           S20 - S23))
```

```
##          Interval Probability
## 1 P(23 <= T <= 30)   0.0116579
## 2  P(23 < T <= 30)   0.0111553
## 3  P(23 <= T < 30)   0.0070845
## 4   P(23 < T < 30)   0.0065820
## 5        P(T > 23)   0.9984975
## 6       P(T >= 23)   0.9990000
## 7       P(T <= 23)   0.0015025
## 8        P(T < 23)   0.0010000
## 9        P(T = 23)   0.0005025
```

7.6.4 Median survival time

The **median survival time** is the time at which 50% of the individuals have experienced the event and 50% have not. On the survival function, $S(\text{median survival time}) = 0.50$. In Example 7.1, the probability of "survival" (not having a preterm birth) never dropped below 0.8716 so the median survival time is not defined. Look at an example for which the probability does go below 0.50.

Example 7.3: The teaching dataset based on the Framingham Heart Study (see Appendix A.6) contains information about whether or not participants experienced hypertension (`HYPERTEN`) and how long (in days) after baseline until they were diagnosed as hypertensive (`TIMEHYP`). If there were no censoring, we could simply compute the median of the event times to compute the median survival time. We must include `na.rm = T` as individuals with prevalent hypertension at baseline have missing values for `TIMEHYPE` and `HYPERTEN`, so analyses of this variable are restricted to individuals who were not hypertensive at baseline.

```
load("Data/fram_time_invar_rmph.rData")

median(fram_time_invar$TIMEHYP, na.rm = T)
```

```
## [1] 5094
```

The median of the event time variable is 5094.5. However, some of the event times are censored (some values of `HYPERTEN`, the event indicator, are 0) so the raw median is an underestimate – it treats censored times as event times when in fact the individual's actual time of diagnosis, if they ever were diagnosed, is later. Use `survfit()` to view the median survival time accounting for censoring, and its 95% confidence interval.

```
surv.ex7.3 <- survfit(Surv(TIMEHYP, HYPERTEN) ~ 1, data = fram_time_invar)
surv.ex7.3
```

```
## Call: survfit(formula = Surv(TIMEHYP, HYPERTEN) ~ 1, data = fram_time_invar)
##
##    1299 observations deleted due to missingness
##           n events median 0.95LCL 0.95UCL
## [1,] 2916   1767   5837    5620    6069
```

To extract the median and confidence interval, use `summary()` on the `survfit` object.

```
summary(surv.ex7.3)$table[c("median", "0.95LCL", "0.95UCL")]
```

```
##   median 0.95LCL 0.95UCL
##     5837    5620    6069
```

For this example, the time is in days, so divide by 365.25 to get the median survival time in years.

```
summary(surv.ex7.3)$table[c("median", "0.95LCL", "0.95UCL")]/365.25
```

```
##   median 0.95LCL 0.95UCL
##    15.98   15.39   16.62
```

The median survival time is 5837 days, or 15.98 years (95% CI = 15.39, 16.62). Since $P(T > \text{median}) = 0.50$, we know that $P(T \leq \text{median}) = 0.50$, so the median survival time (time still free of hypertension) is also the median time of diagnosis.

On the survival function, the median time is the time at which the curve crosses 0.50 (Figure 7.9).

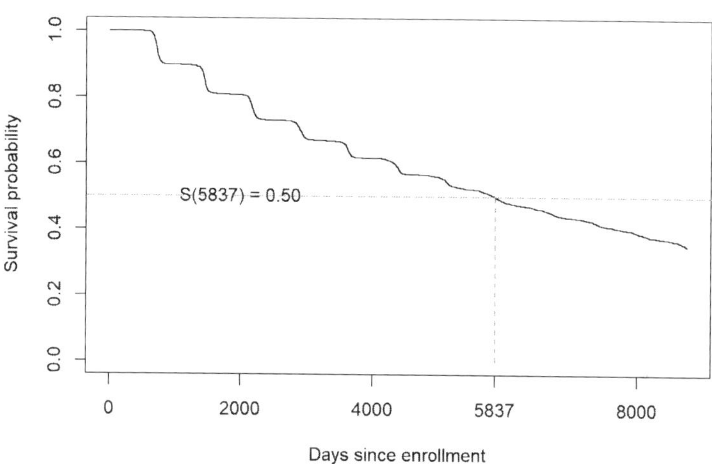

FIGURE 7.9 Median survival time

7.6.5 Comparing groups

So far we have only used the KM estimator to compute the overall survival function for all individuals in the dataset pooled together regardless of how their characteristics differ. We can also estimate separate curves for groups with different characteristics and use the **log-rank test** to test the null hypothesis that the groups' survival functions are the same. In the model formula, place the grouping variable on the right-hand side. Use `survfit()` to facilitate plotting, and `survdiff()` (Harrington and Fleming, 1982) to carry out the log-rank test.

Example 7.4: In the United States, there are large racial disparities in the risk of preterm birth, with Black women having much greater risk than White women (Burris et al., 2019).

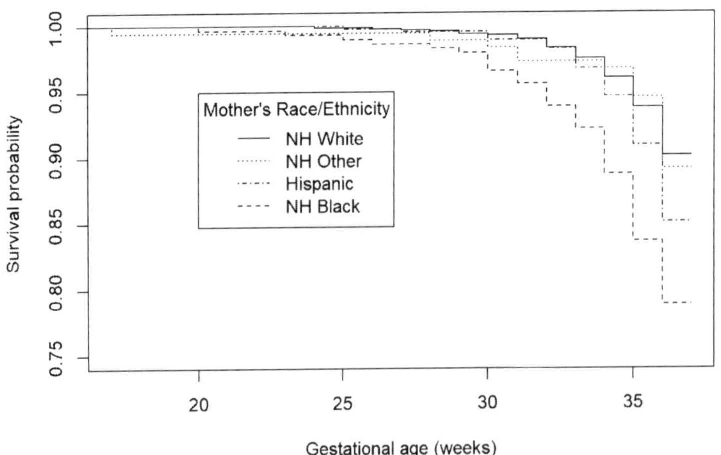

FIGURE 7.10 Survival functions for time to preterm birth by race/ethnicity

Compare the time to preterm birth between mothers of different race/ethnicity (`MRACEHISP`) in the Natality teaching dataset.

First, compute the KM estimate of survival.

```
surv.ex7.4 <- survfit(Surv(gestage37, preterm01) ~ MRACEHISP,
  data = natality)
surv.ex7.4
```

```
## Call: survfit(formula = Surv(gestage37, preterm01) ~ MRACEHISP, data = natality
##     )
##
##    9 observations deleted due to missingness
##                         n events median 0.95LCL 0.95UCL
## MRACEHISP=NH White 1028    100     NA      NA      NA
## MRACEHISP=NH Black  300     62     NA      NA      NA
## MRACEHISP=NH Other  188     20     NA      NA      NA
## MRACEHISP=Hispanic  475     69     NA      NA      NA
```

The median event times are all `NA`, since the proportion of preterm births is always above 0.50 in every group. Figure 7.10 illustrates the survival functions. `plot()` automatically creates separate lines for each group. The previous output shows that there are four groups, so there are four survival functions. The `lty = 1:4` option in `plot()` provides a distinct line type for each group. Note how `c(1,3,4,2)` was used to change the ordering of `lty` and the group labels in `legend()` so the legend order matches the order of the lines in the plot.

```
plot(surv.ex7.4,
  xlab = "Gestational age (weeks)", ylab = "Survival probability",
  xlim = c(17, 37), ylim = c(0.75, 1),
  conf.int = F, mark.time = F,
  lty = 1:4) # 4 groups
# Add c(1,3,4,2) in two places to re-order the legend to match
# the ordering of the lines at 37 weeks
```

```
legend(20, 0.95,
  # Inside c(), the ordering is the same as the order shown in surv.ex7.4
  levels(natality$MRACEHISP)[c(1,3,4,2)],
  lty = c(1,3,4,2),
  title = "Mother's Race/Ethnicity")
```

Similarly, `plot()` automatically plots a separate hazard function for each group (Figure 7.11).

```
# In this case, the default amount of smoothing does not lead to
# an error, so we do not need to specify nk unless we want to
# see what it looks like with less smoothing
haz.ex7.4 <- bshazard::bshazard(Surv(gestage37, preterm01) ~ MRACEHISP,
     data = natality,
     verbose = F)
```

```
## NOTE: entry.status has been set to 0 for all.
```

```
plot(haz.ex7.4,
     xlab = "Gestational age (weeks)",
     ylab = "Hazard of preterm birth",
     conf.int = F, overall = F,
     lty = 1:4, ylim = c(0, 0.08))
legend(10, 0.05,
       levels(natality$MRACEHISP)[c(2,4,3,1)],
       lty = c(2,4,3,1), title = "Mother's Race/Ethnicity")
```

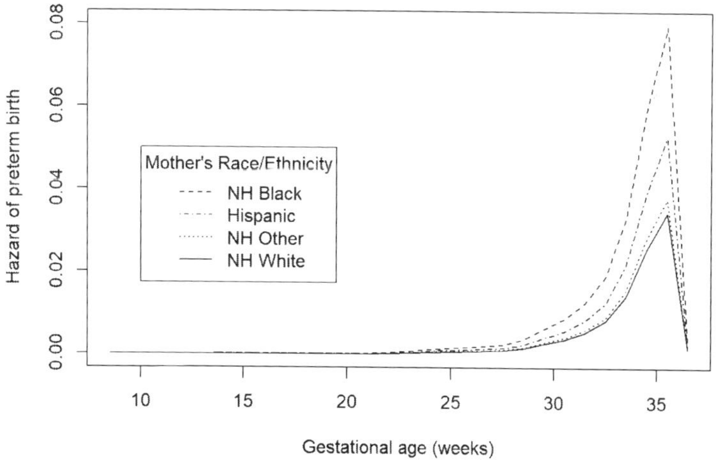

FIGURE 7.11 Hazard functions for time to preterm birth by race/ethnicity

Finally, use `survdiff()` to carry out the log-rank test comparing the groups.

```
survdiff.ex7.4 <- survdiff(Surv(gestage37, preterm01) ~ MRACEHISP,
                           data = natality)
survdiff.ex7.4
```

```
## Call:
## survdiff(formula = Surv(gestage37, preterm01) ~ MRACEHISP, data = natality)
##
## n=1991, 9 observations deleted due to missingness.
##
##                           N Observed Expected (O-E)^2/E (O-E)^2/V
## MRACEHISP=NH White 1028      100    132.0     7.778    16.897
## MRACEHISP=NH Black  300       62     35.4    19.969    23.927
## MRACEHISP=NH Other  188       20     24.0     0.678     0.772
## MRACEHISP=Hispanic  475       69     59.5     1.515     2.044
##
##  Chisq= 30.8  on 3 degrees of freedom, p= 0.0000009
```

Unfortunately, the p-value is not one of the numbers you can extract directly from the `survdiff` object. However, you can compute it directly using `pchisq(, lower.tail=F)` by extracting the `chisq` value and computing the degrees of freedom as one less than the number of levels.

```
pchisq(survdiff.ex7.4$chisq,
       df=length(levels(natality$MRACEHISP)) - 1,
       lower.tail=F)
```

```
## [1] 0.0000009284
```

The racial disparity in preterm birth can be seen in both the survival and hazard function plots – Hispanic and non-Hispanic Black mothers have lower survival (non-preterm birth) probabilities and greater hazards than non-Hispanic White and non-Hispanic Other race mothers. Based on the log-rank test, this difference is statistically significant ($\chi^2 = 30.8$, df = 3, p <.001). In Section 7.7 we will estimate the magnitude of the ratio of the hazards between each pair of race/ethnicity values. While the log-rank test is useful as a hypothesis test, it does not provide an estimate of effect size.

7.7 Cox regression

The terms **Cox regression**, **Cox model**, and **Cox proportional hazards regression** all refer to a semi-parametric method introduced by D. R. Cox in 1972 (Cox, 1972). The method is "semi-parametric" because it makes no assumption about the distribution of the event times (similar to the non-parametric KM method), but it does assume the hazard function depends on a set of parameters (regression coefficients) that define the association between the hazard and a set of predictors.

The Cox model is written as

$$h(t) = h_0(t)e^{\beta_1 X_1 + \ldots + \beta_K X_K} \tag{7.3}$$

where $h_0(t)$ is referred to as the **baseline hazard function** and, as in other forms of regression, there are β terms each multiplied by a predictor. The baseline hazard is similar to the intercept in a linear regression model in that it represents the hazard for individuals whose covariate values are all 0 or at their reference level. The baseline hazard function does not depend on any parameters and drops out completely when estimating the parameters of the model. Thus, Cox regression output does not include an intercept.

How do we interpret the regression coefficients in Equation 7.3? Recall that in logistic regression e^{β} represented an odds ratio (OR). In the Cox model, e^{β} represents a **hazard ratio** (HR) comparing the hazard of experiencing the event at time t between individuals with $X = x + 1$ vs. those with $X = x$ (for a continuous predictor) or between individuals at a specific level of X vs. those at its reference level (for a categorical predictor), holding all other predictors fixed.

To see this, consider the hazards at $X_K = x + 1$ and $X_K = x$ (holding other predictors fixed):

$$\begin{aligned} h(t|X_1 = x_1, ..., X_K = x_K + 1) &= h_0(t)e^{\beta_1 x_1 + ... + \beta_K (x_K + 1)} \\ h(t|X_1 = x_1, ..., X_K = x_K) &= h_0(t)e^{\beta_1 x_1 + ... + \beta_K x_K} \end{aligned}$$

Taking the ratio of these, everything cancels out except e^{β_K} which is, therefore, the HR for X_K. Importantly, the HR does not depend on time. This is the **proportional hazards assumption** – that the hazard functions for any two individuals have a constant proportion over time.

Interpret HRs similarly to ORs estimated from logistic regression (see **Interpreting an OR** in Section 6.4, substituting "HR" for "OR" and "hazard" for "odds"). For example, HR = 1.20 implies that one group has a 20% greater hazard of the event than another.

7.8 Fitting the Cox regression model

7.8.1 Unadjusted

To illustrate the syntax and interpretation of the output, we will fit two unadjusted Cox regression models as examples, one with a categorical predictor and one with a continuous predictor.

Categorical predictor

Example 7.4 (continued): Earlier we plotted the KM estimates of the survival functions for preterm birth for mothers of different race/ethnicities (Figure 7.10). Visually, there were clear disparities. Use Cox regression, via the `coxph()` function, to numerically assess the magnitude of racial disparities by estimating the hazard ratios comparing these groups. As before, use the `Surv(TIME, EVENT) ~ X` syntax for the model formula.

```
cox.ex7.4 <- coxph(Surv(gestage37, preterm01) ~ MRACEHISP,
                   data = natality)
summary(cox.ex7.4)
```

```
## Call:
## coxph(formula = Surv(gestage37, preterm01) ~ MRACEHISP, data = natality)
##
##   n= 1991, number of events= 251
##      (9 observations deleted due to missingness)
```

```
##
##                       coef exp(coef) se(coef)    z    Pr(>|z|)
## MRACEHISPNH Black 0.8483     2.3357   0.1617 5.25 0.00000015 ***
## MRACEHISPNH Other 0.0964     1.1012   0.2449 0.39     0.6940
## MRACEHISPHispanic 0.4332     1.5422   0.1565 2.77     0.0056 **
## ---
## Signif. codes:  0 '***' 0.001 '**' 0.01 '*' 0.05 '.' 0.1 ' ' 1
##
##                   exp(coef) exp(-coef) lower .95 upper .95
## MRACEHISPNH Black      2.34      0.428     1.701      3.21
## MRACEHISPNH Other      1.10      0.908     0.681      1.78
## MRACEHISPHispanic      1.54      0.648     1.135      2.10
##
## Concordance= 0.587  (se = 0.018 )
## Likelihood ratio test= 27.6  on 3 df,   p=0.000004
## Wald test            = 29.4  on 3 df,   p=0.000002
## Score (logrank) test = 30.7  on 3 df,   p=0.000001
```

The output tells us that there were 1991 observations (9 were removed because they had a missing value for `MRACEHISP`) and among these there were 251 events (preterm births), with the remaining event times being censored. The `coef` column contains the estimates of the βs, and the `exp(coef)` column contains e^{β} for each term which is the HR comparing that level to the reference level (non-Hispanic White). For example, non-Hispanic Black mothers have 2.34 times the hazard of preterm birth of non-Hispanic White mothers. The `exp(-coef)` column displays the inverse of the HR, which reverses the direction of comparison. For example, the hazard for non-Hispanic White mothers is 0.43 times that of non-Hispanic Black mothers. The `lower .95` and `upper .95` columns display the 95% CI for the HR (`exp(coef)`, not the inverse).

Use the following code to extract just the regression coefficient table.

```
summary(cox.ex7.4)$coef
```

```
##                      coef exp(coef) se(coef)      z     Pr(>|z|)
## MRACEHISPNH Black 0.84830     2.336   0.1617 5.2472 0.0000001544
## MRACEHISPNH Other 0.09637     1.101   0.2449 0.3934 0.6940061841
## MRACEHISPHispanic 0.43321     1.542   0.1565 2.7680 0.0056401474
```

Use the following code to display the HRs and their 95% confidence intervals and p-values.

```
# HRs, 95% CIs, p-values
cbind("HR"      = exp(summary(cox.ex7.4)$coef[, "coef"]),
                  exp(confint(cox.ex7.4)),
      "p-value" = summary(cox.ex7.4)$coef[, "Pr(>|z|)"])
```

```
##                      HR  2.5 % 97.5 %      p-value
## MRACEHISPNH Black 2.336 1.7014  3.206 0.0000001544
## MRACEHISPNH Other 1.101 0.6813  1.780 0.6940061841
## MRACEHISPHispanic 1.542 1.1348  2.096 0.0056401474
```

Use `car::Anova()` to get multiple degree of freedom Wald tests for categorical variables with more than two levels.

```
car::Anova(cox.ex7.4, type = 3, test.statistic = "Wald")
```

```
## Analysis of Deviance Table (Type III tests)
##
## Response: Surv(gestage37, preterm01)
##            Df Chisq Pr(>Chisq)
## MRACEHISP   3  29.4  0.0000019 ***
## ---
## Signif. codes:  0 '***' 0.001 '**' 0.01 '*' 0.05 '.' 0.1 ' ' 1
```

Conclusion: We estimate from the Natality teaching dataset that there are significant racial disparities in the hazard of preterm birth (p <.001 from the Type III test). Non-Hispanic Black mothers have 2.34 times the hazard of preterm birth of non-Hispanic White mothers (HR = 2.34; 95% CI = 1.70, 3.21; p <.001). Hispanic mothers have 1.54 times the hazard of preterm birth of non-Hispanic White mothers (HR = 1.54; 95% CI = 1.13, 2.10; p = .006). Non-Hispanic Other race mothers had greater observed hazard than non-Hispanic White mothers, but this difference was small and not statistically significant (HR = 1.10; 95% CI = 0.68, 1.78; p = .694).

Continuous predictor

Example 7.5: Using the Framingham Heart Study teaching dataset, assess the association between time to angina pectoris (recurrent chest pain) and the age of the participant at enrollment (`AGE`). The time and event variables are `TIMEAP` and `ANGINA`, respectively. The syntax is the same as the categorical predictor setting, but the interpretation of the HR is slightly different. Rather than being a comparison between a level and the reference level, it is a comparison between individuals who differ by one unit (in this case, by 1 year of age).

```
cox.ex7.5 <- coxph(Surv(TIMEAP, ANGINA) ~ AGE,
                   data = fram_time_invar)
summary(cox.ex7.5)
```

```
## Call:
## coxph(formula = Surv(TIMEAP, ANGINA) ~ AGE, data = fram_time_invar)
##
##   n= 4215, number of events= 558
##
##        coef exp(coef) se(coef)    z            Pr(>|z|)
## AGE 0.03429   1.03488  0.00499 6.87 0.0000000000066 ***
## ---
## Signif. codes:  0 '***' 0.001 '**' 0.01 '*' 0.05 '.' 0.1 ' ' 1
##
##     exp(coef) exp(-coef) lower .95 upper .95
## AGE      1.03      0.966      1.02      1.05
##
## Concordance= 0.595  (se = 0.012 )
## Likelihood ratio test= 46.6  on 1 df,   p=0.000000000009
## Wald test            = 47.1  on 1 df,   p=0.000000000007
## Score (logrank) test = 47.8  on 1 df,   p=0.000000000005
```

```
# HR, 95% CI, p-value
cbind("HR"      = exp(summary(cox.ex7.5)$coef[, "coef"]),
                  exp(confint(cox.ex7.5)),
      "p-value" = summary(cox.ex7.5)$coef[, "Pr(>|z|)"])
```

```
##        HR 2.5 % 97.5 %           p-value
## AGE 1.035 1.025  1.045 0.000000000006608
```

Conclusion: Age at enrollment is significantly associated with time to angina pectoris (p <.001, from the regression coefficient table – since this is just testing one term we do not need `car::Anova()`, although it would have given the same answer). Individuals who were one year older at enrollment have 3.5% greater hazard of angina pectoris (HR = 1.035; 95% CI = 1.025, 1.045; p <.001).

HR associated with other than a 1-unit difference

Similar to the issue with ORs estimated from logistic regression discussed in Section 6.6.1, if you want to know the HR associated with more than a 1-unit difference, you cannot simply multiply the HR by the new amount of units. Instead, multiply the regression coefficient (`coef`) (and the 95% confidence interval) by the units *before* exponentiating. For example, individuals who were 10 years older at enrollment have 41% greater hazard of angina pectoris (not 10 × the % associated with a 1-unit difference which would be 35%).

```
# Multiply by 10 inside of exp()
# Do NOT multiply the p-value
cbind("HR"      = exp(10*summary(cox.ex7.5)$coef[, "coef"]),
                  exp(10*confint(cox.ex7.5)),
      "p-value" = summary(cox.ex7.5)$coef[, "Pr(>|z|)"])
```

```
##        HR 2.5 % 97.5 %           p-value
## AGE 1.409 1.278  1.554 0.000000000006608
```

7.8.2 Adjusted

When adjusting for other predictors, refer to hazard ratios as **adjusted hazard ratios** (AHR) and, when interpreting them, specify what you have adjusted for.

Example 7.6: Using the Natality teaching dataset, estimate the association between time to preterm birth and previous preterm birth (`RF_PPTERM`), adjusted for mother's age (`MAGER`), race/ethnicity (`MRACEHISP`), and marital status (`DMAR`). For comparison, fit the unadjusted model first. To make the two models comparable (based on the same individuals), use a complete case analysis.

```
# Complete-case dataset
natality.complete <- natality %>%
  select(gestage37, preterm01, RF_PPTERM, MAGER, MRACEHISP, DMAR) %>%
  drop_na()

# Unadjusted
cox.ex7.6.unadj <- coxph(Surv(gestage37, preterm01) ~ RF_PPTERM,
                         data = natality.complete)

# AHR, 95% CI, p-value
cbind("HR"      = exp(summary(cox.ex7.6.unadj)$coef[, "coef"]),
                  exp(confint(cox.ex7.6.unadj)),
      "p-value" = summary(cox.ex7.6.unadj)$coef[, "Pr(>|z|)"])
```

```
##                 HR 2.5 % 97.5 %      p-value
## RF_PPTERMYes 3.285 2.074  5.204 0.0000004018
```

```
# Adjusted
cox.ex7.6 <- coxph(Surv(gestage37, preterm01) ~ RF_PPTERM + MAGER +
                     MRACEHISP + DMAR, data = nataliy.complete)

# AHRs, 95% CIs, p-values
cbind("AHR"       = exp(summary(cox.ex7.6)$coef[, "coef"]),
                    exp(confint(cox.ex7.6)),
      "p-value"   = summary(cox.ex7.6)$coef[, "Pr(>|z|)"])
```

```
##                       AHR  2.5 % 97.5 %     p-value
## RF_PPTERMYes       2.8919 1.8188  4.598 0.000007191
## MAGER              1.0318 1.0077  1.056 0.009317466
## MRACEHISPNH Black 1.7485 1.2346  2.476 0.001649379
## MRACEHISPNH Other 0.9292 0.5207  1.658 0.803844986
## MRACEHISPHispanic 1.2972 0.9132  1.843 0.146213569
## DMARUnmarried      1.7950 1.3291  2.424 0.000135888
```

```
# Multiple df p-values
car::Anova(cox.ex7.6, type = 3, test.statistic = "Wald")
```

```
## Analysis of Deviance Table (Type III tests)
##
## Response: Surv(gestage37, preterm01)
##           Df Chisq Pr(>Chisq)
## RF_PPTERM  1 20.14  0.0000072 ***
## MAGER      1  6.76    0.00932 **
## MRACEHISP  3 10.79    0.01292 *
## DMAR       1 14.56    0.00014 ***
## ---
## Signif. codes:  0 '***' 0.001 '**' 0.01 '*' 0.05 '.' 0.1 ' ' 1
```

Before adjusting for confounding, those with a previous preterm birth have an estimated 3.29 times the hazard of experiencing a preterm birth as those without a previous preterm birth. After adjusting for confounding, the estimated association is still large, but somewhat attenuated. After adjusting for mother's age, race/ethnicity, marital status, and education, those with a previous preterm birth have an estimated 2.89 times the hazard of experiencing a preterm birth (AHR = 2.89; 95% CI = 1.82, 4.60; p <.001) as those without a previous preterm birth.

NOTE: Parametric survival regression models

The KM method is a non-parametric estimator of the survival function. That implies that it makes no assumption about the shape of the function. As mentioned above, the Cox model is a semi-parametric method – it uses an unspecified baseline hazard function but includes regression parameters. There are various parametric survival regression models that do assume an underlying shape of the survival function, resulting from an assumption about the distribution of the survival times. Examples of assumed distributions include exponential, Weibull, Gompertz, and gamma. When the distributional assumption is met, these models can be more powerful than Cox regression. However, Cox regression does not depend on a distributional assumption and also easily incorporates time-varying predictors (Allison, 2010) (see Section 7.14). In this text, we focus on the Cox proportional hazards regression model. See the `flexsurvreg()` function in the `flexsurv` package (Jackson, 2023) for information on fitting parametric survival models in R.

7.9 Visualizing hazard ratios

In addition to reporting the numeric results of a Cox regression, it is helpful to create a **forest plot** to visualize the AHRs and their 95% CIs. You can easily create a forest plot using `sjPlot::plot_model()` (Lüdecke, 2023). See `?plot_model` for customization options.

Example 7.6 (continued): Create a forest plot displaying the AHRs for the model. To put the AHR for age on a similar scale as the categorical predictors, use the AHR and 95% CI for a 8-year difference in age (the interquartile range).

By way of reminder, the following code was used to fit the model.

```
cox.ex7.6 <- coxph(Surv(gestage37, preterm01) ~ RF_PPTERM + MAGER +
                     MRACEHISP + DMAR, data = natality.complete)
```

For the purpose of creating a forest plot, changing the age scale must be made via the method that transforms age prior to fitting the model (see Section 6.6.1). Using the interquartile range results in an AHR that compares the adjusted hazard of the outcome between individuals at the 75th and 25th percentiles of the predictor distribution.

```
# Compute the interquartile range
IQR(natality.complete$MAGER)
```

```
## [1] 8
```

```
# Rescale age by dividing by the IQR
fpdat <- natality.complete %>%
  mutate(MAGER8 = MAGER/8)

# Re-fit the model replacing age with the rescaled age variable
fit.fp <- coxph(Surv(gestage37, preterm01) ~ RF_PPTERM + MAGER8 +
                  MRACEHISP + DMAR, data = fpdat)

# Store the CIs for use in setting the axis limits
CI <- exp(confint(fit.fp))
```

The forest plot is shown in Figure 7.12.

```
# Forest plot
sjPlot::plot_model(fit.fp,
                   axis.lim = c(min(CI), max(CI)),
                   auto.label = F)
```

7.10 Prediction

In linear regression, we estimated the mean outcome at $X = x$, $E(Y|X = x)$. In logistic regression, we estimated the probability of the outcome, $P(Y = 1|X = x)$. In Cox regression,

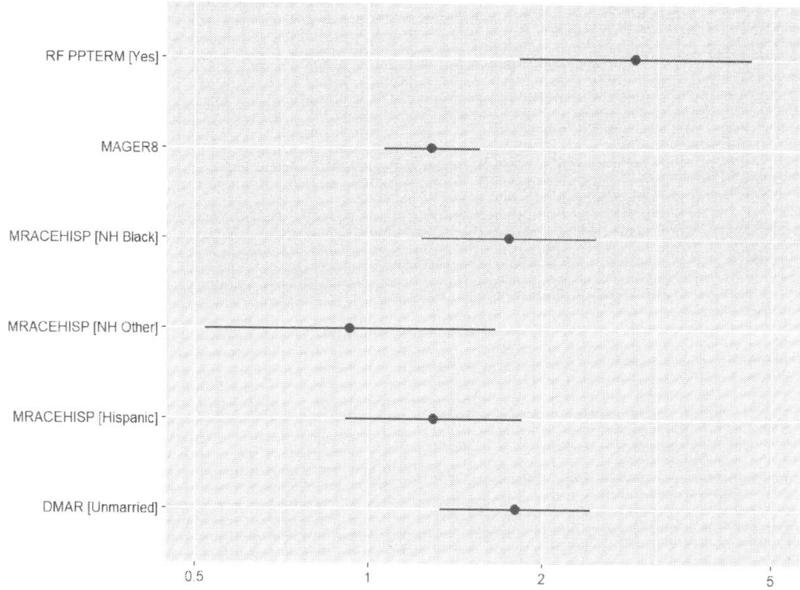

FIGURE 7.12 Forest plot of AHRs and their 95% CIs for the Cox regression analysis of preterm birth

we can estimate the survival probability at a specific time, $S(t|X = x)$, as well as the hazard ratio for an individual relative to a reference individual, $h(t|X = x) / h(t|X = x_{\text{ref}})$.

Use `predict()` to compute these estimates. Along with specifying values for the predictors in the model at which you want to compute the estimate, also specify t, the time at which you want to compute the estimate. For predicted survival, the choice of t matters. However, because of the proportional hazards assumption, a predicted HR will not depend on time, although in the syntax below you still need to specify a time for the code to work (but the HR will be the same regardless of what time is chosen).

Unfortunately, `predict()` only returns the standard error of the estimate when applied to a `coxph` object, not a confidence interval. For the estimated survival probability, an alternative is to use `survfit()` on the fitted Cox model, which does return a confidence interval.

Example 7.6 (continued): Using the Natality teaching dataset, estimate $S(30)$, the probability that a preterm birth has not occurred as of 30 weeks, for a married, non-Hispanic Black, 28-year-old woman with a previous preterm birth. Also, estimate the hazard of preterm birth for this individual relative to a reference individual.

```
# Check the spelling of each level
# (results not shown)
levels(natality$DMAR)
levels(natality$MRACEHISP)
levels(natality$RF_PPTERM)
```

```
# NOTE: For the event indicator variable (preterm), you must specify
#       a value or predict() will not work, but it does not matter which
#       value you specify, you will get the same answer.
NEWDAT <- data.frame(preterm01 = 1,
                     gestage37 = 30,
                     DMAR      = "Married",
```

```
                          MRACEHISP = "NH Black",
                          MAGER     = 28,
                          RF_PPTERM = "Yes")

predict(cox.ex7.6, NEWDAT, type = "survival", se.fit = T)
```

```
## $fit
## [1] 0.9628
##
## $se.fit
## [1] 0.01235
```

We can get the same answer using a `summary()` of `survfit()` with the `times` option. However, `summary()` will also include a confidence interval (via the `conf.int` option). With this method, there is no need to specify values for the event indicator and time inside of `NEWDAT`, although leaving them in `NEWDAT` will not change the answer.

```
unlist(summary(survfit(cox.ex7.6,
                       NEWDAT,
                       se.fit =T, conf.int = 0.95),
       times=30)[c("surv", "std.err", "lower", "upper")])
```

```
##    surv std.err   lower   upper
## 0.96282 0.01235 0.93892 0.98732
```

We conclude that the probability of no preterm birth through 30 weeks is 0.963 (95% CI = 0.939, 0.987).

Use `predict()` with `type = "risk"` to estimate an AHR comparing the hazard for this individual to that of a reference individual. Different `reference` specifications lead to different results.

- `reference = "zero"`: The reference individual has continuous predictor values of 0, and categorical predictors that are each at their reference level. This is not a good choice if $X = 0$ is not a plausible value for a continuous predictor X.
- `reference = "sample"`: The reference individual has continuous predictors that are each at their respective sample mean, and categorical predictors that are each at their reference level. This is equivalent to centering each continuous variable at its mean.
- `reference = "strata"` (the default): This leads to the same answer as `reference = "sample"` unless the model includes a `strata()` term (see Section 7.16.4), in which case continuous variables will be set to their within-stratum means.

Our model contained a single continuous predictor, mother's age, for which a reference value of 0 does not make sense. As we see below, the AHR comparing our individual to a reference individual with age 0 is very large, since the model is extrapolating and estimating a very small hazard for a newborn woman to have a preterm birth (a nonsense statement) and that small number is the denominator of the HR.

```
predict(cox.ex7.6, NEWDAT, type = "risk", reference = "zero", se.fit = T)
```

```
## $fit
##     1
## 12.15
##
## $se.fit
##     1
## 1.487
```

If a value of 0 does not make sense for any of the continuous predictors in your model, then use `reference = "strata"` instead (equivalent to `"sample"` if there are no strata variables, and more sensible if there are, since it makes sense to have the hazard for the individual in `NEWDAT` be compared to an individual in their own strata).

```
predict(cox.ex7.6, NEWDAT, type = "risk", reference = "strata", se.fit = T)
```

```
## $fit
##     1
## 4.906
##
## $se.fit
##      1
## 0.6329
```

Because of the proportional hazards assumption, the hazard ratio is constant over time. You will get the same answer for any `gestage37` value you enter into `NEWDAT`, or even if you omit `gestage37` altogether (the event indicator, `preterm01`, can also be omitted).

```
NEWDAT <- data.frame(DMAR       = "Married",
                     MRACEHISP  = "NH Black",
                     MAGER      = 28,
                     RF_PPTERM  = "Yes")
predict(cox.ex7.6, NEWDAT, type = "risk", reference = "strata")
```

```
##     1
## 4.906
```

Conclusion: This individual has 4.91 times the hazard of preterm birth of an individual at the reference level of each categorical predictor and the mean of each continuous predictor.

To verify that the reference individual has reference level values and mean values for the predictors, the code below demonstrates that the predicted HR for an individual with these characteristics is 1.

```
NEWDAT <- data.frame(DMAR       = levels(natality.complete$DMAR)[1],
                     MRACEHISP  = levels(natality.complete$MRACEHISP)[1],
                     MAGER      = mean(natality.complete$MAGER),
                     RF_PPTERM  = levels(natality.complete$RF_PPTERM)[1])
predict(cox.ex7.6, NEWDAT, type = "risk", reference = "strata")
```

```
## 1
## 1
```

NOTE: `reference = "strata"` is the default value and, for the reasons mentioned above, is a reasonable choice. Thus, you can typically omit the option and type, for example, `predict(cox.ex7.6, NEWDAT, type = "risk")`.

7.11 Plotting the estimated survival function

By computing predicted survival probabilities over a range of times, we can plot the survival function estimated by Cox regression for any set of predictor values. This is especially useful if we want to visually compare the estimated survival functions between groups.

NOTE: The estimated curves plotted here serve the specific purpose of illustrating *relative* differences between groups under the assumptions of the model and the form of the predictors included. The estimated *absolute* survival probabilities depend on the specific predictor values chosen and only represent individuals with those values rather than the average survival of a cohort of individuals with different predictor values. See Therneau et al. (2015) for a discussion of options for estimating the adjusted survival function for a reference cohort, rather than for one specific set of predictor values as is done here, and, optionally, with less restrictive model assumptions. Some of these options are implemented in the `ggadjustedcurves()` function in the `survminer` package (Kassambara et al., 2021).

Example 7.6 (continued): Plot the estimated survival functions for those with and without a previous preterm birth. Set the other predictors at their mean or reference level.

First, create a `data.frame` for each level of the variable we are interested in (`RF_PRETERM`) containing specific values for the other predictors and a range of times (so we can estimate a curve rather than survival at a single time).

```
# DF for RF_PPTERM = "Yes"
PPTERM_Yes <- data.frame(preterm01 = 1,
                         gestage37 = 17:36,
                         DMAR      = "Married",
                         MRACEHISP = "NH White",
                         MAGER     = mean(natality.complete$MAGER),
                         RF_PPTERM = "Yes")

# For the other level(s) of the variable of interest, copy and change the value
PPTERM_No <- PPTERM_Yes
PPTERM_No$RF_PPTERM <- "No"
```

Next, compute the estimates at these values.

```
PRED_PPTERM_Yes <- predict(cox.ex7.6, PPTERM_Yes, type = "survival")
PRED_PPTERM_No  <- predict(cox.ex7.6, PPTERM_No,  type = "survival")
```

Finally, plot the estimates vs. time (Figure 7.13).

```
# Create a blank plot with the correct axis labels and limits
plot(0, 0, col = "white",
     xlab = "Gestational Age (weeks)", ylab = "P(Not yet preterm)",
     xlim = c(17, 36),
     ylim = c(0.70, 1), # ylim = c(0, 1) if you need a larger range
     font.axis = 2, font.lab = 2, cex.axis = 1.25, cex.lab = 1.25)
# Add a line for each group
# NOTE: For lines() the arguments are x, y not y ~ x
lines(PPTERM_No$gestage37,  PRED_PPTERM_No,  lwd = 2, lty = 1)
lines(PPTERM_Yes$gestage37, PRED_PPTERM_Yes, lwd = 2, lty = 2)
legend(17, 0.85, c("No", "Yes"), title = "Previous preterm birth",
       cex=1.2, lty=c(1,2), lwd=c(2,2))
```

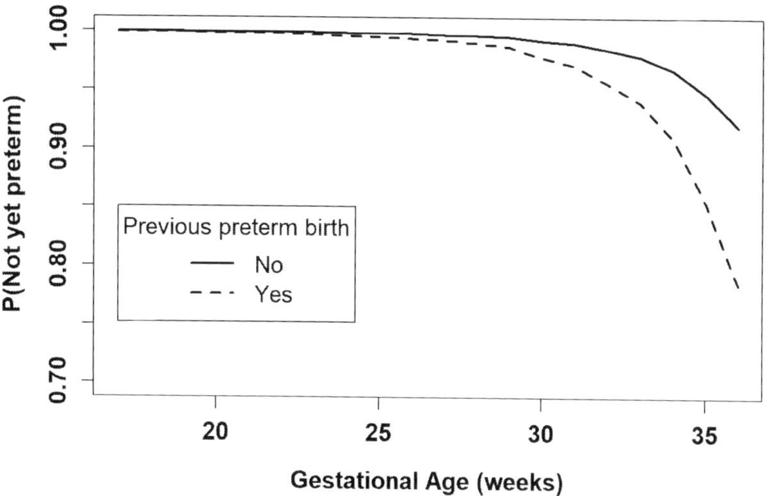

FIGURE 7.13 Estimated survival function by previous preterm birth

If you plot $1 - S(t)$ instead, the vertical axis will display the probability of preterm birth (Figure 7.14).

```
plot(0, 0, col = "white",
     xlab = "Gestational Age (weeks)", ylab = "P(Preterm by this time)",
     xlim = c(17, 36),
     ylim = c(0, 0.3), # ylim = c(0, 1) if you need a larger range
     font.axis = 2, font.lab = 2, cex.axis = 1.25, cex.lab = 1.25)
# Since this is 1 - S(t), reverse the order of the lines
# NOTE: For lines() the arguments are x, y not y ~ x
lines(PPTERM_Yes$gestage37, 1 - PRED_PPTERM_Yes, lwd = 2, lty = 2)
lines(PPTERM_No$gestage37,  1 - PRED_PPTERM_No,  lwd = 2, lty = 1)
legend(17, 0.2, c("Yes", "No"), title = "Previous preterm birth",
       cex=1.2, lty=c(2,1), lwd=c(2,2))
```

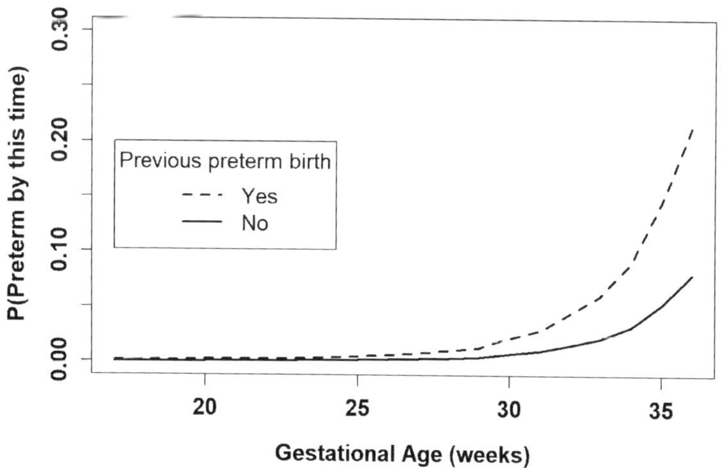

FIGURE 7.14 Estimated 1 – survival function by previous preterm birth

7.12 Interactions

As with linear and logistic regression, include interactions in a Cox model to assess effect modification. Including an interaction allows you to assess if the association between a risk factor and time to an event depends on another variable, and to estimate the HR for one variable at different levels of another.

Example 7.7: Using the Natality teaching dataset, see if the association between previous preterm birth and time to preterm birth (adjusted for age, race/ethnicity, and marital status) depends on whether or not the mother had pre-pregnancy hypertension (`RF_PHYPE`). As with any model with an interaction, include both the main effects and the interaction term.

```
cox.ex7.7.int <- coxph(Surv(gestage37, preterm01) ~ MAGER + MRACEHISP + DMAR +
                        RF_PPTERM + RF_PHYPE +   # Main effects
                        RF_PPTERM:RF_PHYPE,      # Interaction
                      data = natality)

round(
  cbind("AHR"       = exp(summary(cox.ex7.7.int)$coef[, "coef"]),
                      exp(confint(cox.ex7.7.int)),
        "p-value"   = summary(cox.ex7.7.int)$coef[, "Pr(>|z|)"])
  , 3)
```

```
##                                AHR 2.5 % 97.5 % p-value
## MAGER                        1.032 1.008  1.057   0.009
## MRACEHISPNH Black            1.699 1.194  2.416   0.003
## MRACEHISPNH Other            0.929 0.520  1.657   0.802
## MRACEHISPHispanic            1.304 0.918  1.852   0.139
## DMARUnmarried                1.785 1.320  2.413   0.000
## RF_PPTERMYes                 2.660 1.634  4.328   0.000
## RF_PHYPEYes                  1.287 0.526  3.148   0.580
## RF_PPTERMYes:RF_PHYPEYes     6.774 1.195 38.392   0.031
```

The interaction is statistically significant (the p-value for the `RF_PPTERMYes:RF_PHYPEYes` row is .031). Had either of the terms in the interaction been categorical with more than two levels, we would have used `car::Anova(cox.ex7.7.int, type = 3, test = "Wald")` to get the interaction p-value, as it would have been a multiple degree of freedom test.

7.12.1 Overall test of a predictor involved in an interaction

As shown in Section 5.9.11, we can also carry out an overall test of previous preterm birth by comparing the full model to the model with both the main effect (`RF_PPTERM`) and interaction (`RF_PPTERM:RF_PHYPE`) removed. We did not remove missing data prior to fitting the full model, so we need to create a dataset with no missing data first so both models can be fit to the same observations, or use the `subset` option. As with MLR, use `anova()` to compare the models but for a Cox regression we must specify `test = "Chisq"` to get a p-value. There is no option to specify a Wald test in this case; `anova()` uses a likelihood ratio test when comparing Cox models.

```
# Fit the reduced model using the subset option
# to remove missing values for the variable
# being removed to create the reduced model
fit0 <- coxph(Surv(gestage37, preterm01) ~
                  MAGER + MRACEHISP + DMAR +
                  RF_PHYPE,
              data = natality,
              subset = complete.cases(RF_PPTERM))
# Compare to full model
anova(fit0, cox.ex7.7.int, test = "Chisq")
```

```
## Analysis of Deviance Table
##  Cox model: response is  Surv(gestage37, preterm01)
##  Model 1: ~ MAGER + MRACEHISP + DMAR + RF_PHYPE
##  Model 2: ~ MAGER + MRACEHISP + DMAR + RF_PPTERM + RF_PHYPE + RF_PPTERM:RF_
  PHYPE
##    loglik Chisq Df Pr(>|Chi|)
## 1  -1587
## 2  -1578  19.1  2   0.000071 ***
## ---
## Signif. codes:  0 '***' 0.001 '**' 0.01 '*' 0.05 '.' 0.1 ' ' 1
```

The 2 df test (2 df because the models differ in two terms – the main effect and the interaction) results in a p-value that is very small. Thus, after adjusting for age, race/ethnicity, marital status, and pre-pregnancy hypertension, previous preterm birth is significantly associated with time to preterm birth.

7.12.2 Estimating the HR at each level of the other variable

In addition to the test of significance, we also would like to know what is the magnitude of the AHR for previous preterm birth among those with and without pre-pregnancy hypertension. The main effect for `RF_PPTERM` provides the AHR for previous preterm birth among those at the reference level of `RF_PHYPE` ("No"). For linear and logistic regression models, we used `gmodels::estimable()` to estimate the AHR at non-reference levels of the other predictor in the interaction. Unfortunately, `gmodels::estimable()` does not work for a `coxph` object. Instead, re-fit the model after changing the reference level of `RF_PHYPE` to "Yes" which will result in a main effect for `RF_PPTERM` at the new reference level. In this way, we can get the AHR, 95% CI, and p-value for `RF_PPTERM` at each level of `RF_PHYPE`.

First, using the original model, extract the results for the main effect `RF_PPTERM`, which is the effect at `RF_PHYPE` = "No" (the reference level for `RF_PHYPE`).

```
round(
  cbind("AHR"     = exp(summary(cox.ex7.7.int)$coef[, "coef"]),
                    exp(confint(cox.ex7.7.int)),
        "p-value" = summary(cox.ex7.7.int)$coef[, "Pr(>|z|)"])["RF_PPTERMYes", ]
  , 3)
```

```
##      AHR   2.5 %  97.5 % p-value
##    2.660   1.634   4.328   0.000
```

Next, re-level `RF_PHYPE` (not `RF_PPTERM`), re-fit the model, and extract the main effect results again to get the `RF_PPTERM` effect at `RF_PHYPE` = "Yes".

```
natality_b <- natality %>%
  mutate(RF_PHYPE = relevel(RF_PHYPE, ref = "Yes"))

# Re-fit with data = natality_b
cox.ex7.7.int_b <- coxph(Surv(gestage37, preterm01) ~
                         MAGER + MRACEHISP + DMAR +
                         RF_PPTERM + RF_PHYPE +
                         RF_PPTERM:RF_PHYPE,
                         data = natality_b)

round(
  cbind("AHR"       = exp(summary(cox.ex7.7.int_b)$coef[, "coef"]),
                        exp(confint(cox.ex7.7.int_b)),
        "p-value"   = summary(cox.ex7.7.int_b)$coef[, "Pr(>|z|)"])["RF_PPTERMYes",]
  , 3)
```

```
##     AHR   2.5 %  97.5 % p-value
##  18.016   3.406  95.289   0.001
```

The AHR for previous preterm birth is greater among those with pre-pregnancy hypertension (18.016) than among those without (2.660) (interaction p-value = .031, a test of the null hypothesis that these two AHRs are the same).

The AHR among those with pre-pregnancy hypertension is quite large and has a very wide CI. Why? It turns out that there were not many cases with pre-pregnancy hypertension, resulting in high variability when estimating the AHR in this subgroup and a wide CI.

```
# Number with hypertension
table(natality$RF_PHYPE)
```

```
##
##   No  Yes
## 1965   34
```

Also, among those with pre-pregnancy hypertension, only two mothers had a previous preterm birth and both experienced a preterm birth, which explains the large `RF_PPTERM` AHR in this group.

```
# Previous preterm vs. preterm
# among those with hypertension
table(natality$RF_PPTERM[natality$RF_PHYPE == "Yes"],
      natality$preterm01[natality$RF_PHYPE == "Yes"])
```

```
##
##         0  1
##   No   27  5
##   Yes   0  2
```

You may have noticed that there is a 0 in this table. With logistic regression, a 0 in the two-way table comparing a categorical predictor to the outcome was an indication that logistic regression would have convergence problems due to separation (Section 6.10). In Section 7.13 we will discuss this same issue in the context of a Cox regression model. It turns out this particular 0 does not cause a problem with separation, but others will.

7.12.3 Visualizing an interaction

Plot the survival functions illustrating the previous preterm birth effect among those without and with pre-pregnancy hypertension. The code is similar to that in Section 7.11, but now we need a `data.frame` for each combination of levels of `RF_PPTERM` and `RF_PHYPE`.

```
# RF_PPTERM = "Yes", RF_PHYPE   = "Yes"
DAT_PPTERM_Yes_PHYPE_Yes <- data.frame(
            preterm01 = 1,
            gestage37 = 17:36,
            DMAR      = "Married",
            MRACEHISP = "NH White",
            MAGER     = mean(natality.complete$MAGER),
            RF_PPTERM = "Yes",
            RF_PHYPE  = "Yes")

# Other three combinations
# Copy first data.frame
DAT_PPTERM_Yes_PHYPE_No <- DAT_PPTERM_Yes_PHYPE_Yes
DAT_PPTERM_No_PHYPE_Yes <- DAT_PPTERM_Yes_PHYPE_Yes
DAT_PPTERM_No_PHYPE_No  <- DAT_PPTERM_Yes_PHYPE_Yes

# Update values of the 2 terms in the interaction
DAT_PPTERM_Yes_PHYPE_No$RF_PPTERM <- "Yes"
DAT_PPTERM_Yes_PHYPE_No$RF_PHYPE  <- "No"

DAT_PPTERM_No_PHYPE_Yes$RF_PPTERM <- "No"
DAT_PPTERM_No_PHYPE_Yes$RF_PHYPE  <- "Yes"

DAT_PPTERM_No_PHYPE_No$RF_PPTERM <- "No"
DAT_PPTERM_No_PHYPE_No$RF_PHYPE  <- "No"
```

When computing the predictions, be careful to use the original fit for the combinations with `RF_PHYPE` = "No" (the original reference level of the other term in the interaction) and the re-leveled fit for `RF_PHYPE` = "Yes" (the reference level after re-leveling).

```
# Predictions
# Use re-leveled fit when RF_PHYPE = "Yes" (ref level after re-leveling)
PRED_PPTERM_Yes_PHYPE_Yes <-
  predict(cox.ex7.7.int_b, DAT_PPTERM_Yes_PHYPE_Yes, type = "survival")

# Use original fit when RF_PHYPE = "No" (original ref level)
PRED_PPTERM_Yes_PHYPE_No  <-
  predict(cox.ex7.7.int,   DAT_PPTERM_Yes_PHYPE_No,  type = "survival")

# Use re-leveled fit when RF_PHYPE = "Yes" (ref level after re-leveling)
PRED_PPTERM_No_PHYPE_Yes  <-
  predict(cox.ex7.7.int_b, DAT_PPTERM_No_PHYPE_Yes,  type = "survival")

# Use original fit when RF_PHYPE = "No" (original ref level)
PRED_PPTERM_No_PHYPE_No   <-
  predict(cox.ex7.7.int,   DAT_PPTERM_No_PHYPE_No,   type = "survival")
```

Finally, plot the predictions.

```
# RF_PHYPE = "No"
par(mfrow=c(1,2))
plot(0, 0, col = "white",
     xlab = "Gestational Age (weeks)", ylab = "P(Not yet preterm)",
```

```
     xlim = c(17, 36),
     ylim = c(0, 1),
     font.axis = 2, font.lab = 2,
     main = "Without hypertension")
# NOTE: For lines() the arguments are x, y not y ~ x
lines( DAT_PPTERM_No_PHYPE_No$gestage37,
       PRED_PPTERM_No_PHYPE_No,  lwd = 2, lty = 1)
lines( DAT_PPTERM_Yes_PHYPE_No$gestage37,
       PRED_PPTERM_Yes_PHYPE_No, lwd = 2, lty = 2)
legend(17, 0.55, c("No", "Yes"), title = "Previous preterm birth",
       lty=c(1,2), lwd=c(2,2), bty = "n")

# RF_PHYPE = "Yes"
plot(0, 0, col = "white",
     xlab = "Gestational Age (weeks)", ylab = "P(Not yet preterm)",
     xlim = c(17, 36),
     ylim = c(0, 1),
     font.axis = 2, font.lab = 2,
     main = "With hypertension")
# NOTE: For lines() the arguments are x, y not y ~ x
lines(DAT_PPTERM_No_PHYPE_Yes$gestage37,
       PRED_PPTERM_No_PHYPE_Yes,  lwd = 2, lty = 1)
lines(DAT_PPTERM_Yes_PHYPE_Yes$gestage37,
       PRED_PPTERM_Yes_PHYPE_Yes, lwd = 2, lty = 2)
legend(17, 0.55, c("No", "Yes"), title = "Previous preterm birth",
       lty=c(1,2), lwd=c(2,2), bty = "n")
```

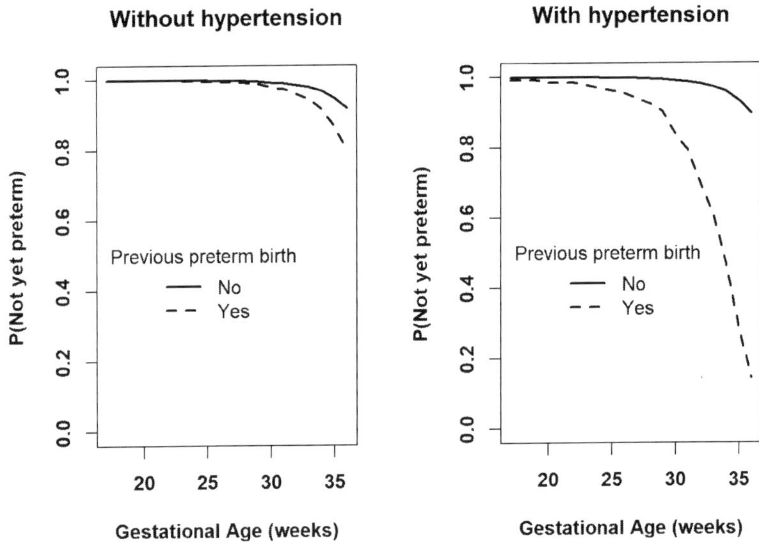

FIGURE 7.15 Comparing the survival functions between those without and with previous preterm birth at each level of pre-pregnancy hypertension

It is a good idea to compare the visualization of an interaction to the estimated AHR at each level of the other variable in the interaction. If they are not consistent, then the estimation and/or plotting code have an error. In Section 7.12.2, we estimated that the AHR for `RF_PPTERM` at `RF_PHYPE` = "No" is 2.660 and at `RF_PHYPE` = "Yes" is 18.016. These are consistent with Figure 7.15. Both AHRs are greater than 1 and in both plots the survival curve for RF_PPTERM = "Yes" is lower than that for "No". Also, the AHR for RF_PPTERM at RF_PHYPE = "Yes" is much larger than at RF_PHYPE = "No" and the lines in the right panel are much further apart than those in the left panel.

The right panel in Figure 7.15 shows that, based on this dataset, we estimate that those with both a previous preterm birth and pre-pregnancy hypertension (and other predictors as specified above) have over 85% probability of preterm birth (the dashed survival function drops below 0.15). Be wary of this estimate, however, since, as seen above, it is based on a very small sample size. Using the methods from Section 7.10, compute a 95% CI for this quantity to get an idea of how imprecise it is.

```
unlist(summary(survfit(cox.ex7.7.int_b,
   DAT_PPTERM_Yes_PHYPE_Yes[DAT_PPTERM_Yes_PHYPE_Yes$gestage37 == 36,],
   se.fit =T, conf.int = 0.95),
 times=36)[c("surv", "std.err", "lower", "upper")])
```

```
##      surv  std.err    lower    upper
## 0.141915 0.204637 0.008407 1.000000
```

Conclusion: After adjusting for mother's age, race/ethnicity, marital status, and pre-pregnancy hypertension, previous preterm birth is significantly associated with time to preterm birth (p <.001 – from the test comparing the model with an interaction to the model with no RF_PPTERM main effect or interaction). The association between previous preterm birth and time to preterm birth differs significantly between those without and with pre-pregnancy hypertension (interaction p = .031). Among mothers who did not have pre-pregnancy hypertension, those with a previous preterm birth have 2.7 times the hazard of preterm birth of those who did not (AHR = 2.66; 95% CI = 1.63, 4.33; p <.001). Among those who did have pre-pregnancy hypertension, the hazard is even larger (AHR = 18.0; 95% CI = 3.4, 95.3; p <.001).

7.13 Separation

As with logistic regression (Section 6.10), Cox regression can have a problem with separation. However, unlike logistic regression, the outcome in a Cox regression is not a simple binary variable, but rather a combination of a binary event indicator and an event time. In logistic regression, separation occurs when the binary outcome is always 0 *or* always 1 within a level of a predictor. In Cox regression, however, separation only occurs when the binary event indicator is always 0 within a level of a predictor (all event times censored). If all event times within a level are uncensored (everyone experienced the event), there is no separation.

Example 7.7 (continued): Consider the subset of mothers with pre-pregnancy hypertension. We saw earlier that this is a small sample, and there was a zero in the two-way table of RF_PPTERM vs. the event indicator.

```
subdat1 <- natality %>%
  filter(RF_PHYPE == "Yes")

table(subdat1$RF_PPTERM, subdat1$preterm01)
```

```
##
##        0  1
##    No  27  5
##    Yes  0  2
```

This zero does *not* cause a problem with separation – `coxph()` runs with no errors and we get an estimated AHR and standard error that seem reasonable (yes, the AHR and upper confidence bound are large, but as we saw earlier this is due to the small sample size).

```
fit1 <- coxph(Surv(gestage37, preterm01) ~ RF_PPTERM,
              data = subdat1)

cbind("HR"        = exp(summary(fit1)$coef[, "coef"]),
                    exp(confint(fit1)),
      "p-value" = summary(fit1)$coef[, "Pr(>|z|)"])
```

```
##                    HR 2.5 % 97.5 %  p-value
## RF_PPTERMYes 14.91  2.44  91.12 0.003435
```

However, what if instead of no censored times at `RF_PPTERM == "Yes"` they were all censored?

```
subdat2 <- natality %>%
  filter(RF_PHYPE == "Yes")

# For illustration, change the two times in the "Yes" row to censored
subdat2$preterm01[subdat2$RF_PPTERM == "Yes" &
                  subdat2$preterm01 == 1] <- 0

table(subdat2$RF_PPTERM, subdat2$preterm01)
```

```
##
##        0  1
##    No  27  5
##    Yes  2  0
```

Now the 0 is in the `preterm01 = 1` column indicating that all the times are censored for `RF_PPTERM = "Yes"`. Using this data, `coxph` returns a warning and a HR approaching 0 or ∞.

```
fit2 <- coxph(Surv(gestage37, preterm01) ~ RF_PPTERM,
              data = subdat2)
```

```
## Warning in coxph.fit(X, Y, istrat, offset, init, control, weights = weights, :
## Loglik converged before variable 1 ; coefficient may be infinite.
```

```
cbind("HR"        = exp(summary(fit2)$coef[, "coef"]),
                    exp(confint(fit2)),
      "p-value" = summary(fit2)$coef[, "Pr(>|z|)"])
```

```
##                               HR 2.5 % 97.5 % p-value
## RF_PPTERMYes 0.00000003847     0    Inf   0.999
```

Diagnosis and resolution

Before running a Cox regression, always check for separation using a complete case dataset so the sample size is the same as will be used in the regression. For each categorical predictor, create a two-way table of the predictor vs. the event indicator. For predictors that have any levels at which all observations are censored (zero events), solve the problem using filtering, collapsing, or removing as discussed in Section 6.10.4. An alternative, not covered here, is to use penalized Cox regression. See, for example, the `coxphf` package (Heinze et al., 2023a).

7.14 Time-varying predictors

In all the examples we have looked at, the predictors did not vary over time. Each individual had only one value for each predictor. Since survival data is the result of follow-up over time, it is possible to have predictors that *do* vary over time. For example, in a study of time to heart attack, researchers could record various other conditions' occurrence over time, such as hypertension or angina. In a study of juvenile recidivism, researchers could record how education or employment status change over time. Cox regression is able to handle such **time-varying predictors** (also known as "time-dependent covariates").

The datasets we have analyzed so far contained one row per individual, including their event time, an indicator of whether the event occurred at that time or the time is censored, and time-invariant predictor values. A dataset with time-varying predictors will have multiple rows per individual, with different rows having different values for the time-varying predictors, reflecting how they change over time. Additionally, rather than having a single event time variable, each row will have two time variables indicating the beginning and end of the time interval represented by that row of data.

Example 7.8: The Opioid teaching dataset (see Appendix A.7) contains longitudinal information for 362 individuals who at baseline had used non-prescribed pharmaceutical opioids (NPPO, "pain pills"), but were not dependent on NPPOs and had never used heroin (Carlson et al., 2016). The dataset contains 1853 observations. Each row contains the time variables `START` and `STOP` which define the time interval (years from initiation of NPPO use) associated with that row. Time-invariant variables in the dataset are constant over all rows for the same individual, while time-varying variables can change between rows. Each `(START, STOP]` interval defines a period of time during which no variables changed. Two time-varying variables in the dataset are heroin use (`heroin`) (the event indicator variable) and lifetime opioid dependence (`dep_lifetime`) based on DSM-IV criteria (Forman et al., 2004; Hudziak et al., 1993). Other than heroin use, which is the event of interest, time-dependent variables were lagged to ensure that they temporally preceded the outcome.

Look at the rows of data for a few variables for one individual.

```
load("Data/opioid_rmph.rData")

opioid %>%
  filter(RANDID == 10) %>%
  select(RANDID, wave, START, STOP, heroin, age_at_init, sex, dep_lifetime)
```

```
##    RANDID wave START STOP heroin age_at_init  sex dep_lifetime
## 1     10    1  3.78 4.26      0          19 Male            0
## 2     10    2  4.26 4.78      0          19 Male            1
## 3     10    3  4.78 5.29      0          19 Male            1
## 4     10    4  5.29 5.84      0          19 Male            1
## 5     10    5  5.84 6.27      0          19 Male            1
## 6     10    6  6.27 6.79      1          19 Male            1
```

This individual is male and started using NPPOs at age 19 years. These two variables are time-invariant and so have the same value in every row. His first interview took place 4.26 years after NPPO initiation and asked about his experiences and behaviors going back 6 months. Thus, the time interval for his first row is `(START, STOP]` $= (3.78, 4.26]$. He first reported using heroin at his 6th interview. At baseline (wave $= 0$, but remember time-varying variables were lagged so wave $= 1$ values in the dataset actually are the values from wave $= 0$), he did not meet the criteria for lifetime opioid dependence, but he did at the next interview. Thus, the associated time-varying variable (`dep_lifetime`) changes from the first to second row.

We will work with datasets that are already in this format. If, in your future work, you have a dataset in a different format, see Therneau et al. (2023) for guidance on transforming it into the correct format. Given data set up in this manner, we can fit a Cox regression model with a slight modification to the `coxph()` syntax we have been using. Instead of `Surv(TIME, EVENT)`, use `Surv(START, STOP, EVENT)`. In this example, the `START` and `STOP` variables are named `START` and `STOP`, but if they have different names in a future dataset you are working with then use those names instead.

```
cox.ex7.8 <- coxph(Surv(START, STOP, heroin) ~
                     age_at_init + sex + dep_lifetime,
                   data = opioid)

summary(cox.ex7.8)
```

```
## Call:
## coxph(formula = Surv(START, STOP, heroin) ~ age_at_init + sex +
##     dep_lifetime, data = opioid)
##
##   n= 1853, number of events= 27
##
##                coef exp(coef) se(coef)     z Pr(>|z|)
## age_at_init  -0.393     0.675    0.135 -2.91   0.0036 **
## sexMale       0.151     1.164    0.392  0.39   0.6996
## dep_lifetime  1.058     2.881    0.399  2.65   0.0080 **
## ---
## Signif. codes:  0 '***' 0.001 '**' 0.01 '*' 0.05 '.' 0.1 ' ' 1
##
##              exp(coef) exp(-coef) lower .95 upper .95
## age_at_init      0.675      1.482     0.518     0.879
## sexMale          1.164      0.859     0.539     2.511
## dep_lifetime     2.881      0.347     1.318     6.298
##
## Concordance= 0.716  (se = 0.042 )
## Likelihood ratio test= 15.2  on 3 df,   p=0.002
## Wald test            = 14.5  on 3 df,   p=0.002
## Score (logrank) test = 15.4  on 3 df,   p=0.001
```

```
cbind("AHR"     = exp(summary(cox.ex7.8)$coef[, "coef"]),
                  exp(confint(cox.ex7.8)),
      "p-value" = summary(cox.ex7.8)$coef[, "Pr(>|z|)"])
```

```
##                 AHR  2.5 % 97.5 %  p-value
## age_at_init  0.6749 0.5180 0.8794 0.003597
## sexMale      1.1635 0.5391 2.5111 0.699560
## dep_lifetime 2.8807 1.3176 6.2981 0.008022
```

```
car::Anova(cox.ex7.8, type = 3, test.statistic = "Wald")
```

```
## Analysis of Deviance Table (Type III tests)
##
## Response: Surv(START, STOP, heroin)
##              Df Chisq Pr(>Chisq)
## age_at_init   1  8.48     0.0036 **
## sex           1  0.15     0.6996
## dep_lifetime  1  7.03     0.0080 **
## ---
## Signif. codes:  0 '***' 0.001 '**' 0.01 '*' 0.05 '.' 0.1 ' ' 1
```

Whether a predictor is time-varying or time-invariant, its HR can be interpreted as a comparison of the hazard between groups of individuals with different values of that predictor. Thus, in this example, we could conclude that after adjusting for age at NPPO initiation and sex, those with lifetime opioid dependence have 2.88 times the hazard of using heroin as those who do not (AHR = 2.88; 95% CI = 1.32, 6.30; p = .008). The HR for a time-varying predictor, however, can also be interpreted as the effect of within-individual change on the hazard. If an individual without opioid dependence transitions to dependence, their hazard of transitioning to heroin is multiplied by 2.88.

NOTES:

- If an observation has the same start and end time, `coxph` will drop it from the analysis since there is no time at risk. However, this may not be appropriate. For example, suppose times are recorded as whole number days. An interval could start on a certain day and the event could have happened that same day. So there is time at risk, it is just less than a whole day. In this case, an approximate solution is to add a fractional number to the stop time for observations with `START` = `STOP`, as suggested in Therneau et al. (2023). For example, if `START` and `STOP` are both day 10, change `STOP` to 10.5 (making the assumption that the person was at risk in this interval for 12 hours). Make sure to also increase the `START` time of the next time interval for this person by the same amount to avoid having overlapping intervals.
- As mentioned previously, the event time distribution in this dataset is left-truncated – those who transitioned to heroin before the study started were not eligible for inclusion in the study. Fortunately, the `Surv(START, STOP, EVENT)` syntax automatically handles the left-truncation when, as in this dataset, the start time for the first interval for each individual is the time from the time origin (e.g., initiation of pain pill use) to the start of the study. Having the start time of each individual's first interval be 0 would lead to bias, as it would assume that everyone was under observation since their time origin and that those in the study were sampled from those at risk since the time origin. In fact, participants were only observed after the study started and those who experienced the event before then were excluded.

7.15 Collinearity

Collinearity in a Cox regression leads to the same problems as it does in other forms of regression (unstable parameter estimates, difficulty in interpretation – see Section 5.20), and the solutions to the problem are the same (e.g., remove redundant variables). However, unlike `lm` and `glm`, the `car::vif()` function will not work with a `coxph` object. Instead, first fit a linear regression model with any numeric variable, as the outcome (here we use the event time variable) and compute the VIFs for that model.

```
pretend.lm <- lm(gestage37 ~ RF_PPTERM + MAGER + MRACEHISP + DMAR,
                 data = natality.complete)
car::vif(pretend.lm)
```

```
##              GVIF Df GVIF^(1/(2*Df))
## RF_PPTERM 1.006  1           1.003
## MAGER     1.162  1           1.078
## MRACEHISP 1.146  3           1.023
## DMAR      1.299  1           1.140
```

7.16 Proportional hazards assumption

The reason Cox regression is called Cox "proportional hazards" (PH) regression is that the standard form of the model assumes the hazards for any two individuals have the same proportion at all times. Thus, the PH assumption implies the HR measuring the effect of any predictor is constant over time.

Consider two individuals who differ by one unit in X_K and do not differ in any other predictor in the model. Their Cox model hazard functions are the following:

$$\begin{aligned} h(t|X_1 = x_1, ..., X_K = x_K + 1) &= h_0(t)e^{\beta_1 x_1 + ... + \beta_K (x_K + 1)} \\ h(t|X_1 = x_1, ..., X_K = x_K) &= h_0(t)e^{\beta_1 x_1 + ... + \beta_K x_K} \end{aligned}$$

Taking the ratio of these, everything cancels out except e^{β_K}, the AHR for X_K, a quantity which does not depend on time. In fact, you could plug in *any* two sets of predictor values, differing on any or all of the predictors, and the ratio of the hazards would not depend on time. The hazard $h(t)$ itself can increase or decrease over time because the baseline hazard $h_0(t)$ varies with time. However, the *ratio* of hazards comparing any two individuals is constant because $h_0(t)$ does not vary between individuals and so cancels out in the ratio.

A violation of the PH assumption does not mean we cannot use Cox regression, just that we need to modify it slightly to account for non-proportional hazards, typically by including an interaction between a predictor and time or by stratifying by a predictor.

7.16.1 Checking the proportional hazards assumption

Suppose we think the HR for X_k varies linearly with time. This means that instead of having $\beta_k X_k$ in the model equation, we would have $\beta_k(t) X_k$ where $\beta_k(t) = \beta_{k_1} + \beta_{k_2} t$. More generally, $\beta(t)$ could take on any form. The `cox.zph()` (Grambsch and Therneau, 1994) function checks the PH assumption by comparing the PH model for each predictor to a model that allows that predictor's regression coefficient to vary smoothly with time, where "smoothly" means the alternative model does not assume a specific form for the time trajectory.

Example 7.6 (continued): Check each predictor for non-proportional hazards in the model we fit for Example 7.6.

To be able to screen each component of a categorical predictor that has $L > 2$ levels, we must replace the factor version of the variable with the corresponding $L - 1$ indicator variables. Although not necessary at this step, we will also replace each binary categorical variable with an indicator variable as we may need the indicator variable version when we relax the PH assumption in Section 7.16.3.

When you enter a factor variable into a regression model, R automatically creates indicator variables and includes them in the model, leaving out the one for the reference level. The resulting indicator variables can be extracted from a model using the `model.matrix()` function. Applying this function to our Cox model fit results in a matrix. The following shows the first few rows.

```
head(
  model.matrix(cox.ex7.6)
)
```

```
##   RF_PPTERMYes MAGER MRACEHISPNH Black MRACEHISPNH Other MRACEHISPHispanic
## 1            0    35                 0                 0                 1
## 2            1    28                 0                 0                 0
## 3            0    22                 1                 0                 0
## 4            0    19                 0                 0                 1
## 5            0    30                 0                 0                 0
## 6            0    20                 0                 1                 0
##   DMARUnmarried
## 1             0
## 2             1
## 3             1
## 4             1
## 5             0
## 6             1
```

Extract the indicator variables and assign them to variables inside the dataset and re-fit the model.

NOTES:

- Pay careful attention to how the columns of the model matrix are spelled. In this example, some have spaces (e.g., "MRACEHISPNH Black"). When you create variables in the code below, the strings inside the `[,]`s on the right must be spelled exactly as they appear in the model matrix.
- This method of extracting indicator variables and adding them to a dataset only works correctly with a complete case dataset. If the dataset used to fit the model was not complete, create a complete-case dataset and re-fit the model before proceeding.

```
# Extract indicator variables and add to dataset
natality.complete <- natality.complete %>%
  mutate(Previous_Preterm  = model.matrix(cox.ex7.6)[, "RF_PPTERMYes"],
         race_eth_NHBlack  = model.matrix(cox.ex7.6)[, "MRACEHISPNH Black"],
         race_eth_NHOther  = model.matrix(cox.ex7.6)[, "MRACEHISPNH Other"],
         race_eth_Hispanic = model.matrix(cox.ex7.6)[, "MRACEHISPHispanic"],
         Unmarried         = model.matrix(cox.ex7.6)[, "DMARUnmarried"])

# Re-fit the model with indicator variables instead of factors
cox.ex7.6.indicator <- coxph(Surv(gestage37, preterm01) ~
                   Previous_Preterm + MAGER +
                   race_eth_NHBlack + race_eth_NHOther + race_eth_Hispanic +
                   Unmarried,
                   data = natality.complete)
```

Next, verify that this new fit is equivalent to the old fit. If it is not, then check the code you used to create the indicator variables and re-fit the model.

```
summary(cox.ex7.6)$coef
```

```
##                         coef exp(coef) se(coef)       z  Pr(>|z|)
## RF_PPTERMYes         1.06193    2.8919  0.23662  4.4879 0.000007191
## MAGER                0.03132    1.0318  0.01204  2.6002 0.009317466
## MRACEHISPNH Black    0.55874    1.7485  0.17754  3.1470 0.001649379
## MRACEHISPNH Other   -0.07341    0.9292  0.29555 -0.2484 0.803844986
## MRACEHISPHispanic    0.26023    1.2972  0.17910  1.4530 0.146213569
## DMARUnmarried        0.58499    1.7950  0.15332  3.8155 0.000135888
```

```
summary(cox.ex7.6.indicator)$coef
```

```
##                         coef exp(coef) se(coef)       z  Pr(>|z|)
## Previous_Preterm     1.06193    2.8919  0.23662  4.4879 0.000007191
## MAGER                0.03132    1.0318  0.01204  2.6002 0.009317466
## race_eth_NHBlack     0.55874    1.7485  0.17754  3.1470 0.001649379
## race_eth_NHOther    -0.07341    0.9292  0.29555 -0.2484 0.803844986
## race_eth_Hispanic    0.26023    1.2972  0.17910  1.4530 0.146213569
## Unmarried            0.58499    1.7950  0.15332  3.8155 0.000135888
```

Finally, use `cox.zph()` to check for non-proportional hazards using the model fit with indicator variables.

```
NPH_CHECK <- cox.zph(cox.ex7.6.indicator)
NPH_CHECK
```

```
##                     chisq df      p
## Previous_Preterm    1.958  1 0.1618
## MAGER               0.806  1 0.3693
## race_eth_NHBlack    4.067  1 0.0437
## race_eth_NHOther    0.608  1 0.4355
## race_eth_Hispanic   0.186  1 0.6661
## Unmarried           8.769  1 0.0031
## GLOBAL             12.257  6 0.0565
```

The row for each predictor tests the null hypothesis that the predictor's coefficient does not vary with time (equivalently, that the predictor's AHR does not vary with time). Based on statistical significance of individual terms, we would conclude that the coefficients for `race_eth_NHBlack` and `Unmarried` have significantly non-proportional hazards. However, we must also take into account the fact that we are carrying out multiple tests. The `GLOBAL` row tests the null hypothesis that all the predictors meet the PH assumption. Based on the global test, we would conclude that the PH assumption is sufficiently met for all the variables. Yet we must *also* take into account that, as with all statistical hypothesis tests, the .05 cutoff for significance is arbitrary and conclusions are very dependent on the sample size. Thus, in addition to the statistical test, it is wise to look at a visualization to assess the magnitude of any violation of the PH assumption.

Again, each row in the `cox.zph()` output is testing the null hypothesis that the β coefficient for that predictor does not vary with time. A `plot()` of the output of `cox.zph()` illustrates how the βs vary with time (Figure 7.16).

```
par(mfrow=c(2,3))
plot(NPH_CHECK)
```

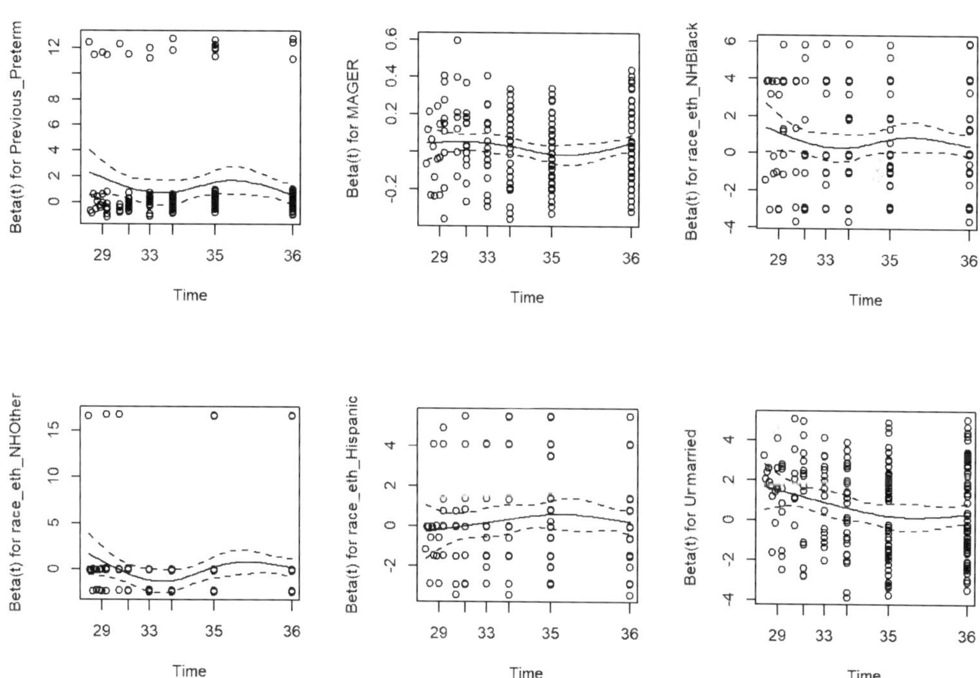

FIGURE 7.16 Visually assessing the proportional hazards assumption

If the null hypothesis is correct (proportional hazards), then the β trajectories will be close to horizontal. This provides a nice visualization for each predictor of both the magnitude of deviation from the PH assumption and the shape of the AHR trajectory. In this example, the AHR trajectories for the first five terms are relatively flat, but the trajectory for `Unmarried` decreases with time.

The following subsections discuss how to relax the PH assumption for a continuous predictor (using `MAGER`, mother's age, for illustration even though it does not significantly violate the

PH assumption in this example) and a categorical predictor (`Unmarried`). Two methods are covered: including an interaction between a predictor and time and stratifying by a categorical predictor.

7.16.2 Adding a time interaction for a continuous predictor

"Interacting a predictor with time" means including that predictor's main effect plus the product of the predictor and some function of time. Which function to use can be based on the shape of the HR trajectory shown when plotting the output of `cox.zph()`. As mentioned previously, modeling the HR for X_k as varying linearly with time is accomplished by replacing $\beta_k X_k$ with $\beta_k(t) X_k$ where $\beta_k(t) = \beta_{k_1} + \beta_{k_2} t$. Equivalently, replace $\beta_k X_k$ with $\beta_{k1} X_k + \beta_{k2} t X_k$. In other words, to model a linearly varying HR, add to the model an interaction between the predictor and t.

We could, instead, assume a non-linear function of time. For example, modeling the HR as varying logarithmically with time corresponds to replacing $\beta_k X_k$ with $\beta_{k1} X_k + \beta_{k2} \ln(t) X_k$, or adding an interaction between the predictor and $\ln(t)$. Either of these approaches, linear or non-linear, relaxes the PH assumption by replacing X's time-invariant regression coefficient with a function of time.

Example 7.6 (continued): Relax the PH assumption for age by assuming the age effect varies linearly with time.

As mentioned above, accomplishing this requires adding, in addition to the age main effect, an interaction between age and linear time. However, we cannot add `t:X` to the model since "time" is the outcome – you cannot include the outcome as part of a predictor. Adding a time interaction in `coxph()` requires adding a special `tt()` function term, as well as a programming statement defining the function. To add a linear interaction with time, use the function $tt(x,t) = x * t$.

```
# The call for cox.ex7.6
cox.ex7.6$call
```

```
## coxph(formula = Surv(gestage37, preterm01) ~ RF_PPTERM + MAGER +
##     MRACEHISP + DMAR, data = natality.complete)
```

```
# Re-fit the model, including tt(x,t) to add the interaction with time
cox.ex7.6.tt <- coxph(Surv(gestage37, preterm01) ~ RF_PPTERM +
                        MRACEHISP + DMAR +
                        MAGER + tt(MAGER),
                     data = natality.complete,
                     tt = function(x,t,...) x*t)
summary(cox.ex7.6.tt)$coef
```

```
##                          coef exp(coef) se(coef)       z    Pr(>|z|)
## RF_PPTERMYes         1.063036    2.8951 0.236600  4.4930 0.000007024
## MRACEHISPNH Black    0.559185    1.7492 0.177545  3.1495 0.001635243
## MRACEHISPNH Other -0.073279    0.9293 0.295551 -0.2479 0.804179208
## MRACEHISPHispanic    0.259949    1.2969 0.179096  1.4514 0.146654688
## DMARUnmarried        0.584711    1.7945 0.153347  3.8130 0.000137290
## MAGER               -0.068712    0.9336 0.137613 -0.4993 0.617559391
## tt(MAGER)            0.002948    1.0030 0.004039  0.7299 0.465467436
```

```
car::Anova(cox.ex7.6.tt, type = 3, test = "Wald")
```

```
## Analysis of Deviance Table (Type III tests)
##
## Response: Surv(gestage37, preterm01)
##             Df Chisq Pr(>Chisq)
## RF_PPTERM   1 20.19   0.000007 ***
## MRACEHISP   3 10.80    0.01283 *
## DMAR        1 14.54    0.00014 ***
## MAGER       1  0.25    0.61756
## tt(MAGER)   1  0.53    0.46547
## ---
## Signif. codes:  0 '***' 0.001 '**' 0.01 '*' 0.05 '.' 0.1 ' ' 1
```

As is the case with any variable involved in an interaction, you can no longer treat the main effect for `MAGER` as the age effect. Instead, it is the age effect at gestational age 0 weeks. The p-value for `tt(MAGER)` ($p = .465$) is a test of the interaction and also a test of the PH assumption (specifically against the alternative of a *linear* time interaction; `cox.zph()` is more general).

In the original PH model, the AHR for `MAGER` was $e^{0.0313} = 1.03$. How does adding the interaction with time change this? Instead of AHR $= e^{\beta_{\text{MAGER}}}$, we now have AHR $= e^{\beta_{\text{MAGER}} + t * \beta_{\text{tt(MAGER)}}}$. In our example, then, the AHR for mother's age is $e^{-0.0687+0.0029*t}$.

- At $t = 0$ (conception), the AHR is $e^{-0.0687} = 0.93$.
- At $t = 30$ weeks, it is $e^{-0.0687+0.0029*30} = 1.02$.
- At $t = 35$ weeks, it is $e^{-0.0687+0.0029*35} = 1.04$.

Compute the time at which the AHR is 1 using the formula $-\beta_X/\beta_{tt(X)}$ (which is the value of t at which $e^{\beta_X + t*\beta_{tt(X)}} = 1$).

```
-1*coef(cox.ex7.6.tt)["MAGER"] /
  coef(cox.ex7.6.tt)["tt(MAGER)"]
```

```
## MAGER
## 23.31
```

This implies that the direction of association changes at 23.3 weeks. If this value is not in the range of the observed event times, then the AHR is either always <1 or always >1. Figure 7.17 illustrates how the AHR changes over time.

```
# Range of observed event times
SUB    <- natality.complete$preterm01 == 1
TIME   <- seq(min(natality.complete$gestage37[SUB]),
              max(natality.complete$gestage37[SUB]), 1)
BETA   <- coef(cox.ex7.6.tt)["MAGER"]
BETATT <- coef(cox.ex7.6.tt)["tt(MAGER)"]
AHR    <- exp(BETA + BETATT*TIME)
plot(AHR ~ TIME, type = "l")
abline(h = 1, lty = 2, col = "darkgray")
abline(v = -1*BETA/BETATT, lty = 2, col = "darkgray")
```

In general, the AHR vs. time plot will be curved. For this continuous predictor, the curve just happens to be approximately linear over this range of weeks.

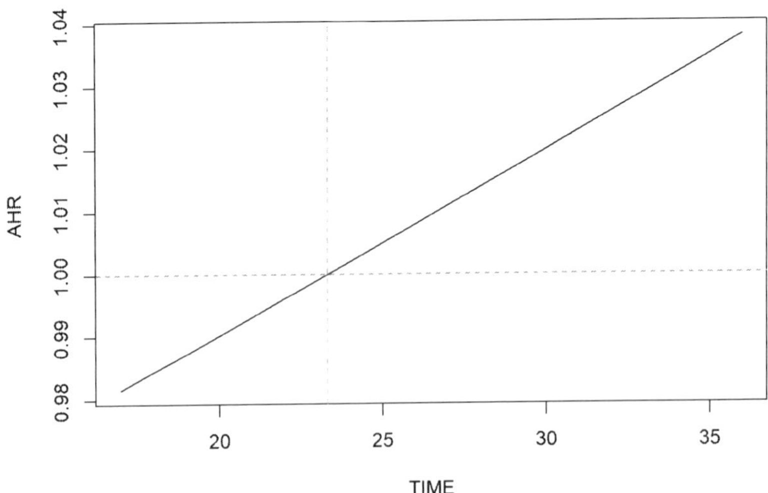

FIGURE 7.17 AHR varying over time when relaxing the proportional hazard assumption

Conclusion: Based on the original PH model, those who are one year older have 3.2% greater hazard of preterm birth (AHR = 1.03; 95% CI = 1.01, 1.06; p = .009). Based on the model that relaxes the PH assumption, this difference varies over time, with younger women having greater hazard of preterm birth earlier in pregnancy and older women greater hazard later in pregnancy (because AHR <1 early, and AHR >1 late). Again, for this predictor, the PH assumption was not actually violated, but we proceeded with relaxing the assumption for illustration.

NOTE: Be careful when interpreting Figure 7.17 – it shows how the hazard *ratio* comparing those who differ by 1-unit in X changes over time, not how the hazard changes over time.

7.16.3 Adding a time interaction for a categorical predictor

Example 7.6 (continued): Relax the PH assumption for marital status by assuming the effect of `Unmarried` varies linearly with time.

The `tt()` function only accepts numeric predictors. We used `model.matrix()` in Section 7.16.1 to create the numeric indicator variable for `DMAR` = "Unmarried". Since that variable is numeric, we can use it here, as well.

The code below includes the interaction with time after replacing the factor variable `DMAR` with its corresponding numeric indicator variable, `Unmarried`. In general, for a categorical predictor with L levels, replace the factor variable with the corresponding $L - 1$ indicator variables and add a `tt()` term for each level that has non-proportional hazards.

```
# Re-fit the model, including tt(x,t) to add the interaction with time
cox.ex7.6.tt2 <- coxph(Surv(gestage37, preterm01) ~ RF_PPTERM + MAGER +
                 MRACEHISP + Unmarried + tt(Unmarried),
                 data = natality.complete,
                 tt=function(x,t,...) x*t)
summary(cox.ex7.6.tt2)$coef
```

```
##                         coef exp(coef) se(coef)      z   Pr(>|z|)
## RF_PPTERMYes         1.05525    2.8727  0.23668  4.459 0.000008253
## MAGER                0.03130    1.0318  0.01205  2.598 0.009369082
## MRACEHISPNH Black    0.55553    1.7429  0.17770  3.126 0.001770542
## MRACEHISPNH Other   -0.07063    0.9318  0.29555 -0.239 0.811110305
## MRACEHISPHispanic    0.26217    1.2998  0.17914  1.464 0.143328318
## Unmarried            7.19074 1327.0797  2.22153  3.237 0.001208583
## tt(Unmarried)       -0.19377    0.8238  0.06466 -2.996 0.002731143
```

```
car::Anova(cox.ex7.6.tt2, type = 3, test = "Wald")
```

```
## Analysis of Deviance Table (Type III tests)
##
## Response: Surv(gestage37, preterm01)
##               Df Chisq Pr(>Chisq)
## RF_PPTERM      1 19.88  0.0000083 ***
## MAGER          1  6.75     0.0094 **
## MRACEHISP      3 10.64     0.0138 *
## Unmarried      1 10.48     0.0012 **
## tt(Unmarried)  1  8.98     0.0027 **
## ---
## Signif. codes:  0 '***' 0.001 '**' 0.01 '*' 0.05 '.' 0.1 ' ' 1
```

Just like for the continuous predictor, we can evaluate how the AHR for DMAR changes with time. In the original PH model, the AHR for DMAR (Unmarried vs. Married) was $e^{0.5850} = 1.79$. How does adding the interaction with time change this? Instead of AHR $= e^{\beta_{\text{DMARUnmarried}}}$, we now have AHR $= e^{\beta_{\text{Unmarried}} + t * \beta_{\text{tt(Unmarried)}}}$. In our example, then, the AHR for mother's age is $e^{7.1907 - 0.1938 * t}$.

- At $t = 0$ (conception), the AHR is $e^{7.1907} = 1327.08$.
- At $t = 30$ weeks, it is $e^{7.1907-0.1938*30} = 3.97$.
- At $t = 35$ weeks, it is $e^{7.1907-0.1938*35} = 1.51$.

Compute the time at which the AHR is 1 using the following formula.

```
-1*coef(cox.ex7.6.tt2)["Unmarried"] /
  coef(cox.ex7.6.tt2)["tt(Unmarried)"]
```

```
## Unmarried
##     37.11
```

Since this is beyond the range of our time variable, the AHR will never be exactly 1 and the direction of association does not change over time. Visualize how the AHR changes over time by computing it over a range of times and plotting (Figure 7.18).

```
SUB    <- natality.complete$preterm01 == 1
TIME   <- seq(min(natality.complete$gestage37[SUB]),
              max(natality.complete$gestage37[SUB]), 1)
BETA   <- coef(cox.ex7.6.tt2)["Unmarried"]
BETATT <- coef(cox.ex7.6.tt2)["tt(Unmarried)"]
AHR    <- exp(BETA + BETATT*TIME)
plot(AHR ~ TIME, type = "l")
abline(h = 1, lty = 2, col = "darkgray")
abline(v = -1*BETA/BETATT, lty = 2, col = "darkgray")
```

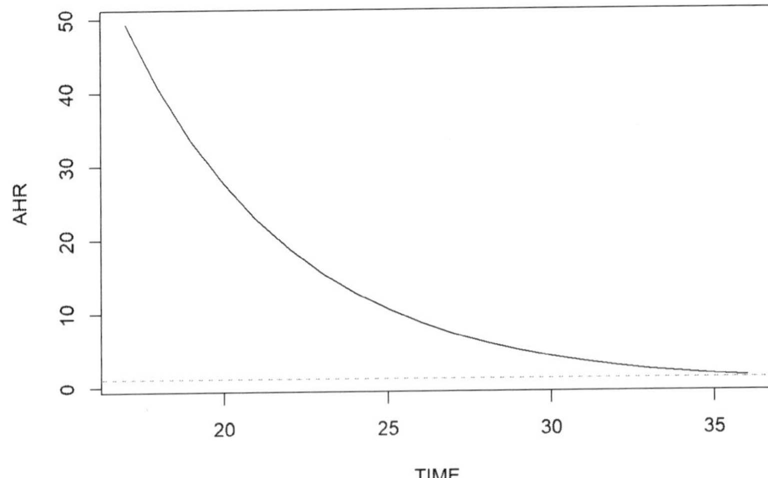

FIGURE 7.18 AHR varying non-linearly over time when relaxing the proportional hazard assumption

Here we clearly see the curved shape of the AHR curve. Why is the AHR so large early in pregnancy? It turns out that almost all of the earliest preterm births occurred among unmarried women. Therefore, in this dataset the hazard of preterm birth early in pregnancy is much greater for unmarried than married women.

```
natality.complete %>%
  filter(preterm01 == 1) %>%
  arrange(gestage37) %>%
  select(gestage37, DMAR) %>%
  slice(1:10)
```

```
## # A tibble: 10 x 2
##    gestage37 DMAR
##        <dbl> <fct>
##  1        17 Unmarried
##  2        20 Unmarried
##  3        23 Unmarried
##  4        24 Unmarried
##  5        25 Unmarried
##  6        26 Unmarried
##  7        27 Unmarried
##  8        27 Married
##  9        28 Unmarried
## 10        28 Married
```

Conclusion: Based on the original PH model, those who are unmarried have 79.5% greater hazard of preterm birth (AHR = 1.79; 95% CI = 1.33, 2.42; p <.001). Based on the model that relaxes the PH assumption, the AHR decreases with time. It is always >1 but is greater earlier in pregnancy.

NOTE: We used $tt(x, t) = xt$ to model how the AHR changes with time. Instead, we could have used a different function of time (e.g., logarithm, polynomial, spline). Had we done so, some of the code above would need to be altered to obtain the correct plot and time at which AHR = 1.

7.16.4 Stratifying by a categorical variable

For a categorical predictor, another method of relaxing the PH assumption is to use **stratification**. In a stratified model, each level of the strata variable (e.g., Married, Unmarried) is assumed to have a different baseline hazard function but the same coefficients for the other predictors in the model. Because the baseline hazard functions in different strata can vary over time differently, the ratio of hazards between individuals in different strata can vary over time. While stratification removes the PH assumption for the strata variable, within levels of the strata variable PH is still assumed for the other predictors.

Example 7.6 (continued): Use stratification to relax the PH assumption for marital status.

Re-fit the model, including replacing `DMAR` with `strata(DMAR)`.

NOTE: When stratifying by `X`, do not include `X` in the model, just `strata(X)`.

```
cox.ex7.6.strata <- coxph(Surv(gestage37, preterm01) ~ RF_PPTERM + MAGER +
                    MRACEHISP + strata(DMAR),
                    data = natality.complete)

summary(cox.ex7.6.strata)$coef
```

```
##                          coef exp(coef) se(coef)       z  Pr(>|z|)
## RF_PPTERMYes          1.05413    2.8695  0.23670  4.4535 0.00000845
## MAGER                 0.03128    1.0318  0.01205  2.5965 0.00941837
## MRACEHISPNH Black     0.55508    1.7421  0.17771  3.1236 0.00178676
## MRACEHISPNH Other    -0.07071    0.9317  0.29555 -0.2393 0.81090523
## MRACEHISPHispanic     0.26203    1.2996  0.17914  1.4627 0.14355676
```

```
car::Anova(cox.ex7.6.strata, type = 3, test = "Wald")
```

```
## Analysis of Deviance Table (Type III tests)
##
## Response: Surv(gestage37, preterm01)
##           Df Chisq Pr(>Chisq)
## RF_PPTERM  1 19.83  0.0000084 ***
## MAGER      1  6.74     0.0094 **
## MRACEHISP  3 10.62     0.0139 *
## ---
## Signif. codes:  0 '***' 0.001 '**' 0.01 '*' 0.05 '.' 0.1 ' ' 1
```

The output looks very similar to an unstratified model, except that now there is no estimate of the effect of the predictor on which you stratified (there is no row for `DMAR` in the output above). If you want to estimate an AHR for `DMAR`, do not use stratification, instead include a time interaction as described previously.

Finally, re-check the PH assumption after stratification to verify that there is no remaining violation of the assumption (Figure 7.19).

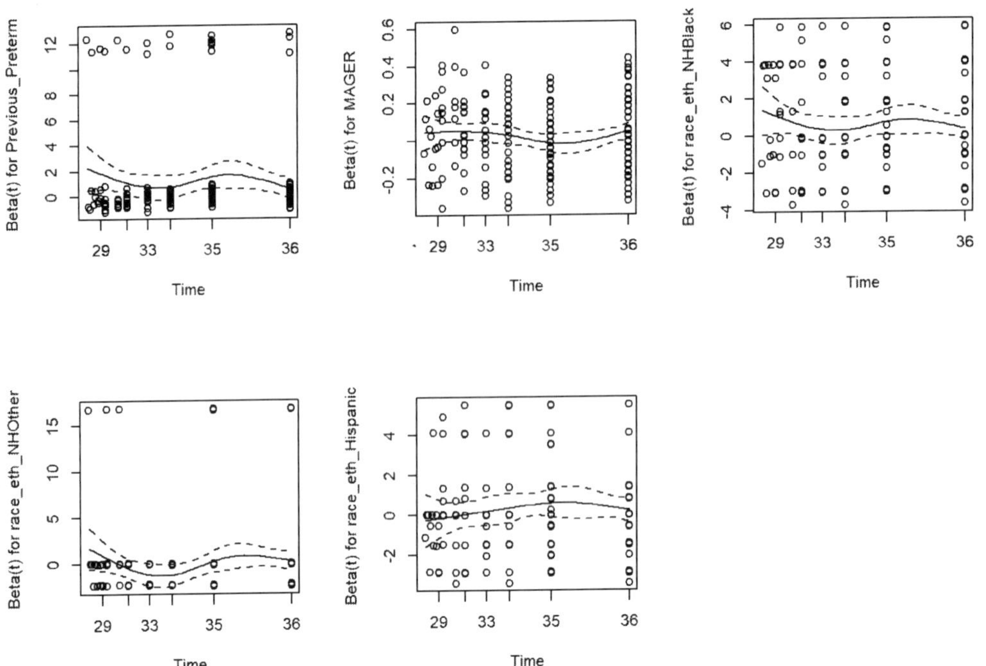

FIGURE 7.19 Rechecing the proportional hazards assumption

```
# Replace factors with indicator variables
cox.ex7.6.strata.indicator <- coxph(Surv(gestage37, preterm01) ~
                    Previous_Preterm + MAGER +
                    race_eth_NHBlack + race_eth_NHOther + race_eth_Hispanic +
                    strata(DMAR),
                  data = natality.complete)

# Verify the fit is the same
summary(cox.ex7.6.strata)$coef
```

```
##                         coef exp(coef) se(coef)       z   Pr(>|z|)
## RF_PPTERMYes         1.05413    2.8695  0.23670  4.4535 0.00000845
## MAGER                0.03128    1.0318  0.01205  2.5965 0.00941837
## MRACEHISPNH Black    0.55508    1.7421  0.17771  3.1236 0.00178676
## MRACEHISPNH Other   -0.07071    0.9317  0.29555 -0.2393 0.81090523
## MRACEHISPHispanic    0.26203    1.2996  0.17914  1.4627 0.14355676
```

```
summary(cox.ex7.6.strata.indicator)$coef
```

```
##                         coef exp(coef) se(coef)       z   Pr(>|z|)
## Previous_Preterm     1.05413    2.8695  0.23670  4.4535 0.00000845
## MAGER                0.03128    1.0318  0.01205  2.5965 0.00941837
## race_eth_NHBlack     0.55508    1.7421  0.17771  3.1236 0.00178676
## race_eth_NHOther    -0.07071    0.9317  0.29555 -0.2393 0.81090523
## race_eth_Hispanic    0.26203    1.2996  0.17914  1.4627 0.14355676
```

```
# Re-check for non-proportional hazards
NPH_CHECK <- cox.zph(cox.ex7.6.strata.indicator)
NPH_CHECK
```

```
##                        chisq df    p
## Previous_Preterm     1.71483  1 0.19
## MAGER                0.00781  1 0.93
## race_eth_NHBlack     1.48254  1 0.22
## race_eth_NHOther     0.28813  1 0.59
## race_eth_Hispanic    0.66920  1 0.41
## GLOBAL               3.41160  5 0.64
```

```
par(mfrow=c(2,3))
plot(NPH_CHECK)
```

7.17 Independence assumption

As with linear and logistic regression, Cox regression assumes the observations are **independent** (see Section 5.14). However, as with logistic regression, the Cox regression model does not include an error term (ϵ). While it is possible to compute various kinds of residuals in a Cox regression, there is no assumption that they be normally distributed or have constant variance.

7.18 Linearity assumption

Cox regression assumes that continuous predictors have a **linear** relationship with the outcome (in this case, the log-hazard of the outcome relative to the baseline hazard). To assess this relationship visually, plot what are called "Martingale residuals" vs. each continuous predictor (Figure 7.20).

```
# Residuals vs. continuous predictor
Y <- resid(cox.ex7.6, type = "martingale")
X <- natality.complete$MAGER
plot(Y ~ X, pch = 20, col = "darkgray",
     ylab = "Martingale Residual",
     xlab = "Mother's Age",
     main = "Residuals vs. Predictor")
abline(h = 0)
lines(smooth.spline(Y ~ X, df = 7), lty = 2, lwd = 2)
```

The residuals vs. mother's age curve appears somewhat non-linear. As with other regression models, to relax the linearity assumption, transform the predictor using a polynomial, logarithm, or other function. The code below adds a quadratic term for maternal age and re-checks the linearity assumption (Figure 7.21).

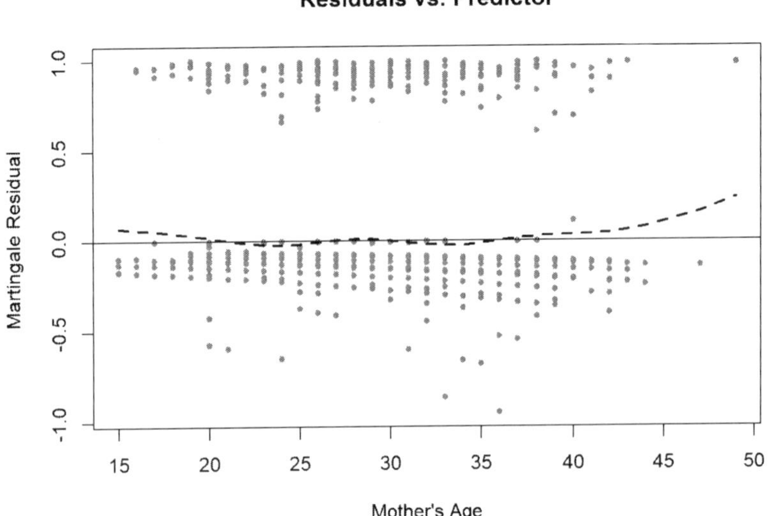

FIGURE 7.20 Visually assessing the linearity assumption of a Cox regression

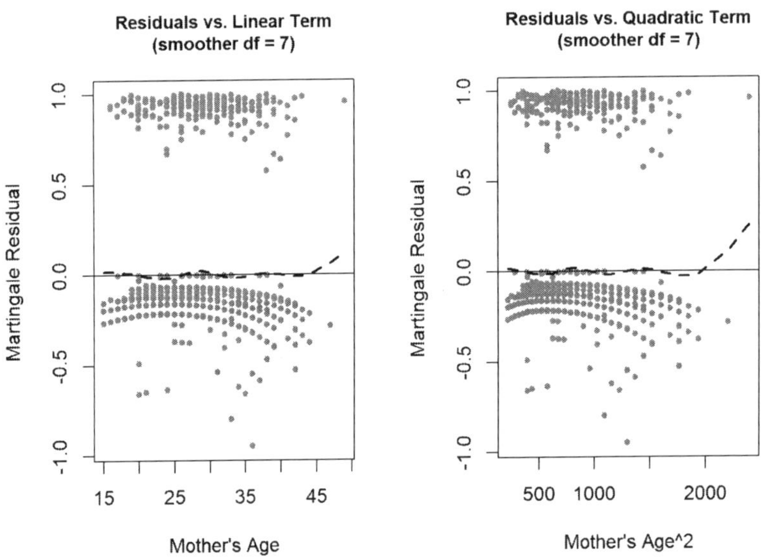

FIGURE 7.21 Re-checking the linearity assumption after adding a quadratic term

As shown in Figure 7.21, adding a quadratic term helps with the non-linearity (the uptick in the curves at older maternal age is due to a single observation with an extreme age value). Inside of `smooth.spline()`, change `df` to a smaller (larger) value to get more (less) smoothing. In Figure 7.22, the plots with `df = 3` appear linear at all ages.

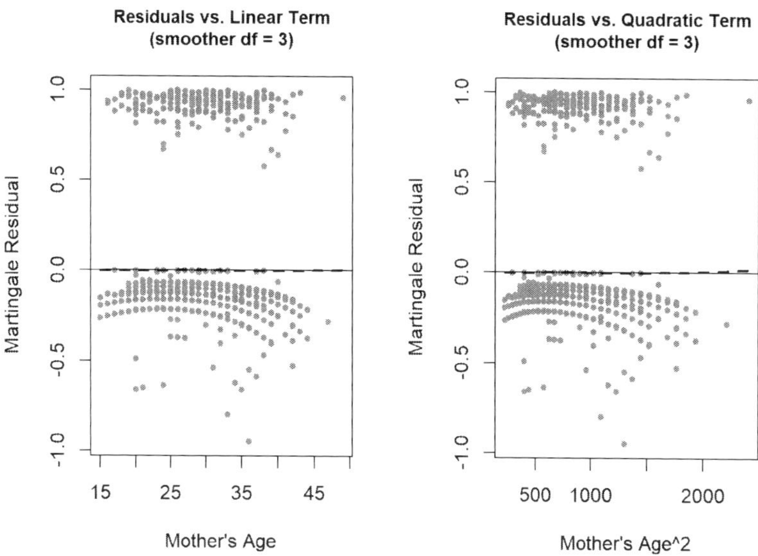

FIGURE 7.22 Changing the amount of smoothing when checking the linearity assumption

In general, be careful when using a small `df` value; make sure to also look at the plot with a larger `df` value, as we did here, to make sure you are not over-smoothing.

7.19 Outliers

As with logistic regression, Cox regression has no normality assumption, so outliers are not a problem from that perspective. However, it is still useful to examine outliers to find observations that are not predicted well by the model. Figure 7.23 illustrates two different kinds of residuals (Martingale and deviance) which can be used to see if any observations stand out from the others.

```
par(mfrow=c(1,2))
plot(resid(cox.ex7.6, type = "martingale"),
     ylab = "Martingale residual", xlab = "Observation")
plot(resid(cox.ex7.6, type = "deviance"),
     ylab = "Deviance residual",   xlab = "Observation")
```

Look at the two observations that have the most negative Martingale residuals. If any of the deviance residuals stood out, you could look at those, as well (although they might be from the same individuals).

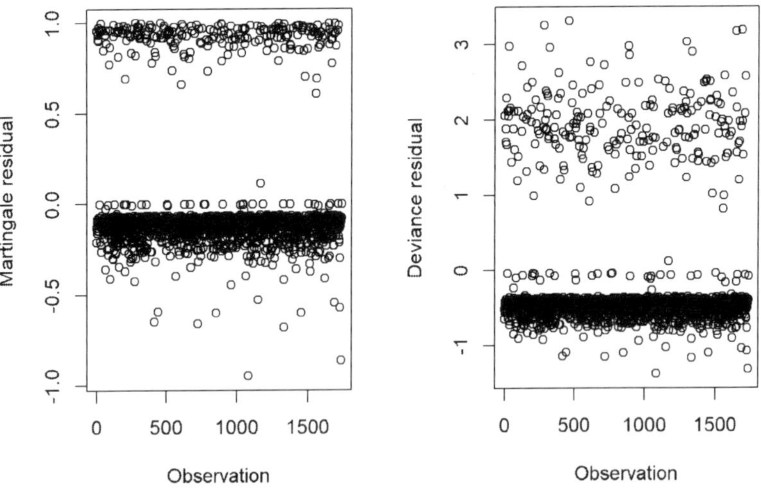

FIGURE 7.23 Martingale and deviance residuals from a Cox regression

```
# Which individuals have the large negative residuals?
# Sort and pick out the first two
sort(resid(cox.ex7.6, type = "martingale"))[1:2]
```

```
##    1080    1738
## -0.9491 -0.8640
```

Had you been looking for individuals with large positive residuals, you would look at the first few observations after sorting on the *negative* of the residual value. For example, `-1*sort(-1*resid(cox.ex7.6, type = "martingale"))[1:2]`.

```
# Examine these two individuals
natality.complete[c("1080", "1738"), ] %>%
  select(gestage37, preterm01, RF_PPTERM, MAGER, MRACEHISP, DMAR)
```

```
## # A tibble: 2 x 6
##   gestage37 preterm01 RF_PPTERM MAGER MRACEHISP DMAR
##       <dbl>     <dbl> <fct>     <dbl> <fct>     <fct>
## 1        37         0 Yes          36 NH Black  Unmarried
## 2        37         0 Yes          33 NH Black  Unmarried
```

```
# "1080" and "1738" are row LABELS
# But to examine the hazard for these two,
# we need row NUMBERS.
# Row labels are not always the same as row numbers
# Check which elements of the prediction correspond to
# these two individuals
which(rownames(natality.complete) %in% c("1080", "1738"))
```

```
## [1] 1080 1738
```

```
# In this case it happened to be the same numbers
# but it may not always be

# What are the estimated hazard ratios for these two individuals
# relative to a reference individual?
predict(cox.ex7.6, type = "risk")[c(1080, 1738)]
```

```
## [1] 11.31 10.30
```

We find that these two individuals, based on their predictor values, are each estimated to have a very high hazard of preterm birth relative to a reference individual, yet neither experienced a preterm birth. This discrepancy resulted in very large negative residuals. As discussed in Section 7.10, the reference individual being compared to is at the mean value for each continuous predictor and the reference level for each categorical predictor.

7.20 Influential observations

Check for influential observations by examining standardized DFBetas (see Section 5.22). Fit the model, plot the DFBetas, and, if there are observations with values that are large, conduct a sensitivity analysis to see if their removal impacts your overall conclusions (see Section 5.25). A cutoff of 0.2 is reasonable (Harrell, 2015, p.504).

```
# Compute DFBETAS
DFBETAS <- resid(cox.ex7.6, type = "dfbetas")
# Examine the first few rows
head(DFBETAS)
```

```
##          [,1]      [,2]      [,3]       [,4]      [,5]      [,6]
## 1 0.0040352 -0.005820 -0.004481 -0.0001244 -0.017555  0.009765
## 2 0.2085628 -0.012956 -0.104859 -0.0335294 -0.091015  0.084854
## 3 0.0048140  0.013115  0.018220  0.0001619  0.001473 -0.003342
## 4 0.0023748  0.014455  0.002487  0.0001607 -0.014082 -0.002793
## 5 0.0012622  0.001298  0.003561  0.0030316  0.003920  0.004394
## 6 0.0006693  0.009121  0.003413 -0.0273620  0.002322 -0.006725
```

```
# Plot
par(mfrow=c(2,3))
plot(DFBETAS[, 1], ylab="RF_PPTERMYes")
abline(h = c(-0.2, 0.2), lty = 2, col = "red")
plot(DFBETAS[, 2], ylab="MAGER")
abline(h = c(-0.2, 0.2), lty = 2, col = "red")
plot(DFBETAS[, 3], ylab="MRACEHISPNH Black")
abline(h = c(-0.2, 0.2), lty = 2, col = "red")
plot(DFBETAS[, 4], ylab="MRACEHISPNH Other")
abline(h = c(-0.2, 0.2), lty = 2, col = "red")
plot(DFBETAS[, 5], ylab="MRACEHISPHispanic")
abline(h = c(-0.2, 0.2), lty = 2, col = "red")
plot(DFBETAS[, 6], ylab="DMARUnmarried")
abline(h = c(-0.2, 0.2), lty = 2, col = "red")
```

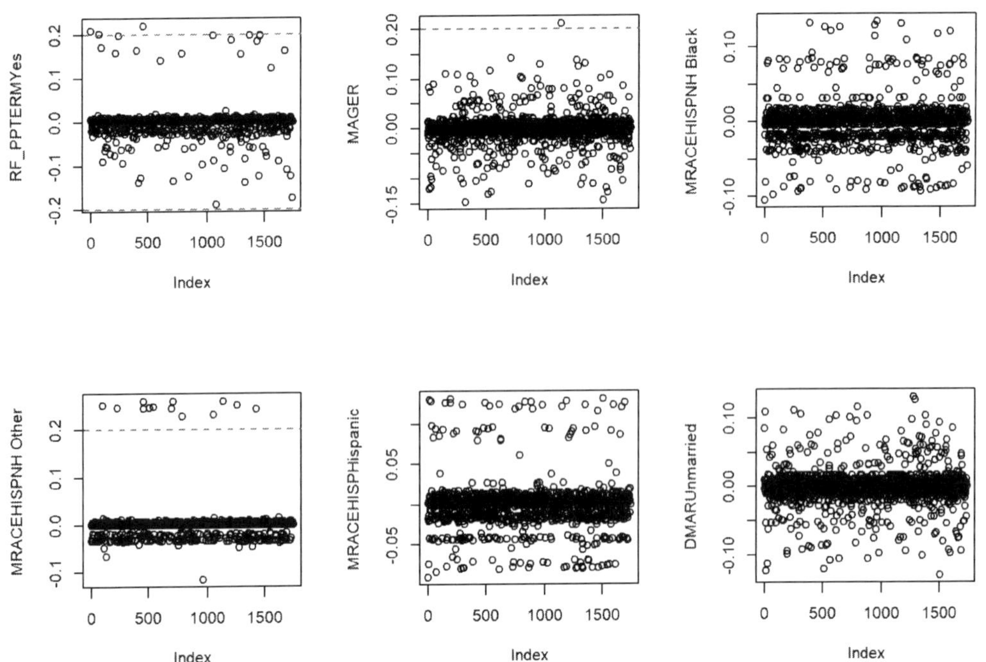

FIGURE 7.24 DFBetas from a Cox regression

There appear to be a number of influential observations (Figure 7.24). For `MRACEHISP` = "Other" (lower left panel), there is a group of observations with standardized DFBeta > 0.2. Upon closer examination, we find that these individuals are the entire group of preterm births among non-Hispanic Other mothers. This group is small enough, and the prevalence of preterm birth is small enough, that there are only a few events in this group. Thus, removing any one of them highly impacts the regression coefficient (they each have a large DFBeta value).

```
# "MRACEHISPNH Other" is the 4th term in the model
SUB <- DFBETAS[, 4] > 0.2

# Race x Preterm among those with large DFBetas
table(natality.complete$MRACEHISP[ SUB],
      natality.complete$preterm01[ SUB])
```

```
##
##              1
##   NH White   0
##   NH Black   0
##   NH Other  13
##   Hispanic   0
```

```
# Race x Preterm among those without large DFBetas
table(natality.complete$MRACEHISP[!SUB],
      natality.complete$preterm01[!SUB])
```

```
##
##                0   1
##     NH White 865  96
##     NH Black 233  58
##     NH Other 128   0
##     Hispanic 299  51
```

7.21 Generalization / overfitting

As with linear (Section 5.26) and logistic (Section 6.15) regression, it is important to limit the number of predictors in a Cox regression model to avoid overfitting and ensure generalizability. The rule of thumb is to have no more than $n_e/15$ predictors, where n_e is the number of events (non-censored event times) (Babyak, 2004; Harrell, 2015, pp.72-73). As with any rule of thumb, this is meant as guidance – there is no requirement that it be strictly applied.

Seen the other way around, if you are designing a study and plan to include K predictors, you need a sample size large enough to result in $n_e \geq 15 \times K$ events. If your background research leads you to believe that the proportion of events will be p_e, then the expected number of events is $n_e = p_e \times n$, where n is the total sample size, and the sample size required for generalizability is $n \geq 15 \times K/p_e$.

As mentioned before, this does not preclude a sample size calculation for the purpose of having enough power. If the sample size is too small, then the Cox regression parameter estimates are unstable, and so they are not generalizable to new cases. This is not the same as lacking in power to detect a real effect; a study can lack power but still be generalizable if the parameter estimates are stable.

For example, suppose you are studying "time to disease after exposure" in a sample in which $n_e = 80$ individuals developed the disease. In a Cox regression, you should limit the number of predictors in your model to no more than $K = n_e/15 = 5.33$ (this is just a rule of thumb, so it is reasonable to round up to 6).

Suppose instead you were designing a study in which you expect 15% of individuals to develop the disease and you would like to include five predictors in your model. For generalizability, you need a sample size of at least $n \geq 15 \times K/p_e = 15 \times 5/0.15 = 500$.

7.22 Likelihood ratio test vs. Wald test

As with logistic regression, we have used the simpler method of using Wald tests for p-values in this chapter. However, as discussed in Section 6.18, likelihood ratio tests are preferred, albeit requiring additional steps to compute for certain parameters. As described in Section 6.18, use the `test.statistic = "LR"` option in `car::Anova()` to obtain likelihood ratio tests.

7.23 Writing it up

This section demonstrates how to write up the Cox regression methods and results for Example 7.6, first for a model assuming PH (`cox.ex7.6`), followed by a model that relaxes the PH assumption (`cox.ex7.6.tt2`). By way of reminder, **Example 7.6** was: Using the Natality teaching dataset, estimate the association between previous preterm birth (`RF_PPTERM`) and time to preterm birth, adjusted for mother's age (`MAGER`), race/ethnicity (`MRACEHISP`), and marital status (`DMAR`).

7.23.1 Writing up Cox regression results (assuming PH)

Below are a few lines of code to extract the information we need for our write-up.

Outcome, predictors, and dataset name

```
cox.ex7.6$call
```

```
## coxph(formula = Surv(gestage37, preterm01) ~ RF_PPTERM + MAGER +
##     MRACEHISP + DMAR, data = natality.complete)
```

Number of observations: If we used a complete case analysis, then the number of rows in the dataset (`nrow(natality.complete)`) is the number of cases used in the analysis. But counting the number of residuals to get the sample size will work even when you have missing data.

```
length(cox.ex7.6$residuals)
```

```
## [1] 1743
```

Selected descriptives: Extract any descriptive summaries you might want to add to your write-up. For example, the distribution of ages or the proportion of individuals in each race/ethnicity group. If you did not remove missing data first before fitting the model, do that here before creating any summary statistics. We want our summary to be based on the same sample as our regression model, and a regression will always carry out a complete case analysis whether you removed the cases with missing values first or not.

```
# Distribution of ages
c(
  min(natality.complete$MAGER),
  max(natality.complete$MAGER),
  mean(natality.complete$MAGER),
  sd(natality.complete$MAGER)
)
```

```
## [1] 15.000 49.000 28.964  5.817
```

```
# Number and proportion for previous preterm
rbind(
  table(natality.complete$RF_PPTERM),
  round(100*prop.table(table(natality.complete$RF_PPTERM)), 1)
)
```

```
##           No  Yes
## [1,] 1683.0 60.0
## [2,]   96.6  3.4
```

```
# Number and proportion for race/ethnicity
rbind(
  table(natality.complete$MRACEHISP),
  round(100*prop.table(table(natality.complete$MRACEHISP)), 1)
)
```

```
##      NH White NH Black NH Other Hispanic
## [1,]    961.0    291.0    141.0    350.0
## [2,]     55.1     16.7      8.1     20.1
```

```
# Number and proportion for marital status
rbind(
  table(natality.complete$DMAR),
    round(100*prop.table(table(natality.complete$DMAR)), 1)
)
```

```
##      Married Unmarried
## [1,]  1088.0     655.0
## [2,]    62.4      37.6
```

```
# Number and proportion of events
rbind(
  table(natality.complete$preterm01),
    round(100*prop.table(table(natality.complete$preterm01)), 1)
)
```

```
##           0     1
## [1,] 1525.0 218.0
## [2,]   87.5  12.5
```

```
# If S(t) dropped below 0.50, you could get the median time to preterm birth
tmp <- survfit(Surv(gestage37, preterm01) ~ 1, data = natality.complete)
summary(tmp)$table[c("median", "0.95LCL", "0.95UCL")]
```

```
##  median 0.95LCL 0.95UCL
##      NA      NA      NA
```

```
# If median survival time is missing, you could instead compute
# the median among those with the event
median(natality.complete$gestage37[natality.complete$preterm01 == 1])
```

```
## [1] 35
```

Individual term regression coefficients, p-values, AHRs, and 95% confidence intervals: The p-value for the regression coefficients for a continuous predictor tests the association between that predictor and the outcome. For categorical predictors, each p-value is for a test comparing the hazard of the outcome between the level shown in that row and the reference level.

```
# Regression coefficients and 95% CIs
cbind("Beta" = summary(cox.ex7.6)$coef[, "coef"],
      confint(cox.ex7.6))
# (results not shown)
```

```
# AHRs, 95% CIs, p-values
cbind("AHR"      = exp(summary(cox.ex7.6)$coef[, "coef"]),
                   exp(confint(cox.ex7.6)),
      "p-value"  = summary(cox.ex7.6)$coef[, "Pr(>|z|)"])
# (results not shown)
```

Type III test p-values: For continuous predictors and categorical predictors with exactly two levels, the Type III Wald test p-values are the same as those in the regression coefficient table. For categorical predictors with more than two levels, Type III tests provide the multiple degree of freedom p-values for the overall test of that predictor's association with the outcome (unless there is an interaction in the model, in which case the main effect test assumes the other term in the interaction is 0 or at its reference level).

```
# Multiple df p-values
car::Anova(cox.ex7.6, type = 3, test.statistic = "Wald")
# (results not shown)
```

Methods: We used data from 1743 births from a teaching dataset based on a modified subset of the 2019 US Natality data, including gestational age and an indicator of preterm birth (gestational age < 37 weeks). Cox proportional hazards regression was used to assess the association between previous preterm birth and the hazard of preterm birth, adjusted for mother's age, race/ethnicity (non-Hispanic White, non-Hispanic Black, non-Hispanic Other, and Hispanic), and marital status (married, unmarried). Gestational ages ≥ 37 weeks were censored at 37 weeks. We assessed the linearity and proportional hazards assumptions and checked for separation, collinearity, outliers, and influential observations (no issues were found). *(In fact, we did find non-proportional hazards, and the corresponding write-up is shown below.)*

Results: There were 218 preterm births (12.5%). Among these, the median gestational age was 35 weeks. Mothers were age 15 to 49 years (mean = 29.0, SD = 5.8), the majority were married (62.4%) and non-Hispanic White (NH White = 55.1%, NH Black = 16.7%, NH Other = 8.1%, Hispanic = 20.1%). A total of 60 (3.4%) mothers had a previous preterm birth. After adjusting for mother's age, race/ethnicity, and marital status, previous preterm birth was significantly positively associated with preterm birth (AHR = 2.89; 95% CI = 1.82, 4.60; p <.001). Those with a previous preterm birth have 2.89 times the hazard of preterm birth. Mother's age, race/ethnicity, and marital status were each also significantly associated with preterm birth, with older, non-Hispanic Black, and unmarried mothers having greater hazard.

See Table 7.2 for full regression results. Once again, use `tbl_regression()` to produce a nice looking table.

```
NOBS <- length(cox.ex7.6$residuals)

library(gtsummary)
# Table of regression coefficients
t1 <- cox.ex7.6 %>%
  tbl_regression(estimate_fun = function(x) style_number(x, digits = 2),
                 label  = list(RF_PPTERM   ~ "Previous Pre-Term",
                               MAGER       ~ "Age (y)",
                               MRACEHISP   ~ "Race/Ethnicity",
                               DMAR        ~ "Marital Status")) %>%
  modify_column_hide(p.value) %>%
  modify_caption(paste("Cox regression results for time to preterm birth vs.
                       previous preterm birth (N = ", NOBS, ")", sep=""))

# Table of hazard ratios
t2 <- cox.ex7.6 %>%
  tbl_regression(exponentiate = T,  # HR = exp(B)
                 estimate_fun = function(x) style_number(x, digits = 2),
                 pvalue_fun   = function(x) style_pvalue(x, digits = 3),
                 label  = list(RF_PPTERM   ~ "Previous Pre-Term",
                               MAGER       ~ "Age (y)",
                               MRACEHISP   ~ "Race/Ethnicity",
                               DMAR        ~ "Marital status")) %>%
  add_global_p(keep = T, test = "Wald")
```

```
TABLE <- tbl_merge(
  tbls = list(t1, t2),
  tab_spanner = c("**Adjusted Coefficient**", "**Adjusted Hazard Ratio**")
  ) %>%
  modify_footnote(everything() ~ NA, abbreviation = TRUE) %>%
  modify_header(update = list(
    estimate_1 = "**log(AHR)**",
    estimate_2 = "**AHR**"
  ))
```

```
TABLE
```

7.23.2 Writing up Cox regression results (relaxing PH)

There was one predictor (`DMAR`) that had both significantly non-proportional hazards and a visualization that indicated a meaningful deviation from PH. The model that included its interaction with time was `cox.ex7.6.tt2`. Extract the components needed from that model for the write-up. Also, re-use the AHR vs. time figure created in Section 7.16.3.

```
# Regression coefficients and 95% CIs
cbind("Beta" = summary(cox.ex7.6.tt2)$coef[, "coef"],
      confint(cox.ex7.6.tt2))
# (results not shown)
```

TABLE 7.2 Cox regression results for time to preterm birth vs. previous preterm birth (N = 1743)

	Adjusted Coefficient		Adjusted Hazard Ratio		
Characteristic	**log(AHR)**	**95% CI**	**AHR**	**95% CI**	**p-value**
Previous Pre-Term					<0.001
No	—	—	—	—	
Yes	1.06	0.60, 1.53	2.89	1.82, 4.60	<0.001
Age (y)	0.03	0.01, 0.05	1.03	1.01, 1.06	0.009
Race/Ethnicity					0.013
NH White	—	—	—	—	
NH Black	0.56	0.21, 0.91	1.75	1.23, 2.48	0.002
NH Other	-0.07	-0.65, 0.51	0.93	0.52, 1.66	0.804
Hispanic	0.26	-0.09, 0.61	1.30	0.91, 1.84	0.146
Marital Status					
Married	—	—			
Unmarried	0.58	0.28, 0.89			
Marital status					<0.001
Married			—	—	
Unmarried			1.79	1.33, 2.42	<0.001

```
# AHRs, 95% CIs, p-values
cbind("AHR"      = exp(summary(cox.ex7.6.tt2)$coef[, "coef"]),
                   exp(confint(cox.ex7.6.tt2)),
      "p-value"  = summary(cox.ex7.6.tt2)$coef[, "Pr(>|z|)"])
# (results not shown)
```

```
car::Anova(cox.ex7.6.tt2, type = 3, test = "Wald")
# (results not shown)
```

Methods: *(Same as above, but with the following addition.)* We checked the PH assumption statistically and numerically. For predictors with a meaningfully large deviation from PH, we relaxed the assumption by including an interaction with time.

Results: *(Same descriptives as above, but regression results are different.)* In the adjusted model, due to a failure of the PH assumption, the effect of marital status was allowed to vary linearly with time. After adjusting for mother's age, race/ethnicity, and marital status, previous preterm birth was significantly positively associated with preterm birth (AHR = 2.87; 95% CI = 1.81, 4.57; p <.001). Those with a previous preterm birth have 2.87 times the hazard of preterm birth. Mother's age and race/ethnicity were each also significantly associated with preterm birth, with older and non-Hispanic Black mothers having greater hazard. Marital status had a significant interaction with time: unmarried mothers had greater hazard of preterm birth than married mothers, even more so earlier in pregnancy.

See Table 7.3 for full regression results.

```
# Table of regression coefficients
t1 <- cox.ex7.6.tt2 %>%
  tbl_regression(estimate_fun = function(x) style_number(x, digits = 2),
                 label  = list(RF_PPTERM  ~ "Previous Pre-Term",
                               MAGER      ~ "Age (y)",
```

```
                  MRACEHISP    ~ "Race/Ethnicity",
                  Unmarried      ~ "Unmarried",
                  # Have to use `name` since the name has a ( in it
                  `tt(Unmarried)` ~ "Unmarried x Time")) %>%
  modify_column_hide(p.value) %>%
  modify_caption(paste("Cox regression results for time to preterm birth vs.
            previous preterm birth (N = ", NOBS, ")", sep=""))

# Table of hazard ratios
t2 <- cox.ex7.6.tt2 %>%
  tbl_regression(exponentiate = T,  # HR = exp(B)
          estimate_fun = function(x) style_number(x, digits = 2),
          pvalue_fun   = function(x) style_pvalue(x, digits = 3),
          label  = list(RF_PPTERM   ~ "Previous Pre-Term",
                  MAGER        ~ "Age (y)",
                  MRACEHISP    ~ "Race/Ethnicity",
                  Unmarried      ~ "Unmarried",
                  `tt(Unmarried)` ~ "Unmarried x Time")) %>%
  add_global_p(keep = T, test = "Wald")
```

```
TABLE <- tbl_merge(
  tbls = list(t1, t2),
  tab_spanner = c("**Adjusted Coefficient**", "**Adjusted Hazard Ratio**")
  ) %>%
  modify_footnote(everything() ~ NA, abbreviation = TRUE) %>%
  modify_header(update = list(
    estimate_1 = "**log(AHR)**",
    estimate_2 = "**AHR**"
  ))
```

```
TABLE
```

TABLE 7.3 Cox regression results for time to preterm birth vs. previous preterm birth (N = 1743)

	Adjusted Coefficient		**Adjusted Hazard Ratio**		
Characteristic	**log(AHR)**	**95% CI**	**AHR**	**95% CI**	**p-value**
Previous Pre-Term					<0.001
No	—	—	—	—	
Yes	1.06	0.59, 1.52	2.87	1.81, 4.57	<0.001
Age (y)	0.03	0.01, 0.05	1.03	1.01, 1.06	0.009
Race/Ethnicity					0.014
NH White	—	—	—	—	
NH Black	0.56	0.21, 0.90	1.74	1.23, 2.47	0.002
NH Other	-0.07	-0.65, 0.51	0.93	0.52, 1.66	0.811
Hispanic	0.26	-0.09, 0.61	1.30	0.91, 1.85	0.143
Unmarried	7.19	2.84, 11.54	1,327.08	17.06, 103,243.47	0.001
Unmarried x Time	-0.19	-0.32, -0.07	0.82	0.73, 0.94	0.003

The very large AHR value for the Unmarried main effect is actually an extrapolation to t = 0 weeks, a time at which there were no events. In fact, there were no events prior to $t = 17$

weeks. Had we centered t, for example at 30 weeks (e.g., `tt = function(x, t, ...) x*(t - 30)`) then the AHR for the main effect (Unmarried) would be much smaller. See Figure 7.25 for a visualization of how the marital status effect changes over time.

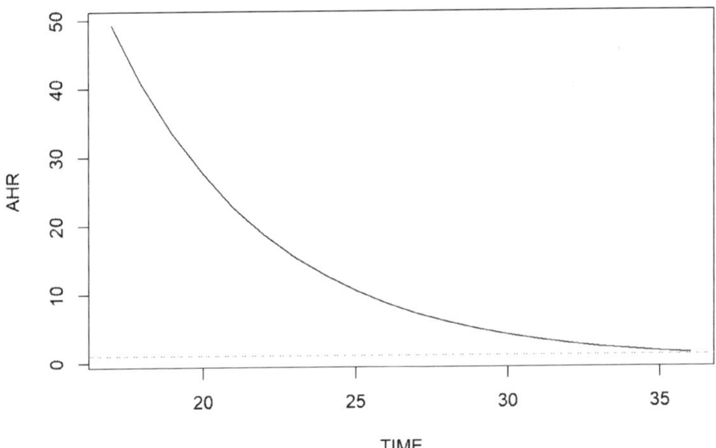

FIGURE 7.25 Preterm birth AHR for marital status (unmarried vs. married) vs. weeks (non-proportional hazards)

7.24 Summary of survival analysis

The steps of carrying out survival analysis using the Kaplan-Meier method and Cox regression are similar to MLR (Section 5.28) with the following differences:

- Considerations when thinking about the outcome.
  - What is the "event" of interest? (e.g., death, preterm birth, heart attack)
  - What is the time origin, the time when an individual could first experience the event? (e.g., enrollment in the study, age of initiation of pain pill use)
  - Which events are censored? Are some individuals' times censored due to loss to follow-up? Are there competing risks that censor times (e.g., death leads to a censored time for other events)? Is there an end-of-study time at which times are censored? (see Section 7.2)
- If the dataset is not already in the correct format (unlike the examples we have been working with), then set up your dataset to fit the structure required for a survival analysis (see Section 7.5).
  - Create a numeric event indicator variable that has the value 1 for events and 0 for censored times.
  - If there are no time-varying predictors, the dataset should have one row per individual.
  - If there are time-varying predictors, the dataset should have one row for every time period in which predictors do not vary (see Section 7.14).

   - Create a numeric time to event variable (or `START` and `STOP` variables if you have time-varying predictors). If you need to work with date variables, see, for example, the chapter "Dates and Times" in **R for Data Science**[3] (Wickham et al., 2023).
- The outcome (time to event) is not typically transformed.
- Start by using the KM method to estimate and plot the survival and hazard functions, ignoring any other variables in your dataset. This will help you get a feel for the data and the extent of censoring (see Section 7.6).
- Use the KM method (and log-rank test) to compare survival functions between groups and plot these comparisons, as well (see Section 7.6.5).
- Check for separation by creating a two-way table for each categorical predictor vs. the event indicator (three-way table if there is an interaction) and look for levels at which all the event times are censored (see Section 7.13). Resolve issues using filtering, collapsing variables, or removing variables (see Section 6.10.4).
   - If you need to re-do the previous steps because the sample size has changed or any categorical variables have changed, make sure to also re-evaluate separation.
- Check the proportional hazards assumption. If necessary, include an interaction with time or stratify (see Section 7.16).
- Evaluate linearity (Section 7.18), outliers (Section 7.19), and influential observations (Section 7.20). In Cox regression, there are no normality or constant variance assumptions.

7.25 Exercises

1. True or false? Time-to-event data can easily be handled using linear regression.

2. Name two reasons why logistic regression may be inappropriate for time-to-event data.

3. During an infectious disease pandemic, researchers follow vaccinated individuals for 1 year post-vaccination and compare the outcome "time to infection" between groups receiving two different vaccines. What type of outcome is this and what regression method would be appropriate for this outcome?

4. In Exercise 3, suppose the researchers only know whether or not each individual experienced the disease during the year after vaccination. What type of outcome is this and what regression method would be appropriate for this outcome?

5. In Exercise 3, suppose some individuals who have not yet experienced infection are lost to follow-up prior to 1 year. What kind (right, left, or interval) and mechanism (Type I, Type II, or random) of censoring is this? Suppose some individuals have not yet experienced infection when the study ends at 1 year. What kind and mechanism of censoring is this?

6. During an infectious disease pandemic, researchers follow vaccinated individuals post-vaccination and record the outcome "time to infection". They stop the study after 100 individuals are infected. For individuals who never experience infection, what kind and mechanism of censoring is this?

7. A retrospective chart review is used to identify patients in a psychiatric clinic who received one of two different treatments for depression and the first visit at which treatment was started (the index visit). Patients were seen every 6 months following the index visit.

[3] https://r4ds.hadley.nz/datetimes

Follow-up charts were reviewed to determine when each patient first no longer met criteria for depression. For the outcome "time to remission", are the times censored and, if so, with what kind of censoring?

8. Suppose you want to carry out a hypothesis test in which the null hypothesis is that survival is the same for three groups of individuals. Name two methods of testing this hypothesis.

9. In the previous exercise, suppose you wanted to carry out the test after adjusting for confounding due to other variables. Would both methods still work?

10. After fitting a Cox regression model, what two assumptions should you check, and what alternatives are possible if each assumption is not met?

11. In the survival function below, approximately what is S(18)?

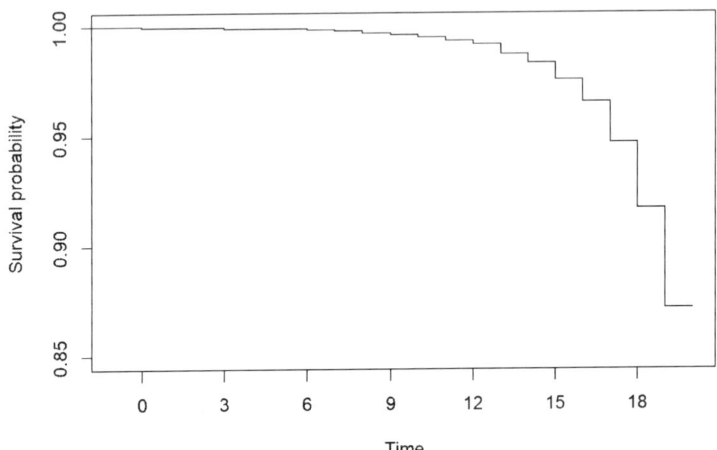

For Exercises 12-17, use the Digitalis teaching dataset (`dig_rmph.rData`, see Appendix A.6) to answer questions related to the time to first hospitalization for any cardiovascular disease (CVD) cause (worsening heart failure, arrythmia, Digoxin toxicity, MI, unstable angina, stroke, coronary revascularization, cardiac transplantation, or other cardiovascular event) (`CVDDAYS` = days from randomization to first CVD hospitalization; `CVD` = event indicator).

12. Check that the time and event indicator variables are in the correct format. If not, modify them so that they are.

13. Compute the Kaplan-Meier estimate of "survival" (in this case, survival is not yet being hospitalized for a CVD event) and answer the following questions.

a. How many are in the risk set at randomization?
b. How many CVD hospitalizations occurred in total?
c. How many CVD hospitalizations occurred exactly 10 days after randomization?
d. What is the estimated survival probability (probability of not yet being hospitalized for a CVD event) and 95% confidence interval 15 days after randomization?
e. What is the probability of CVD hospitalization any time up to and including 500 days?
f. What is the probability of CVD hospitalization strictly between 500 and 750 days after randomization?

14. Plot the survival function, adjusting the X and Y axes to make the curve easy to read and removing the marking of censored times (there are a lot so they clutter up the curve).

15. Plot the hazard function. Based on the shape of the hazard function, describe how the instantaneous risk (hazard) of CVD hospitalization changes over time.

16. What is the median (and 95% CI) days to CVD hospitalization?

17. Compute the Kaplan-Meier estimates of survival comparing treatment groups (`TRTMT`) and answer the following questions.

a. Which group has an earlier median time to CVD hospitalization?
b. Plot the survival and hazard curves comparing treatment groups. Which treatment (placebo or Digoxin) seems to be more effective in preventing CVD hospitalization?
c. Is the difference between the survival functions statistically significant?

For Exercises 18 and 19, use the NHANES examination subset teaching dataset (`nhanes1718_adult_exam_sub_rmph.Rdata`, see Appendix A.1) to answer questions about the association between income (`income`) and age when first told has asthma (`MCQ025`) (the event indicator is `MCQ010` – "Ever been told you have asthma").

18. Do the following:

a. Check that the time and event indicator variables are in the correct format. If not, modify them so that they are. **Hint:** For the event indicator, name the new variable "asthma".
b. "Age when first had asthma" is missing for those who have never been told they have asthma. For a time-to-event analysis, we want those who have never been told they have asthma to have a censored time. It makes sense, in this case, to set their censored "age when first had asthma" to their current age (their event indicator is already 0, so we do not need to change that). Use the following code to accomplish this (assuming your dataset name is `nhanes`).

```
# Create a copy of the age variable
nhanes$asthma_age <- nhanes$MCQ025

# Subset on asthma = 0 and replace the missing asthma ages with current age

# SUB <- nhanes$asthma == 0
# nhanes$asthma_age[SUB] <- nhanes$RIDAGEYR[SUB]
# # Error in nhanes$asthma_age[nhanes$asthma == 0] <- nhanes$RIDAGEYR :
# #   NAs are not allowed in subscripted assignments
# The above error is due to the fact that nhanes$asthma is sometimes missing.
# When subsetting on a logical vector, you have to exclude the missing values.

# Add complete.cases() to the logic
SUB <- complete.cases(nhanes$asthma) & nhanes$asthma == 0

# Use [SUB] on both sides so the number of obs you are changing on the
# left-hand side-is equal to the number of obs you are replacing them with
# from the right hand side
nhanes$asthma_age[SUB] <- nhanes$RIDAGEYR[SUB]

# Check
# For !SUB, these should be the same
summary(nhanes$MCQ025[!SUB])
summary(nhanes$asthma_age[!SUB])
summary(nhanes$MCQ025[!SUB] - nhanes$asthma_age[!SUB])
# For SUB, this is all missing
summary(nhanes$MCQ025[SUB])
```

```
# and these should be the same
summary(nhanes$asthma_age[SUB])
summary(nhanes$RIDAGEYR[SUB])
summary(nhanes$asthma_age[SUB] - nhanes$RIDAGEYR[SUB])
```

c. Use Cox regression to test the association between income and time to asthma. Is the association statistically significant?
d. Ignoring the overall statistical significance or lack thereof, compare the lower, middle, and upper income groups. Report the HR, 95% CI, and p-value for each of the three pairwise comparisons, along with an interpretation of each HR. Based on these results, what is the direction of association?
e. Even if the overall association were statistically significant, it would not necessarily imply a causal relationship (greater income –> less chance of developing asthma). Income here is *current* income, whereas the asthma variable reflects being told sometime in the past that one has asthma. Give a possible alternative explanation for the observed association.

19. Using the same dataset from the previous exercise, test the association between body mass index (BMXBMI) and time to asthma. Answer the question, report the HR, 95% confidence interval, and p-value, and interpret the HR, both for a 1-unit difference in BMI and for a 5-unit difference.

For Exercises 20 and 21, use the Digitalis teaching dataset that you created in Exercise 12.

20. After adjusting for age (`AGE`), sex (`SEX`), minority status (`RACE`), body mass index (`BMI`), and systolic blood pressure (`SYSBP`), is the cause of congestive heart failure (`chfcause`) associated with time to first hospitalization for any CVD cause? `CVDDAYS` is the time variable and the event indicator is the modified version of `CVD` you created in Exercise 12. Answer the question, report the adjusted HRs (AHRs), 95% confidence intervals, and p-values comparing each non-reference cause to the reference level. Order the causes from least to greatest hazard of hospitalization for CVD.

21. Using the model from the previous exercise, what is the predicted probability (and its standard error and 95% confidence interval) of not yet experiencing a CVD hospitalization 100 days post-randomization for a patient with ischemic cause of CHF, who is age 45 years, male, minority, has a BMI of 35 kg/m^2, and has a systolic blood pressure of 140 mmHg?

22. For the patient described in the previous exercise, what is their hazard relative to a reference individual with categorical predictors at their reference level and continuous predictors at the sample mean? Report the AHR and its standard error. Does this patient have more or less hazard of CVD hospitalization than the reference patient?

23. For the patient described in the previous exercise, what is their hazard relative to a reference individual with categorical predictors at their reference level and continuous predictors set to 0? Report the AHR and its standard error. Why is this not a valid prediction? Compared to the predicted AHR in the previous exercise (which was valid), this AHR is not very different. Why would this invalid prediction not be very different from a valid prediction for this particular model?

24. Using the model from the previous exercises, plot the survival functions from the day after randomization (time = 1) to 1781 days for the four CHF cause groups. Assume the other variables are equal to that of the patient for whom you predicted the hazard in the previous exercises.

25. You would like to test the association between the cause of congestive heart failure (`chfcause`) and time to coronary revascularization (`CREV` = event indicator, `CREVDAYS` = time to event) among females, adjusted for age (`AGE`), minority status (`RACE`), body mass index (`BMI`), and systolic blood pressure (`SYSBP`). Knowing this outcome is somewhat rare, you decide to check for separation. Check for separation, resolve any issues, and fit the model to answer the research question. How did separation affect your ability to answer the research question?

For Exercises 26-28, use the teaching dataset based on the Framingham Heart Study (`fram_time_invar_rmph.rData`, see Appendix A.6).

26. Does the association between time to angina pectoris (chest pain) (`TIMEAP` = time variable, `ANGINA` = event indicator) and the age of the participant at enrollment (`AGE`) differ between male and female patients (`SEX`)? Before fitting the model, center age at 50 years. Answer the question, including the appropriate p-value. Estimate the HR, 95% CI, and p-value for age at each level of sex, and interpret the results.

27. In the model you fit in the previous exercise, is sex significantly associated with the outcome?

28. Using the full model from the previous exercises, for each level of sex, create a plot that compares the survival functions between individuals age 40, 50, and 60 years and interpret the results.

For Exercises 29 and 30, use the teaching dataset based on the Framingham Heart Study that contains time-varying predictors for the outcome death (`fram_tv_death_rmph.rData`, see Appendix A.6). The (START, STOP] variables are `tstart` and `tstop` and the event indicator is `DEATH`.

29. Is there an association between smoking status (`CURSMOKE`) and time to death, adjusted for baseline age (`AGE`), sex (`SEX`), and education (`EDUC`)? Answer the question and report the adjusted hazard ratio (AHR), 95% confidence interval, p-value, and interpret the AHR. Also, which of the confounders were statistically significant?

30. The dataset `fram_tv_death_rmph.rData` differs from the dataset `fram_time_invar_rmph.rData` in that the former is set up using the (START, STOP] syntax in order to accommodate time-varying predictors. For analyses that involve only time-invariant predictors, you can use either dataset.

Do the following:

a. Examine each dataset by looking at the rows of data for `RANDID` 6238 and 10552. (**Hint:** Use a filter statement.) Select the variables that are needed for the model fitting. For the time-varying dataset, also include the variable systolic blood pressure (`SYSBP`). How are the two datasets structured differently?
b. Fit a Cox regression model using each dataset to test the association between time to death and baseline age (`AGE`), sex (`SEX`), and education (`EDUC`). For the time-varying dataset, use `tstart` and `tstop` as the time variables. For the time-invariant dataset, use `TIMEDTH`. For both, the event indicator is `DEATH`. You should get the same fit for both datasets. Had there been any time-varying predictors in the model, you would not have been able to use the time-invariant dataset because it only has one row per person.

For Exercises 31-36, use the Digitalis teaching dataset that you created in Exercise 12.

31. Numerically and visually assess the proportional hazards assumption for the predictors in the model you fit in Exercise 20. **Hint:** If the lines are difficult to see, use `lwd=3` and `col="red"` inside your call to `plot()` to make the lines stand out. If the scale is too small, use the `ylim` option to zoom in. For example, `ylim = c(-1, 1)`.

32. Continuing the previous exercise, if there is a variable that violates the proportional hazards assumption, add a time interaction and re-fit the model. Compute the time at which the AHR is 1 and plot the AHR vs. time. Interpret the results.

33. Instead of using a time interaction as in the previous exercise, relax the proportional hazards assumption using stratification. Re-fit the model and re-check the PH assumption. Any remaining problems?

34. Check the linearity assumption for the continuous predictors for the model you fit in Exercise 20 (before relaxing the proportional hazards assumption). Any serious problems? **Hint:** Do not worry about non-linearities that are in areas with just a few points; those are just artifacts of smoothing.

35. Check for outliers for the model you fit in Exercise 20 (before relaxing the proportional hazards assumption). Which individual has the largest positive residual? Examine this individual's predictor values and the output of the model and explain why this individual has the largest residual.

36. Check for influential observations for the model you fit in Exercise 20 (before relaxing the proportional hazards assumption).

37. You are designing a study for which you wish to fit a Cox regression model with four predictors. Assuming the proportion of events (non-censored event times) is expected to be 0.23, what is the minimum sample size you need to avoid overfitting? What if the proportion of events is expected to be 0.10? What if the proportion of events is expected to be 0.80? What can you say about the relationship between the expected proportion of events and the minimum sample size for a given number of predictors?

38. Suppose you have a dataset with $n = 581$ observations. In order to avoid overfitting, what is the maximum number of predictors you should include in a Cox regression model if the observed number of non-censored event times is 20, 100, or 300? What can you say about the relationship between the observed number of non-censored events and the number of predictors for a given sample size?

8

Analyzing Complex Survey Data

In this chapter, you will learn:

- Complex survey concepts and terminology; and
- How to incorporate a complex survey design to obtain estimates of population quantities using the following statistical analysis methods:
  - Descriptive statistics;
  - Linear regression;
  - Binary logistic regression;
  - Kaplan-Meier estimate of the survival function;
  - Log-rank test to compare survival functions between groups; and
  - Cox proportional hazards regression.

This chapter assumes that you have read the previous chapters which discuss the unweighted versions of these statistical analysis methods. In places, the presentation in this chapter is brief, only highlighting what is new when analyzing data from a complex survey.

To use the code in this chapter, first load the `tidyverse` and `survey` (Lumley, 2004, 2023) libraries, set two global `survey` options, and load the file `Functions_rmph.R` (downloadable from RMPH Resources[1]).

```
library(tidyverse)
library(survey)
options(survey.lonely.psu = "adjust")
options(survey.adjust.domain.lonely = TRUE)
source("Functions_rmph.R")
```

8.1 Introduction

In research we are often interested in quantifying some characteristic of a population, such as the prevalence of a condition, the average of a measurement, or the association between an exposure and a disease. Typically, we measure the characteristic in a sample of **units** from the population and use that information to make inference about the population characteristic. Units could be individual people, neighborhoods, hospitals, counties, or even nations.

A **census** attempts to select every unit in a population. A **sample** is a subset of units from a population. The process by which the sample is selected is called the **sampling design**. A **non-probability sample** consists of units selected based on some known or unknown

[1] https://github.com/rwnahhas/RMPH_Resources

non-random method (e.g., a convenience sample). A **probability sample** consists of units selected randomly with known probabilities of selection (not necessarily equal). Probability sampling requires a **sampling frame** – a listing of all the units in the population and their associated selection probabilities.

Whatever the sampling design, the goal is to use information from the sample to infer something about the population. For example, we might use the sample mean blood pressure among individuals as an estimate of the population mean blood pressure. If the sample was obtained using probability sampling, the data analyst can take into account the survey design to obtain results that are representative of the population. The default assumption of most standard statistical methods is that every group of size n in the population has the same probability of being selected. As a consequence, every single unit in the population has the same probability of being selected. Such a design is known as **simple random sampling**. With this design, many standard statistics (e.g., sample mean, sample regression slope) are unbiased estimates of their population counterparts.

Many surveys, however, use a **complex sampling design**, not simple random sampling. There are various reasons for this. For example, if constructing a sampling frame listing every unit in the population is difficult or likely to result in errors, one could use **multistage sampling** to sample larger, easier to list, groups of units followed by surveying some or all units within each group, where an accurate sampling frame can be constructed on site. In multistage sampling, you first sample **primary sampling units** (PSUs) (e.g., households). Then, you sample units within each PSU (e.g., individuals within a household). There could, of course, be more than two stages of sampling. The units from earlier stages form **clusters**.

Another reason to use a complex sampling design is that a simple random sample may result in small sample sizes among some subgroups of interest. For example, if race/ethnicity-specific mean blood pressure is of interest, a researcher may want a sampling design that increases the sample size within smaller subgroups. A simple random sample will likely result in a much larger sample size for the majority race/ethnicity and smaller sample sizes for minority groups. Rather than increase the overall sample size to ensure sufficient sizes in the smaller groups, it is more cost-effective to under-sample large groups and over-sample small groups using **unequal probability sampling**.

One form of unequal probability sampling that is seen in some multistage sampling designs is sampling with **probability proportional to size** (PPS), in which larger PSUs have a greater probability of being selected. Another is **stratified random sampling** in which the population is first non-randomly split into **strata** (e.g., geographic region) within each of which a simple random sample is drawn. Stratification into unequal size strata followed by simple random sampling within strata results in unequal probability sampling because individuals in smaller strata have a greater probability of being selected than individuals in larger strata.

A unit's **design weight** is the inverse of its probability of selection. In a simple random sample, each sampling unit has the same probability of selection, so it has the same weight when measurements are combined to form a statistic. For example, the sample mean of a variable X in a sample of size n is $(x_1 + x_2 + ... + x_n)/n$. Each unit in a simple random sample is equally weighted and we refer to the sample mean as an **unweighted** statistic. But what if, as in many complex survey designs, units in the population have different probabilities of being selected? In that case, the unweighted sample mean is a biased estimate of the population mean. The **weighted** mean, however, is an unbiased estimate. Additional complexities can arise that must be accounted for, as well, such as non-response, in which individuals are selected but refuse to participate. Methods exist to combine the design weights with the other complexities to produce **sampling weights** which, ideally, correspond for each unit to the number of population units represented by that unit.

There are three main consequences to incorrectly treating a complex survey design as a simple random sample. First, standard sample statistics computed from data sampled using a complex survey design may be biased estimates of population statistics. Second, estimates of the variation of sample statistics may be incorrect, resulting also in incorrect confidence intervals and p-values. Finally, ignoring the complex survey design leads to a violation of the independence assumption required for standard statistical methods (Hahs-Vaughn et al., 2011).

In this chapter, we use the `survey` package (Lumley, 2004, 2023) to account for complex survey designs when computing descriptive statistics and carrying out regression analyses. Full documentation can be found at `help(package="survey")` and Analysis of Complex Survey Samples[2] (accessed February 7, 2023). For further reading about sampling design and analysis in general, see, for example, Skinner and Wakefield (2017) and Lohr (2021). For further reading about sampling design and analysis in R specifically, see Lumley (2010).

8.1.1 NHANES survey design

An example of a survey with a complex design is the National Health and Nutrition Examination Survey (NHANES) (Centers for Disease Control and Prevention (CDC). National Center for Health Statistics (NCHS), 2017).

> The NHANES samples are not simple random samples. Rather, a complex, multistage, probability sampling design is used to select participants representative of the civilian, non-institutionalized US population. Oversampling of certain population subgroups is also done to increase the reliability and precision of health status indicator estimates for these particular subgroups. Researchers need to take this into account in their analyses by appropriately specifying the sampling design parameters.
>
> — NHANES Tutorial: Sample Design[3] (accessed February 3, 2023).

Briefly, NHANES has a stratified four-stage sampling design. First, strata are (non-randomly) constructed based on census regions and other geographic information. Within each strata, U.S. counties (the PSUs) are randomly selected, with larger counties having a greater probability of selection. Within counties, city blocks are selected, also proportional to size. Within blocks, households are randomly selected, with certain age, ethnic, and income groups oversampled (higher probability of selection). Finally, within households, individuals are randomly selected. For a full description of the NHANES complex survey design, see NHANES Tutorial: Sample Design[4] (accessed February 3, 2023).

[2] https://cran.r-project.org/web/packages/survey/survey.pdf
[3] https://wwwn.cdc.gov/nchs/nhanes/tutorials/sampledesign.aspx
[4] https://wwwn.cdc.gov/nchs/nhanes/tutorials/sampledesign.aspx

The NHANES website provides sample R code[5] (accessed February 3, 2023) for analyzing NHANES data using the `survey` package, as well as some special considerations when analyzing NHANES data.

In this text, we use the 2017-2018 NHANES cycle, so the information given below is from that cycle. The following variables are included in the dataset to account for the sampling design.

- **Stratum** (`SDMVSTRA`): There were 15 strata.
- **Primary sampling unit** (`SDMVPSU`): This variable takes on only two values (1 or 2). This does not mean that only two counties were selected, rather that two counties were selected within each stratum. Thus, in total, there were 30 PSUs.
- **Interview sampling weight** (`WTINT2YR`): Every participant was interviewed so the interview sampling weight is >0 for every individual (n = 9254). Interviewers used questionnaires to collect self-reported information.
- **Examination sampling weight** (`WTMEC2YR`): Most participants (n = 8704) were also examined at a mobile examination center (MEC). Examinations collected objective measures using, for example, anthropometrics (e.g., height, weight), blood draws (e.g., lipids), and other instrumentation (e.g., dual-energy x-ray absorptiometry (DXA) scans to assess body composition). The 550 participants who were not examined have an examination sampling weight of 0.
- **Fasting subsample sampling weight** (`WTSAF2YR`): A subset of 2711 participants aged 12 years and older also had blood measurements taken using blood drawn after fasting for 8-24 hours. The 6218 participants not in this subsample have a missing (`NA`) fasting subsample weight. The remaining 325 participants were selected for this subsample but were not able to provide an appropriate blood draw. These individuals have fasting subsample sampling weights of zero. See the NHANES documentation[6] (accessed February 8, 2023) for more information.

There were other subsamples, as well, and their corresponding weight variables are noted in the Analytic Notes for certain NHANES variables (e.g., Perfluoroalkyl and Polyfluoroalkyl Substances[7], accessed February 3, 2023). When using NHANES data, always consult the appropriate data documentation and codebooks[8] (accessed February 3, 2023) to ensure you use the appropriate sampling weights.

Which NHANES sampling weight to use

In general, "use the weight of the smallest subpopulation that includes all the variables you want to include in your analysis" (NHANES Tutorial: Weighting[9] (accessed February 3, 2023)). For example, if you only have variables collected in the interview, examination, or on the fasting subsample:

- If any of the variables in your analysis were collected only in the fasting subsample, then use the fasting subsample sampling weights.
- Otherwise, if any were collected only in the examination subsample, then use the examination sampling weights.
- Otherwise, if all variables were collected in the interview, use the interview sampling weights.

[5] https://wwwn.cdc.gov/nchs/nhanes/tutorials/samplecode.aspx
[6] https://wwwn.cdc.gov/Nchs/Nhanes/2017-2018/GLU_J.htm
[7] https://wwwn.cdc.gov/Nchs/Nhanes/2017-2018/PFAS_J.htm
[8] https://wwwn.cdc.gov/nchs/nhanes/continuousnhanes/default.aspx?BeginYear=2017
[9] https://wwwn.cdc.gov/nchs/nhanes/tutorials/weighting.aspx

Combining data over multiple NHANES cycles

Sample statistics based on a single NHANES cycle, while unbiased estimates of U.S. population characteristics, can have large variability due to the fact that not very many PSUs are sampled in any given cycle (NHANES Tutorial: Sample Design[10], accessed February 3, 2023). For example, NHANES 2017-2018 sampled only 30 counties. However, information can easily be combined over multiple cycles. When doing so, you must create a new sampling weight variable, as well as consider the possibility of trends over time. Instructions for how to combine weights over cycles can be found in NHANES Tutorial: Weighting[11] (accessed February 3, 2023).

8.1.2 NSDUH survey design

Another example of a survey with a complex design is the National Survey of Drug Use and Health (NSDUH), which incorporated geographic stratification followed by multistage sampling (U.S. Department of Health and Human Services, Substance Abuse and Mental Health Services Administration, Center for Behavioral Health Statistics and Quality, 2019). See the 2019 NSDUH Public Use File Codebook[12] (accessed February 3, 2023) for detailed information about the sampling design.

The variables needed to account for the complex survey design of the 2019 NSDUH are the following:

- **Stratum** (`vestr`): There were 50 strata.
- **Primary sampling unit** (`verep`): As with the NHANES PSU variable, this variable takes on only two values (1 or 2), nested within strata. Thus, in total, there were 100 PSUs.
- **Final analysis weight** (`ANALWT_C`): These sampling weights are positive for all participants.

8.2 Specifying the survey design

The first step when using the `survey` package is to specify the variables in the dataset that define the components of the complex survey design (e.g., strata, PSUs, sampling weights). This information is needed by all the other survey analysis functions and is stored in a `survey.design` object which is a required argument in all the `survey` functions.

Example 8.1: Specify the survey design for the full NHANES 2017-2018 dataset (`nhanes1718_rmph.Rdata`). This dataset has all the participants, not just a random subsample as was used in earlier chapters, although it still has just a subset of all the NHANES variables. Ultimately, our goal with this example is, as we did in Chapter 5, to carry out a linear regression analysis for the outcome fasting glucose (`LBDGLUSI`) with the predictors waist circumference (`BMXWAIST`), smoking status (`smoker`), age (`RIDAGEYR`), gender (`RIAGENDR`), race/ethnicity (`RIDRETH3`), and

[10] https://wwwn.cdc.gov/nchs/nhanes/tutorials/sampledesign.aspx

[11] https://wwwn.cdc.gov/nchs/nhanes/tutorials/weighting.aspx

[12] https://www.datafiles.samhsa.gov/sites/default/files/field-uploads-protected/studies/NSDUH-2019/NSDUH-2019-datasets/NSDUH-2019-DS0001/NSDUH-2019-DS0001-info/NSDUH-2019-DS0001-info-codebook.pdf

income (`income`). Unlike in Chapter 5, this analysis will use the entire NHANES 2017-2018 dataset and incorporate the complex survey design in order to produce unbiased population estimates with appropriate estimates of variation. Since this analysis uses fasting glucose, use the fasting subsample weights, `WTSAF2YR`.

```
load("Data/nhanes1718_rmph.Rdata")
nrow(nhanes)
```

```
## [1] 9254
```

When incorporating a complex survey design, always use the full dataset. Do not exclude any observations, not even to apply inclusion/exclusion criteria or to remove cases with missing data. The `survey` procedures need to have access to the survey design variable values for all observations in order to obtain valid estimates, even estimates in subgroups. However, we can tell R that we want the final analysis to apply only to a subgroup by creating a subpopulation indicator variable and subsetting the design (not the dataset) based on that variable at a later step. In this example, we will subset on those with non-missing data and positive weights. More generally, you can subset to any subpopulation, as discussed in Section 8.5.

For this example, create a complete case indicator variable called `nomiss` that also conditions on sampling weight > 0. This makes sure descriptive statistics generalize to the same subpopulation as the regression analysis, as well as avoids an error when running some `survey` functions which do not accept zero weights. If doing a regression without a "Table 1" of descriptive statistics (see Section 3.3), you only need to condition on sampling weight $>$ 0. If using a design in which all sampling weights are positive, however, there is no need to condition on sampling weight > 0.

Also, carry out the following data management steps:

- Collapse the race/ethnicity variable as we have done with other analyses.
- Set missing fasting subsample weights to 0 (`svydesign()` returns an error if there are any missing weights).

```
nhanes.mod <- nhanes %>%
  mutate(# Collapse race/ethnicity variable
         race_eth = fct_collapse(RIDRETH3,
          "Hispanic" = c("Mexican American", "Other Hispanic"),
          "Non-Hispanic Other" = c("Non-Hispanic Asian", "Other/Multi")),

         # Set missing fasting subsample weights to 0 since
         # svydesign() returns an error if there are any missing weights.
         WTSAF2YR = replace_na(WTSAF2YR, 0),

         # Complete case and positive weight indicator
         # (a vector of TRUE and FALSE values)
         nomiss = complete.cases(LBDGLUSI, BMXWAIST, smoker, RIDAGEYR,
                                 RIAGENDR, race_eth, income) &
                  WTSAF2YR > 0)
```

```
# Verify the derivations

# Subset on missing weights
SUB <- is.na(nhanes$WTSAF2YR)
# Old NAs
summary(nhanes$WTSAF2YR[SUB])
```

```
# are now 0 in the new dataset
summary(nhanes.mod$WTSAF2YR[SUB])
# For those with non-missing weights, new - old = 0 (no change)
summary(nhanes$WTSAF2YR[!SUB] - nhanes.mod$WTSAF2YR[!SUB])
# (results not shown)
```

Next, use `svydesign()` to create the `survey.design` object. For each of the `strata`, `id`, and `weights` arguments, input a formula (e.g., `~X`) specifying the variable that defines the survey strata, PSUs, and sampling weights, respectively.

For NHANES 2017-2018, specify the following three designs: (1) Interview, (2) Examination, and (3) Fasting subsample. Each design has the same strata and PSU variable names, but a different sampling weight variable name. See Section 8.1.1 for definitions of the variables used. As mentioned in that section, only use the interview weights if *all* the variables in the analysis were included in the interview; if any were from the examination, use the examination weights unless any were only measured in the fasting subsample in which case use the fasting subsample weights.

```
library(survey)
options(survey.lonely.psu = "adjust")
options(survey.adjust.domain.lonely = TRUE)

design.INT <- svydesign(strata=~SDMVSTRA, id=~SDMVPSU, weights=~WTINT2YR,
                        nest=TRUE, data=nhanes.mod)

design.MEC <- svydesign(strata=~SDMVSTRA, id=~SDMVPSU, weights=~WTMEC2YR,
                        nest=TRUE, data=nhanes.mod)

design.FST <- svydesign(strata=~SDMVSTRA, id=~SDMVPSU, weights=~WTSAF2YR,
                        nest=TRUE, data=nhanes.mod)

# View basic information about the design
design.FST
```

```
## Stratified 1 - level Cluster Sampling design (with replacement)
## With (30) clusters.
## svydesign(strata = ~SDMVSTRA, id = ~SDMVPSU, weights = ~WTSAF2YR,
##     nest = TRUE, data = nhanes.mod)
```

```
# View more information about the design
summary(design.FST)
# (results not shown)
```

Finally, for Example 8.1, use the `subset()` function to create a design object to be used with analyses of cases with `nomiss = TRUE`, those with complete data and positive weights.

```
design.FST.nomiss <- subset(design.FST, nomiss)
```

Filtering out cases with missing values is not actually necessary for the regression analysis itself but will ensure that the descriptive statistics and the regression analysis generalize to the same subpopulation. Had we used `design.FST` instead of `design.FST.nomiss`, the subpopulation represented by the descriptive statistics for a variable would be those who would have had non-missing values for *that variable* had they been sampled, regardless of whether or not other variables were missing. With `design.FST.nomiss`, descriptive statistics for all variables and the regression analysis all generalize to those in the population who would have had non-missing values for *all the variables* had they been sampled.

NOTES:

- The global options, `survey.lonely.psu = "adjust"` and `survey.adjust.domain.lonely = TRUE` (which we set at the beginning of this chapter) specify how variance estimation will be handled in cases where a stratum contains only one PSU. These options "center the stratum at the population mean rather than the stratum mean" and make the same adjustment within subgroup analyses (see `?surveyoptions`). These options only need to be set once per R session.
- The `survey` package assumes the `id` variable has a unique value for every unique PSU. However, in many datasets, including NHANES and NSDUH, the `id` variable values are nested within `strata`, duplicated across `strata`, requiring the use of the the `nest=TRUE` option in `svydesign()`. If you leave this option out, but the `id` values are actually nested, R will return the warning `Clusters not nested in strata at top level; you may want nest=TRUE`.
- `svydesign()` does not accept missing weights. That is why we set missing weights to 0 before specifying the design for the fasting subsample.
- Some functions (e.g., `svycoxph()`) return an error if there are any weights of 0. That is why we conditioned on sampling weight > 0 in the code above. See also Section 8.7.3.2.
- `lm()`, `glm()`, and `coxph()` all have a `weights` argument that can be used to specify a weighted regression model. However, passing sampling weights to that argument will not produce the correct results for a complex survey design since `weights` in the context of those functions refers to a different kind of weight. Instead, use the methods described in this chapter, using functions from the `survey` library, to correctly incorporate the survey design into your analysis.

8.2.1 Design degrees of freedom

Estimated confidence intervals and p-values depend on the degrees of freedom. Both NHANES and NSDUH documentation recommended using the **design degrees of freedom** (DF), the number of PSUs less the number of strata, to obtain correct measures of variation (NHANES Tutorial: Variance Estimation[13], accessed February 7, 2023; 2019 NSDUH Public Use File Codebook[14], page H-2, accessed February 7, 2023).

Where does this quantity come from? "Degrees of freedom" are related to the number of independent pieces of information in a sample. For a simple random sample, the DF is the sample size minus the number of population quantities being estimated. For example, the sample mean has DF equal to the sample size minus 1. However, in many complex survey designs, there are fewer independent pieces of information. Where does the "design DF" come from? Here is one way to think about it. After splitting the population up into strata, NHANES and NSDUH sampled PSUs within strata, that is, they conducted a simple random sample of PSUs within each stratum. Thus, the sample mean of a PSU-level variable within a single stratum has DF = (number of PSUs within the stratum) − 1. Summing that up over strata results in DF = number of PSUs less the number of strata.

[13] https://wwwn.cdc.gov/nchs/nhanes/tutorials/VarianceEstimation.aspx

[14] https://www.datafiles.samhsa.gov/sites/default/files/field-uploads-protected/studies/NSDUH-2019/NSDUH-2019-datasets/NSDUH-2019-DS0001/NSDUH-2019-DS0001-info/NSDUH-2019-DS0001-info-codebook.pdf

For the NHANES 2017-2018 data, the design DF are 15 because there are 30 PSUs and 15 strata. For the 2019 NSDUH data, the design DF are 50 because there are 100 PSUs and 50 strata.

Different statistical software may use different default DF. In R, most `survey` functions do not use the design DF by default but do have an option to change the DF. The code `degf(design)` computes the design DF for a given `design`. In a subgroup analysis (see Section 8.5) the correct design DF is based on the numbers of PSUs and strata in the subsample representing the subgroup of interest, either of which might be smaller than in the full sample. Fortunately, `degf()` modifies the DF appropriately for an analysis on a subgroup with fewer PSUs and/or strata. See the "Details" section in `?summary.svyglm` for more information.

8.3 Weighted descriptive statistics

There are a number of `survey` functions for computing weighted descriptive statistics, as well as a `gtsummary` (Sjoberg et al., 2021, 2023) function to conveniently create a "Table 1". We will compute these statistics overall and by exposure or outcome.

8.3.1 Overall

Example 8.1 (continued): Compute weighted descriptive statistics that estimate values for the population represented by NHANES 2017-2018 participants with non-missing values on all our analysis variables. Since the outcome variable is fasting glucose, use the design corresponding to the fasting subsample with complete data and positive weights (`design.FST.nomiss`).

Examples are shown first for two variables, one continuous and one categorical. Later, we will use `gtsummary` to create a Table 1 of weighted descriptive statistics for all the variables.

For **continuous** variables, use `svymean()`, `svyvar()`, and `svyquantile()` to compute the weighted mean, standard deviation, median, and interquartile range, and `svyhist()` and `svyboxplot()` to plot weighted histograms and boxplots (Figures 8.1 and 8.2, respectively). When using `confint()` to get a 95% confidence interval, add `df = degf(design)` to use the design DF.

NOTE: If not using a complete case analysis and a variable has missing values, add `na.rm = T` to each `svymean` or `svyvar` call to avoid returning a missing value. When using `confint(svymean())`, the `na.rm=T` must go in the `svymean()` call, not the `confint()` call (e.g., `confint(svymean(~X, design, na.rm=T), df = degf(design))`).

```
# Weighted mean, standard deviation, and 95% CI
cbind(
  "wMEAN" = svymean(     ~LBDGLUSI, design.FST.nomiss),
  "wSD"   = sqrt(svyvar(~LBDGLUSI, design.FST.nomiss)[1]),
        confint(svymean(~LBDGLUSI, design.FST.nomiss), df=degf(design.FST.nomiss))
)
```

```
##          wMEAN   wSD 2.5 % 97.5 %
## LBDGLUSI 6.106 1.763 5.976  6.236
```

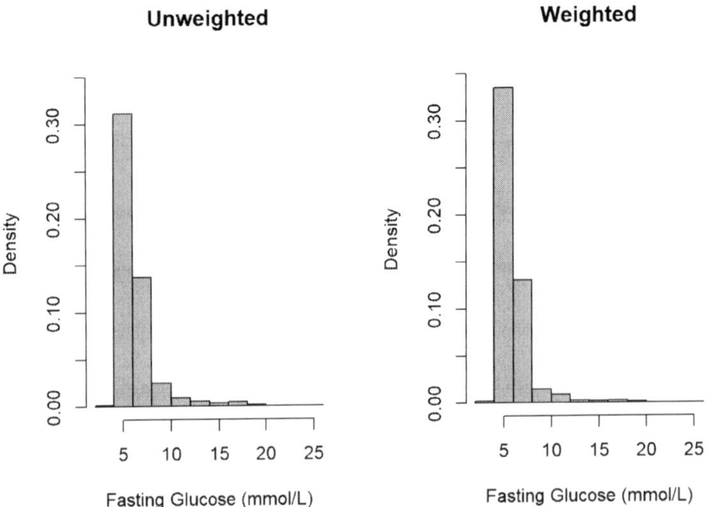

FIGURE 8.1 Unweighted vs. weighted histograms

```
# Weighted median, 95% CI, and standard error
# df = degf(design) is already the default for svyquantile
svyquantile(~LBDGLUSI, design.FST.nomiss, 0.50)[[1]]
```

```
##     quantile ci.2.5 ci.97.5      se
## 0.5     5.72   5.66    5.83 0.03988
```

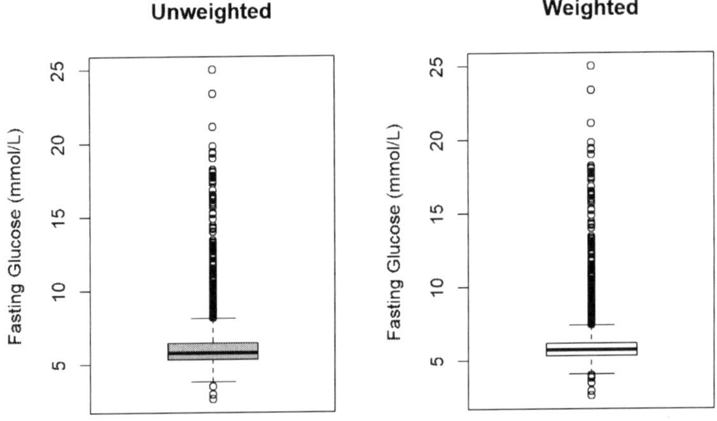

FIGURE 8.2 Unweighted vs. weighted boxplots

```
# Weighted interquartile range
c("wIQR" =
    svyquantile(~LBDGLUSI, design.FST.nomiss, 0.75)[[1]][1, "quantile"] -
    svyquantile(~LBDGLUSI, design.FST.nomiss, 0.25)[[1]][1, "quantile"])
```

```
## wIQR
## 0.83
```

```
par(mfrow=c(1,2))
# Unweighted probability histogram
hist(nhanes.mod$LBDGLUSI[nhanes.mod$nomiss], probability = T,
     xlab = "Fasting Glucose (mmol/L)", main = "Unweighted",
     ylim = c(0,0.35))
# Weighted probability histogram
svyhist(   ~LBDGLUSI, design.FST.nomiss,
     xlab = "Fasting Glucose (mmol/L)", main = "Weighted",
     ylim = c(0,0.35))
```

```
par(mfrow=c(1,2))
# Unweighted boxplot
boxplot(nhanes.mod$LBDGLUSI[nhanes.mod$nomiss],
     ylab = "Fasting Glucose (mmol/L)",
     main = "Unweighted")
# Weighted boxplot
svyboxplot(~LBDGLUSI ~ 1, design.FST.nomiss, all.outliers = T,
     ylab = "Fasting Glucose (mmol/L)",
     main = "Weighted")
```

For **categorical** variables, use `svytotal()` or `svytable()` to estimate population totals. In a sample we compute sample frequencies, but the corresponding population quantity is a population total. Use `svymean()`, `svytable()` or `svyciprop()` to estimate population proportions. Use `barplot()` on the output of `svytable()` to plot a weighted barchart (Figure 8.3).

NOTE: If not using a complete case analysis and a variable has missing values, add `na.rm = T` to each `svytotal()` or `svymean()` call to avoid returning a missing value.

```
# Weighted total with SE of total
svytotal(~race_eth, design.FST.nomiss)
```

```
##                                  total       SE
## race_ethHispanic              32712402  4552659
## race_ethNon-Hispanic White 137704704 10603571
## race_ethNon-Hispanic Black  23063852  3388626
## race_ethNon-Hispanic Other  21072543  2998573
```

```
# Weighted total (no SE)
svytable(~race_eth, design.FST.nomiss)
```

```
## race_eth
##           Hispanic Non-Hispanic White Non-Hispanic Black Non-Hispanic Other
##           32712402          137704704           23063852           21072543
```

```
# Weighted proportion using svymean()
svymean(~race_eth, design.FST.nomiss)
```

```
##                               mean   SE
## race_ethHispanic            0.1525 0.02
## race_ethNon-Hispanic White 0.6418 0.03
## race_ethNon-Hispanic Black 0.1075 0.02
## race_ethNon-Hispanic Other 0.0982 0.01
```

```
# Weighted proportion using svytable by normalizing the total to 1
svytable(~race_eth, design.FST.nomiss, Ntotal=1)
```

```
## race_eth
##          Hispanic Non-Hispanic White Non-Hispanic Black Non-Hispanic Other
##           0.15247            0.64182            0.10750            0.09822
```

```
# Weighted proportion with 95% CI
# df = degf(design) is already the default for svyciprop
svyciprop(~I(race_eth=="Hispanic"), design.FST.nomiss)
```

```
##                                  2.5% 97.5%
## I(race_eth == "Hispanic") 0.152 0.113   0.2
```

```
# (results for other 3 levels not shown)
svyciprop(~I(race_eth=="Non-Hispanic White"), design.FST.nomiss)
svyciprop(~I(race_eth=="Non-Hispanic Black"), design.FST.nomiss)
svyciprop(~I(race_eth=="Non-Hispanic Other"), design.FST.nomiss)
```

```
par(mfrow=c(2,1))
# Unweighted barplot
barplot(prop.table(table(nhanes.mod$race_eth[nhanes.mod$nomiss])),
        ylab = "Proportion", xlab = "Race/Ethnicity",
        main = "Unweighted", cex.names = 0.65)
# Weighted barplot
mybar <- svytable(~race_eth, design.FST.nomiss, Ntotal=1)
barplot(mybar,
        ylab = "Proportion", xlab = "Race/Ethnicity",
        main = "Weighted", cex.names = 0.65)
```

NHANES over-sampled minority groups in order to have sufficient sample sizes for subgroup analyses. Thus, there are large differences between the unweighted and weighted sample distributions of race/ethnicity, as shown in Figure 8.3. Unweighted proportions only reflect the composition of the sample, whereas weighted proportions estimate the population distribution.

Finally, use `tbl_svysummary()` from the `gtsummary` library to produce a Table 1 of weighted descriptive statistics. Instead of starting with a dataset as we did in unweighted analyses, start with the design object (`design.FST.nomiss`). For categorical variables, N and n values in the table are estimated population totals – when we did unweighted analyses, these were sample sizes.

The results are shown in Table 8.1.

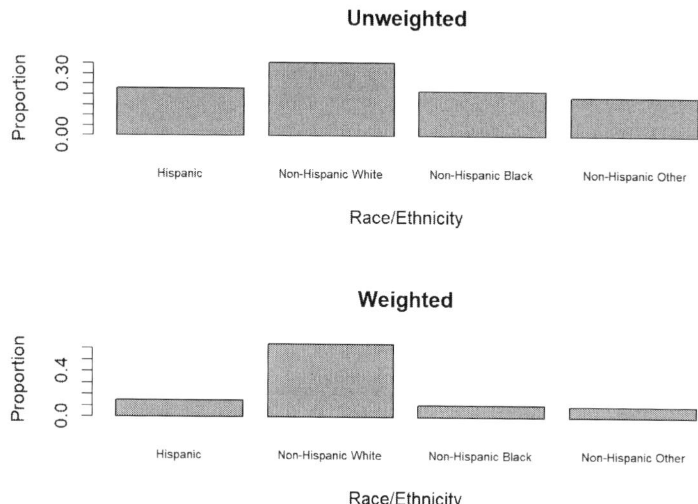

FIGURE 8.3 Unweighted vs. weighted barcharts

TABLE 8.1 Weighted descriptive statistics

Variable	N = 214,553,501
Fasting Glucose (mmol/L)	6.11 (1.76)
Waist Circumference (cm)	100.2 (17.1)
Age (years)	47.0 (17.5)
Smoking Status	
Never	123,589,210 (57.6%)
Past	55,919,877 (26.1%)
Current	35,044,414 (16.3%)
Gender	
Male	104,286,253 (48.6%)
Female	110,267,248 (51.4%)
Race/Ethnicity	
Hispanic	32,712,402 (15.2%)
Non-Hispanic White	137,704,704 (64.2%)
Non-Hispanic Black	23,063,852 (10.7%)
Non-Hispanic Other	21,072,543 (9.8%)
Annual Income	
<$25,000	38,141,114 (17.8%)
$25,000 to <$55,000	57,342,659 (26.7%)
$55,000+	119,069,729 (55.5%)

[1] Mean (SD); n (%)

```
library(gtsummary)
TABLE <- design.FST.nomiss %>%
  tbl_svysummary(
    # Use include to select variables
    include = c(LBDGLUSI, BMXWAIST, RIDAGEYR,
                smoker, RIAGENDR, race_eth, income),
    statistic = list(all_continuous()  ~ "{mean} ({sd})",
                     all_categorical() ~ "{n}    ({p}%)"),
    digits = list(LBDGLUSI ~ c(2, 2),
                  BMXWAIST ~ c(1, 1),
                  RIDAGEYR ~ c(1, 1),
                  all_categorical() ~ c(0, 1)),
    label  = list(LBDGLUSI ~ "Fasting Glucose (mmol/L)",
                  BMXWAIST ~ "Waist Circumference (cm)",
                  RIDAGEYR ~ "Age (years)",
                  smoker   ~ "Smoking Status",
                  RIAGENDR ~ "Gender",
                  race_eth ~ "Race/Ethnicity",
                  income   ~ "Annual Income")
  ) %>%
  modify_header(label = "**Variable**") %>%
  modify_caption("Weighted descriptive statistics") %>%
  bold_labels()
```

```
TABLE
```

8.3.2 By exposure or outcome

Example 8.1 (continued): Compute weighted descriptive statistics by smoking status.

For `svymean()`, `svyvar()`, `svyquantile()`, `svytotal()`, and `svyciprop()`, use the wrapper function `svyby()` to compute statistics at each level of another variable. For `svytable()`, however, add the by variable to the formula.

```
# Continuous variables, by smoking status
# Weighted mean, standard deviation, and 95% CI
WTD <- cbind(
  svyby(~LBDGLUSI, ~smoker, design.FST.nomiss, svymean)[, 1:2],
  svyby(~LBDGLUSI, ~smoker, design.FST.nomiss, svyvar)[ ,   2],
  confint(svyby(~LBDGLUSI, ~smoker, design.FST.nomiss, svymean),
          df = degf(design.FST.nomiss))
)
WTD[, 3] <- sqrt(WTD[, 3])
names(WTD)[3] <- paste(names(WTD)[2], "wSD")
names(WTD)[2] <- paste(names(WTD)[2], "wMEAN")
WTD
```

```
##          smoker LBDGLUSI wMEAN LBDGLUSI wSD 2.5 % 97.5 %
## Never     Never          5.965        1.543 5.796  6.135
## Past       Past          6.457        2.066 6.297  6.617
## Current Current          6.041        1.894 5.798  6.285
```

```
# Weighted median, standard error, and 95% CI
# df = degf(design) is already the default for svyquantile
#      but not for confint
cbind(
  svyby(~LBDGLUSI, ~smoker, design.FST.nomiss, svyquantile, 0.50),
  confint(svyby(~LBDGLUSI, ~smoker, design.FST.nomiss, svyquantile, 0.50),
          df = degf(design.FST.nomiss))
)
```

```
##          smoker LBDGLUSI se.LBDGLUSI 2.5 % 97.5 %
## Never     Never     5.66     0.03753 5.580  5.740
## Past       Past     5.83     0.03988 5.745  5.915
## Current Current     5.66     0.06568 5.520  5.800
```

```
# Weighted interquartile range
Q75 <- as.data.frame(
  svyby(~LBDGLUSI, ~smoker, design.FST.nomiss, svyquantile, 0.75)
  )[,1:2]
Q25 <- as.data.frame(
  svyby(~LBDGLUSI, ~smoker, design.FST.nomiss, svyquantile, 0.25)
  )[,1:2]
names(Q75)[2] <- paste(names(Q75)[2], "wQ75")
names(Q25)[2] <- paste(names(Q25)[2], "wQ25")
WIQR <- merge(Q75, Q25)
WIQR$wIQR <- WIQR[,2] - WIQR[,3]
WIQR
```

```
##    smoker LBDGLUSI wQ75 LBDGLUSI wQ25 wIQR
## 1 Current          6.11          5.27 0.84
## 2   Never          6.11          5.27 0.84
## 3    Past          6.49          5.50 0.99
```

```
# Categorical variables, by smoking status
# Weighted total
svytable(~race_eth + smoker, design.FST.nomiss)
```

```
##                      smoker
## race_eth                 Never     Past  Current
##   Hispanic            20524606  7861212  4326583
##   Non-Hispanic White 75120526 40106079 22478099
##   Non-Hispanic Black 15137153  3487312  4439386
##   Non-Hispanic Other 12806924  4465273  3800346
```

```
# Weighted total with SE of total
svyby(~race_eth, ~smoker, design.FST.nomiss, svytotal)
# (results not shown)
```

Be careful when normalizing `svytable()` to get proportions. Using `Ntotal=1` as we did previously normalizes the frequencies to the overall total. Rather, we want to normalize to the column totals to get proportions within each level of smoking status. This is accomplished using `prop.table(, margin = 2)`. Optionally, use `addmargins(, 1)` to confirm that each column sums to 1.

```
# Weighted proportion
addmargins(
  prop.table(svytable(~race_eth + smoker, design.FST.nomiss), margin = 2)
  , 1)
```

```
##                          smoker
## race_eth                    Never    Past Current
##   Hispanic               0.16607 0.14058 0.12346
##   Non-Hispanic White 0.60782 0.71721 0.64142
##   Non-Hispanic Black 0.12248 0.06236 0.12668
##   Non-Hispanic Other 0.10362 0.07985 0.10844
##   Sum                    1.00000 1.00000 1.00000
```

```
# Weighted proportion with SE
svyby(~I(race_eth=="Hispanic"), ~smoker, design.FST.nomiss, svyciprop)
```

```
##           smoker I(race_eth == "Hispanic")
## Never      Never                    0.1661
## Past        Past                    0.1406
## Current Current                    0.1235
##         se.as.numeric(I(race_eth == "Hispanic"))
## Never                                    0.02650
## Past                                     0.01976
## Current                                  0.02512
```

```
# (results for other levels not shown)
svyby(~I(race_eth=="Non-Hispanic White"), ~smoker, design.FST.nomiss, svyciprop)
svyby(~I(race_eth=="Non-Hispanic Black"), ~smoker, design.FST.nomiss, svyciprop)
svyby(~I(race_eth=="Non-Hispanic Other"), ~smoker, design.FST.nomiss, svyciprop)
```

Finally, use `tbl_svysummary()` from the `gtsummary` library to produce a Table 1 of weighted descriptive statistics by smoking status. The results are shown in Table 8.2.

```
library(gtsummary)
TABLE <- design.FST.nomiss %>%
  tbl_svysummary(
    by = smoker,
    # Use include to select variables
    include = c(LBDGLUSI, BMXWAIST, RIDAGEYR,
                RIAGENDR, race_eth, income),
    statistic = list(all_continuous()  ~ "{mean} ({sd})",
                     all_categorical() ~ "{n}    ({p}%)"),
    digits = list(LBDGLUSI ~ c(2, 2),
                  BMXWAIST ~ c(1, 1),
                  RIDAGEYR ~ c(1, 1),
                  all_categorical() ~ c(0, 1)),
    label  = list(LBDGLUSI ~ "Fasting Glucose (mmol/L)",
                  BMXWAIST ~ "Waist Circumference (cm)",
                  RIDAGEYR ~ "Age (years)",
                  RIAGENDR ~ "Gender",
                  race_eth ~ "Race/Ethnicity",
                  income   ~ "Annual Income")
  ) %>%
  modify_header(label = "**Variable**",
   all_stat_cols() ~ "**{level}**<br>N = {n} ({style_percent(p, digits=1)}%)") %>%
  modify_caption("Weighted descriptive statistics, by smoking status") %>%
  bold_labels()
```

TABLE

TABLE 8.2 Weighted descriptive statistics, by smoking status

Variable	Never N = 123589210 (57.6%)	Past N = 55919877 (26.1%)	Current N = 35044414 (16.3%)
Fasting Glucose (mmol/L)	5.97 (1.54)	6.46 (2.07)	6.04 (1.89)
Waist Circumference (cm)	98.6 (17.0)	104.0 (16.6)	99.9 (17.4)
Age (years)	45.4 (17.7)	51.5 (18.0)	45.4 (14.4)
Gender			
Male	51,169,518 (41.4%)	34,805,126 (62.2%)	18,311,610 (52.3%)
Female	72,419,692 (58.6%)	21,114,752 (37.8%)	16,732,805 (47.7%)
Race/Ethnicity			
Hispanic	20,524,606 (16.6%)	7,861,212 (14.1%)	4,326,583 (12.3%)
Non-Hispanic White	75,120,526 (60.8%)	40,106,079 (71.7%)	22,478,099 (64.1%)
Non-Hispanic Black	15,137,153 (12.2%)	3,487,312 (6.2%)	4,439,386 (12.7%)
Non-Hispanic Other	12,806,924 (10.4%)	4,465,273 (8.0%)	3,800,346 (10.8%)
Annual Income			
<$25,000	18,081,806 (14.6%)	9,269,756 (16.6%)	10,789,552 (30.8%)
$25,000 to <$55,000	31,806,525 (25.7%)	15,168,200 (27.1%)	10,367,933 (29.6%)
$55,000+	73,700,879 (59.6%)	31,481,921 (56.3%)	13,886,929 (39.6%)

[1] Mean (SD); n (%)

8.4 Weighted linear regression

8.4.1 Fitting the model

To carry out a linear regression that incorporates a survey design, use `svyglm()` with `family =gaussian()`. "Gaussian" means "normally distributed" so this is specifying a model with an outcome that, given the predictors, is normally distributed, which is equivalent to specifying a linear regression with normally distributed errors. See, for example, Lumley and Scott (2017) and for further reading about fitting regression models to complex survey data.

After fitting the model, use `summary()`, `confint()`, and `car::Anova(, type = 3, test.statistic = "F")` to view the regression coefficients, their 95% confidence intervals, and p-values. Each of these functions has an option to change the DF to the design DF. Finally, table the regression results using `tbl_regression()`. There is an option to change the DF in `tbl_regression()`, but it only applies to the overall p-values for categorical predictors, so the confidence intervals and remaining p-values in the table are based on default DF rather than the design DF. See the "Details" section in `?summary.svyglm` for information regarding the default DF.

Example 8.1 (continued): Regress the outcome fasting glucose (`LBDGLUSI`) on the predictors waist circumference (`BMXWAIST`), smoking status (`smoker`), age (`RIDAGEYR`), gender (`RIAGENDR`), race/ethnicity (`RIDRETH3`), and income (`income`), correctly incorporating the NHANES survey design to obtain estimates representative of the population. Use the NHANES 2017-2018 fasting subset survey design we have been working with in this chapter (`design.FST.nomiss`).

```
fit.ex8.1 <- svyglm(LBDGLUSI ~ BMXWAIST + smoker + RIDAGEYR +
                      RIAGENDR + race_eth + income,
                 family=gaussian(), design=design.FST.nomiss)
```

When calling `summary()`, `confint()`, and `car::Anova()` to summarize the results, use the `df.resid`, `ddf`, and `error.df` options, respectively, to specify the design DF for confidence intervals and hypothesis tests. To ensure you are using the same design as was used in the regression, extract the design from the `svyglm` object as `fit$survey.design`.

```
round(
  cbind(
    summary(fit.ex8.1, df.resid = degf(fit.ex8.1$survey.design))$coef,
    confint(fit.ex8.1, ddf      = degf(fit.ex8.1$survey.design))
  )
, 4)
```

```
##                              Estimate Std. Error t value Pr(>|t|)    2.5 %  97.5 %
## (Intercept)                    3.3890     0.3542  9.5670   0.0000  2.6340  4.1441
## BMXWAIST                       0.0221     0.0031  7.0544   0.0000  0.0154  0.0288
## smokerPast                     0.1914     0.0961  1.9916   0.0649 -0.0134  0.3963
## smokerCurrent                  0.0055     0.1103  0.0499   0.9609 -0.2297  0.2407
## RIDAGEYR                       0.0210     0.0020 10.2867   0.0000  0.0166  0.0253
## RIAGENDRFemale                -0.3331     0.0684 -4.8694   0.0002 -0.4789 -0.1873
## race_ethNon-Hispanic White    -0.3448     0.1461 -2.3599   0.0322 -0.6561 -0.0334
## race_ethNon-Hispanic Black    -0.2577     0.1565 -1.6470   0.1203 -0.5912  0.0758
## race_ethNon-Hispanic Other    -0.1095     0.1381 -0.7928   0.4402 -0.4040  0.1849
## income$25,000 to <$55,000     -0.1437     0.1414 -1.0162   0.3256 -0.4452  0.1577
## income$55,000+                -0.1224     0.0895 -1.3674   0.1917 -0.3132  0.0684
```

```
car::Anova(fit.ex8.1, type = 3, test.statistic = "F",
           error.df = degf(fit.ex8.1$survey.design))
```

```
## Analysis of Deviance Table (Type III tests)
##
## Response: LBDGLUSI
##             Df      F      Pr(>F)
## (Intercept)  1  91.53 0.000000089 ***
## BMXWAIST     1  49.77 0.000003909 ***
## smoker       2   2.09      0.1584
## RIDAGEYR     1 105.82 0.000000034 ***
## RIAGENDR     1  23.71      0.0002 ***
## race_eth     3   3.00      0.0636 .
## income       2   1.03      0.3793
## Residuals   15
## ---
## Signif. codes:  0 '***' 0.001 '**' 0.01 '*' 0.05 '.' 0.1 ' ' 1
```

To table these results, use the `tbl_regression()` function in `gtsummary` after adding `test.statistic = "F"` in `add_global_p()`. The results are shown in Table 8.3.

Warning: There is no option in `tbl_regression()` to change the DF to make the confidence intervals and p-values for the individual regression coefficients be based on the design DF. There is an option in `add_global_p()` to change the DF for the Type III tests (commented out in the code below); however, using that option will lead to p-values that do not match between the overall (Type III) test for a binary predictor and the test for the single term in the model for that binary predictor. The code below uses R's default DF throughout. If, instead, you want to use the design DF throughout, uncomment the `error.df` row below and export the table (see Sections 3.3.2 and 3.3.3). Then, in a word processing program, edit the confidence intervals and p-values for each regression coefficient, replacing them with those you computed above using the design DF.

```
library(gtsummary)
TABLE <- fit.ex8.1 %>%
  tbl_regression(intercept = T,
                 estimate_fun = function(x) style_sigfig(x, digits = 3),
                 pvalue_fun   = function(x) style_pvalue(x, digits = 3),
                 label  = list(BMXWAIST ~ "Waist Circumference (cm)",
                              smoker   ~ "Smoking Status",
                              RIDAGEYR ~ "Age (years)",
                              RIAGENDR ~ "Gender",
                              race_eth ~ "Race/Ethnicity",
                              income   ~ "Annual Income")) %>%
  add_global_p(keep = T, test.statistic = "F"
  # Uncomment to use design df for Type III test p-values
  # (but then they will not match the regression term p-values for
  #  binary predictors)
  #            , error.df = degf(fit.ex8.1$survey.design)
                                ) %>%
  modify_caption(
    "Weighted linear regression results for fasting glucose (mmol/L)")
```

```
TABLE
```

TABLE 8.3 Weighted linear regression results for fasting glucose (mmol/L)

Characteristic	Beta	95% CI	p-value
(Intercept)	3.39	2.48, 4.30	<0.001
Waist Circumference (cm)	0.022	0.014, 0.030	<0.001
Smoking Status			0.219
Never	—	—	
Past	0.191	-0.056, 0.438	0.103
Current	0.006	-0.278, 0.289	0.962
Age (years)	0.021	0.016, 0.026	<0.001
Gender			0.005
Male	—	—	
Female	-0.333	-0.509, -0.157	0.005
Race/Ethnicity			0.134
Hispanic	—	—	
Non-Hispanic White	-0.345	-0.720, 0.031	0.065
Non-Hispanic Black	-0.258	-0.660, 0.145	0.160
Non-Hispanic Other	-0.110	-0.465, 0.246	0.464
Annual Income			0.421
<$25,000	—	—	
$25,000 to <$55,000	-0.144	-0.507, 0.220	0.356
$55,000+	-0.122	-0.353, 0.108	0.230

Conclusion: After adjusting for age, gender, race/ethnicity, income, and each other, waist circumference is significantly associated with fasting glucose (B = 0.022; 95% CI = 0.014, 0.030; p <.001) but smoking status is not (p = .219).

8.4.2 Prediction

As was the case with `glm()`, `predict()` applied to a `svyglm()` object does not return a CI. Instead, use `svycontrast_design_df()` (in `Functions_rmph.R`, which you loaded at the beginning of this chapter), a modified version of `svycontrast()` with a DF option, which is similar to `gmodels::estimable()`. See Section 6.8 for how to specify the second argument to get the estimate of interest.

Example 8.1 (continued): What is the estimated population mean fasting glucose among individuals with a waist circumference of 120 cm, who have never smoked, are age 35 years, female, and Hispanic, and have an annual income from \$25,000 to <\$55,000?

```
# Always include the intercept for estimation
# Specify a 1 for the intercept, a # for each continuous predictor
# and a 1 for each non-reference level of a categorical variable.
# If a predictor is at its reference level, it should be left out.
svycontrast_design_df(fit.ex8.1,
                      c("(Intercept)"                = 1,
                        "BMXWAIST"                   = 120,
                        "RIDAGEYR"                   = 35,
                        "RIAGENDRFemale"             = 1,
                        "income$25,000 to <$55,000" = 1))
```

```
##     est lower upper
## 1 6.302 6.018 6.585
```

Conclusion: The estimated population mean fasting glucose among such individuals is 6.30 mmol/L (95% CI = 6.02, 6.59).

8.4.3 Interactions

The syntax for including an interaction in `svyglm()` is exactly as in `lm()`: add `X:Z` to include an interaction between `X` and `Z`.

Example 8.2: Does the association between the outcome fasting glucose (`LBDGLUSI`) and waist circumference (`BMXWAIST`) depend on gender (`RIAGENDR`), adjusted for smoking status, age, race, and income? Correctly incorporate the NHANES survey design to obtain an estimate of the waist circumference slope for each gender that is generalizable to the population.

```
fit.ex8.2 <- svyglm(LBDGLUSI ~ BMXWAIST + smoker + RIDAGEYR +
                      RIAGENDR + race_eth + income + BMXWAIST:RIAGENDR,
                    family=gaussian(), design=design.FST.nomiss)


round(
  cbind(
    summary(fit.ex8.2,
            df.resid=degf(fit.ex8.2$survey.design))$coef,
    confint(fit.ex8.2,
            ddf=degf(fit.ex8.2$survey.design))
  )
, 4)
```

```
##                                 Estimate Std. Error t value Pr(>|t|)    2.5 %  97.5 %
## (Intercept)                       3.2130     0.4739  6.7793   0.0000  2.2028  4.2231
## BMXWAIST                          0.0239     0.0048  4.9532   0.0002  0.0136  0.0342
## smokerPast                        0.1937     0.0991  1.9546   0.0695 -0.0175  0.4050
## smokerCurrent                     0.0102     0.1151  0.0886   0.9305 -0.2351  0.2555
## RIDAGEYR                          0.0208     0.0023  9.2184   0.0000  0.0160  0.0256
## RIAGENDRFemale                   -0.0256     0.5868 -0.0437   0.9657 -1.2765  1.2252
## race_ethNon-Hispanic White       -0.3429     0.1447 -2.3705   0.0316 -0.6513 -0.0346
## race_ethNon-Hispanic Black       -0.2495     0.1573 -1.5866   0.1335 -0.5848  0.0857
## race_ethNon-Hispanic Other       -0.1085     0.1374 -0.7898   0.4420 -0.4014  0.1843
## income$25,000 to <$55,000        -0.1447     0.1409 -1.0269   0.3207 -0.4450  0.1556
## income$55,000+                   -0.1257     0.0911 -1.3795   0.1880 -0.3198  0.0685
## BMXWAIST:RIAGENDRFemale          -0.0031     0.0055 -0.5592   0.5843 -0.0147  0.0086
```

The p-value for `BMXWAIST:RIAGENDRFemale` is >0.05, so the interaction is not statistically significant.

To estimate the waist circumference slope at each level of gender, use `svycontrast_design_df()` (in Functions_rmph.R, which you loaded at the beginning of this chapter). See Section 5.9.7 for how to specify the second argument to get the estimates of interest.

```
# Estimate slope at each level of other variable, along with a CI
# using the design DF (the default)
rbind(
  "BMXWAIST @ Male"   = svycontrast_design_df(fit.ex8.2,
                            c("BMXWAIST"                 = 1)),
  "BMXWAIST @ Female" = svycontrast_design_df(fit.ex8.2,
                            c("BMXWAIST"                 = 1,
                              "BMXWAIST:RIAGENDRFemale" = 1))
)
```

```
##                         est   lower   upper
## BMXWAIST @ Male   0.02393 0.01363 0.03423
## BMXWAIST @ Female 0.02088 0.01336 0.02840
```

Conclusion: After incorporating the NHANES complex survey design, the association between waist circumference and fasting glucose was not significantly different between males and females (p = .584; slope for males = 0.024; slope for females = 0.021).

8.4.4 Visualize the weighted unadjusted relationships

Use `svyplot()` to visualize the weighted unadjusted relationships between predictors and the outcome using bubble plots, with points representing more individuals in the population plotted with larger circles. This function works for both continuous and categorical predictors (Figures 8.4 and 8.5, respectively). For a continuous predictor, use `svysmooth()` to superimpose an estimate of the mean outcome vs. the predictor that does not assume linearity.

```
# Plot for continuous predictor
svyplot(LBDGLUSI ~ BMXWAIST, design.FST.nomiss,
        ylab = "Fasting Glucose (mmol/L)",
        xlab = "Waist Circumference (cm)")
# Fit unadjusted model
fit.wc <- svyglm(LBDGLUSI ~ BMXWAIST,
                 family=gaussian(), design=design.FST.nomiss)
```

```
# Add line
abline(fit.wc, col = "red", lwd = 2)
# Add smoother
lines(svysmooth(fit.wc$formula,
                fit.wc$survey.design,
                df = 5),
      col = "blue", lwd = 2, lty = 2)
legend("topright", c("Regression line", "Smoother"),
       title = "Weighted estimate",
       col = c("red", "blue"), lwd = c(2,2), lty = c(1,2),
       bty = "n", seg.len = 5)
```

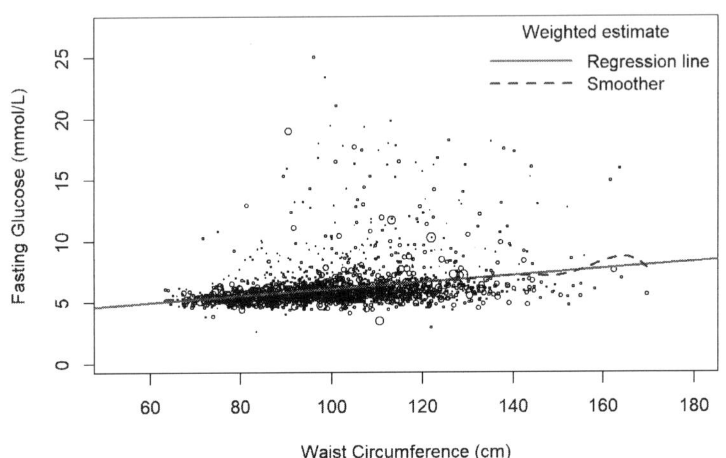

FIGURE 8.4 Weighted scatterplot of outcome vs. a continuous predictor

```
# Plot for categorical predictor
svyplot(LBDGLUSI ~ smoker, design.FST.nomiss,
        ylab = "Fasting Glucose (mmol/L)",
        xlab = "Smoking Status",
        xaxt = "n")
axis(1, at = 1:3, labels = levels(nhanes.mod$smoker))
# Fit unadjusted model
fit.smoker <- svyglm(LBDGLUSI ~ smoker,
                  family=gaussian(), design=design.FST.nomiss)
# Compute means
MEANS <- as.numeric(
  predict(fit.smoker,
          data.frame(smoker=levels(nhanes.mod$smoker)))
)
# Add points and lines
points(1:length(MEANS), MEANS, col = "red", pch = 20, cex = 2)
lines( 1:length(MEANS), MEANS, col = "red")
```

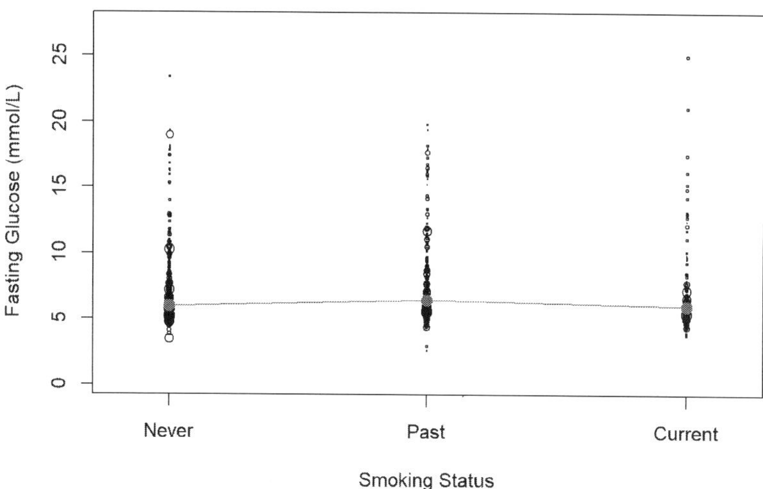

FIGURE 8.5 Weighted plot of outcome vs. a categorical predictor

8.5 Domain (subgroup) analysis

Sometimes we are interested in estimating quantities among subgroups of the population, such as "females age 45 years and older". This is called a **domain analysis** or **subgroup analysis**. In an unweighted analysis, one would subset the *dataset* to filter out those who do not meet the criteria and then run the analysis on the resulting subset of observations. In a weighted analysis, however, removing observations from the dataset results in incorrect standard errors, confidence intervals, and p-values. The information about the sample design found in the excluded cases is needed to incorporate the sampling design correctly. Instead, use the `subset()` function to subset the design, that is, to specify the domain of interest *in the design object.* Rather than actually removing observations, this retains all observations and lets `survey` know what subset you are interested in. This is exactly what we did earlier, using a complete case indicator variable (`nomiss`) to subset on those with non-missing values. However, here, we show how to apply this method more generally to any subpopulation.

Example 8.1 (continued): Repeat the creation of a Table 1 of descriptive statistics and the linear regression from Example 8.1 among females age 45 years and older. Due to the inclusion criteria, gender must be removed from the analysis since it no longer varies between individuals.

Previously, we used `subset()` to specify a domain with no missing values and positive weights. Here, we add additional conditions to limit to females age 45 years and older.

```
design.FST.domain <- subset(design.FST,
                            nomiss & RIAGENDR == "Female" & RIDAGEYR >= 45)
```

NOTE: If the process of specifying your domain of interest involves creating new variables, you must re-create your design object to make those variables available to the design before using `subset()` on the `survey.design` object.

Use `tbl_svysummary()` to create a table of descriptive statistics. The results are shown in Table 8.4.

```
library(gtsummary)
TABLE <- design.FST.domain %>%
  tbl_svysummary(
    by = smoker,
    # Use include to select variables
    include = c(LBDGLUSI, BMXWAIST, RIDAGEYR, race_eth, income),
    statistic = list(all_continuous()  ~ "{mean} ({sd})",
                     all_categorical() ~ "{n}    ({p}%)"),
    digits = list(LBDGLUSI ~ c(2, 2),
                  BMXWAIST ~ c(1, 1),
                  RIDAGEYR ~ c(1, 1),
                  all_categorical() ~ c(0, 1)),
    label  = list(LBDGLUSI ~ "Fasting Glucose (mmol/L)",
                  BMXWAIST ~ "Waist Circumference (cm)",
                  RIDAGEYR ~ "Age (years)",
                  race_eth ~ "Race/Ethnicity",
                  income   ~ "Annual Income")
  ) %>%
  modify_header(label = "**Variable**",
    all_stat_cols() ~ "**{level}**<br>N = {n} ({style_percent(p, digits=1)}%)")
    %>%
  modify_caption("Weighted descriptive statistics, by smoking status
                (Females age 45y and older)") %>%
  bold_labels()
```

```
TABLE
```

TABLE 8.4 Weighted descriptive statistics, by smoking status (Females age 45y and older)

Variable	**Never** N = 40923599 (65.8%)	**Past** N = 13054149 (21.0%)	**Current** N = 8207032 (13.2%)
Fasting Glucose (mmol/L)	6.12 (1.55)	6.46 (2.38)	6.20 (1.42)
Waist Circumference (cm)	98.1 (15.1)	104.4 (18.0)	98.6 (17.1)
Age (years)	60.9 (10.5)	62.0 (10.3)	57.0 (7.9)
Race/Ethnicity			
Hispanic	4,701,252 (11.5%)	1,219,766 (9.3%)	841,551 (10.3%)
Non-Hispanic White	28,446,162 (69.5%)	9,779,559 (74.9%)	5,853,934 (71.3%)
Non-Hispanic Black	4,125,440 (10.1%)	1,217,709 (9.3%)	829,841 (10.1%)
Non-Hispanic Other	3,650,745 (8.9%)	837,114 (6.4%)	681,706 (8.3%)
Annual Income			
<$25,000	5,158,320 (12.6%)	3,323,765 (25.5%)	3,168,536 (38.6%)
$25,000 to <$55,000	10,585,563 (25.9%)	3,199,300 (24.5%)	1,753,277 (21.4%)
$55,000+	25,179,717 (61.5%)	6,531,083 (50.0%)	3,285,219 (40.0%)

[1] Mean (SD); n (%)

Fit the model using the subsetted domain design.

```
fit.ex8.1.domain <- svyglm(LBDGLUSI ~ BMXWAIST + smoker + RIDAGEYR +
                               race_eth + income,
                           family=gaussian(), design=design.FST.domain)
```

Finally, use `tbl_regression()` to create a table of regression coefficients. The results are shown in Table 8.5.

```
TABLE <- fit.ex8.1.domain %>%
  tbl_regression(intercept = T,
                 estimate_fun = function(x) style_sigfig(x, digits = 3),
                 pvalue_fun   = function(x) style_pvalue(x, digits = 3),
                 label  = list(BMXWAIST ~ "Waist Circumference (cm)",
                               smoker    ~ "Smoking Status",
                               RIDAGEYR ~ "Age (years)",
                               race_eth ~ "Race/Ethnicity",
                               income    ~ "Annual Income")) %>%
  add_global_p(keep = T, test.statistic = "F"
  # Uncomment to use design df for Type III test p-values
  # (but then they will not match the regression term p-values for
  #  binary predictors)
  #             , error.df = degf(fit.ex8.1$survey.design)
                               ) %>%
  modify_caption("Weighted linear regression results for fasting glucose (mmol/L)
                 (Females age 45y and older)")
```

```
TABLE
```

TABLE 8.5 Weighted linear regression results for fasting glucose (mmol/L) (Females age 45y and older)

Characteristic	Beta	95% CI	p-value
(Intercept)	3.15	1.85, 4.46	0.001
Waist Circumference (cm)	0.031	0.015, 0.047	0.003
Smoking Status			0.731
Never	—	—	
Past	0.118	-0.367, 0.604	0.573
Current	0.037	-0.559, 0.632	0.885
Age (years)	0.008	-0.012, 0.028	0.362
Race/Ethnicity			0.017
Hispanic	—	—	
Non-Hispanic White	-0.480	-0.924, -0.036	0.038
Non-Hispanic Black	-0.391	-1.02, 0.234	0.177
Non-Hispanic Other	0.071	-0.476, 0.618	0.762
Annual Income			0.436
<$25,000	—	—	
$25,000 to <$55,000	-0.124	-0.501, 0.252	0.450
$55,000+	-0.255	-0.706, 0.196	0.216

Conclusion: Among females age 45 years and older, after adjusting for age, gender, race/ethnicity, income, and each other, waist circumference is significantly associated with fasting glucose (B = 0.031; 95% CI = 0.015, 0.047; p = .003), but smoking status is not (p = .731).

8.6 Weighted binary logistic regression

8.6.1 Fitting the model

To carry out a binary logistic regression that incorporates a survey design, use `svyglm()` with `family=quasibinomial()`. This produces the same results as `family=binomial()` but avoids a warning about non-integer numbers of successes. As with `glm()`, `svyglm()` models the probability that the outcome is at the non-reference level, if the outcome is a factor, or the probability that the outcome is 1, if the outcome is numeric with values 0 and 1 (see Section 6.6.2). Although not discussed in this text, survey-weighted ordinal logistic regression can be carried out using `svyolr()`.

Example 8.3: Using data from adult participants in the 2019 NSDUH (`nsduh2019_rmph.RData`), what is the association between lifetime marijuana use (`mj_lifetime`) and age at first use of alcohol (`alc_agefirst`), adjusted for age (`demog_age_cat6`), sex (`demog_sex`), and income (`demog_income`)?

Since the question states "among adult participants", use a domain analysis for this example. Our domain definition is based on one of the variables in the model (`demog_age_cat6`) and excludes individuals at certain levels of that variable. This does not affect any of the analyses below, but does lead to an issue with `tbl_regression()`, as the excluded levels show up in the table and show up multiple times. To avoid this problem, create a new version of this variable with missing values at the excluded levels and use `fct_drop()` to drop the unused levels. This must be done *before* specifying and subsetting the design so that the new variable is available to the design.

```
load("Data/nsduh2019_rmph.RData")

# Check the levels of the age variable
table(nsduh$demog_age_cat6)
```

```
##
## 12-17 18-25 26-34 35-49 50-64   65+
## 13397 14226  8601 11134  4880  3898
```

```
nsduh <- nsduh %>%
  mutate(domain = demog_age_cat6 %in%
           c("18-25", "26-34", "35-49", "50-64", "65+"),
         # So tbl_regression works correctly:
         # Set non-adult age group to missing
         age_new = na_if(demog_age_cat6, "12-17"),
         # Drop unused levels
         age_new = fct_drop(age_new))

# Check derivation
table(nsduh$age_new, nsduh$demog_age_cat6, useNA = "ifany")
```

```
##
##          12-17 18-25 26-34 35-49 50-64   65+
##    18-25     0 14226     0     0     0     0
##    26-34     0     0  8601     0     0     0
##    35-49     0     0     0 11134     0     0
```

```
##    50-64      0     0     0     0  4880     0
##    65+        0     0     0     0     0  3898
##    <NA>   13397     0     0     0     0     0
```

Next, specify the survey design and subset the design.

```
library(survey)
options(survey.lonely.psu = "adjust")
options(survey.adjust.domain.lonely = TRUE)

design.NSDUH <- svydesign(strata=~vestr, id=~verep, weights=~ANALWT_C,
                        nest=TRUE, data=nsduh)

design.NSDUH.adults <- subset(design.NSDUH, domain)
```

Use `svyglm()` with `family=quasibinomial()` to fit the model, and the same `summary()`, `confint()`, and `car::Anova()` calls used for weighted linear regression to view the output.

```
fit.ex8.3 <- svyglm(mj_lifetime ~ alc_agefirst +
                      age_new + demog_sex + demog_income,
                    family = quasibinomial(),
                    design = design.NSDUH.adults)
round(
  cbind(
    summary(fit.ex8.3,
            df.resid=degf(fit.ex8.3$survey.design))$coef,
    confint(fit.ex8.3,
            ddf=degf(fit.ex8.3$survey.design))
  )
, 4)
```

```
##                                  Estimate Std. Error t value Pr(>|t|)   2.5 %
## (Intercept)                        4.8352     0.1433  33.743   0.0000  4.5474
## alc_agefirst                      -0.2427     0.0074 -32.711   0.0000 -0.2576
## age_new26-34                       0.2048     0.0456   4.490   0.0000  0.1132
## age_new35-49                      -0.1694     0.0429  -3.949   0.0002 -0.2556
## age_new50-64                      -0.1030     0.0516  -1.996   0.0514 -0.2067
## age_new65+                        -0.7953     0.0565 -14.078   0.0000 -0.9088
## demog_sexFemale                   -0.0672     0.0310  -2.169   0.0348 -0.1294
## demog_income$20,000 - $49,999     -0.1559     0.0362  -4.302   0.0001 -0.2287
## demog_income$50,000 - $74,999     -0.0804     0.0559  -1.438   0.1567 -0.1928
## demog_income$75,000 or more       -0.1254     0.0430  -2.919   0.0053 -0.2117
##                                   97.5 %
## (Intercept)                       5.1230
## alc_agefirst                     -0.2278
## age_new26-34                      0.2964
## age_new35-49                     -0.0832
## age_new50-64                      0.0007
## age_new65+                       -0.6819
## demog_sexFemale                  -0.0050
## demog_income$20,000 - $49,999 -0.0831
## demog_income$50,000 - $74,999  0.0319
## demog_income$75,000 or more     -0.0391
```

```
car::Anova(fit.ex8.3, type = 3, test.statistic = "F",
           error.df = degf(fit.ex8.3$survey.design))
```

```
## Analysis of Deviance Table (Type III tests)
##
## Response: mj_lifetime
##                Df       F               Pr(>F)
## (Intercept)     1 1138.61 < 0.0000000000000002 ***
## alc_agefirst    1 1070.01 < 0.0000000000000002 ***
## age_new         4   92.48 < 0.0000000000000002 ***
## demog_sex       1    4.71              0.03483 *
## demog_income    3    6.57              0.00079 ***
## Residuals      50
## ---
## Signif. codes:  0 '***' 0.001 '**' 0.01 '*' 0.05 '.' 0.1 ' ' 1
```

Compute the AORs and their 95% confidence intervals.

```
# Compute odds ratios and their 95% confidence intervals
OR.CI <- cbind("AOR" = exp(   coef(fit.ex8.3)),
                       exp(confint(fit.ex8.3,
                           df.resid=degf(fit.ex8.3$survey.design))))[-1,]
round(OR.CI, 3)
```

```
##                                      AOR 2.5 % 97.5 %
## alc_agefirst                       0.784 0.773  0.796
## age_new26-34                       1.227 1.119  1.346
## age_new35-49                       0.844 0.774  0.921
## age_new50-64                       0.902 0.813  1.001
## age_new65+                         0.451 0.403  0.506
## demog_sexFemale                    0.935 0.878  0.995
## demog_income$20,000 - $49,999 0.856 0.795  0.921
## demog_income$50,000 - $74,999 0.923 0.824  1.033
## demog_income$75,000 or more   0.882 0.809  0.962
```

To table these results, use the `tbl_regression()` function in `gtsummary` after adding `test .statistic = "F"` in `add_global_p()`. As mentioned in Section 8.4.1, there is no option in `tbl_regression()` to change the DF to make the confidence intervals and p-values for the individual regression coefficients be based on the design DF. See the warning there for how to resolve this issue.

The results are shown in Table 8.6.

```
library(gtsummary)
t1 <- fit.ex8.3 %>%
  tbl_regression(intercept = T,
                 # Use style_number to round to 2 digits (e.g., 1.27)
                 estimate_fun = function(x) style_number(x, digits = 2),
                 # Use style_sigfig to keep 2 significant digits (e.g., 1.3)
                 # estimate_fun = function(x) style_sigfig(x, digits = 2),
                 label  = list('(Intercept)' ~ "Intercept",
                               alc_agefirst ~ "Age of First Alcohol Use (years)",
                               age_new ~ "Age (years)",
                               demog_sex      ~ "Sex",
                               demog_income   ~ "Income")) %>%
  modify_column_hide(p.value) %>%
  modify_caption("Weighted logistic regression results for
  lifetime marijuana use vs. age at first alcohol use (years)")

t2 <- fit.ex8.3 %>%
```

```
  tbl_regression(intercept = T,
                 exponentiate = T,  # OR = exp(B)
                 # Use style_number to round to 2 digits (e.g., 1.27)
                 estimate_fun = function(x) style_number(x, digits = 2),
                 # Use style_sigfig to keep 2 significant digits (e.g., 1.3)
                 pvalue_fun   = function(x) style_pvalue(x, digits = 3),
                 label  = list('(Intercept)' ~ "Intercept",
                               alc_agefirst ~ "Age of First Alcohol Use (years)",
                               age_new ~ "Age (years)",
                               demog_sex        ~ "Sex",
                               demog_income     ~ "Income")) %>%
  add_global_p(keep = T, test.statistic = "F"
  # Uncomment to use design df for Type III test p-values
  # (but then they will not match the regression term p-values for
  #  binary predictors)
  #              , error.df = degf(fit.ex8.1$survey.design)
                              )
```

```
TABLE <- tbl_merge(
  tbls = list(t1, t2),
  tab_spanner = c("**Adjusted Coefficient**", "**Adjusted Odds Ratio**")
  ) %>%
  modify_footnote(everything() ~ NA, abbreviation = TRUE) %>%
  modify_header(update = list(
    estimate_1 = "**log(AOR)**",
    estimate_2 = "**AOR**"
))
```

```
TABLE
```

TABLE 8.6 Weighted logistic regression results for lifetime marijuana use vs. age at first alcohol use (years)

	Adjusted Coefficient		Adjusted Odds Ratio		
Characteristic	**log(AOR)**	**95% CI**	**AOR**	**95% CI**	**p-value**
Intercept	4.84	4.55, 5.12	125.86	94.24, 168.10	<0.001
Age of First Alcohol Use (years)	-0.24	-0.26, -0.23	0.78	0.77, 0.80	<0.001
Age (years)					<0.001
18-25	—	—	—	—	
26-34	0.20	0.11, 0.30	1.23	1.12, 1.35	<0.001
35-49	-0.17	-0.26, -0.08	0.84	0.77, 0.92	<0.001
50-64	-0.10	-0.21, 0.00	0.90	0.81, 1.00	0.053
65+	-0.80	-0.91, -0.68	0.45	0.40, 0.51	<0.001
Sex					0.036
Male	—	—	—	—	
Female	-0.07	-0.13, 0.00	0.94	0.88, 1.00	0.036
Income					<0.001
Less than $20,000	—	—	—	—	
$20,000 - $49,999	-0.16	-0.23, -0.08	0.86	0.80, 0.92	<0.001
$50,000 - $74,999	-0.08	-0.19, 0.03	0.92	0.82, 1.03	0.158
$75,000 or more	-0.13	-0.21, -0.04	0.88	0.81, 0.96	0.006

Conclusion: After adjusting for age, sex, and income, lifetime marijuana use is significantly negatively associated with age at first use of alcohol (AOR = 0.78; 95% CI = 0.77, 0.80; p <.001). Those who start using alcohol one year later have 22% lower odds of lifetime marijuana use.

8.6.2 Prediction

Use `svycontrast_design_df()` to compute estimated log-odds from a `svyglm()` object fit using `family = quasibinomial()` and use `ilogit()` to convert the log-odds to a probability.

Example 8.3 (continued): What is the estimated probability (and 95% CI) of lifetime marijuana use for someone who first used alcohol at age 13 years, is currently age 30 years, male, and has an income of < $20,000?

```
# Always include the intercept for estimation
# Specify a 1 for the intercept, a # for each continuous predictor
# and a 1 for each non-reference level of a categorical variable.
# If a predictor is at its reference level, specify a 0 or exclude it.
ilogit(
  svycontrast_design_df(fit.ex8.3,
                        c("(Intercept)"  = 1,
                          "alc_agefirst" = 13,
                          "age_new26-34" = 1))
)
```

```
##       est  lower  upper
## 1 0.8682 0.8554 0.8799
```

Conclusion: The predicted probability of lifetime marijuana use among such individuals is 86.8% (95% CI = 85.5%, 88.0%).

8.6.3 Interactions

Including an interaction in the model and estimating the effect of one variable at levels of another proceeds as described for weighted linear regression above with the addition of exponentiating to obtain AORs.

Example 8.3 (continued): After adjusting for age and income, does the association between lifetime marijuana use and age at first use of alcohol differ by sex? What are the AORs for age at first alcohol use for males and females?

```
fit.ex8.3.int <- svyglm(mj_lifetime ~ alc_agefirst +
                          age_new + demog_sex + demog_income +
                          alc_agefirst:demog_sex,
                        family =quasibinomial(),
                        design = design.NSDUH.adults)


round(
  cbind(
    summary(fit.ex8.3.int,
            df.resid=degf(fit.ex8.3.int$survey.design))$coef,
    confint(fit.ex8.3.int,
            ddf=degf(fit.ex8.3.int$survey.design))
  )
, 4)
```

```
##                                       Estimate Std. Error t value Pr(>|t|)    2.5 %
## (Intercept)                             4.4769     0.1719  26.048   0.0000   4.1317
## alc_agefirst                           -0.2219     0.0096 -23.228   0.0000  -0.2411
## age_new26-34                            0.2063     0.0457   4.511   0.0000   0.1145
## age_new35-49                           -0.1658     0.0429  -3.868   0.0003  -0.2519
## age_new50-64                           -0.0941     0.0512  -1.838   0.0720  -0.1970
## age_new65+                             -0.7821     0.0565 -13.843   0.0000  -0.8955
## demog_sexFemale                         0.6840     0.3188   2.145   0.0368   0.0436
## demog_income$20,000 - $49,999          -0.1586     0.0359  -4.419   0.0001  -0.2308
## demog_income$50,000 - $74,999          -0.0846     0.0564  -1.500   0.1399  -0.1978
## demog_income$75,000 or more            -0.1303     0.0431  -3.021   0.0040  -0.2170
## alc_agefirst:demog_sexFemale           -0.0432     0.0181  -2.380   0.0212  -0.0796
##                                        97.5 %
## (Intercept)                            4.8222
## alc_agefirst                          -0.2027
## age_new26-34                           0.2982
## age_new35-49                          -0.0797
## age_new50-64                           0.0087
## age_new65+                            -0.6686
## demog_sexFemale                        1.3243
## demog_income$20,000 - $49,999         -0.0865
## demog_income$50,000 - $74,999          0.0287
## demog_income$75,000 or more           -0.0437
## alc_agefirst:demog_sexFemale          -0.0067
```

```
rbind(
  "alc_agefirst @ Male" = svycontrast_design_df(fit.ex8.3.int,
      c("alc_agefirst"                 = 1),
      EXP = T),        # Exponentiate to get OR
  "alc_agefirst @ Female" = svycontrast_design_df(fit.ex8.3.int,
      c("alc_agefirst"                 = 1,
        "alc_agefirst:demog_sexFemale" = 1),
      EXP = T)
)
```

```
##                          est  lower  upper
## alc_agefirst @ Male   0.8010 0.7858 0.8165
## alc_agefirst @ Female 0.7671 0.7461 0.7887
```

Conclusion: After adjusting for age and income, the association between lifetime marijuana use and age at first use of alcohol significantly differs between sexes (p = .021, the p-value for the interaction term `alc_agefirst:demog_sexFemale`). For both males and females, those who start using alcohol at an older age have lower odds of lifetime marijuana use, but this effect is stronger (AOR further from 1) in females (AOR = 0.77) than in males (AOR = 0.80).

8.7 Weighted survival analysis

To incorporate a complex survey design into a survival analysis, use

- `svykm()` to compute a weighted Kaplan-Meier estimate of the survival function,

- `svylogrank()` to carry out a weighted log-rank test to compare survival curves between groups, and
- `svycoxph()` to carry out weighted Cox regression.

8.7.1 Weighted Kaplan-Meier estimate of the survival function

Example 8.4: Using the NHANES 2017-2018 data (`nhanes1718_rmph.Rdata`), estimate the weighted Kaplan-Meier estimate of the survival function for time to first diagnosis of asthma. The time variable is age when first had asthma (`MCQ025`) and the event indicator is "Ever been told you have asthma" (`MCQ010`).

We need an event indicator that is 1 for those who have been told they have asthma and 0 otherwise, and a time variable that is the age when told one first had asthma for those who have been told and current age otherwise (censored asthma age). Format the event indicator variable and then fill in current age for those without asthma.

NOTE: This step is needed for the Kaplan-Meier estimator, log-rank test, and Cox regression.

```
# Format the event indicator
levels(nhanes.mod$MCQ010)
```

```
## [1] "Yes" "No"
```

```
nhanes.mod$asthma <- as.numeric(nhanes.mod$MCQ010 == "Yes")
table(nhanes.mod$MCQ010, nhanes.mod$asthma, useNA = "ifany")
```

```
##
##           0    1 <NA>
##   Yes     0 1325    0
##   No   7563    0    0
##   <NA>    0    0  366
```

```
# Create a copy of the age variable
nhanes.mod$asthma_age <- nhanes.mod$MCQ025

# For those with asthma = 0, set asthma age to current age
SUB <- complete.cases(nhanes.mod$asthma) & nhanes.mod$asthma == 0
nhanes.mod$asthma_age[SUB] <- nhanes.mod$RIDAGEYR[SUB]
```

```
# Check derivation
# For !SUB, the difference between these should be 0 or NA
# (asthma_age is MCQ025)
summary(nhanes.mod$MCQ025[    !SUB] -
        nhanes.mod$asthma_age[!SUB])
# For SUB, this should be all missing
summary(nhanes.mod$MCQ025[SUB])
# For !SUB, the difference between these should be 0
# (asthma_age is RIDAGEYR)
summary(nhanes.mod$RIDAGEYR[  SUB] -
        nhanes.mod$asthma_age[SUB])
# (results not shown)
```

Next, re-create the design object so that these new variables are available for weighted analyses. The asthma variables were part of the NHANES interview, and no other variables are included in this analysis, so use the interview design weights.

```
library(survey)
options(survey.lonely.psu = "adjust")
options(survey.adjust.domain.lonely = TRUE)

design.INT <- svydesign(strata=~SDMVSTRA, id=~SDMVPSU, weights=~WTINT2YR,
                        nest=TRUE, data=nhanes.mod)
```

Use `svykm()` and `plot()` to estimate and plot the survival function (Figure 8.6).

```
km.ex8.4 <- svykm(Surv(asthma_age, asthma) ~ 1,
                  design = design.INT)
```

```
plot(km.ex8.4, ylim = c(0.65, 1),
     xlab = "Age (years)",
     ylab = "Proportion without asthma")
```

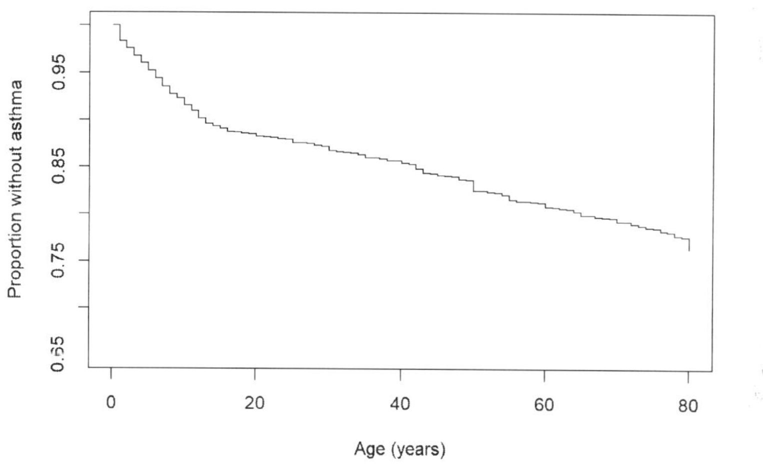

FIGURE 8.6 Weighted Kaplan-Meier survival function

Conclusion: We see that the proportion without asthma drops rapidly before age 20 years, indicating the greatest risk of an asthma diagnosis is in childhood. However, as the survival curve continues to decrease over the lifespan, adults who have not yet been diagnosed with asthma are still at risk.

8.7.2 Weighted log-rank test for comparing groups

To compare groups, use `svykm` with a factor variable on the right-hand side of the formula, and then `svylogrank` to carry out the log-rank test.

NOTE: If you have not already done so, format the event indicator and time variable (see Section 8.7.1).

Example 8.4 (continued): Using the NHANES 2017-2018 data (`nhanes1718_rmph.Rdata`), use a weighted log-rank test to compare time to first diagnosis of asthma between race/ethnicity groups. The race/ethnicity variable was also in the interview, so we can continue using the design object that includes the interview weights.

```
km.ex8.4b <- svykm(Surv(asthma_age, asthma) ~ race_eth,
                   design = design.INT)
```

Figure 8.7 illustrates the comparison of weighted survival functions between race/ethnicity groups.

```
# When plotting svykm objects, the lty option must be in a list.
# When including lty, ylim must also be in a list
# (without lty, ylim works without being in a list)
plot(km.ex8.4b,
     xlab = "Age (years)",
     ylab = "Proportion without asthma",
     pars = list(lty=1:4), ylim = list(c(0.65, 1)))
legend(45, 1, levels(nhanes.mod$race_eth)[c(2,4,1,3)],
        lty = (1:4)[c(2,4,1,3)])
```

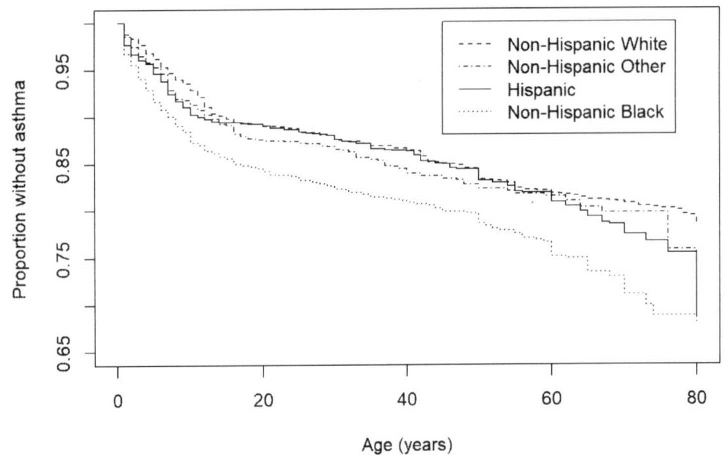

FIGURE 8.7 Weighted Kaplan-Meier survival function by race/ethnicity

```
# WARNING: svylogrank() will fail if there are missing values.
# Create a domain with only complete cases
nhanes.mod <- nhanes.mod %>%
  mutate(domain_asthma = complete.cases(asthma_age) & complete.cases(asthma) &
                         complete.cases(race_eth))

# Re-create the design to include the new domain variable
design.INT <- svydesign(strata=~SDMVSTRA, id=~SDMVPSU, weights=~WTINT2YR,
                        nest=TRUE, data=nhanes.mod)
```

```
# Subset the design to do a domain analysis
design_asthma <- subset(design.INT, domain_asthma)

# Log-rank test
svylogrank(Surv(asthma_age, asthma) ~ race_eth, design = design_asthma)[[2]]
```

```
##       Chisq           p
## 29.330856009  0.000001908
```

Conclusion: Time to first diagnosis of asthma differs significantly between race/ethnicity groups ($p < .001$). As shown in Figure 8.7, non-Hispanic Black individuals have the greatest proportion of individuals with asthma at any given age, starting from childhood and persisting across the lifespan.

8.7.3 Weighted Cox regression

8.7.3.1 Fitting the model

Use `svycoxph()` to fit a weighted Cox proportional hazards regression. The current functions do not have an option to change the DF from the default. An exception is the function `svycontrast_design_df()` (in `Functions_rmph.R`) which uses the design DF by default, with an option to use the residual DF.

NOTE: If you have not already done so, format the event indicator and time variable (see Section 8.7.1).

Example 8.5: Using the NHANES 2017-2018 data (`nhanes1718_rmph.Rdata`), compare time to first diagnosis of asthma between race/ethnicity groups, adjusted for gender (`RIAGENDR`), using a Cox proportional hazards regression.

The results are shown in Table 8.7.

```
cox.ex8.5 <- svycoxph(Surv(asthma_age, asthma) ~ race_eth + RIAGENDR,
                      design = design.INT)
```

```
car::Anova(cox.ex8.5, type = 3)
summary(cox.ex8.5)
cbind("AHR" = exp(summary(cox.ex8.5)$coef[, "coef"]),
              exp(confint(cox.ex8.5)),
      "p"   = summary(cox.ex8.5)$coef[, "Pr(>|z|)"])
# (results not shown)
```

```
library(gtsummary)
# Table of regression coefficients
t1 <- cox.ex8.5 %>%
  tbl_regression(estimate_fun = function(x) style_number(x, digits = 3),
                 label   = list(race_eth ~ "Race/Ethnicity",
                                RIAGENDR ~ "Gender")) %>%
  modify_column_hide(p.value) %>%
  modify_caption("Weighted Cox regression results for time to asthma
                 vs. race/ethnicity, adjusted for gender")
```

```
## Stratified 1 - level Cluster Sampling design (with replacement)
## With (30) clusters.
## svydesign(strata = ~SDMVSTRA, id = ~SDMVPSU, weights = ~WTINT2YR,
##     nest = TRUE, data = nhanes.mod)
```

```
# Table of hazard ratios
t2 <- cox.ex8.5 %>%
  tbl_regression(exponentiate = T,  # HR = exp(B)
                 estimate_fun = function(x) style_number(x, digits = 3),
                 pvalue_fun   = function(x) style_pvalue(x, digits = 3),
                 label  = list(race_eth ~ "Race/Ethnicity",
                               RIAGENDR ~ "Gender")) %>%
  add_global_p(keep = T)
```

```
## Stratified 1 - level Cluster Sampling design (with replacement)
## With (30) clusters.
## svydesign(strata = ~SDMVSTRA, id = ~SDMVPSU, weights = ~WTINT2YR,
##     nest = TRUE, data = nhanes.mod)
```

```
TABLE <- tbl_merge(
  tbls = list(t1, t2),
  tab_spanner = c("**Regression Coefficient**", "**Adjusted Hazard Ratio**")
  ) %>%
  modify_footnote(everything() ~ NA, abbreviation = TRUE) %>%
  modify_header(update = list(
    estimate_1 = "**log(AHR)**",
    estimate_2 = "**AHR**"
  ))
```

```
TABLE
```

TABLE 8.7 Weighted Cox regression results for time to asthma vs. race/ethnicity, adjusted for gender

	Regression Coefficient		**Adjusted Hazard Ratio**		
Characteristic	**log(AHR)**	**95% CI**	**AHR**	**95% CI**	**p-value**
Race/Ethnicity					<0.001
Hispanic	—	—	—	—	
Non-Hispanic White	-0.083	-0.316, 0.151	0.921	0.729, 1.162	0.487
Non-Hispanic Black	0.314	0.071, 0.557	1.369	1.074, 1.745	0.011
Non-Hispanic Other	0.028	-0.391, 0.446	1.028	0.677, 1.562	0.897
Gender					0.117
Male	—	—	—	—	
Female	0.138	-0.035, 0.310	1.148	0.966, 1.364	0.117

Conclusion: After adjusting for gender, time to first diagnosis of asthma is significantly different between race/ethnicity groups ($p < .001$). Non-Hispanic Black participants have 36.9% greater hazard of asthma diagnosis than Hispanic participants. Note that the AHR for non-Hispanic White participants is < 1, so the disparity between non-Hispanic Black and non-Hispanic White is even greater than 36.9%. To estimate it, divide the two AHRs,

1.369/0.921 = 1.486 to see that non-Hispanic Black participants have 48.6% greater hazard of an asthma diagnosis than non-Hispanic White participants. If you also want a 95% confidence interval and a p-value to go along with this estimate, then re-level the race/ethnicity variable to make non-Hispanic White the reference level, refit the model, and then examine the effect for non-Hispanic Black, which would then be a comparison with non-Hispanic White.

8.7.3.2 Exclude cases with zero weights

Unlike the other functions we have used in this chapter, `svycoxph()` returns an error if there are any cases with 0 weight. To avoid this, use a domain analysis to exclude these cases. If you were already using a domain analysis, modify that domain to also exclude these cases. For example, in the NHANES dataset, there are some cases with exam weights of 0 (those not in the examination subsample).

```
design.MEC <- svydesign(strata=~SDMVSTRA, id=~SDMVPSU, weights=~WTMEC2YR,
                        nest=TRUE, data=nhanes.mod)

design.MEC.pos.weights <- subset(design.MEC, WTMEC2YR > 0)
```

8.7.3.3 Interactions

Including an interaction in the model and estimating effects for one variable at levels of another proceeds as described for weighted binary logistic regression above.

Example 8.5 (continued): Does the effect of body mass index (`BMXBMI`) on time to asthma differ between race/ethnicity groups, adjusted for gender? Estimate the BMI effect at each level of race/ethnicity.

Since BMI was only collected in the examination subsample, we must use the examination weights when specifying the sample design (and use a domain analysis that excludes those with zero weights; see Section 8.7.3.2).

```
cox.ex8.5.int <- svycoxph(Surv(asthma_age, asthma) ~ race_eth + RIAGENDR +
                          BMXBMI + race_eth:BMXBMI,
                          design = design.MEC.pos.weights)
car::Anova(cox.ex8.5.int, type = 3)
```

```
## Analysis of Deviance Table (Type III tests)
##
## Response: Surv(asthma_age, asthma)
##                 Df Chisq Pr(>Chisq)
## race_eth         3  6.73      0.081 .
## RIAGENDR         1  3.55      0.060 .
## BMXBMI           1  1.53      0.216
## race_eth:BMXBMI  3  5.09      0.165
## ---
## Signif. codes:  0 '***' 0.001 '**' 0.01 '*' 0.05 '.' 0.1 ' ' 1
```

```
summary(cox.ex8.5.int)
# (results not shown)
```

```
rbind(
  "BMI @ Hispanic" = svycontrast_design_df(cox.ex8.5.int,
    c("BMXBMI" = 1),
    EXP=T),
  "BMI @ NH White" = svycontrast_design_df(cox.ex8.5.int,
    c("BMXBMI"                             = 1,
      "race_ethNon-Hispanic White:BMXBMI" = 1),
    EXP=T),
  "BMI @ NH Black" = svycontrast_design_df(cox.ex8.5.int,
    c("BMXBMI"                             = 1,
      "race_ethNon-Hispanic Black:BMXBMI" = 1),
    EXP=T),
  "BMI @ NH Other" = svycontrast_design_df(cox.ex8.5.int,
    c("BMXBMI"                             = 1,
      "race_ethNon-Hispanic Other:BMXBMI" = 1),
    EXP=T)
)
```

```
##                   est  lower upper
## BMI @ Hispanic 0.9823 0.9525 1.013
## BMI @ NH White 1.0047 0.9808 1.029
## BMI @ NH Black 0.9945 0.9785 1.011
## BMI @ NH Other 1.0211 0.9962 1.047
```

Conclusion: After adjusting for gender, the effect of body mass index on time to asthma does not significantly differ between race/ethnicity groups (p = .165).

8.8 Likelihood ratio test vs. Wald test

As with previous chapters, Wald tests for p-values were used in this chapter. However, likelihood ratio (LR) tests are, in general, more powerful. True LR tests are not possible with `svyglm()` objects since they were not fit using maximum likelihood (Lumley, 2010). However, the function `regTermTest()` can be used to carry out a "working" LR test for weighted linear, logistic, or Cox regression models (the Rao-Scott LR test) (Rao and Scott, 1984; Lumley and Scott, 2013, 2014) to compare any two nested models, similar to `anova()` which we used in previous chapters.

`regTermTest()` can therefore obtain an overall Type 3 Wald or working LR test for a categorical predictor with more than two levels. To get a test for a single level of a categorical predictor, first create indicator variables for the levels of that predictor as described in Section 6.18.

Example 8.1 (continued): Use a working LR test to test the overall significance of race/ethnicity in the weighted adjusted linear regression model for fasting glucose. For comparison, also compute the Wald test.

```
# Model fit previously
fit.ex8.1 <- svyglm(LBDGLUSI ~ BMXWAIST + smoker + RIDAGEYR +
                      RIAGENDR + race_eth + income,
                 family=gaussian(), design=design.FST.nomiss)
```

```
# Working LR test for race_eth
regTermTest(fit.ex8.1,
            test.terms = ~ race_eth,
            df = degf(fit.ex8.1$survey.design),
            method = "LRT")
```

```
## Working (Rao-Scott+F) LRT for race_eth
##  in svyglm(formula = LBDGLUSI ~ BMXWAIST + smoker + RIDAGEYR + RIAGENDR +
##     race_eth + income, design = design.FST.nomiss, family = gaussian())
## Working 2logLR =  11.29 p= 0.042
## (scale factors:  1.8 0.93 0.31 );  denominator df= 15
```

```
# Wald test for race_eth
regTermTest(fit.ex8.1,
            test.terms = ~ race_eth,
            df = degf(fit.ex8.1$survey.design),
            method = "Wald")
```

```
## Wald test for race_eth
##  in svyglm(formula = LBDGLUSI ~ BMXWAIST + smoker + RIDAGEYR + RIAGENDR +
##     race_eth + income, design = design.FST.nomiss, family = gaussian())
## F =  3.003  on  3  and  15  df: p= 0.064
```

Conclusion: Based on the likelihood ratio test, race/ethnicity is significantly associated with fasting glucose, after adjusting for the other variables in the model (p = .042). As previously mentioned, LRTs are generally more powerful than Wald tests, which means lower p-values. In this example, that is the case, with the Wald test p-value being .064.

8.9 Summary of special cases

There were a number of places in this chapter where there was a special case regarding how to handle a situation. The list below serves as a reminder.

- Replace `NA` weights with 0 in a data management step before specifying the survey design.
- Use `subset()` to create a design for a domain (subgroup) analysis. This was needed for the following situations:
  - Subsetting on complete cases when creating a Table 1 to correspond to your regression analysis. But if you are not creating a Table 1, there is no need to use a complete case analysis.
  - Filtering out zero weights to avoid an error in some `survey` functions (e.g., `svycoxph()`).
  - Subgroup analyses.
- If you add/remove/change any variables, re-specify the design (and subset, if any).
- `tbl_regression()` does not have an option to change the DF for other than the Type III tests. If you want to use the design DF throughout, create the table, export, and replace CIs and p-values with values from functions that used the design DF.

8.10 Exercises

1. True or false? In a simple random sample, every unit in the population has an equal probability of selection.

2. True or false? In multistage sampling, you first non-randomly split the population into groups, and then randomly sample within each group.

3. What is the difference between a sample and a census?

4. What kind of sampling first non-randomly splits the population into groups before simple random sampling withing groups?

5. In multistage sampling, what are the units called that are sampled first?

6. What do we call the set of numbers that are related to how many population units are represented by each sampled unit?

For Exercises 7 to 15, use the 2019 NSDUH data (`nsduh2019_rmph.RData`, see Appendix A.5).

7. Estimate the population mean, standard deviation, and 95% confidence interval for the mean for the age of first cigarette use (`cig_agefirst`), and plot a weighted histogram.

8. Estimate the population totals and proportions for the levels of total family income (`demog_income`) and plot a barplot displaying the weighted proportions.

9. Estimate the population totals and proportions for the levels of total family income (`demog_income`) at each level of employment status (`demog_employ`).

10. Create a Table 1 of weighted descriptive statistics (mean and SD for continuous variables, total and proportion for categorical variables) for the following variables, using a `subset()` of the design so the statistics are based on a complete case analysis. The variables to include are: sex (`demog_sex`), health status (`demog_health`), education (`demog_educ_cat4`), age of first alcohol use (`alc_agefirst`), and days of alcohol use in the past year (`alc_pastly`). The table should contain an "Overall" column and columns by sex.

11. Visualize the estimated population relationship between the predictor age at first alcohol use (`alc_agefirst`) and the outcome age at first heroin use (`her_agefirst`), along with the estimated population regression line and smoother.

12. Is the predictor age at first alcohol use (`alc_agefirst`) associated with the outcome age at first heroin use (`her_agefirst`), after adjusting for sex (`demog_sex`) and education (`demog_educ_cat4`)? Estimate the population association, 95% confidence interval, and p-value, and interpret the results.

13. Using the model from the previous exercise, what is the estimated age of first heroin use (and 95% confidence interval) for someone who first used alcohol at age 15, is male, and has a college degree?

14. After adjusting for sex (`demog_sex`) and education (`demog_educ_cat4`), does the association between the predictor age at first cigarette use (`cig_agefirst`) and the outcome age at first heroin use (`her_agefirst`) differ between those who live in Nonmetro, Small Metro, and Large Metro areas (`demog_urban`)? Answer the question, along with the appropriate p-value, and estimate the population association and its 95% confidence interval at each level of urbanicity. At which levels of urbanicity is the association significant?

15. Repeat Exercise 12 within the subgroup females age 18-34 years (`demog_age_cat`). Regardless of statistical significance, interpret the regression coefficient. How do these results compare to the results from Exercise 12? **Hint:** You will have to remove `demog_sex` from the model since it no longer has any variation.

For Exercises 16 to 21, use the NHANES 2017-2018 data (`nhanes1718_rmph.Rdata`, see Appendix A.1).

16. Estimate the population proportion of individuals not covered by health insurance (`HIQ011 = "No"`). Use the interview weights (`WTINT2YR`) when specifying the design.

17. Create a complete case indicator for the variables covered by health insurance (`HIQ011`), education (`DMDEDUC2`), and age (`RIDAGEYR`). For the subpopulation with non-missing values for these variables, estimate the proportion of individuals not covered by health insurance (`HIQ011 = "No"`). Use the interview weights (`WTINT2YR`) when specifying the design.

18. Using the design you specified in the previous exercise, is there an association between age and not being covered by health insurance after adjusting for education? Answer the question, including the adjusted odds ratio, 95% confidence interval, and p-value. Also, interpret the association in terms of the change in odds associated with a 5-year age difference.

19. Using the model from the previous exercise, what is the estimated population proportion with no health insurance among those with a HS/GED education who are 21, 30, and 45 years of age?

20. Plot the weighted Kaplan-Meier estimate of the survival function for age when first tried marijuana at each level of education (`DMDEDUC2`). Are the survival curves significantly different? The time variable is age when first tried marijuana (DUQ210) and the event indicator is "Ever used marijuana or hashish" (`DUQ200`). Use the interview weights (`WTINT2YR`) when specifying the design. Just like for the asthma example in this chapter, first format the event indicator and fill in current age for the time variable for those who did not experience the event.

21. Using the design you specified in the previous exercise, estimate the hazard ratios for marijuana use comparing each level of education to those with less than a 9th grade education.

9

Multiple Imputation of Missing Data

In this chapter, you will learn:

- Missing data concepts and terminology;
- How to create multiply imputed datasets; and
- How to carry out the following analyses accounting for missing data via multiple imputation:
  - Descriptive statistics;
  - Linear regression;
  - Binary logistic regression; and
  - Cox proportional hazards regression.

This chapter assumes that you have read the chapters on these statistical analysis methods.

To use the code in this chapter, first load the `tidyverse`, `mice` (van Buuren and Groothuis-Oudshoorn, 2011, 2023), and `miceadds` (Robitzsch et al., 2023) libraries along with the file `Functions_rmph.R` (downloadable from RMPH Resources[1]).

```
library(tidyverse)
library(mice)
library(miceadds)
source("Functions_rmph.R")
```

9.1 Introduction

Missing data are common in public health research. Whatever method is used to select individuals for a survey or other study, there may be **unit non-response** – individuals may decline to participate, or the researcher may not succeed in contacting some individuals. Additionally, among those who do participate, some may drop out of the study early resulting in **loss to follow-up**. Among individuals who do participate, there may be **item non-response** – individuals may leave some survey items unanswered, whether due to refusal or lack of knowledge. Similarly, when physical measurements are involved, some individuals may decline to be measured.

For some survey designs, statistical methods can be used to account for unit non-response or loss to follow-up via survey weights. In any case, it is helpful for any research report to include a statement about the response rate (the proportion of individuals selected who participated in the research overall and at each study visit, if longitudinal), characteristics

[1] https://github.com/rwnahhas/RMPH_Resources

of the non-responders (if known), and how the respondent characteristics differ between those with complete responses and those with item non-response (see Section 9.4.5).

This chapter addresses item non-response, which is referred to from this point forward as **missing data**. The default approach for handling missing data in regression methods is **listwise deletion** – remove any case (individual) from the dataset that has a missing value for the outcome or for any predictor. This **complete case analysis** approach is the method used in previous chapters. Listwise deletion can be done either explicitly by the data analyst or implicitly by the software. In the previous chapters of this text, cases with missing data were explicitly removed prior to computing descriptive statistics so that these statistics would be based on the same sample as the regression.

It would be preferable, however, to use a method that retains *all* cases in the dataset, making use of their non-missing information, and to do so in a way that leads to **consistent** estimates of regression coefficients that have the correct estimated precision. "Consistency" is similar to "unbiased", but not exactly. A consistent estimator is one that gets closer and closer to being unbiased in larger and larger sample sizes.

This chapter introduces a method that is almost always better than a complete case analysis. First introduced by Donald Rubin (Rubin, 1987, 1996), **multiple imputation** has become widely used and recognized as an effective method for handling missing data that is applicable to a wide variety of analyses.

Imputation refers to replacing a missing value with a guess of its true value. When this is done for every missing value, an incomplete dataset becomes a complete dataset. **Mean imputation** replaces a missing value for a variable with the mean of that variable's non-missing values. This, however, ignores relationships between variables and, as a result, distorts those relationships. **Conditional mean imputation** improves upon this by replacing a missing value for a variable with its predicted value using a regression model with that variable as the outcome and the other variables in the dataset as predictors.

Conditional mean imputation is intuitively appealing – it makes sense to replace a missing value with the best guess given the non-missing information. Unfortunately, doing so actually results in a complete dataset that does not accurately reflect reality because the *variation* in the data is understated. An improvement can be made by **adding randomness** to the guess – instead of replacing the missing value with the best guess, replace it with the best guess plus or minus some random noise. The correct way to do this is to take a random draw from the proper distribution, one that uses the relationships between all the observed (non-missing) data to predict the missing value, taking into account all sources of uncertainty in this prediction.

Such a **single imputation** method, however, results in treating imputed values as if they were known and overstating the precision of regression results (too small standard errors, too narrow confidence intervals, and too small p-values). This limitation is overcome by using **multiple imputation** (MI) – whatever your single imputation method, do it multiple times. Multiple imputation consists of taking, for each missing value, M random draws from the proper distribution. The result is M complete datasets.

Next, given the M complete datasets, fit the regression model of interest M times, once with each dataset, to get M sets of estimated regression coefficients and their estimated variances. Finally, **Rubin's rules** are used to (1) compute the final estimated regression coefficients by averaging the estimates over the M sets and (2) compute the final estimated variation of the final estimates by averaging the estimated variances over the M sets and adding a bit more based on how variable the M estimates were between imputations.

The second Rubin's rule corrects the overstatement of precision that results from single imputation. Consider a dataset with N observations, n of which have no missing data. A complete case analysis (via listwise deletion) would have a sample size of n, and computation of standard errors would be based on that sample size. Since that ignores the non-missing information in the excluded cases, these standard errors will, in general, be larger than if we had made use of this information. Had we used imputation to complete the data, but then pretended as if there never were any missing data (as in single imputation), standard errors would be based on a sample size of N, and would be too small. MI plus Rubin's rules lead us to the correct compromise between the two extremes of excluding the missing data altogether and imputing values for it but pretending it was never missing.

The modest goal of this chapter is to present the basics of MI and how to apply them using the `mice` package (van Buuren and Groothuis-Oudshoorn, 2011, 2023), which includes the `mice()` function to create imputed datasets and additional functions to visualize imputed values, carry out analyses on each imputed dataset, and pool results over imputations. Some additional functions will be used from the `miceadds` package (Robitzsch et al., 2023), as well.

While the functions in `mice` have reasonable default settings, there are many nuances that go into doing MI well. This chapter covers the basics but is simply an introduction. After becoming comfortable with the basics here, you are highly encouraged to learn more and/or seek help from an experienced statistician before carrying out analyses utilizing MI. Also, see Flexible Imputation of Missing Data[2] for a thorough and accessible treatment of MI (van Buuren, 2018). An excellent reference for the topic of missing data in general is Little and Rubin (2019). For an overview of R methods for handling missing data, see CRAN Task View: Missing Data[3] (accessed March 7, 2023). Also, in this chapter the examples use the NHANES and NSDUH teaching datasets. These examples are only for illustrating the use of multiple imputation to handle missing data. Multiple imputation of missing complex survey data is more complicated; for example, see Liu et al. (2016) and Quartagno et al. (2020).

9.2 Missing data mechanism

Missing data mechanisms are typically classified as one of the following (Rubin, 1976):

- MCAR: Missing completely at random,
- MAR: Missing at random, or
- MNAR: Missing not at random.

Missing data are **MCAR** if the probability of missingness is independent of the data. In other words, the data are MCAR if the reason for missing values in the outcome or predictors has nothing to do with the data values themselves, whether observed or missing. Missing data are **MAR** if the probability of missingness is independent of the missing values given the observed data. In other words, under MAR how likely a value is to be missing can be estimated based on the non-missing data. Missing data are **MNAR** if, even given all the observed information, the probability of missingness depends on the unobserved missing values themselves.

[2] https://stefvanbuuren.name/fimd/

[3] https://cran.r-project.org/web/views/MissingData.html

To illustrate the difference between these three missing data mechanisms, consider a dataset with the following three variables: disease status, level of exposure, and age. Suppose, for some individuals, exposure is missing (they do have an exposure value, but we do not know what it is). What do each of the three missing data mechanisms imply in this setting?

- **MCAR:** Any two individuals, regardless of their values of disease status, level of exposure, and age, have the same probability of having a missing value for exposure.
- **MAR:** Any two individuals with the same disease status and age have the same chance of having a missing exposure value, regardless of how large or small their actual level of exposure.
- **MNAR:** Even among individuals with the same disease status and age, the chance of having a missing exposure value depends on the level of exposure.

MCAR is generally unrealistic. Individuals who skip a question or refuse to be measured may be very different than those who comply, and they may have systematically different outcome and/or predictor values. MAR is at least somewhat plausible – it allows for systematic differences, as long as those differences are predictable based on observed information. Under MAR (and another condition called ignorability that is beyond the scope of this text), MI leads to consistent estimates with correct standard errors.

MNAR is often plausible, as well, but is more complicated to handle since the distinction between MAR and MNAR depends on unknown information. As a result, a common approach is to assume the data are MAR and use MI to handle the missing data. The plausibility of MAR is improved by including in the imputation model any variable that could be related to the chance of missingness. Evaluating the potential impact of a violation of the MAR assumption (in other words, MNAR) involves using advanced methods to posit a missing data model that depends on the unknown information. This model, by necessity, requires strong assumptions, so MNAR analyses typically include a sensitivity analysis in which the assumptions are varied, resulting in a range of possible conclusions (van Buuren, 2018). For example, see `mice`: An approach to sensitivity analysis[4] (accessed April 4, 2022).

When the data are MCAR, a complete case analysis will yield unbiased estimates, although MI will be more efficient in that the effective sample size will increase since no cases need to be discarded. Under MAR and MNAR, however, a complete case analysis will typically yield biased estimates. MI, however, can provide consistent estimates under MAR (assuming the imputation model does not leave out important variables and is correctly specified), but not under MNAR. These differences are illustrated in Figure 9.1, which is explained in the following.

In Chapter 6, our 2019 NSDUH teaching dataset was used to estimate the adjusted odds ratio (AOR) assessing the association between age of first alcohol use and lifetime marijuana use, adjusted for age, sex, and income. Figure 9.1 displays the results of the following simulation steps:

- **Create a "no missing data" dataset:** Remove all individuals with a missing age of first alcohol use. These individuals never used alcohol, so we do not want to impute ages for them. There were no other missing values in the dataset for the variables of interest.
- **Create a MCAR dataset:** Randomly set about 5% of each variable's values to missing, independent of any of the data values.
- **Create a MAR dataset:** Randomly set each variable's values to missing with a probability of missingness that depends on the values of all the other variables, but not that variable.

[4]https://www.gerkovink.com/miceVignettes/Sensitivity_analysis/Sensitivity_analysis.html

- **Create a MNAR dataset:** Randomly set each variable's values to missing with a probability of missingness that depends on the values of all the variables, including that variable.
- For each of the three datasets that have missing values, estimate the regression coefficient for each term in the model using a complete case analysis and using MI. Within each of these six analyses, average over 25 runs of the simulation, and then exponentiate the average to obtain the estimated AOR for each term.
- Plot the AORs for each missing data analysis vs. the AORs from the "no missing data" dataset analysis.

The top left panel in Figure 9.1 illustrates that if the missing data are MCAR a complete case analysis results in unbiased estimates (this is a simulation with 25 iterations, so the points have some variation about the 45-degree line of equivalence just due to random noise). MI is consistent under MCAR, as shown in the top right panel, with the benefit of improved precision (not shown). The middle row illustrates that under MAR a complete case analysis yields biased estimates (the points are not on the line), but MI yields consistent estimates. The bottom row illustrates that, under MNAR, while MI reduces the bias found in a complete case analysis, it does not eliminate it.

9.3 The imputation model

MI works by drawing values randomly from a distribution that uses the relationships between all the variables, estimated from the observed values, to predict the values for cases with missing data, taking into account all sources of uncertainty in this prediction. `mice` stands for **multivariate imputation by chained equations** (van Buuren and Groothuis-Oudshoorn, 2011). The `mice` algorithm starts by predicting one variable's missing values using a model where that variable is the outcome and all the other variables are the predictors. Given those predictions, the algorithm moves on to the next variable and predicts its missing values based on all the others. These prediction models are strung together in a "chain", hence the name.

Suppose you have three variables, X, Y, and Z. Roughly, the `mice` algorithm is as follows:

- Randomly impute missing X values using the model $X \sim Y + Z +$ randomness.
- Given those predictions, impute Y using the model $Y \sim X + Z +$ randomness.
- Given those predictions, impute Z using the model $Z \sim X + Y +$ randomness.
- Iterate this "chain" many times; often enough that, ideally, the results converge (are not changing much with additional iterations).

For each variable, the model used depends on the type of that variable. For example, for a continuous variable, a linear regression model could be used. It turns out, however, that using a more flexible method that does not depend on normality so heavily is more generally applicable. The default in the `mice()` imputation function is to use **predictive mean matching** (`pmm`) for continuous variables, logistic regression for binary variables (`logreg`), multinomial (polytomous) logistic regression for unordered (nominal) categorical variables (`polyreg`), and ordinal (proportional odds) logistic regression for ordered categorical variables (`polr`).

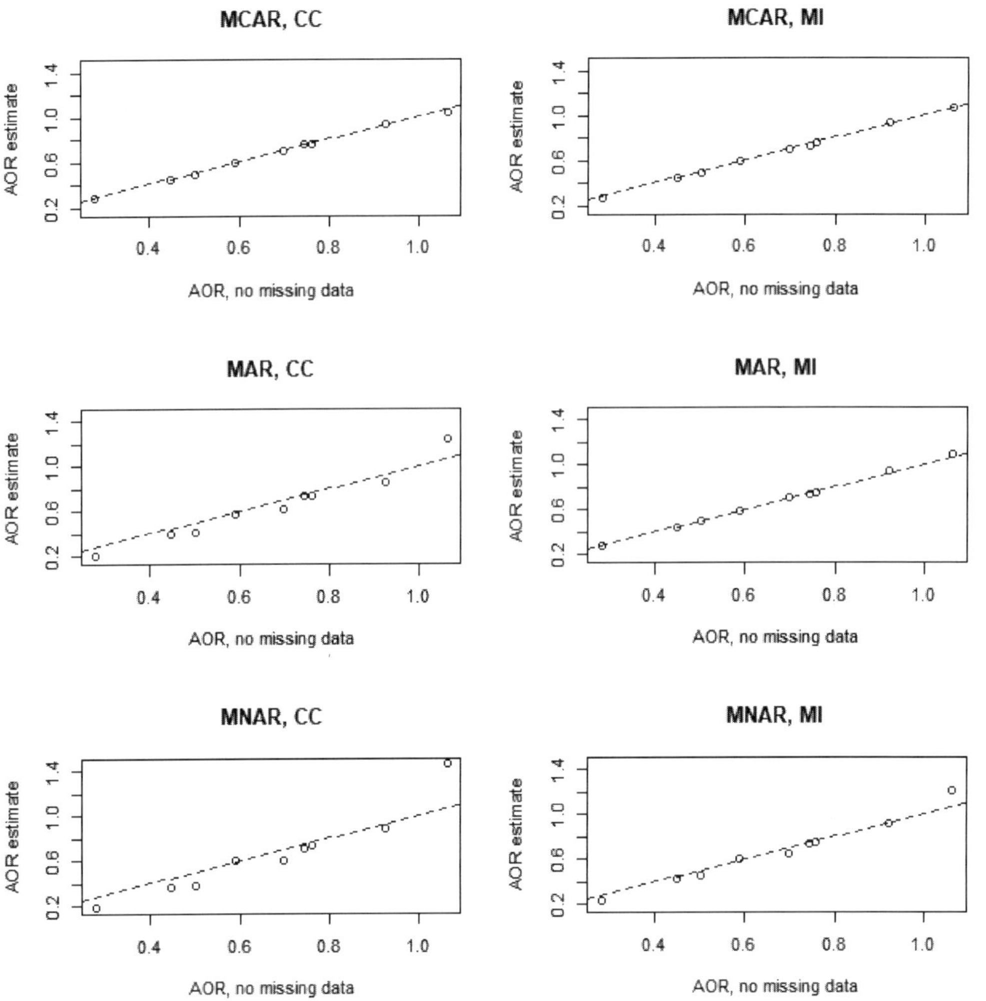

FIGURE 9.1 Bias in adjusted odds ratios (AORs) estimated using complete case analysis (CC) and multiple imputation (MI) under MCAR, MAR, and MNAR missing data mechanisms

If we were to use linear regression to predict a missing value, we would fit the model, compute the predicted mean value, add some random noise to account for the uncertainty in the prediction and in the model itself, and impute using that value. Predictive mean matching differs in that, instead of replacing a missing value with the *prediction* from the model, it replaces it with a randomly chosen *observed value*, where the choice is among those with predicted values closest to the missing value's predicted value. Thus, an imputation based on predictive mean matching is always equal to one of the observed values in the dataset.

NOTE: The terms "outcome" and "predictor" have different meanings between the regression model to be fit after MI and the imputation model used to fill in the missing values. In the regression model, there is one outcome variable and the remaining variables are the predictors. In the imputation model, each variable takes a turn at being the outcome in each iteration of the `mice` algorithm and acts as a predictor in the models for which a different variable is the outcome.

9.4 Fitting the imputation model

Example 9.1: As an example, return to the linear regression of fasting glucose (mmol/L) (`LBDGLUSI`) on waist circumference (`BMXWAIST`) and smoking status (`smoker`; Never, Past, Current) in the NHANES 2017-2018 fasting subset teaching dataset (see Appendix A.1), adjusted for age (`RIDAGEYR`), gender (`RIAGENDR`; Male, Female), race/ethnicity (`RIDRETH3`; Mexican American, Other Hispanic, Non-Hispanic White, Non-Hispanic Black, Non-Hispanic Asian, Other/Multi), and `income` (<$25,000, $25,000 to <$55,000, $55,000+). Previously, in Chapter 5, a complete case analysis was carried out using listwise deletion. Now MI will be used to handle the missing data.

9.4.1 What variables to include

Include in the imputation model all variables that will be used in the analysis, including those with no missing values. The underlying principle is that the imputation model should be at least as complex as the analysis to be done later. Otherwise, information about the relationships between variables will be lost.

For example, if the analysis is a regression, then include the regression outcome and all the predictors *in the same form they will be used in the regression model.* This strategy is referred to as **transform-then-impute** (von Hippel, 2009). Specifically, if any variable in the regression model is collapsed, transformed, or modified in any way, include the modified version in the imputation model instead of the original version. If a predictor has both linear and quadratic terms in the regression model, include both in the imputation model (see Section 9.6.3). If predictors are involved in an interaction in the regression model, make sure the imputation model reflects that interaction either via stratification or by explicitly including an interaction term in the imputation model (see Section 9.6.4).

Additionally, if there are **auxiliary variables** available, variables not in the regression model that are suspected to be related to the probability of missingness and/or to the values of incomplete variables in the imputation model, include them in the imputation model, as well.

Finally, do not impute values for variables that are impossible. For example, do not impute values for "age of first alcohol use" among those who have never used alcohol. Instead of imputing values for these individuals, exclude such individuals and recognize that the scope of the analysis would be "among those who have ever used alcohol". An exception to excluding these individuals would be if "age of first alcohol use" is the time-to-event outcome in a survival analysis. In that case, prior to fitting the imputation model, set each missing value to that individual's chronological age (or to missing if their chronological age is unknown) and the event indicator to 0 (to indicate censoring).

9.4.2 Transform-then-impute vs. impute-then-transform

If you need a variable transformation for the regression model, create it *before* carrying out multiple imputation and include it in the imputation model in that form. For example, if you need a Box-Cox outcome transformation (demonstrated in Section 9.4.3) or a quadratic term (demonstrated in Section 9.6.3), create the transformed variable or quadratic term and add it to the dataset *before* carrying out multiple imputation, not after.

This method is referred to as **transform-then-impute** and preserves the relationships between variables in the regression model, as opposed to **impute-then-transform** which can distort the relationships between variables. Transform-then-impute is the better method and leads to consistent results (von Hippel, 2009).

9.4.3 Pre-processing

Prior to fitting the imputation model, do all the same data checking and cleaning steps you would for any regression analysis. For example, set missing value codes to `NA`, collapse sparse categorical variable levels, code categorical variables as factors, and compute any needed variable transformations.

NOTE: The methods in this chapter assume that the dataset contains only variables that are in the imputation model. Therefore, `select()` only those variables prior to fitting the imputation model.

Example 9.1 (continued): Imitate the pre-processing steps used when this analysis was done in Chapter 5. First, collapse the sparse race/ethnicity levels. Second, transform fasting glucose using the Box-Cox outcome transformation computed in Section 5.18, as that is the form of the outcome used in the analysis. Finally, `select()` just the variables to be included in the imputation model.

```
# Load data
load("Data/nhanes1718_adult_fast_sub_rmph.Rdata")
```

```
# Pre-processing
nhanes <- nhanes_adult_fast_sub %>%
    mutate(
      # (1) Collapse a sparse categorical variable
      race_eth = fct_collapse(RIDRETH3,
      "Hispanic" = c("Mexican American", "Other Hispanic"),
      "Non-Hispanic Other" = c("Non-Hispanic Asian", "Other/Multi")),
      # (2) Box-Cox transformation for our regression
      LBDGLUSI_trans = -1*LBDGLUSI^(-1.5)
      ) %>%
  # Select the variables to be included in the imputation model.
  # If any variables were modified, include only the modified version.
  # Exceptions (see Sections 9.6.3 and 9.6.4):
  #   If including a quadratic, also include the linear term.
  #   If including an interaction, also include the main effects.
  select(LBDGLUSI_trans, BMXWAIST, smoker, RIDAGEYR, RIAGENDR, race_eth, income)

# Shorten labels for race_eth so output fits on the page
levels(nhanes$race_eth) <- c("Hispanic", "NH White", "NH Black", "NH Other")
```

NOTES:

- `mice()` by default checks for collinearity when fitting the imputation model. If a variable is perfectly collinear, it is removed and not imputed. If there is severe approximate collinearity, all variables are imputed, but highly collinear variables are removed from some of the chained models.
- Some methods in this chapter will fail if a non-categorical variable is class `integer`. You can check the class using `class(dat$X)` and if `integer` is returned, then convert to

numeric using `mutate(X = as.numeric(X))` during pre-processing.

9.4.4 Visualize the missing data pattern

Visualize the missing data pattern using `md.pattern()`, as shown in Figure 9.2. Each row is a block of cases with missing values on the same variables, the numbers on the left are the number of cases in each block, the numbers on the right are the number of variables with missing values in that block, and the numbers on the bottom are the number of cases with missing values for each variable individually. For example, 857 cases have no missing data, 15 cases have missing data on both waist circumference and income, the fourth block has two variables with missing values, and there are 123 cases with missing income.

```
PATTERN <- md.pattern(nhanes, rotate.names = T)
```

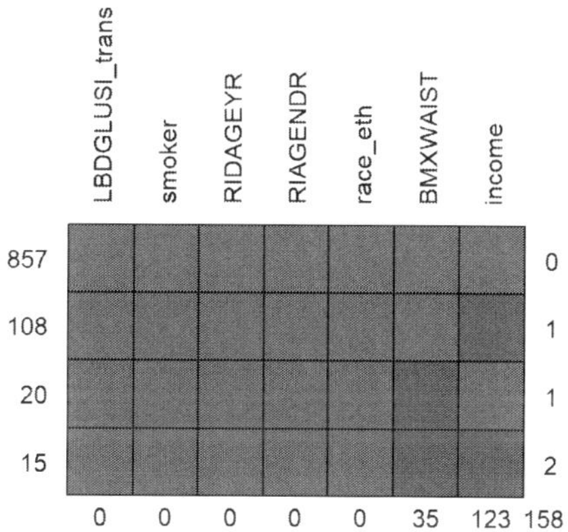

FIGURE 9.2 Missing data pattern

```
# To see a tabular view of the same information
PATTERN
```

```
##     LBDGLUSI_trans smoker RIDAGEYR RIAGENDR race_eth BMXWAIST income
## 857              1      1        1        1        1        1      1   0
## 108              1      1        1        1        1        1      0   1
## 20               1      1        1        1        1        0      1   1
## 15               1      1        1        1        1        0      0   2
##                  0      0        0        0        0       35    123 158
```

NOTE: The number in the lower right (158) is the sum of the number of missing values across the variables. This is not necessarily the total number of incomplete cases, as some individuals may have a missing value for more than one variable. In this example, the number of incomplete cases is `sum(!complete.cases(nhanes))` = 143, not 158.

9.4.5 Compare those with and without missing data

As mentioned previously, it is helpful for any research report to include a description of how the respondent characteristics differ between those with complete responses and those with item non-response. Split the sample once for each variable that has missing values and compare the other variables between each of the resulting pairs of samples.

If the distributions of the other variables are about the same in the complete and incomplete subgroups, that lends support to an assumption of MCAR and the use of a complete-case analysis. However, even if each other variable is similar between those with and without missing values for a variable, their joint distributions might not be similar. Additionally, if the number of cases with incomplete data is not small, a complete case analysis will result in lower precision due to the loss in sample size. Therefore, ultimately, this step of comparing those with and without missing data is not the basis for the decision regarding whether or not to use MI. If there is only a small fraction of missing values, a complete case analysis and MI will produce about the same results. If the fraction is larger, MI is superior whether the missing data mechanism is MCAR or not. This step is useful, however, as part of a complete description of the data and the nature of the missingness.

Example 9.1 (continued): For the NHANES teaching dataset, compare the descriptive statistics between those with and without missing data. This requires creating a missingness indicator for each variable with missing values to use as a "by" variable in the descriptive statistics table.

Look back at the visualization of missing data in the previous section, or use `summary()`, to see which variables have missing values in the dataset.

```
summary(nhanes)
```

```
## LBDGLUSI_trans        BMXWAIST           smoker         RIDAGEYR        RIAGENDR
## Min.    :-0.2372   Min.    : 63.2   Never  :579   Min.    :20.0   Male  :457
## 1st Qu.:-0.0813   1st Qu.: 88.3   Past   :264   1st Qu.:34.0   Female:543
## Median :-0.0731   Median : 97.8   Current:157   Median :47.0
## Mean    :-0.0715   Mean    :100.5                 Mean    :47.9
## 3rd Qu.:-0.0645   3rd Qu.:111.0                 3rd Qu.:61.0
## Max.    :-0.0121   Max.    :169.5                 Max.    :80.0
##                     NA's    :35
##       race_eth                   income
## Hispanic:191    <$25,000             :164
## NH White:602    $25,000 to <$55,000:224
## NH Black:115    $55,000+             :489
## NH Other: 92    NA's                 :123
##
##
##
```

Since there are only two variables with missing values, we can create two tables and merge them together. If there were more than a few variables with missing values, such a merged table might be too large, in which case just look at each table individually.

NOTE: Below, `missing` is created only for the purpose of creating the table. Do not permanently add `missing` to the dataset, as that variable should not be included in the imputation model later.

The results are shown in Table 9.1.

```
library(gtsummary)
t1 <- nhanes %>%
  mutate(missing = factor(is.na(BMXWAIST),
                          levels = c(FALSE, TRUE),
                          labels = c("Not Missing",
                                     "Missing"))) %>%
  tbl_summary(
    by = missing,
    statistic = list(all_continuous()  ~ "{mean} ({sd})",
                     all_categorical() ~ "{n}    ({p}%)"),
    digits = list(all_continuous()  ~ c(2, 2),
                  LBDGLUSI_trans    ~ c(4, 4),
                  all_categorical() ~ c(0, 1)),
    label = list(LBDGLUSI_trans ~ "Transformed FG",
                 BMXWAIST ~ "WC (cm)",
                 smoker ~ "Smoking",
                 RIDAGEYR ~ "Age (y)",
                 RIAGENDR ~ "Gender",
                 race_eth ~ "Race/Ethnicity",
                 income ~ "Income")
  ) %>%
  modify_header(
    label = "**Variable**",
    all_stat_cols() ~ "**{level}**<br>N = {n} ({style_percent(p, digits=1)}%)"
  ) %>%
  modify_caption("Participant characteristics, by missingness") %>%
  bold_labels()

t2 <- nhanes %>%
  mutate(missing = factor(is.na(income),
                          levels = c(FALSE, TRUE),
                          labels = c("Not Missing",
                                     "Missing"))) %>%
  tbl_summary(
    by = missing,
    statistic = list(all_continuous()  ~ "{mean} ({sd})",
                     all_categorical() ~ "{n}    ({p}%)"),
    digits = list(all_continuous()  ~ c(2, 2),
                  LBDGLUSI_trans    ~ c(4, 4),
                  all_categorical() ~ c(0, 1)),
    label = list(LBDGLUSI_trans ~ "Transformed FG",
                 BMXWAIST ~ "WC (cm)",
                 smoker ~ "Smoking",
                 RIDAGEYR ~ "Age (y)",
                 RIAGENDR ~ "Gender",
                 race_eth ~ "Race/Ethnicity",
                 income ~ "Income")
  ) %>%
  modify_header(
    label = "**Variable**",
    all_stat_cols() ~ "**{level}**<br>N = {n} ({style_percent(p, digits=1)}%)"
  )

TABLE <- tbl_merge(
  tbls = list(t1, t2),
  tab_spanner = c("**Waist Circumference**", "**Income**")
)
```

```
TABLE
```

TABLE 9.1 Participant characteristics, by missingness

Variable	Waist Circumference		Income	
	Not Missing N = 965 (96.5%)	Missing N = 35 (3.50%)	Not Missing N = 877 (87.7%)	Missing N = 123 (12.3%)
Transformed FG	-0.0715 (0.0173)	-0.0734 (0.0174)	-0.0714 (0.0176)	-0.0727 (0.0148)
WC (cm)	100.48 (16.95)	NA (NA)	100.82 (17.15)	97.73 (15.03)
Unknown	0	35	20	15
Smoking				
Never	557 (57.7%)	22 (62.9%)	511 (58.3%)	68 (55.3%)
Past	255 (26.4%)	9 (25.7%)	228 (26.0%)	36 (29.3%)
Current	153 (15.9%)	4 (11.4%)	138 (15.7%)	19 (15.4%)
Age (y)	47.70 (16.46)	52.63 (20.90)	47.84 (16.55)	48.12 (17.41)
Gender				
Male	443 (45.9%)	14 (40.0%)	398 (45.4%)	59 (48.0%)
Female	522 (54.1%)	21 (60.0%)	479 (54.6%)	64 (52.0%)
Race/Ethnicity				
Hispanic	180 (18.7%)	11 (31.4%)	157 (17.9%)	34 (27.6%)
NH White	589 (61.0%)	13 (37.1%)	544 (62.0%)	58 (47.2%)
NH Black	111 (11.5%)	4 (11.4%)	100 (11.4%)	15 (12.2%)
NH Other	85 (8.8%)	7 (20.0%)	76 (8.7%)	16 (13.0%)
Income				
<$25,000	157 (18.3%)	7 (35.0%)	164 (18.7%)	0 (NA%)
$25,000 to <$55,000	221 (25.8%)	3 (15.0%)	224 (25.5%)	0 (NA%)
$55,000+	479 (55.9%)	10 (50.0%)	489 (55.8%)	0 (NA%)
Unknown	108	15	0	123

[1] Mean (SD); n (%)

Summary:

- Those with missing waist circumference have, on average, lower fasting glucose and older age, are less likely to smoke, and are more likely to be female, Hispanic or non-Hispanic Other race, and earn <$25,000 per year.
- Those with missing income have, on average, lower fasting glucose and smaller waist circumference, and are more likely to be Hispanic or non-Hispanic Other race.

9.4.6 Number of imputations

If you need to try out different options in the `mice()` function, then use a small number of imputations (e.g., 5), as that will speed up processing while you work out the details of your code. When you are ready to fit your final imputation model, however, use 20 imputations. That is actually the first step in the method of von Hippel (2020). However, it is easier to learn other aspects of multiple imputation first, so the details of that method will be left to Section 9.9.

9.4.7 mice()

Use the `mice()` function to fit the imputation model. Since MI involves random draws from distributions, each time you run `mice()` you will get slightly different results. To ensure reproducibility, use the `seed` option to set the random seed to any number you choose (use `seed = 3` to reproduce the results in this text). The `m` option specifies the number of imputations, and `print = F` suppresses a listing of the iterations. The first argument to the function is the dataset containing **only** the variables to be included in the imputation model; **any variable that is not part of the imputation model should be excluded from this dataset.**

Example 9.1 (continued): Fit the imputation model and examine the output.

```
imp.nhanes <- mice(nhanes,
                   seed  = 3,
                   m     = 20,
                   print = F)
imp.nhanes
```

```
## Class: mids
## Number of multiple imputations:  20
## Imputation methods:
## LBDGLUSI_trans        BMXWAIST          smoker        RIDAGEYR        RIAGENDR
##             ""           "pmm"              ""              ""              ""
##       race_eth          income
##             ""       "polyreg"
## PredictorMatrix:
##                LBDGLUSI_trans BMXWAIST smoker RIDAGEYR RIAGENDR race_eth income
## LBDGLUSI_trans              0        1      1        1        1        1      1
## BMXWAIST                    1        0      1        1        1        1      1
## smoker                      1        1      0        1        1        1      1
## RIDAGEYR                    1        1      1        0        1        1      1
## RIAGENDR                    1        1      1        1        0        1      1
## race_eth                    1        1      1        1        1        0      1
```

The output states that the object named `imp.nhanes` is of class `mids` (multiply imputed dataset), the number of imputations (20), the `Imputation method` used for each variable, and which variables were used to predict which other variables (`PredictorMatrix`). By default, `mice()` used predictive mean matching (`pmm`) for waist circumference (a continuous variable) and multinomial logistic regression (`polyreg`) for income (less restrictive than using ordinal logistic regression). The `PredictorMatrix` will generally have 0s on the diagonal and 1s elsewhere indicating that each variable is predicted using all the others.

Some methods used later in this chapter operate directly on a `mids` object, but others require first extracting the individual imputed datasets. `complete()` converts a `mids` object to either a single complete dataset or all the completed datasets. There are a number of different ways of doing this. The two used in this chapter are returning the first imputed dataset and returning the original data and all imputed datasets all in one stacked dataset.

```
# Return the first imputed dataset
imp.nhanes.dat <- complete(imp.nhanes)
# Same dimensions as original data
dim(nhanes)
```

```
## [1] 1000    7
```

```
dim(imp.nhanes.dat)
```

```
## [1] 1000    7
```

```
# Return all the imputed datasets in "long" form (they will be stacked vertically)
# including the original (unimputed) data
imp.nhanes.dat <- complete(imp.nhanes, "long", include = TRUE)
# imp.nhanes$m*(n+1) rows
dim(imp.nhanes.dat)
```

```
## [1] 21000     9
```

```
names(imp.nhanes.dat)
```

```
## [1] ".imp"            ".id"             "LBDGLUSI_trans" "BMXWAIST"
## [5] "smoker"          "RIDAGEYR"        "RIAGENDR"        "race_eth"
## [9] "income"
```

```
# A new variable called ".imp" is created that indexes the imputations
# .imp = 0 corresponds to the original data
# .imp = 1 to m corresponds to the imputed data
range(imp.nhanes.dat$.imp) # 0 to m
```

```
## [1]  0 20
```

```
# A new variable called ".id"  is created that indexes the observations
range(imp.nhanes.dat$.id)  # 1 to number of cases
```

```
## [1]    1 1000
```

9.4.8 Examine the imputed values

After fitting the imputation model, examine the imputations by plotting the observed and imputed data together. Ideally, the imputed values are plausible compared to the observed values. Imputed values that are far away from the distribution of the observed values (not possible with `pmm` but possible with other methods) may indicate a problem with the imputation model, or perhaps that an MNAR model is needed. Imputed values which only span a subset of the distribution of the observed values are interesting in that they provide some information about the nature of the missing values that may assist in reducing missing data in future studies.

`stripplot()` plots the observed and missing values for **continuous variables** (Figure 9.3). The observed data are plotted (labeled as 1 on the x-axis) as well as the observed and imputed data together for each completed dataset (labeled as 2 to the number of imputations + 1 on the x-axis). The points are "jittered" to provide some spread, so it is easier to see the imputed values superimposed over the observed values. By default, `stripplot()` will make plots even for variables with no missing data, but a formula can be included to specify just a variable with missing data (in our example, `BMXWAIST ~ .imp`, where `.imp` is a special variable referring to the imputation number). The `col`, `pch`, and `cex` options specify the color, plotting character, and point size, respectively. For each, specify two values, the first for the observed values, the second for the imputed values.

```
stripplot(imp.nhanes, BMXWAIST ~ .imp,
          col = c("gray", "black"),
          pch = c(21, 20),
          cex = c(1, 1.5))
```

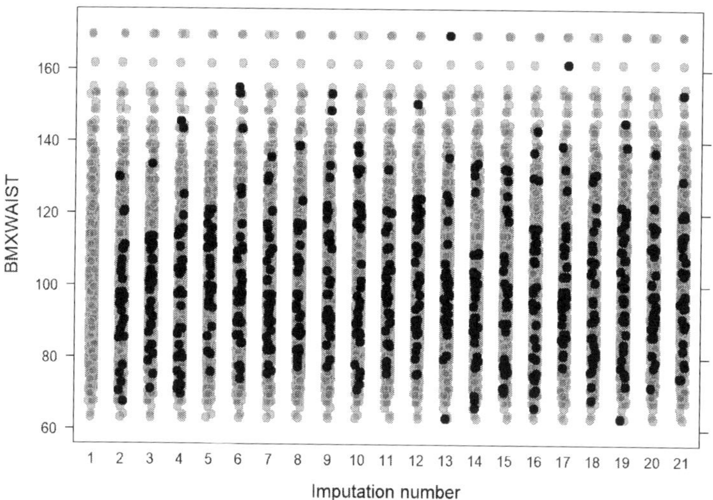

FIGURE 9.3 All observed and imputed values, by imputation

Alternatively, you can plot the imputed values by another variable.

```
# vs. a categorical variable (results not shown)
stripplot(imp.nhanes, BMXWAIST ~ RIAGENDR,
          col = c("gray", "black"),
          pch = c(21, 20),
          cex = c(1, 1.5))

# vs. a continuous variable (results not shown)
stripplot(imp.nhanes, BMXWAIST ~ LBDGLUSI_trans,
          col = c("gray", "black"),
          pch = c(21, 20),
          cex = c(1, 1.5))
```

If the number of missing values is large, `stripplot()` might not be very informative since imputed values are plotted on top of observed values. An alternative is to use `bwplot()` to plot side-by-side boxplots (Figure 9.4). With this function, the observed data are labeled as imputation 0 on the x-axis, and the boxplots for each imputation only plot the distribution of imputed values.

```
bwplot(imp.nhanes, BMXWAIST ~ .imp)
```

For a categorical variable, create a two-way table comparing the distribution of the variable between the complete data and the imputed data. This requires using the "long" form of the data, including the original data.

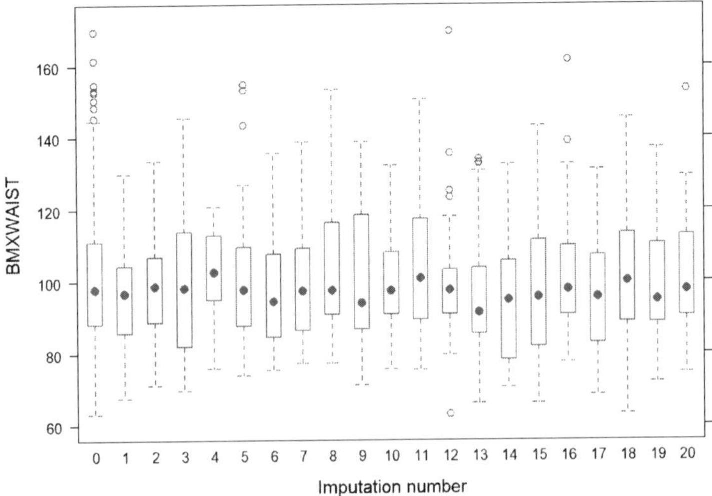

FIGURE 9.4 Boxplots of observed and imputed values, by imputation

```
imp.nhanes.dat <- complete(imp.nhanes, "long", include = TRUE)

imp.nhanes.dat <- imp.nhanes.dat %>%
  mutate(imputed = .imp > 0,
         imputed = factor(imputed,
                          levels = c(F,T),
                          labels = c("Observed", "Imputed")))

prop.table(table(imp.nhanes.dat$income,
                 imp.nhanes.dat$imputed),
           margin = 2)
```

```
##
##                        Observed Imputed
##   <$25,000               0.1870  0.1889
##   $25,000 to <$55,000    0.2554  0.2584
##   $55,000+               0.5576  0.5527
```

The following code creates a table that includes a separate column for each imputation.

```
prop.table(table(imp.nhanes.dat$income,
                 imp.nhanes.dat$.imp),
           margin = 2)
# (results not shown)
```

9.4.9 For descriptive statistics only: back-transformation and derived variables

If there are variables you transformed for the imputation and regression models, but want to include on their original scale for descriptive statistics, then you must back-transform them after imputation. Similarly, if there are new variables, not in the imputation model, that you would like to derive based on the variables in the imputation model, derive them after imputation and add them to the imputed datasets; **however, it is critical that you do not include the post-imputation back-transformed or derived variables in the regression model; they are only to be used for descriptive statistics.**

Very important: To re-iterate, back-transformation is not to be used to derive variables for inclusion in a regression model; doing so would be impute-then-transform which results in a distortion of the relationship between the variables in the regression model (as described in Section 9.4.2). If you need a variable for the regression model, create it in the pre-processing step and include it in the imputation model in that form (Section 9.4.3). **Back-transformation is only to be used to un-transform or derive variables for the purpose of descriptive statistics, not for inclusion in the regression model.**

Example 9.1 (continued): For fasting glucose, we need the transformed variable for the regression model. However, we would like the original variable for computing descriptive statistics. Additionally, we would like to compute descriptive statistics by waist circumference, but since it is a continuous variable we need to derive a median split version for use as our "by" variable.

The code below demonstrates how to back-transform a variable, create a derived variable, and include these in the imputed datasets. First, extract the imputed datasets in "long" form from the `mids` object. Second, manipulate the long-form dataset (e.g., using a `mutate()` statement). Finally, convert the long-form dataset back into a `mids` object.

```
# Use complete() to extract all the imputed datasets, with
# include = TRUE to also extract the original incomplete dataset
imp.nhanes.dat <- complete(imp.nhanes, "long", include = TRUE)

# Inverse Box-Cox transformation
# If Z = -1*Y^(-1.5),  then Y = (-Z)^(1/-1.5)
imp.nhanes.dat <- imp.nhanes.dat %>%
  mutate(LBDGLUSI = (-1*LBDGLUSI_trans)^(1/-1.5))

# Median split
MEDIAN <- median(imp.nhanes.dat$BMXWAIST, na.rm = T)
LABEL0 <- paste("WC <",  MEDIAN, sep = "")
LABEL1 <- paste("WC >=", MEDIAN, sep = "")

imp.nhanes.dat <- imp.nhanes.dat %>%
  mutate(wc_median_split = as.numeric(BMXWAIST >= MEDIAN),
         wc_median_split = factor(wc_median_split,
                                  levels = 0:1,
                                  labels = c(LABEL0, LABEL1)))

# Convert back to a `mids` object
imp.nhanes.new <- as.mids(imp.nhanes.dat)
```

As always, do a before vs. after check to make sure any back-transformations or derivations did what you expected. In particular, make sure the back-transformed variable is on the same scale as the corresponding original variable, but do not be concerned if the `summary()` before vs. after is not *exactly* identical since any previously missing values are now imputed. This check is especially important when back-transforming a Box-Cox transformed variable, as it is easy to put a parenthesis in the wrong place resulting in the wrong formula.

```
# Check back-transformation vs. original scale
summary(nhanes_adult_fast_sub$LBDGLUSI) # Original
```

```
##    Min. 1st Qu.  Median    Mean 3rd Qu.    Max.
##    2.61    5.33    5.72    6.09    6.22   19.00
```

```
summary(imp.nhanes.dat$LBDGLUSI) # Back-transformed
```

```
##    Min. 1st Qu.  Median    Mean 3rd Qu.    Max.
##    2.61    5.33    5.72    6.09    6.22   19.00
```

```
# Check derivation of median split
tapply(imp.nhanes.dat$BMXWAIST,
       imp.nhanes.dat$wc_median_split, range)
```

```
## $`WC <97.8`
## [1] 63.2 97.7
##
## $`WC >=97.8`
## [1]  97.8 169.5
```

9.5 Descriptive statistics after MI

After MI, compute descriptive statistics within each complete dataset and use Rubin's rules to pool statistics and compute their standard errors. Rubin's rules comprise two parts – one for computing the statistic and one for its variation. For the statistic, simply average the values over the imputations. For the variation of the statistic, combine the average variance over imputations with the variation of the statistic between imputations. Additionally, if the sampling distribution of the statistic is not close to normal, then it is recommended to transform the statistic to normality before applying Rubin's rules.

Statistics for which we can assume normality (in a large sample) include the mean and standard deviation of a continuous variable and the frequencies and proportions of levels of a categorical variable. A list of appropriate transformations for other statistics (e.g., correlation, odds ratio) can be found at Scalar inference of non-normal quantities[5] (van Buuren, 2018). The strategy in that case is to transform, apply Rubin's rules, and then back-transform to the original scale.

Later, this section will introduce some convenient functions for computing descriptive statistics after MI. But first, we demonstrate how to use Rubin's rules to compute the mean and standard error of the mean waist circumference from Example 9.1. To do this, extract the multiply imputed datasets, compute the sample mean and variance of the sample mean for waist circumference in each, and then combine them using Rubin's rules to get the mean and standard error pooled over the imputations. Recall that the variance of the sample mean is the sample variance divided by the sample size.

[5]https://stefvanbuuren.name/fimd/sec-pooling.html#sec:poolnon

```
# Multiple imputed datasets stacked together
# but NOT including the original data
imp.nhanes.dat <- complete(imp.nhanes.new, action = "long")

# Sample mean in each complete imputed dataset
MEAN      <- tapply(imp.nhanes.dat$BMXWAIST, imp.nhanes.dat$.imp, mean)

# Variance of the sample mean in each imputed dataset
# Make sure to use imp.nhanes.new (the mids object)
# inside nrow!
VAR.MEAN <- tapply(imp.nhanes.dat$BMXWAIST,
                   imp.nhanes.dat$.imp,
                   var) / nrow(imp.nhanes.new$data)
```

The following shows the means and variances by imputation, in a table and visually.

```
cbind("MEAN"     = MEAN,
      "VAR.MEAN" = VAR.MEAN)
```

```
##       MEAN VAR.MEAN
## 1  100.3   0.2857
## 2  100.4   0.2834
## 3  100.4   0.2903
## 4  100.6   0.2826
## 5  100.5   0.2910
## 6  100.4   0.2860
## 7  100.4   0.2865
## 8  100.6   0.2896
## 9  100.5   0.2908
## 10 100.4   0.2835
## 11 100.6   0.2876
## 12 100.5   0.2882
## 13 100.3   0.2892
## 14 100.3   0.2894
## 15 100.4   0.2908
## 16 100.5   0.2878
## 17 100.4   0.2875
## 18 100.5   0.2888
## 19 100.4   0.2842
## 20 100.5   0.2866
```

```
par(mfrow=c(1,2))
plot(MEAN ~ I(1:imp.nhanes$m),
     ylab = "Mean", xlab = "Imputation",
     main = "Mean\nby imputation")
abline(h=mean(MEAN), col="red", lty=2, lwd=2)
plot(VAR.MEAN ~ I(1:imp.nhanes$m),
     ylab = "Variance of Mean", xlab = "Imputation",
     main = "Variance of Mean\nby imputation")
abline(h=mean(VAR.MEAN), col="red", lty=2, lwd=2)
```

The average of the 20 within-imputation sample means is 100.44 cm (the horizontal dashed line in the left panel of Figure 9.5) and is the correct estimate of the mean waist circumference. The average of the 20 within-imputation variances of the sample mean is 0.2875. This turns out to be an underestimate of the variance of the sample mean waist circumference. It needs to be increased based on how variable the means are between imputations. The Rubin's rule

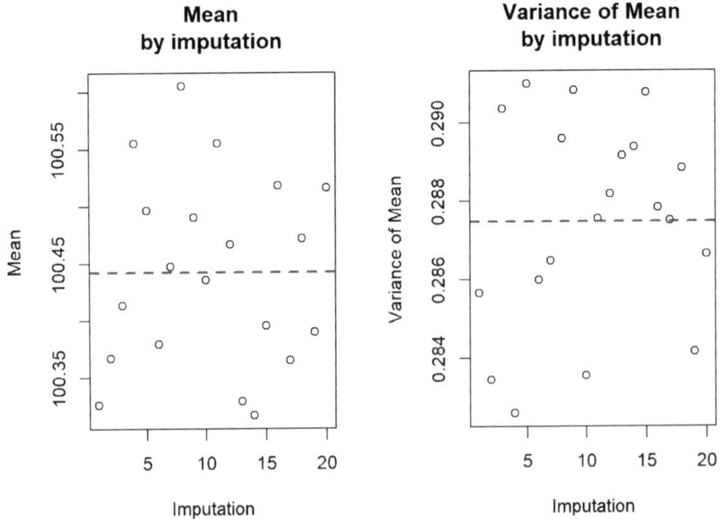

FIGURE 9.5 Mean and variance of waist circumference across multiple complete (imputed) datasets

formula for the pooled variance adds the average variance to a factor times the variance of the means. Below, the `pool.scalar()` function does the computations for us and the pooled mean and variance of the mean are manually computed, as well. The latter is just to illustrate what `pool.scalar()` is doing; in future, just use `pool.scalar()`.

```
# Rubin's rules for the mean and variance of the mean
# Using pool.scalar()
POOLED <- pool.scalar(MEAN,
                      VAR.MEAN,
                      nrow(imp.nhanes.new$data))
data.frame(pooled.mean = POOLED$qbar,
           mean.var    = POOLED$ubar,
           btw.var     = POOLED$b,
           pooled.var  = POOLED$t)
```

```
##   pooled.mean mean.var  btw.var pooled.var
## 1       100.4   0.2875 0.007014     0.2948
```

```
# Manually (just for illustration)
data.frame(pooled.mean = mean(MEAN),
           mean.var    = mean(VAR.MEAN),
           btw.var     = var(MEAN),
           pooled.var  = mean(VAR.MEAN) + (1 + 1/length(MEAN))*var(MEAN))
```

```
##   pooled.mean mean.var  btw.var pooled.var
## 1       100.4   0.2875 0.007014     0.2948
```

In this example, there was not much missing data, so there was not much variation in the means (the between-imputation variance, `btw.var` is small), so the pooled variance is not that much larger than the mean of the variances.

Finally, to get the pooled standard error of the mean, take the square root of the pooled variance of the mean.

```
sqrt(POOLED$t)
```

```
## [1] 0.543
```

NOTE: This is the pooled *standard error of the mean.* If you want the pooled *standard deviation* of waist circumference, average the within-imputation standard deviations over the imputations.

The steps of computing descriptive statistics (mean, standard error of the mean, and standard deviation for continuous variables; frequency and proportion for categorical variables) in each dataset and pooling using Rubin's rules have been coded into the functions `mi.mean.se.sd()`, `mi.n.p()`, `mi.mean.se.sd.by()`, and `mi.n.p.by()` (found in `Functions_rmph.R` which you loaded at the beginning of this chapter). These operate directly on a `mids` object so there is no need to use `complete()` to convert to a `data.frame` first.

NOTES:

- For variables with no missing data, these `mi.` functions will return the same values as standard descriptive statistic functions.
- For frequencies of categorical variables that have missing values, the results after MI may not be whole numbers, as they are averages over imputations.
- If a "by" variable has missing values, then even frequencies for variables that did not have missing values may not be whole numbers, as the split between groups will differ between imputations.

Each function call returns a `data.frame` and `rbind()` stacks the results together.

```
# Means and standard deviations
rbind(mi.mean.se.sd(imp.nhanes.new, "LBDGLUSI"),
      mi.mean.se.sd(imp.nhanes.new, "BMXWAIST"),
      mi.mean.se.sd(imp.nhanes.new, "RIDAGEYR"))
```

```
##             mean      se     sd
## LBDGLUSI   6.093 0.05096  1.611
## BMXWAIST 100.442 0.54299 16.955
## RIDAGEYR  47.876 0.52657 16.652
```

```
# Frequencies and proportions
rbind(mi.n.p(imp.nhanes.new, "smoker"),
      mi.n.p(imp.nhanes.new, "RIAGENDR"),
      mi.n.p(imp.nhanes.new, "race_eth"),
      mi.n.p(imp.nhanes.new, "income"))
```

```
##                            n      p
## smoker: Never          579.0 0.5790
## smoker: Past           264.0 0.2640
## smoker: Current        157.0 0.1570
## RIAGENDR: Male         457.0 0.4570
## RIAGENDR: Female       543.0 0.5430
## race_eth: Hispanic     191.0 0.1910
## race_eth: NH White     602.0 0.6020
## race_eth: NH Black     115.0 0.1150
```

```
## race_eth: NH Other            92.0 0.0920
## income: <$25,000             188.9 0.1889
## income: $25,000 to <$55,000 258.4 0.2584
## income: $55,000+             552.7 0.5527
```

```
# Means and standard deviations by another variable
rbind(mi.mean.se.sd.by(imp.nhanes.new, "LBDGLUSI", BY = "wc_median_split"),
      mi.mean.se.sd.by(imp.nhanes.new, "RIDAGEYR", BY = "wc_median_split"))
```

```
##           mean.1    se.1  sd.1 mean.2    se.2   sd.2
## LBDGLUSI   5.701 0.05341  1.173   6.48 0.08404  1.871
## RIDAGEYR 44.594 0.76423 16.727  51.11 0.72326 15.943
```

```
# Frequencies and proportions by another variable
rbind(mi.n.p.by(imp.nhanes.new, "smoker",   BY = "wc_median_split"),
      mi.n.p.by(imp.nhanes.new, "RIAGENDR", BY = "wc_median_split"),
      mi.n.p.by(imp.nhanes.new, "race_eth", BY = "wc_median_split"),
      mi.n.p.by(imp.nhanes.new, "income",   BY = "wc_median_split"))
```

```
##                                n.1    p.1    n.2     p.2
## smoker: Never               314.50 0.6336 264.50 0.52522
## smoker: Past                110.15 0.2219 153.85 0.30550
## smoker: Current              71.75 0.1445  85.25 0.16928
## RIAGENDR: Male              212.20 0.4275 244.80 0.48610
## RIAGENDR: Female            284.20 0.5725 258.80 0.51390
## race_eth: Hispanic           91.60 0.1845  99.40 0.19738
## race_eth: NH White          292.40 0.5890 309.60 0.61477
## race_eth: NH Black           57.15 0.1151  57.85 0.11487
## race_eth: NH Other           55.25 0.1113  36.75 0.07297
## income: <$25,000             83.55 0.1683 105.35 0.20919
## income: $25,000 to <$55,000 132.20 0.2663 126.20 0.25060
## income: $55,000+            280.65 0.5654 272.05 0.54021
```

9.6 Linear regression after MI

Use `with()` to fit a regression model on a **multiply imputed dataset** (`mids`) object to produce a **multiple imputation repeated analysis** (`mira`) object. Afterward, use `pool()` to create a **multiple imputation pooled results** (`mipo`) object. Returning to Example 9.1, the following code fits a linear regression model after MI.

NOTE: Only include in the regression model variables that were included in the dataset prior to multiple imputation (see Section 9.4.2). Some can be left out, but no new ones should be added – the regression model must not be more complex than the imputation model.

```
fit.imp.lm <- with(imp.nhanes.new,
                   lm(LBDGLUSI_trans ~ BMXWAIST + smoker + RIDAGEYR +
                      RIAGENDR + race_eth + income))
```

`with()` fits the regression model separately for each complete dataset. The following extracts a `list` of regression results for each imputation.

```
fit.imp.lm$analyses
# (results not shown)
```

For example, the results for the first analysis are:

```
# Full summary
# summary(fit.imp.lm$analyses[[1]])
# Just the coefficients
round(summary(fit.imp.lm$analyses[[1]])$coef, 5)
```

```
##                           Estimate Std. Error   t value Pr(>|t|)
## (Intercept)               -0.11247    0.00352 -31.94801  0.00000
## BMXWAIST                   0.00031    0.00003  10.63284  0.00000
## smokerPast                 0.00133    0.00116   1.13986  0.25462
## smokerCurrent              0.00071    0.00141   0.50642  0.61267
## RIDAGEYR                   0.00031    0.00003  10.08568  0.00000
## RIAGENDRFemale            -0.00492    0.00098  -5.02719  0.00000
## race_ethNH White          -0.00395    0.00132  -2.99556  0.00281
## race_ethNH Black          -0.00298    0.00183  -1.62541  0.10439
## race_ethNH Other          -0.00196    0.00194  -1.00794  0.31373
## income$25,000 to <$55,000  0.00090    0.00152   0.59550  0.55165
## income$55,000+             0.00010    0.00137   0.06974  0.94441
```

Use `summary(pool())` to apply Rubin's rules to obtain the final regression coefficients and 95% confidence intervals.

```
summary(pool(fit.imp.lm), conf.int=T)
```

The numbers in the resulting table may have a lot of digits, and unfortunately the `round()` function does not work directly on a `mipo.summary` object. To round the results, use the function `round.summary()` (found in `Functions_rmph.R` which you loaded at the beginning of this chapter) which takes a `mira` object, pools the results, and rounds the output to a specified number of `digits`.

```
round.summary(fit.imp.lm, digits = 4)
```

```
##                           estimate std.error statistic    df p.value   2.5 %
## (Intercept)                -0.1117    0.0036  -30.9663 889.2  0.0000 -0.1188
## BMXWAIST                    0.0003    0.0000    9.9470 846.6  0.0000  0.0002
## smokerPast                  0.0014    0.0012    1.1532 978.1  0.2491 -0.0009
## smokerCurrent               0.0007    0.0014    0.5238 980.7  0.6005 -0.0020
## RIDAGEYR                    0.0003    0.0000   10.0374 979.4  0.0000  0.0002
## RIAGENDRFemale             -0.0048    0.0010   -4.8885 981.2  0.0000 -0.0068
## race_ethNH White           -0.0038    0.0013   -2.8704 975.3  0.0042 -0.0064
## race_ethNH Black           -0.0028    0.0019   -1.5040 974.6  0.1329 -0.0064
## race_ethNH Other           -0.0019    0.0020   -0.9904 970.2  0.3222 -0.0058
## income$25,000 to <$55,000   0.0011    0.0016    0.6740 803.5  0.5005 -0.0020
## income$55,000+              0.0004    0.0014    0.2469 622.9  0.8051 -0.0025
##                           97.5 %
## (Intercept)              -0.1046
## BMXWAIST                  0.0004
## smokerPast                0.0037
## smokerCurrent             0.0035
## RIDAGEYR                  0.0004
## RIAGENDRFemale           -0.0029
## race_ethNH White         -0.0012
## race_ethNH Black          0.0008
```

```
## race_ethNH Other           0.0019
## income$25,000 to <$55,000  0.0041
## income$55,000+             0.0032
```

9.6.1 Multiple degree of freedom tests

`mi.anova()` (Grund et al., 2016) from the `miceadds` package (Robitzsch et al., 2023) carries out multiple degree of freedom Type III tests pooled over imputations and computes the pooled R^2 value.

```
mi.anova(imp.nhanes.new,
         "LBDGLUSI_trans ~ BMXWAIST + smoker + RIDAGEYR + RIAGENDR +
         race_eth + income",
         type = 3)
```

```
## Univariate ANOVA for Multiply Imputed Data (Type 3)
##
## lm Formula:  LBDGLUSI_trans ~ BMXWAIST + smoker + RIDAGEYR + RIAGENDR +
##          race_eth + income
## R^2=0.1953
## ..........................................................................
## ANOVA Table
##              SSQ df1    df2  F value  Pr(>F)     eta2 partial.eta2
## BMXWAIST 0.02434   1   5894  97.9628 0.00000 0.08455      0.09507
## smoker   0.00033   2 363903   0.6861 0.50355 0.00114      0.00141
## RIDAGEYR 0.02374   1 836530 100.8680 0.00000 0.08247      0.09296
## RIAGENDR 0.00563   1 813144  23.8936 0.00000 0.01954      0.02371
## race_eth 0.00201   3 250497   2.8327 0.03677 0.00699      0.00861
## income   0.00017   2   5274   0.2740 0.76038 0.00058      0.00072
## Residual 0.23166  NA     NA       NA      NA      NA           NA
```

`mi.anova()` only works, however, for linear regression. A more generally applicable method, which works for linear and other forms of regression, is `D1()` which carries out a Wald test comparing two nested models. To get a multiple degree of freedom test for a categorical variable with more than two levels, fit a reduced model that omits that variable, and use `D1()` to compare the full and reduced models. The first time you run `D1()`, you will be prompted to install the `mitml` package (Grund et al., 2023) if you have not already installed it.

For example, to get a Type III test for `smoker` in Example 9.1:

```
# Reduced model omitting smoker
fit.imp.lm.smoker   <- with(imp.nhanes.new,
  lm(LBDGLUSI_trans ~ BMXWAIST + RIDAGEYR + RIAGENDR + race_eth + income))

# Compare full and reduced model
summary(D1(fit.imp.lm, fit.imp.lm.smoker))
```

```
##
## Models:
##  model
    formula
##      1 LBDGLUSI_trans ~ BMXWAIST + smoker + RIDAGEYR + RIAGENDR + race_eth +
    income
##      2          LBDGLUSI_trans ~ BMXWAIST + RIDAGEYR + RIAGENDR + race_eth +
    income
```

```
##
## Comparisons:
##    test statistic dfl    df2 dfcom p.value      riv
##  1 ~~ 2     0.6891    2 985.9    989  0.5023 0.005758
##
## Number of imputations:  20   Method D1
```

To carry out a likelihood ratio test instead, use `D3()`.

```
summary(D3(fit.imp.lm, fit.imp.lm.smoker))
```

```
##
## Models:
##  model
   formula
##       1 LBDGLUSI_trans ~ BMXWAIST + smoker + RIDAGEYR + RIAGENDR + race_eth +
   income
##       2            LBDGLUSI_trans ~ BMXWAIST + RIDAGEYR + RIAGENDR + race_eth +
   income
##
## Comparisons:
##    test statistic df1      df2 dfcom p.value      riv
##  1 ~~ 2     0.6972    2 1101109    989    0.498 0.005294
##
## Number of imputations:  20   Method D3
```

9.6.2 Predictions

To obtain predictions following MI, estimate within each analysis and pool the results over imputations using Rubin's rules. First, create a `data.frame` containing the values at which you want to predict. Second, compute the estimate (and its standard error) for each analysis. Since the `mira` object, `fit.imp.lm`, contains the results of each analysis in a list (`fit.imp.lm$analyses`), use `lapply()` to apply the `predict()` function to each analysis (each element of the list). Finally, extract the predictions and their variances and use Rubin's rules to pool the results via `pool.scalar()`.

Example 9.1 (continued): Predict (transformed) fasting glucose for an individual with a waist circumference of 130 cm who is a current smoker, age 50 years, male, non-Hispanic Black, and earns at least $55,000 annually.

```
# Prediction data.frame
NEWDAT <- data.frame(
  BMXWAIST = 130,
  smoker   = "Current",
  RIDAGEYR = 50,
  RIAGENDR = "Male",
  race_eth = "NH Black",
  income   = "$55,000+")
# The code below assumes NEWDAT has only 1 row.
# To make predictions at different values,
# run again with a different NEWDAT.

# Get prediction for each analysis
PREDLIST <- lapply(fit.imp.lm$analyses, predict, newdata=NEWDAT, se.fit=T)
```

```
# Compute the mean and variance in each analysis
MEAN <- VAR <- rep(NA, length(fit.imp.lm$analyses))
for(i in 1:length(MEAN)) {
  MEAN[i] <-  PREDLIST[[i]]$fit
  VAR[ i] <- (PREDLIST[[i]]$se.fit)^2
}

# Apply Rubin's rules
PRED.POOLED <- pool.scalar(MEAN,
                           VAR,
                           nrow(imp.nhanes.new$data))

# Extract the pooled mean
PRED.POOLED$qbar
```

```
## [1] -0.05931
```

```
# Extract the pooled standard error of the mean
sqrt(PRED.POOLED$t)
```

```
## [1] 0.002225
```

If the outcome was transformed and you would like the prediction on the original scale, then back-transform (this works for the mean, but not the standard error). In Example 9.1, a Box-Cox transformation with λ = -1.5 was used.

```
(-1*PRED.POOLED$qbar)^(1/-1.5)
```

```
## [1] 6.575
```

9.6.3 Polynomial predictor transformations

If you are going to include non-linear functions of variables in the regression model (e.g., Box-Cox, log, quadratic), your imputation model should include those non-linearities. This ensures that your imputation model includes the same level of complexity as your regression model. Otherwise, you will be imputing values under the assumption of linearity.

For functions that convert X into a single term $f(x)$ (e.g., $\log(X)$), simply create a transformed version of the variable prior to imputation. This was done for the outcome fasting glucose in Example 9.1. What about a transformation that converts a predictor X into multiple terms, such as a quadratic transformation which includes both X and X^2? Should the square term be created prior to imputation (**transform-then-impute**) or after (**impute-then-transform**)? As described in Section 9.4.2, create the term prior to imputation (transform-then-impute).

NOTES:

- When using a polynomial transformation, avoid collinearity issues in the imputation model by first centering (e.g., $cX = X-$ mean of X) and then creating higher order terms as powers of the centered variable (e.g., $cX2 = (cX)^2, cX3 = (cX)^3$, etc.) prior to imputation.
- In both the imputation and regression model, include higher order terms using the derived version (e.g., include the square term as $cX2$, not $I(cX^2)$).

- In both the imputation and regression model, include *all* the polynomial terms (e.g., centered linear, quadratic, cubic, etc.) in your polynomial, not just the highest order term (e.g., for a quadratic transformation, include both cX and $cX2$, not just $cX2$).

Example 9.1 (continued): Regress (transformed) fasting glucose on waist circumference, adjusted for smoking status, age, gender, race/ethnicity, and income, and include a quadratic transformation for waist circumference.

First, update the pre-processed dataset to include a centered version of waist circumference and a squared version of that centered variable.

```
# Compute the mean waist circumference for centering
mean(nhanes_adult_fast_sub$BMXWAIST, na.rm=T)
```

```
## [1] 100.5
```

```
nhanesb <- nhanes %>%
  mutate(
    # Centered, quadratic version of waist circumference
    cBMXWAIST  =  BMXWAIST - 100,
    cBMXWAIST2 = cBMXWAIST^2
  ) %>%
  # Select the variables to be included in the imputation model.
  # If including a quadratic, also include the linear term.
  select(LBDGLUSI_trans, cBMXWAIST, cBMXWAIST2,
         smoker, RIDAGEYR, RIAGENDR, race_eth, income)
```

Next, fit the imputation model.

```
imp.quad <- mice(nhanesb,
                 seed = 3,
                 m = 20,
                 print = F)
```

Finally, fit the regression model with both the centered linear and quadratic waist circumference terms, with the quadratic term included as `cBMXWAIST2` not `I(cBMXWAIST^2)`.

```
fit.imp.quad <- with(imp.quad,
                lm(LBDGLUSI_trans ~ cBMXWAIST + cBMXWAIST2 + smoker + RIDAGEYR +
                   RIAGENDR + race_eth + income))
round.summary(fit.imp.quad, digits = 8)[, c("estimate", "2.5 %",
                                             "97.5 %", "p.value")]
```

```
##                         estimate      2.5 %      97.5 %    p.value
## (Intercept)          -0.08249436 -0.0870761 -0.07791260 0.00000000
## cBMXWAIST             0.00028633  0.0002205  0.00035212 0.00000000
## cBMXWAIST2            0.00000081 -0.0000016  0.00000322 0.50834469
## smokerPast            0.00136314 -0.0009372  0.00366350 0.24516547
## smokerCurrent         0.00078328 -0.0020152  0.00358175 0.58294854
## RIDAGEYR              0.00031310  0.0002507  0.00037544 0.00000000
## RIAGENDRFemale       -0.00485763 -0.0068005 -0.00291479 0.00000109
## race_ethNH White     -0.00395902 -0.0065948 -0.00132328 0.00327873
## race_ethNH Black     -0.00291613 -0.0065572  0.00072493 0.11634728
## race_ethNH Other     -0.00214190 -0.0059891  0.00170533 0.27486657
## income$25,000 to <$55,000  0.00134174 -0.0017960  0.00447953 0.40140845
## income$55,000+        0.00058526 -0.0022808  0.00345135 0.68849725
```

9.6.4 Interactions

If you want to include an interaction in your regression model, make sure to include the interaction in your imputation model, as well. This ensures that your imputation model includes at least the same level of complexity as your regression model. Otherwise, you will be imputing values under the assumption of no interaction.

For an interaction involving a categorical variable, this can be done by stratifying the dataset by the categorical variable, fitting the imputation model separately in each stratum, and then combining the results. This will only work if the categorical variable has no missing values (or, if there are not many missing values, if you first exclude cases with missing values on the categorical variable). An alternative, which is more generally applicable, is to use transform-then-impute (Section 9.4.2); create the interaction term as a distinct variable in your pre-processing step and include it in the imputation model (von Hippel, 2009).

The stratification and transform-then-impute methods will not, in general, lead to the same imputed values. The stratification method leads to a more complex imputation model because it effectively includes an interaction between the categorical variable and *every* other variable, whereas the transform-then-impute method only includes whatever interactions you explicitly include. However, transform-then-impute is more generally applicable; if the categorical variable has missing data, or if both terms in the interaction are continuous, the stratification method is not possible.

Example 9.1 (continued): Regress (transformed) fasting glucose on waist circumference, adjusted for smoking status, age, gender, race/ethnicity, and income, and include an interaction to see if the waist circumference effect depends on gender. Use MI to handle missing data.

9.6.4.1 Interaction via stratification

First, check to see if gender has any missing values. If yes, and the number is very small, then remove the cases with missing gender from the dataset. If the number is large, or you do not want to lose even a small number of cases, use the interaction via transform-then-impute method described in the next section.

```
table(nhanes$RIAGENDR, useNA = "ifany")
```

```
##
##   Male Female
##    457    543
```

Gender has no missing values. To use stratification to include an interaction in the imputation model, do the following: split the dataset into separate datasets for each level of gender, use `mice()` on each dataset, and combine the resulting `mids` objects using `rbind()`. Make sure to use the same number of imputations for all strata, the same number you would use for the full dataset if you were not using stratification.

```
# Split data by gender
nhanes_F <- nhanes %>%
  filter(RIAGENDR == "Female")

nhanes_M <- nhanes %>%
  filter(RIAGENDR == "Male")
```

```
# Impute separately by gender
imp_F <- mice(nhanes_F,
              seed   = 3,
              m      = 20,
              print = F)

imp_M <- mice(nhanes_M,
              seed   = 3,
              m      = 20,
              print = F)

# Combine the imputations across strata
imp.int <- rbind(imp_F, imp_M)
```

Just to check, table gender on the first imputation. It should have all its values and no missing values. If it does not, you may have stratified on a variable with missing values (which you should not do).

```
# Checking the first imputation
# (should have all levels of gender and no missing values)
table(complete(imp.int)$RIAGENDR, useNA = "ifany")
```

```
##
##   Male Female
##    457    543
```

Next, fit the model with the interaction and pool the results.

```
fit.imp.int <- with(imp.int,
                    lm(LBDGLUSI_trans ~ BMXWAIST + smoker + RIDAGEYR +
                         RIAGENDR + race_eth + income +
                         BMXWAIST:RIAGENDR))
round.summary(fit.imp.int, digits = 5)[, c("estimate", "2.5 %",
                                           "97.5 %", "p.value")]
```

```
##                            estimate    2.5 %   97.5 % p.value
## (Intercept)                -0.11386 -0.12345 -0.10426 0.00000
## BMXWAIST                    0.00032  0.00023  0.00041 0.00000
## smokerPast                  0.00142 -0.00089  0.00374 0.22889
## smokerCurrent               0.00088 -0.00193  0.00368 0.53975
## RIDAGEYR                    0.00031  0.00025  0.00037 0.00000
## RIAGENDRFemale             -0.00027 -0.01233  0.01178 0.96466
## race_ethNH White           -0.00388 -0.00650 -0.00125 0.00383
## race_ethNH Black           -0.00272 -0.00636  0.00092 0.14342
## race_ethNH Other           -0.00203 -0.00588  0.00182 0.30049
## income$25,000 to <$55,000   0.00104 -0.00205  0.00413 0.51008
## income$55,000+              0.00040 -0.00241  0.00320 0.78212
## BMXWAIST:RIAGENDRFemale    -0.00005 -0.00016  0.00007 0.45021
```

In this example, the interaction is only one term, but if there were multiple terms, use `D1()` to get a multiple degree of freedom Wald test of significance by comparing the model with an interaction to the model without (or `D3()` to get a likelihood ratio test).

```
# Fit reduced model with no interaction
fit.imp.noint <- with(imp.int,
                      lm(LBDGLUSI_trans ~ BMXWAIST + smoker + RIDAGEYR +
                         RIAGENDR + race_eth + income))

# Type 3 Wald test of interaction
summary(D1(fit.imp.int, fit.imp.noint))
```

```
##
## Models:
##  model
##      1
##      2
##
             formula
##  LBDGLUSI_trans ~ BMXWAIST + smoker + RIDAGEYR + RIAGENDR + race_eth + income +
    BMXWAIST:RIAGENDR
##                     LBDGLUSI_trans ~ BMXWAIST + smoker + RIDAGEYR + RIAGENDR +
    race_eth + income
##
## Comparisons:
##    test statistic df1   df2 dfcom p.value    riv
##  1 ~~ 2    0.5706   1 911.4   988  0.4502 0.0326
##
## Number of imputations:  20   Method D1
```

9.6.4.2 Interaction via transform-then-impute

A more general approach to including an interaction, which works even when a categorical variable in the interaction has missing values or when both variables are continuous, is to explicitly compute the interaction terms during pre-processing followed by imputation including the computed terms. If both terms in the interaction are continuous, simply create (during pre-processing) a variable that is their product (transform-then-impute). If either are categorical, the process involves a few additional steps.

First, fit the regression model, including the interaction, to the original pre-imputation dataset to confirm what terms comprise the interaction.

```
fit.pre <- lm(LBDGLUSI_trans ~ BMXWAIST + smoker + RIDAGEYR +
                    RIAGENDR + race_eth + income +
               BMXWAIST:RIAGENDR, data = nhanes)
names(coef(fit.pre))
```

```
##  [1] "(Intercept)"              "BMXWAIST"
##  [3] "smokerPast"               "smokerCurrent"
##  [5] "RIDAGEYR"                 "RIAGENDRFemale"
##  [7] "race_ethNH White"         "race_ethNH Black"
##  [9] "race_ethNH Other"         "income$25,000 to <$55,000"
## [11] "income$55,000+"           "BMXWAIST:RIAGENDRFemale"
```

There is one interaction term – `BMXWAIST:RIAGENDRFemale`. To explicitly compute this term, derive a variable that is `BMXWAIST` multiplied by an indicator variable corresponding to `RIAGENDER == Female` (1 when this condition is true, 0 when false). In general, first create an indicator variable for each non-reference level of the categorical variable and compute the product of each indicator variable and the other term in the interaction.

NOTE: Make sure to drop the old categorical variable from the dataset.

```
nhanes_int <- nhanes %>%
  mutate(RIAGENDRFemale = as.numeric(RIAGENDR == "Female"),
         BMXWAIST_RIAGENDRFemale = BMXWAIST*RIAGENDRFemale) %>%
  select(-RIAGENDR)
```

```
# Checking
table(nhanes$RIAGENDR, nhanes_int$RIAGENDRFemale, useNA = "ifany")
SUB <- nhanes$RIAGENDR == "Female"
# Should all be 0
summary(nhanes_int$BMXWAIST[SUB] -
        nhanes_int$BMXWAIST_RIAGENDRFemale[SUB])
summary(nhanes_int$BMXWAIST_RIAGENDRFemale[!SUB])
# (results not shown)
```

Next, fit the imputation model.

```
imp.int <- mice(nhanes_int,
            seed  = 3,
            m     = 20,
            print = F)
```

Finally, fit the regression model using `with()`, including the new gender indicator variable and the derived interaction term (which is entered just like any other term in the model, not with a :).

```
fit.imp.int <- with(imp.int,
                lm(LBDGLUSI_trans ~ BMXWAIST + smoker + RIDAGEYR +
                     RIAGENDRFemale + race_eth + income +
                     BMXWAIST_RIAGENDRFemale))
round.summary(fit.imp.int, digits = 5)
```

```
##                             estimate std.error statistic    df p.value     2.5 %
## (Intercept)                 -0.11339   0.00487  -23.2773 943.7 0.00000 -0.12295
## BMXWAIST                     0.00032   0.00005    6.9489 915.3 0.00000  0.00023
## smokerPast                   0.00138   0.00117    1.1777 982.6 0.23921 -0.00092
## smokerCurrent                0.00078   0.00143    0.5491 977.9 0.58308 -0.00201
## RIDAGEYR                     0.00031   0.00003    9.8267 977.1 0.00000  0.00024
## RIAGENDRFemale              -0.00122   0.00613   -0.1992 894.9 0.84215 -0.01326
## race_ethNH White            -0.00390   0.00133   -2.9242 971.4 0.00353 -0.00652
## race_ethNH Black            -0.00292   0.00185   -1.5780 979.0 0.11490 -0.00655
## race_ethNH Other            -0.00207   0.00196   -1.0584 976.6 0.29012 -0.00592
## income$25,000 to <$55,000    0.00082   0.00164    0.5002 462.1 0.61717 -0.00240
## income$55,000+               0.00019   0.00145    0.1333 610.8 0.89402 -0.00266
## BMXWAIST_RIAGENDRFemale     -0.00004   0.00006   -0.5837 882.4 0.55956 -0.00015
##                               97.5 %
## (Intercept)                 -0.10383
## BMXWAIST                     0.00041
## smokerPast                   0.00369
## smokerCurrent                0.00358
## RIDAGEYR                     0.00037
## RIAGENDRFemale               0.01082
## race_ethNH White            -0.00128
## race_ethNH Black             0.00071
```

```
## race_ethNH Other               0.00177
## income$25,000 to <$55,000      0.00404
## income$55,000+                 0.00305
## BMXWAIST_RIAGENDRFemale        0.00008
```

9.6.4.3 Estimate the effect of one variable at levels of the other

To estimate the effect of one variable at levels of the other, use `gmodels::estimable()` on each imputation separately and pool the results using Rubin's rules via `pool.scalar()` (with some additional code to compute p-values).

```
# Compute mean and variance from each analysis
# Initialize vector for level 1 (Male)
MEAN1 <- VAR1 <- rep(NA, length(fit.imp.int$analyses))
# Initialize vector for level 2 (Female)
MEAN2 <- VAR2 <- rep(NA, length(fit.imp.int$analyses))

for(i in 1:length(MEAN1)) {
  EST1 <- gmodels::estimable(fit.imp.int$analyses[[i]],
                             c("BMXWAIST" = 1))

  EST2 <- gmodels::estimable(fit.imp.int$analyses[[i]],
                             c("BMXWAIST" = 1,
                               "BMXWAIST_RIAGENDRFemale" = 1))

  MEAN1[i] <-  as.numeric(EST1["Estimate"])
  VAR1[ i] <- (as.numeric(EST1["Std. Error"])^2)
  MEAN2[i] <-  as.numeric(EST2["Estimate"])
  VAR2[ i] <- (as.numeric(EST2["Std. Error"])^2)
}

# Apply Rubin's rules
POOLED1 <- pool.scalar(MEAN1,
                       VAR1,
                       nrow(imp.int$data))
POOLED2 <- pool.scalar(MEAN2,
                       VAR2,
                       nrow(imp.int$data))

# Extract the pooled mean and standard error
DF <- data.frame(gender = c("Male", "Female"),
                 est    = c(POOLED1$qbar, POOLED2$qbar),
                 se     = sqrt(c(POOLED1$t, POOLED2$t)),
                 df     = c(POOLED1$df, POOLED2$df))

# Compute test statistic and p-value
DF$t <- DF$est/DF$se
DF$p <- 2*(pt(abs(DF$t), DF$df, lower.tail = F))
DF
```

```
##   gender       est         se    df     t                 p
## 1   Male 0.0003173 0.00004567 925.1 6.949 0.000000000006944
## 2 Female 0.0002822 0.00003910 841.6 7.219 0.000000000001175
```

9.7 Logistic regression after MI

The syntax for logistic regression after MI is similar to linear regression after MI, except that now `glm()` is used instead of `lm()`.

Example 9.2: Using the NSDUH 2019 teaching dataset (see Appendix A.5), what is the association between lifetime marijuana use (`mj_lifetime`) and age at first use of alcohol (`alc_agefirst`), adjusted for age (`demog_age_cat6`), sex (`demog_sex`), and income (`demog_income`)? Use MI to handle missing data.

NOTE: This example uses the MAR dataset created for the illustration in Section 9.2 (`nsduh_mar_rmph.RData`). Those who never used alcohol were removed from the dataset, since we do not want to impute missing ages for them. Also, a random subset of the values for each variable were set to missing, with the probability of missingness for each variable depending on all the other variables but not the variable itself.

First, load the data and view the extent of the missing data.

```
load("Data/nsduh_mar_rmph.RData")
summary(nsduh_mar)
```

```
##  mj_lifetime  alc_agefirst  demog_age_cat6  demog_sex              demog_income
##  No  :343    Min.   : 3.0   18-25: 98       Male  :380   Less than $20,000:117
##  Yes :491    1st Qu.:15.0   26-34:120       Female:392   $20,000 - $49,999:230
##  NA's:  9    Median :17.0   35-49:216       NA's  : 71   $50,000 - $74,999:118
##              Mean   :17.6   50-64:195                    $75,000 or more  :345
##              3rd Qu.:20.0   65+  :173                    NA's             : 33
##              Max.   :45.0   NA's : 41
##              NA's   :49
```

Next, fit the imputation model.

```
imp.nsduh <- mice(nsduh_mar,
                  seed  = 3,
                  m     = 20,
                  print = F)
imp.nsduh
```

```
## Class: mids
## Number of multiple imputations:  20
## Imputation methods:
##    mj_lifetime   alc_agefirst demog_age_cat6      demog_sex   demog_income
##       "logreg"          "pmm"      "polyreg"       "logreg"      "polyreg"
## PredictorMatrix:
##                mj_lifetime alc_agefirst demog_age_cat6 demog_sex demog_income
## mj_lifetime              0            1              1         1            1
## alc_agefirst             1            0              1         1            1
## demog_age_cat6           1            1              0         1            1
## demog_sex                1            1              1         0            1
## demog_income             1            1              1         1            0
```

Finally, fit the logistic regression on the imputed datasets and pool the results.

NOTE: Only include in the regression model variables that were included in the dataset prior to multiple imputation (see Section 9.4.2).

```
fit.imp.glm <- with(imp.nsduh,
                glm(mj_lifetime ~ alc_agefirst + demog_age_cat6 +
                    demog_sex + demog_income, family = binomial))
# summary(pool(fit.imp.glm), conf.int = T)
round.summary(fit.imp.glm, digits=3)
```

```
##                                   estimate std.error statistic    df p.value  2.5 %
## (Intercept)                          6.022     0.588    10.239 683.1   0.000  4.867
## alc_agefirst                        -0.265     0.028    -9.361 594.4   0.000 -0.320
## demog_age_cat626-34                 -0.252     0.334    -0.754 716.0   0.451 -0.908
## demog_age_cat635-49                 -0.824     0.298    -2.769 800.8   0.006 -1.408
## demog_age_cat650-64                 -0.743     0.301    -2.473 773.5   0.014 -1.334
## demog_age_cat665+                   -1.391     0.307    -4.532 788.9   0.000 -1.993
## demog_sexFemale                      0.012     0.172     0.067 478.7   0.947 -0.327
## demog_income$20,000 - $49,999       -0.457     0.271    -1.689 746.0   0.092 -0.989
## demog_income$50,000 - $74,999       -0.032     0.310    -0.103 760.3   0.918 -0.642
## demog_income$75,000 or more         -0.325     0.259    -1.254 727.3   0.210 -0.834
##                                   97.5 %
## (Intercept)                        7.177
## alc_agefirst                      -0.209
## demog_age_cat626-34                0.404
## demog_age_cat635-49               -0.240
## demog_age_cat650-64               -0.153
## demog_age_cat665+                 -0.789
## demog_sexFemale                    0.350
## demog_income$20,000 - $49,999      0.074
## demog_income$50,000 - $74,999      0.577
## demog_income$75,000 or more        0.184
```

Add `exponentiate = T` to `summary()` or `round.summary()` to compute odds ratios, but be careful interpreting the results, as only the estimate and confidence interval are exponentiated; the standard error in the table is still the standard error of the un-exponentiated regression coefficient. As such, to avoid confusion, extract just the estimate, confidence interval, and p-value when exponentiating (and drop the first row since the exponentiated intercept is not of interest).

```
# summary(pool(fit.imp.glm), conf.int = T,
#    exponentiate = T)[-1, c("term", "estimate", "2.5 %", "97.5 %", "p.value")]
round.summary(fit.imp.glm, digits = 3,
        exponentiate = T)[-1, c("estimate", "2.5 %", "97.5 %", "p.value")]
```

```
##                               estimate 2.5 % 97.5 % p.value
## alc_agefirst                     0.767 0.726  0.811   0.000
## demog_age_cat626-34              0.777 0.403  1.498   0.451
## demog_age_cat635-49              0.439 0.245  0.787   0.006
## demog_age_cat650-64              0.475 0.263  0.858   0.014
## demog_age_cat665+                0.249 0.136  0.455   0.000
## demog_sexFemale                  1.012 0.721  1.419   0.947
## demog_income$20,000 - $49,999    0.633 0.372  1.077   0.092
## demog_income$50,000 - $74,999    0.968 0.526  1.782   0.918
## demog_income$75,000 or more      0.722 0.434  1.202   0.210
```

9.7.1 Multiple degree of freedom tests

`mi.anova()` does not work for logistic regression models, but `D1()` and `D3()` do. Fit reduced models, each omitting one categorical variable that has more than two levels and then use `D1()` to compare the full and reduced models using a Wald test or `D3()` for a likelihood ratio test.

For example, to get a Type 3 Wald test and LR test for `demog_age_cat6` in Example 9.2:

```
# Fit reduced model
fit.imp.glm.age    <- with(imp.nsduh,
  glm(mj_lifetime ~ alc_agefirst +
        demog_sex + demog_income, family = binomial))

# Compare full and reduced models
# Wald test
summary(D1(fit.imp.glm, fit.imp.glm.age))
# LR test
summary(D3(fit.imp.glm, fit.imp.glm.age))
# (results not shown)
```

9.7.2 Predictions

Prediction proceeds in almost the same way as described in Section 9.6.2 – compute estimates within each imputed dataset and then pool the results using Rubin's rules. The one change is that `type = "response"` is added to get estimates that are probabilities.

Example 9.2 (continued): What is the estimated probability (and standard error) of lifetime marijuana use for someone who first used alcohol at age 13 years, is currently age 30 years, male, and has an annual income of <$20,000?

```
# Prediction data.frame
NEWDAT <- data.frame(alc_agefirst    = 13,
                     demog_age_cat6 = "26-34",
                     demog_sex      = "Male",
                     demog_income   = "Less than $20,000")

# Estimate for each analysis
PREDLIST <- lapply(fit.imp.glm$analyses, predict, newdata=NEWDAT,
                   se.fit=T, type = "response")

# Extract mean and variance from each analysis
MEAN <- VAR <- rep(NA, length(fit.imp.glm$analyses))
for(i in 1:length(MEAN)) {
  MEAN[i] <-  PREDLIST[[i]]$fit
  VAR[ i] <- (PREDLIST[[i]]$se.fit)^2
}

# Apply Rubin's rules
PRED.POOLED <- pool.scalar(MEAN,
                           VAR,
                           nrow(imp.nsduh$data))

# Extract the pooled mean
PRED.POOLED$qbar
```

```
## [1] 0.911
```

```
# Extract the pooled standard error
sqrt(PRED.POOLED$t)
```

```
## [1] 0.02772
```

9.8 Cox regression after MI

MI with time-to-event data that includes censoring is more complex than what we have seen so far. The issue is that the outcome is not just a single variable, but a pair of variables, the event time and the event indicator, with the event time being censored for those with event indicator = 0. The obvious solution is to include both the event indicator and time in the imputation model. However, a better approach is to include in the imputation model the event indicator and, instead of the event time, the cumulative baseline hazard, and possibly also the interaction between the event indicator and the cumulative baseline hazard (White and Royston, 2009). The hazard function was discussed in Section 7.4. The baseline hazard is the hazard function when all predictors are at zero or their reference level, and the cumulative baseline hazard sums the baseline hazard up to a given time.

The strategy is to compute each individual's cumulative baseline hazard at their event time using the `mice` function `nelsonaalen()` and add it to the dataset in place of the event time, carry out MI using `mice()`, and put the event time back into the imputed datasets before fitting the Cox model on each imputed dataset and pooling the results.

Example 9.3: Using the Natality teaching dataset (see Appendix A.3), estimate the association between previous preterm birth (`RF_PPTERM`) and time to preterm birth, adjusted for mother's age (`MAGER`), race/ethnicity (`MRACEHISP`), and marital status (`DMAR`). The event time and event indicator variables are `gestage37` and `preterm01`, respectively. Handle missing data using MI.

```
load("Data/natality2018_rmph.Rdata")

# Compute cumulative baseline hazard and add to the dataset
natality$cumhaz <- nelsonaalen(natality, gestage37, preterm01)
```

What is this actually doing? Each individual has an event time. We *could* include that time in the imputation model. Doing so would mean that we think the predictors are associated linearly with the event time. What makes more sense, however, is to assume that the predictors are associated with the *risk* of the event at a given time. It turns out that the appropriate term to put into the imputation model is the cumulative baseline hazard up to that time. This replaces linear time with a non-linear function of time. In Figure 9.6, we see that below about 26 weeks the cumulative baseline hazard is relatively constant, and then increases non-linearly. Using linear time in the imputation model would be equivalent to assuming the accumulation of risk is constant over time, which is generally not realistic.

Next, fit the imputation model, including `cumhaz` instead of the time variable.

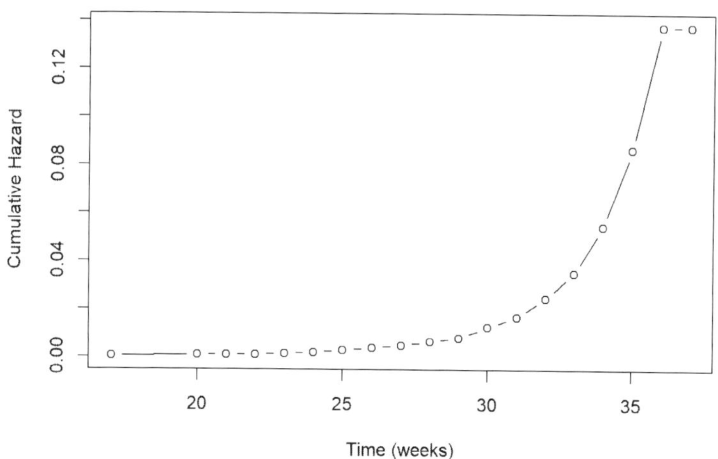

FIGURE 9.6 Cumulative baseline hazard vs. time

```
# Select the variables for inclusion in the imputation model
# Include the cumulative baseline hazard
# Include the event indicator variable
# Do NOT include the event time variable
natality_for_imp <- natality %>%
  select(cumhaz, preterm01, RF_PPTERM, MAGER, MRACEHISP, DMAR)

imp.natality <- mice(natality_for_imp,
                     seed  = 3,
                     m     = 20,
                     print = F)
```

Next, put the time variable back into the dataset for use in `coxph()`. If there were any missing time values, replace them with imputed values. This is done by creating a look-up table (`bhz`) and finding the time value that corresponds to each imputed `cumhaz` value. This works because the baseline cumulative hazard depends only on time, not on any of the predictors in the model.

NOTE: The code below only works if predictive mean matching (the default) was used to impute missing cumulative baseline hazard values.

```
# Imputed datasets in long form
imp.natality.dat <- complete(imp.natality, "long", include = TRUE)

# Repeat time variable m + 1 times since impdat
# includes the original data as well as m imputations
imp.natality.dat$gestage37 <- rep(natality$gestage37, imp.natality$m + 1)

# Replace missing time values with time corresponding
# to the imputed cumulative hazard value
# (only needed if there were missing event times,
#  but no harm in running if there were not)
#  .imp > 0 prevents replacing missing values in the original data
```

```
SUB <- imp.natality.dat$.imp > 0 & is.na(imp.natality.dat$gestage37)
if(sum(SUB) > 0) {
  # Create a look-up table with the event times
  # and corresponding cumulative hazards
  bhz <- data.frame(time   = natality$gestage37,
                    cumhaz = natality$cumhaz)
  # Sort and remove duplicates
  bhz <- bhz[order(bhz$time),]
  bhz <- bhz[!duplicated(bhz) & complete.cases(bhz$time),]

  # The following only works if pmm (the default) was used
  # to impute missing cumhaz values
  # (because it relies on the imputed values being values
  #  present in the non-missing values)
  for(i in 1:sum(SUB)) {
    # Use max since last 2 times have the same cumhaz
    imp.natality.dat$gestage37[SUB][i] <-
      max(bhz$time[bhz$cumhaz == imp.natality.dat$cumhaz[SUB][i]], na.rm = T)
  }
}

# Convert back to a mids object
imp.natality.new <- as.mids(imp.natality.dat)
```

Finally, fit the Cox regression on the imputed datasets and display the pooled results.

NOTE: Only include in the regression model variables that were included in the dataset prior to multiple imputation (see Section 9.4.2).

```
library(survival)
fit.imp.cox <- with(imp.natality.new,
                    coxph(Surv(gestage37, preterm01) ~
                          RF_PPTERM + MAGER + MRACEHISP + DMAR))
# Do NOT include the -1 here since a Cox model has no intercept
# summary(pool(fit.imp.cox), conf.int = T,
#         exponentiate = T)[, c("term", "estimate", "2.5 %", "97.5 %", "p.value")]
round.summary(fit.imp.cox, digits = 3,
              exponentiate = T)[, c("estimate", "2.5 %", "97.5 %", "p.value")]
```

```
##                      estimate 2.5 % 97.5 % p.value
## RF_PPTERMYes            3.213 2.079  4.966   0.000
## MAGER                   1.031 1.008  1.054   0.007
## MRACEHISPNH Black       1.858 1.322  2.613   0.000
## MRACEHISPNH Other       1.111 0.685  1.801   0.668
## MRACEHISPHispanic       1.328 0.965  1.828   0.082
## DMARUnmarried           1.822 1.339  2.479   0.000
```

9.8.1 Multiple degree of freedom tests

Neither `mi.anova()` nor `D3()` works for Cox regression models, but `D1()` does. For example, to get a Type 3 Wald test for `MRACEHISP` in Example 9.3:

```
fit.imp.cox.race <- with(imp.natality.new,
                    coxph(Surv(gestage37, preterm01) ~
                          RF_PPTERM + MAGER + DMAR))

summary(D1(fit.imp.cox, fit.imp.cox.race))
# (results not shown)
```

9.8.2 Predictions

See Section 7.10 for how to compute predictions from a Cox model. Prediction after MI proceeds as in Sections 9.6.2 and 9.7.2 – estimate within each of the models fit with the imputed datasets and pool the results using Rubin's rules.

9.9 Number of imputations revisited

In Section 9.4.6 you were instructed to simply use 20 imputations. That approach was taken so as to not complicate the initial presentation of the other aspects of multiple imputation. However, the correct number of imputations needed is actually a function of information that is obtained *after* fitting the regression model. von Hippel (2020) suggests to first fit the imputation model 20 times, then fit your regression model, and then recompute the number of imputations using the `how_many_imputations()` function in the `howManyImputations` library (Errickson, 2023). If that number is larger than 20, then re-fit the imputation model with the larger number of imputations. See also von Hippel (2019).

For example:

```
imp <- mice(nhanes,
            seed  = 3,
            m     = 20,
            print = F)

# Temporary fit, just to get the number of imputations
fit <- with(imp,
            lm(LBDGLUSI_trans ~ BMXWAIST + smoker + RIDAGEYR +
                                RIAGENDR + race_eth + income))

howManyImputations::how_many_imputations(fit)
```

```
## [1] 7
```

Since we already fit the model with `m = 20` imputations, and having more imputations is better than fewer, there is no need to re-fit the imputation and regression models. However, if `how_many_imputations()` had returned a number larger than 20, then we would re-fit the imputation model with that larger number and then re-fit the regression model with the updated `mids` object.

9.10 Regression diagnostics after MI

In previous chapters, various assumption checking and diagnostics were discussed for each type of regression model. How do we diagnose the fit of a model after MI? Any visual diagnostic plot can be examined within each imputation individually and compared between imputations. Also, diagnostic statistics can be computed within each imputation and pooled, for example, using `pool.scalar()`, `micombine.chisquare()`, or `micombine.F()` (the latter two from the `miceadds` package (Robitzsch et al., 2023)).

Two examples will be demonstrated here: visually checking linearity for a continuous predictor in a linear regression model and carrying out a Hosmer-Lemeshow goodness-of-fit test for a logistic regression model.

9.10.1 Example: Examining a diagnostic plot across imputations

As described in Section 5.16, use a CR plot to check the linearity assumption of a linear regression model. Display the plot for each imputation to get a visual assessment of the adequacy of the linearity assumption. For the linear regression model fit in Section 9.6, there were 20 imputations.

In practice, view all the plots for all continuous predictors. Here, for brevity, only the linearity assumption for one of the continuous predictors from Example 9.1 (`BMXWAIST`) is assessed and only for the first four imputations. As shown in Figure 9.7, the linearity assumption appears to be consistently met.

```
# To view for all imputations
# for(i in 1:length(fit.imp.lm$analyses)) {
#     Insert car::crPlots call here
# }

# Just the first four imputations
par(mfrow=c(2,2))
for(i in 1:4) {
  car::crPlots(fit.imp.lm$analyses[[i]],
               terms = ~ BMXWAIST,
               pch=20, col="gray",
               smooth = list(smoother=car::gamLine),
               ylab = "Comp + Resid",
               # In general, no need to specify ylim.
               # It is included here because outliers
               # make it harder to see patterns on the
               # the original scale, so zooming in helps.
               ylim = c(-0.05, 0.05))
}
```

9.10.2 Example: Pooling a diagnostic test over imputations

The Hosmer-Lemeshow goodness-of-fit test computes a chi-squared statistic. Compute this statistic on the fit to each imputed dataset, and then use `micombine.chisquare()` (Li et al., 1991; Enders, 2010) to pool over imputations. Although, in this example, the individual tests are chi-square tests, after pooling the over imputations the test turns out to be an F test.

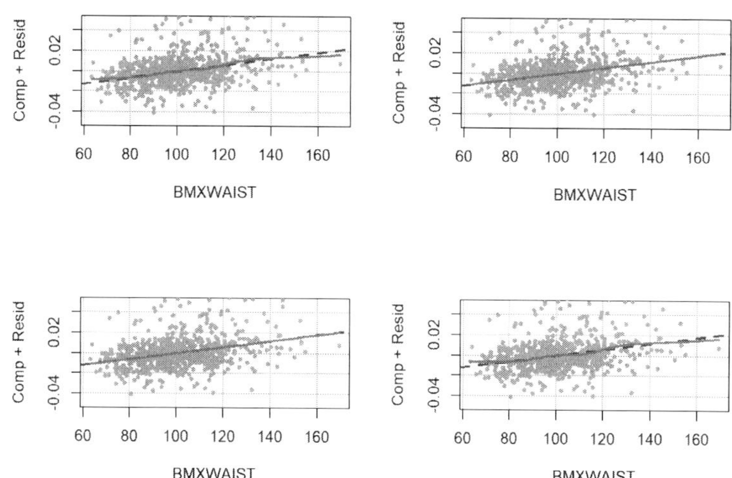

FIGURE 9.7 Checking the linearity assumption within each imputation

```
# Initialize a vector to store the chi-squared values
X2 <- rep(NA, length(fit.imp.glm$analyses))

# Compute chi-squared statistic on each imputation
# (and store the degrees of freedom, as well)
for(i in 1:length(X2)) {
  HL <- ResourceSelection::hoslem.test(fit.imp.glm$analyses[[i]]$y,
                                       fit.imp.glm$analyses[[i]]$fitted.values)
  X2[i] <- HL$statistic # Chi-squared value
  DF    <- HL$parameter # Degrees of freedom
  rm(HL)
}

# Input the chi-squared vector and the degrees of freedom
# to pool over imputations
micombine.chisquare(X2, DF)
```

```
## Combination of Chi Square Statistics for Multiply Imputed Data
## Using 20 Imputed Data Sets
## F(8, 163.64)=0.321     p=0.9571
```

9.11 Exercises

1. True or false? Listwise deletion is an effective method for handling missing data that is applicable to a wide variety of analyses.

2. True or false? “Rubin’s Rules” are used to process the imputation-specific estimates to produce a single estimate and appropriately account for both the within- and between-imputation variation.

3. Explain the distinction between unit and item non-response. Which type of non-response can be handled with multiple imputation (as discussed in this chapter)?

4. What type of missing data method is used when carrying out a complete case analysis, and is the default for most regression software?

5. When imputing missing values, what is wrong with imputing a missing value with the mean of the observed values? How does conditional mean imputation help, and why is it still inadequate?

6. Suppose you have a dataset with two variables, Y = hospital readmission within one year of a stroke and X = annual income, where the readmission information is complete but income is missing for many individuals. For each of the following statements, is the missing data mechanism MCAR, MAR, or MNAR? NOTE: The notation $P(\text{Income is missing}|\text{Readmission, Income})$ is read "the probability that income is missing for a patient given their readmission and income values". It may seem strange to talk about the probability that income is missing given income. But think of it as "how likely is it that, given the actual but possibly unknown value of income, income is missing". Perhaps those with higher or lower income values are more or less likely to not report their income in a survey.

- $P(\text{Income missing}|\text{Readmission, Income}) = P(\text{Income missing}|\text{Readmission})$. Given the observed information (readmission status), the probability that income is missing is independent of its value. All patients with the same readmission status have the same chance of having not reported their income, regardless of their actual income.
- $P(\text{Income missing}|\text{Readmission, Income}) = P(\text{Income missing})$. The probability that income is missing is independent of the observed information (readmission status) and the incomplete information (income). All patients have the same chance of having not reported their income.
- $P(\text{Income missing}|\text{Readmission, Income}) = P(\text{Income missing}|\text{Readmission, Income})$. Even after controlling for the observed information (readmission status), the probability that income is missing depends on the (possibly unknown) income value. Even among observations with the same readmission status, the chance that income is missing depends on its value.

7. When using multiple imputation, how many imputations do you need?

8. When using multiple imputation, how should you handle a variable that will be transformed in your analysis (e.g., a categorical variable that will be collapsed; a continuous variable that will be transformed using some function)?

For Exercises 9 to 13, use the Natality teaching dataset (`natality2018_rmph.Rdata`, see Appendix A.3).

9. Prior to imputation, compute descriptive statistics for father's race/Hispanic origin (`FRACEHISP`), education (`FEDUC`), and age (`FAGECOMB`), including the number of missing values. Use all available data for each variable (rather than a complete case analysis).

10. Next, we would like to compute descriptive statistics for father's race/Hispanic origin (`FRACEHISP`), education (`FEDUC`), and age (`FAGECOMB`) after using multiple imputation. There are a number of auxiliary variables that can be included in the imputation model. The auxiliary variables are non-analysis variables in the dataset that may be correlated with the father's characteristics and/or correlated with the chance that the father's characteristics are missing. The following is the full list of analysis and auxiliary variables.

- Father's race/Hispanic origin (FRACEHISP)
- Father's education (FEDUC)
- Father's age (FAGECOMB)
- Mother's race/Hispanic origin (MRACEHISP)
- Mother's education (MEDUC)
- Mother's age (MAGER)
- Marital status (DMAR)
- Prior births now living (PRIORLIVE)
- Birthweight (g) (DBWT)
- Month prenatal care began (PRECARE)
- WIC (WIC)
- Risk factors reported (risks)
- Preterm birth (preterm)

11. Fit the imputation model with five imputations and examine the output. What method was used to impute each variable?

12. Visualize the imputations for father's age.

13. Compute the descriptive statistics after MI. How do these compare to the descriptives before using MI (computed in Question 11) and what does that say about the nature of the missing data?

For Exercises 14 to 20, use the 2020 UN Human Development Data (`unhdd2020.rmph.Rdata`, see Appendix A.2).

14. After handling missing data using multiple imputation, fit a regression model to test the association between the outcome "child under 5y mortality (2018, per 1,000 live births)" (`mort_lt5`) and the predictors "female population with at least some secondary education (2015-2019, % ages 25 and older)" (`educ_f`), "child malnutrition - stunting (moderate or severe) (2010-2019, % under age 5)" (`stunt`), and "infants exclusively breastfed (2010-2019, % ages 0-5 months)" (`breast`). Assume no changes need to be made to any of these variables – you will explore other aspects of this analysis in subsequent questions.

15. Check the normality, linearity, and constant variance assumptions for the model you fit in the previous exercise. What assumptions are violated?

16. Examine a histogram of the outcome. Use a Box-Cox outcome transformation (based on the original data, before imputation), re-fit the imputation model, re-fit the regression model, and re-check the normality, linearity, and constant variance assumptions. Do any problems remain?

17. Starting with the model you fit in the previous exercise, relax the linearity assumption for `stunt` and `breast` using polynomial transformations (e.g., quadratic, cubic, or higher order). Remember to center each variable prior to transformation. Then re-check the normality, linearity, and constant variance assumptions. Do any problems remain?

18. Re-do the previous exercise, this time also including a quadratic for `educ_f`. Then re-check the normality, linearity, and constant variance assumptions. Do any problems remain?

19. Using the final model from the previous exercise, predict child mortality for a nation with 40% child malnutrition, 25% infants exclusively breastfed, and 35% female population with at least some secondary education. Compare this to the prediction when these values are 10%, 70%, and 90%. Hint: In your prediction `data.frame`, enter a value for every term in the model, and the terms in this model were centered and some where squared or cubed.

20. Using the final model from Exercise 18, expand the imputation and regression models to assess if the quadratic association between female education and mortality depends on countries' HDI group (`hdi_group`). Use the transform-then-impute method for including an interaction.

For Exercises 21 to 23, use the NHANES 2017-2018 fasting subsample teaching dataset (`nhanes1718_adult_fast_sub_rmph.Rdata`, see Appendix A.1). Create a dichotomous version of PHQ-9 representing "at least mild depression" (PHQ-9 $\geq$ 5) using the following code.

```
load("Data/nhanes1718_adult_fast_sub_rmph.Rdata")

# Create dichotomized PHQ-9
# "PHQ-9 scores of 5, 10, 15, and 20 represented
#  mild, moderate, moderately severe, and
#  severe depression, respectively"
#  (https://www.ncbi.nlm.nih.gov/pmc/articles/PMC1495268/)

nhanes <- nhanes_adult_fast_sub %>%
  mutate(depression = factor(phq9 >= 5,
                             levels = c(F, T),
                             labels = c("No", "Yes")))

# Check derivation
table(nhanes$phq9, nhanes$depression, useNA = "ifany")
tapply(nhanes$phq9, nhanes$depression, range, na.rm=T)
```

21. After handling missing data using multiple imputation, fit a regression model to test if the outcome "at least mild depression" is significantly associated with trouble sleeping (`SLQ050`) after adjusting for age (`RIDAGEYR`), gender (`RIAGENDR`), income (`income`), and days someone engages in vigorous recreational activities (`PAQ655`)? Answer the question and report the AOR, 95% confidence interval, and p-value. Also, which other predictors are significantly associated with "at least mild depression"?

22. Expand the model you fit in the previous exercise to assess whether the association between "at least mild depression" and trouble sleeping depends on gender. Use the stratification method for imputing an interaction. Regardless of the statistical significance of the interaction, estimate the sleep effect at each level of gender.

23. Assess the goodness-of-fit of the model from the previous exercise using the Hosmer-Lemeshow test, as well as calibration plots.

24. For this exercise, use the teaching dataset based on the Framingham Heart Study (`fram_time_invar_rmph.rData`, see Appendix A.6). After handling missing data using multiple imputation, fit a regression model to test if time to angina differs between participants with different levels of education (`EDUC`), adjusted for age (`AGE`) and sex (`SEX`). The time variable is `TIMEAP` and the event indicator is `ANGINA`.

A

Datasets

This text uses datasets derived from freely available public health datasets for examples and exercises. In many cases the original datasets were modified in some way (e.g., taking a random subset, deriving variables).

These datasets are suitable for teaching only. The author makes no claim or implication that any inferences derived from these teaching datasets are valid. If you wish to use a dataset for publishable research, obtain the data and documentation directly from each source. Each has a unique design, and it is vital to understand the details before using the data for research.

For each teaching dataset, the dataset itself, instructions regarding how to download the dataset, and/or R code for creating the dataset are available at RMPH Resources[1]. To use the code in this text as-is, create an R project with a subfolder called `Data`, place all the datasets in that subfolder, and run R from within your project.

A.1 NHANES (2017-2018)

Background

The National Health and Nutrition Examination Survey (NHANES)[2] is a survey designed to "assess the health and nutritional status of adults and children in the United States". The study began in the 1960s and since 1999 has has been conducted in 2-year cycles (e.g, 1999-2000, 2001-2002). This nationally representative survey includes both interviews and physical examinations. NHANES is conducted by the National Center for Health Statistics (NCHS), which is part of the Centers for Disease Control and Prevention (CDC) (National Center for Health Statistics, 2021a,b).

Documentation

A description of the 2017-2018 survey cycle target population, objectives, and data collection procedures can be found at NHANES 2017-2018 Overview[3]. See, in particular, the section on "Guidance for NHANES Data Users". NHANES data includes demographics, chronic conditions, and risk factors. See NHANES 2017-2018 Data Details[4] for a full list of datasets and their documentation. Individual datasets (e.g., Demographics, Body Measures,

[1] https://github.com/rwnahhas/RMPH_Resources
[2] https://www.cdc.gov/nchs/nhanes
[3] https://wwwn.cdc.gov/nchs/nhanes/continuousnhanes/overview.aspx?BeginYear=2017
[4] https://wwwn.cdc.gov/nchs/nhanes/continuousnhanes/default.aspx?BeginYear=2017

Cholesterol – Total) are freely downloadable and can be merged (within a cycle) on the variable SEQN. Data from different survey cycles are from different individuals, however, not longitudinal. See also the NCHS Data User Agreement[5].

Teaching Datasets

Any analyses, interpretations, or conclusions reached herein are only for the purpose of illustrating regression methods and are credited to the author, not to NCHS, which is responsible only for the initial data. The author makes no claim or implication that any inferences derived from these teaching datasets are valid.

The teaching datasets used in this text were merged from multiple NHANES 2017-2018 data files and include a random subset of 1000 observations from adults in the examination data (`nhanes1718_adult_exam_sub_rmph.Rdata`), and a random subset of 1000 observations from adults in the fasting subsample (`nhanes1718_adult_fast_sub_rmph.Rdata`). In each case, sampling was done with replacement using the appropriate subsample weights in order to approximate a nationally representative distribution. This sampling method is solely for the purpose of creating a teaching dataset to illustrate regression methods. Chapter 8 discusses analyzing data using the survey weights appropriately using the full dataset (`nhanes1718_rmph.Rdata`).

In these NHANES teaching datasets, variable names in CAPS are coded as in the original dataset (with the exception of missing value codes being set to NA and some cases assigned values based on skip patterns). Variable names in lowercase were derived from other variables (e.g., `smoker`).

Creating the Teaching Datasets

To create the teaching datasets, do the following:

- Download the R script file `NHANES 2017 2018 Process.R` from RMPH Resources[6].
- Run the R script file `NHANES 2017 2018 Process.R` to download and process the raw NHANES data. There is no need to download the NHANES data directly from NCHS as it will be downloaded automatically when you run the script.
- The script will create the following teaching datasets:
  - `nhanes1718_rmph.Rdata`
  - `nhanes1718_adult_exam_sub_rmph.Rdata`
  - `nhanes1718_adult_fast_sub_rmph.Rdata`
  - `nhanes_CC_rmph.Rdata` (an artificial matched case-control dataset used to illustrate conditional logistic regression)
  - `nhanesf.complete.50_rmph.Rdata` (a subsample of size 50 used for a small sample size example)
  - `nhanesf.complete.30_rmph.Rdata` (a subsample of size 30 used for a small sample size exercise)
- Place these `.Rdata` files in your “Data” folder.

Rows and Columns

These files have the following numbers of rows and columns:

```
load("Data/nhanes1718_rmph.Rdata")
dim(nhanes)
```

[5] https://www.cdc.gov/nchs/data_access/restrictions.htm
[6] https://github.com/rwnahhas/RMPH_Resources

```
## [1] 9254   90
```

```
load("Data/nhanes1718_adult_exam_sub_rmph.Rdata")
dim(nhanes_adult_exam_sub)
```

```
## [1] 1000   85
```

```
load("Data/nhanes1718_adult_fast_sub_rmph.Rdata")
dim(nhanes_adult_fast_sub)
```

```
## [1] 1000   86
```

```
load("Data/nhanes_CC_rmph.Rdata")
dim(nhanes_CC)
```

```
## [1] 890   6
```

```
load("Data/nhanesf.complete.50_rmph.Rdata")
dim(nhanesf.complete.50)
```

```
## [1] 50 12
```

```
load("Data/nhanesf.complete.30_rmph.Rdata")
dim(nhanesf.complete.30)
```

```
## [1] 30   4
```

A.2 United Nations Human Development Data (2020)

Background

The 2020 United Nations Human Development Data (UNHDD) contain nation-level measures, including the Human Development Index (HDI) and various economic and health indicators.

"The entire series of Human Development Index (HDI) values and rankings are recalculated every year using the same the most recent (revised) data and functional forms. The HDI rankings and values in the 2014 Human Development Report cannot therefore be compared directly to indices published in previous Reports." Please see hdr.undp.org[7] for more information.

"The HDI was created to emphasize that people and their capabilities should be the ultimate criteria for assessing the development of a country, not economic growth alone. The HDI can also be used to question national policy choices, asking how two countries with the same

[7] http://hdr.undp.org/

level of GNI per capita can end up with different human development outcomes". (United Nations Development Programme, 2020)

Documentation

See the Human Development Report 2020: Reader's Guide[8] for details about the data. The sidebar on that page has links to additional documentation. See the use license at Copyright and Terms of Use[9].

Teaching Dataset

The licensor in no way endorses the author or the author's use of this data. Any analyses, interpretations, or conclusions reached herein are only for the purpose of illustrating regression methods and are credited to the author, not to the licensor. The author makes no claim or implication that any inferences derived from this teaching dataset are valid.

The raw data were downloaded from the UNHDD Download Data page[10] on September 8, 2021, and processed to format variables as factors to create the teaching dataset `unhdd2020.rmph.rData`. No other changes were made to the data, although not all available variables were included in the teaching dataset. All included variables represent 2019 data unless noted otherwise in the variable labels. The `.rData` file is not labelled; however, the file `UNHDD 2020.csv` (available from RMPH Resources) has labels in the first row.

Creating the Teaching Dataset

To create the teaching dataset, do the following:

- Download `unhdd2020.rmph.rData` from RMPH Resources[11].
- Place this `.Rdata` file in your "Data" folder.

Rows and Columns

This file has the following numbers of rows and columns:

```
load("Data/unhdd2020.rmph.rData")
dim(unhdd)
```

```
## [1] 189  70
```

A.3 U.S. Natality (2018)

Background

As required by federal law in the United States, the 2018 United States Birth Data[12] were compiled from information on birth certificates by the National Vital Statistics System, part of the National Center for Health Statistics, in cooperation with states (National Center for

[8] https://hdr.undp.org/en/content/human-development-report-2020-readers-guide
[9] http://hdr.undp.org/en/content/copyright-and-terms-use
[10] http://hdr.undp.org/en/content/download-data
[11] https://github.com/rwnahhas/RMPH_Resources
[12] https://www.cdc.gov/nchs/nvss/births.htm

Health Statistics, 2022). Information collected includes gestational age, birthweight, maternal and paternal demographic information, risk factors, and characteristics of the labor and delivery.

Documentation

Documentation can be found in the User Guide to the 2018 Natality Public Use File[13]. See also the NCHS Data User Agreement[14].

Teaching Datasets

Any analyses, interpretations, or conclusions reached herein are only for the purpose of illustrating regression methods and are credited to the author, not to NCHS, which is responsible only for the initial data. The author makes no claim or implication that any inferences derived from these teaching datasets are valid.

The teaching dataset `natality2018_rmph.Rdata` is a simple random sample of 2000 births intended only for illustrating regression methods. In the teaching dataset, variable names in CAPS are coded as in the original dataset (with the exception of missing value codes being set to NA and some cases assigned values based on skip patterns). Variable names in lowercase were derived from other variables. The gestational age variable `COMBGEST` was modified to create the variable `gestage37` for use in survival analysis in which gestational ages >37 weeks were censored at 37 weeks and a random subset of gestational ages were censored at times <37 weeks. Thus, in addition to being only a small sample of U.S. births, the data are slightly modified for teaching purposes.

Creating the Teaching Datasets

To create the teaching datasets, do the following:

- Download the .zip file containing the 2018 CSV file[15] found at Vital Statistics Natality Birth Data[16].
- Extract the CSV file `natl2018us.csv` from the .zip file.
- Download the R script file `Natality 2018 Process.R` from RMPH Resources[17].
- Run the R script file `Natality 2018 Process.R` to process the raw data and create the following teaching datasets:
  - `natality2018_rmph.Rdata`
  - `natality_CC_rmph.Rdata` (an artificial matched case-control dataset used to illustrate conditional logistic regression)
- Place these `.Rdata` files in your "Data" folder.

Rows and Columns

These files have the following numbers of rows and columns:

```
load("Data/natality2018_rmph.Rdata")
dim(natality)
```

```
## [1] 2000   39
```

[13]https://data.nber.org/natality/2018/UserGuide2018-508.pdf
[14]https://www.cdc.gov/nchs/data_access/restrictions.htm
[15]https://data.nber.org/natality/2018/natl2018.csv.zip
[16]https://www.nber.org/research/data/vital-statistics-natality-birth-data
[17]https://github.com/rwnahhas/RMPH_Resources

```
load("Data/natality_CC_rmph.Rdata")
dim(natality_CC)
```

```
## [1] 1570    4
```

A.4 COVID-19 county-level data

Background

The COVID Severity Forecasting data include county-level demographics and risk factors related to COVID-19 (Altieri et al., 2021). The USA Facts case, death, and county population data include county-level COVID-19 case and death counts, as well as population sizes (Source: USAFacts[18]).

Documentation

Data documentation is available at COVID Severity Forecasting data[19] and USA Facts case, death, and county population data[20]. The COVID Severity Forecasting data are distributed under the MIT License[21]. The USA Facts data are distributed under a Creative Commons Attribution-ShareAlike 4.0 (or higher) International Public License (the "CC BY-SA 4.0 License"). See How to Cite USAFacts[22] and Terms and Conditions[23] for more information. Data derived herein from USA Facts are made available under the same license as the original USA Facts data.

Teaching Dataset

Any analyses, interpretations, or conclusions reached herein are only for the purpose of illustrating regression methods and are credited to the author, not to the licensors. The author makes no claim or implication that any inferences derived from this teaching dataset are valid.

The dataset `covid_20210908_rmph.rData` was created by merging the COVID Severity Forecasting data with USA Facts case, death, and county population data, excluding counties with FIPS that did not match between the two datasets. The COVID Severity Forecasting data were downloaded on September 9, 2021, and the USA Facts data were downloaded on September 10, 2021 and contained data collected through September 8, 2021.

After merging, the following code was used to derive additional variables (no need to run, just shown here for your information).

[18] https://usafacts.org/
[19] https://github.com/Yu-Group/covid19-severity-prediction
[20] https://usafacts.org/visualizations/coronavirus-covid-19-spread-map/
[21] https://github.com/Yu-Group/covid19-severity-prediction/blob/master/LICENSE
[22] https://usafacts.org/how-to-cite-usafacts/
[23] https://usafacts.org/terms-and-conditions/

```
covid <- covid %>%
  mutate(State                    = factor(State),
         Rural.UrbanContinuumCode2013 = factor(Rural.UrbanContinuumCode2013),
         hospitals_per_100k = 100000*X.Hospitals / PopulationEstimate2018,
         icu_beds_per_100k  = 100000*X.ICU_beds  / PopulationEstimate2018,
         icu_beds_per_hosp  =        X.ICU_beds  / X.Hospitals,
         fte_hosp_per_100k  = 100000*X.FTEHospitalTotal2017 /
                                     PopulationEstimate2018,
         fte_hosp_per_hosp  =        X.FTEHospitalTotal2017 / X.Hospitals,
         mds_per_100k       = 100000*TotalM.D..s.TotNon.FedandFed2017 /
                                     PopulationEstimate2018,
         mds_per_hosp       =        TotalM.D..s.TotNon.FedandFed2017 /
                                     X.Hospitals,
         fte_hosp_per_hosp  = na_if(fte_hosp_per_hosp, Inf),
         mds_per_hosp       = na_if(mds_per_hosp,      Inf))
```

Additionally, labels were added to the variables. To view the labels, run the following code:

```
load("Data/covid_20210908_rmph.rData")
Hmisc::label(covid)
```

Creating the Teaching Dataset

To create the teaching dataset, do the following:

- Download `covid_20210908_rmph.rData` from RMPH Resources[24].
- Place this `.Rdata` file in your "Data" folder.

Rows and Columns

This file has the following numbers of rows and columns:

```
load("Data/covid_20210908_rmph.rData")
dim(covid)
```

```
## [1] 3074   99
```

A.5 NSDUH (2019)

Background

The National Survey on Drug Use and Health (NSDUH)[25], a product of the Substance Abuse and Mental Health Services Administration (SAMHSA) under the U.S. Department of Health and Human Services, measures the use of illegal substances, the use and misuse of prescribed substances, substance use disorder and treatment, and mental health outcomes (U.S. Department of Health and Human Services, Substance Abuse and Mental Health Services Administration, Center for Behavioral Health Statistics and Quality, 2019).

[24] https://github.com/rwnahhas/RMPH_Resources
[25] https://www.samhsa.gov/data/data-we-collect/nsduh-national-survey-drug-use-and-health

Documentation

Downloadable data and documentation[26] are freely available from SAMHSA (U.S. Department of Health and Human Services, Substance Abuse and Mental Health Services Administration, Center for Behavioral Health Statistics and Quality, 2019) for research and statistical purposes. Documentation for the 2019 data can be found in the 2019 NSDUH Public Use File Codebook[27]. See also, Policies[28].

Teaching Datasets

SAMHSA bears no responsibility for use of the data or for interpretations or inferences based upon such uses. Any analyses, interpretations, or conclusions reached herein are only for the purpose of illustrating regression methods and are credited to the author, not to SAMHSA. The author makes no claim or implication that any inferences derived from these teaching datasets are valid.

The teaching dataset `nsduh2019_adult_sub_rmph.RData` includes a random subset of 1000 observations of adults, and variables that have been renamed for clarity. Sampling was done with replacement using sampling weights in order to approximate a nationally representative distribution. This sampling method is solely for the purpose of creating a teaching dataset to illustrate regression methods. Chapter 8 discusses analyzing data using the survey weights appropriately using the full dataset (`nsduh2019_rmph.RData`). Chapter 9 uses the dataset `nsduh_mar_rmph.RData`, derived from `nsduh2019_adult_sub_rmph.RData` with some cases removed and some data values randomly set to missing, in an illustration of multiple imputation.

Creating the Teaching Datasets

To create the teaching datasets, do the following:

- Download the .zip file containing the 2019 R dataset[29] found at 2019 Population Data[30].
- Extract the .RData file `NSDUH_2019.RData` from the .zip file.
- Download the R script files `NSDUH_2019 Process.R` and `NSDUH_2019 MI Simulation.R` from RMPH Resources[31].
- Run the R script file `NSDUH_2019 Process.R` to process the raw data and create the following teaching datasets:
  - `nsduh2019_rmph.RData`
  - `nsduh2019_adult_sub_rmph.RData`
- Place these `.Rdata` files in your "Data" folder.
- Run the R script file `NSDUH_2019 MI Simulation.R` to process the raw data and create the following teaching datasets:
  - `nsduh_mar_rmph.RData`
- Place this `.Rdata` file in your "Data" folder.

---

[26] https://www.datafiles.samhsa.gov/dataset/national-survey-drug-use-and-health-2019-nsduh-2019-ds0001

[27] https://www.datafiles.samhsa.gov/sites/default/files/field-uploads-protected/studies/NSDUH-2019/NSDUH-2019-datasets/NSDUH-2019-DS0001/NSDUH-2019-DS0001-info/NSDUH-2019-DS0001-info-codebook.pdf

[28] https://www.datafiles.samhsa.gov/about-us/policies

[29] https://www.datafiles.samhsa.gov/sites/default/files/field-uploads-protected/studies/NSDUH-2019/NSDUH-2019-datasets/NSDUH-2019-DS0001/NSDUH-2019-DS0001-bundles-with-study-info/NSDUH-2019-DS0001-bndl-data-r.zip

[30] https://www.datafiles.samhsa.gov/dataset/national-survey-drug-use-and-health-2019-nsduh-2019-ds0001

[31] https://github.com/rwnahhas/RMPH_Resources

Rows and Columns

These files have the following numbers of rows and columns:

```
load("Data/nsduh2019_rmph.RData")
dim(nsduh)
```

```
## [1] 56136    57
```

```
load("Data/nsduh2019_adult_sub_rmph.RData")
dim(nsduh_adult_sub)
```

```
## [1] 1000   54
```

```
load("Data/nsduh_mar_rmph.RData")
dim(nsduh_mar)
```

```
## [1] 843   5
```

A.6 BioLINCC teaching datasets

Background

BioLINCC[32] is the National Heart, Lung, and Blood Institute (NHLBI) Biologic Specimen and Data Repository (National Heart, Lung and Blood Institute, 2022). They have made available the following teaching datasets: the Childhood Asthma Management Program (CAMP) dataset, the Digitalis Investigation Group (DIG) dataset, and the Framingham Heart Study dataset.

Documentation

- Digitalis Investigation Group (DIG)[33]
- Childhood Asthma Management Program (CAMP)[34]
- Framingham Heart Study[35]

These teaching datasets, as provided by BioLINCC, are not appropriate for publication purposes. Each has been rendered anonymous through the application of certain statistical processes such as permutations and/or random visit selection.

Teaching Datasets

Any analyses, interpretations, or conclusions reached herein are only for the purpose of illustrating regression methods and are credited to the author, not

[32]https://biolincc.nhlbi.nih.gov/about/
[33]https://biolincc.nhlbi.nih.gov/media/teachingstudies/digdoc.pdf
[34]https://biolincc.nhlbi.nih.gov/media/teachingstudies/CAMP_Documentation.pdf
[35]https://biolincc.nhlbi.nih.gov/media/teachingstudies/FHS_Teaching_Longitudinal_Data_Documentation_2021a.pdf

to BioLINCC. The author makes no claim or implication that any inferences derived from these teaching datasets are valid.

For the versions of these teaching datasets used in this text, the data were not changed but some variables were converted to factors, and only a subset of variables were retained. For the CAMP dataset, only data for the main study and 48-month follow-up were retained. For the Framingham study, the dataset `fram_time_invar_rmph.rData` contains only time-invariant predictors and excludes individuals with prevalent coronary heart disease, angina pectoris, myocardial infarction, or stroke at baseline. Additionally, variables describing hypertension were set to missing for those with prevalent hypertension at baseline. Additional datasets containing both time-invariant and time-varying information were created for a set of outcomes.

Creating the Teaching Datasets

- Request the DIG, CAMP, and Framingham datasets from the Request a teaching dataset[36] link at the NHLBI Teaching Datasets[37] site.
- After your request has been granted, download the datasets `DIG.csv`, `camp_teach.csv`, and `frmgham2.csv` from the links provided by NHLBI.
- Download the R script files `DIG Process.R`, `CAMP Process.R`, and `Framingham Process.R` from RMPH Resources[38].
- Run the R scripts `DIG Process.R`, `CAMP Process.R`, and `Framingham Process.R` to process the raw data and create the following teaching datasets:
  - `dig_rmph.rData`
  - `camp_0_48_rmph.rData`
  - `fram_time_invar_rmph.rData`
  - `fram_tv_angina_rmph.rData`, `fram_tv_mi_rmph.rData`, `fram_tv_mi_fchd_rmph.rData`, `fram_tv_anychd_rmph.rData`, `fram_tv_stroke_rmph.rData`, `fram_tv_cvd_rmph.rData`, `fram_tv_hyperten_rmph.rData`, and `fram_tv_death_rmph.rData`

Rows and Columns

These files have the following numbers of rows and columns:

```
load("Data/dig_rmph.rData")
dim(dig)
```

```
## [1] 6800   71
```

```
load("Data/camp_0_48_rmph.rData")
dim(camp_0_48)
```

```
## [1] 629  15
```

```
load("Data/fram_time_invar_rmph.rData")
dim(fram_time_invar)
```

```
## [1] 4215   20
```

[36]https://biolincc.nhlbi.nih.gov/login/?next=/requests/teaching-dataset-request/
[37]https://biolincc.nhlbi.nih.gov/teaching/
[38]https://github.com/rwnahhas/RMPH_Resources

A.7 Opioid

Background

The Opioid teaching dataset contains de-identified data from a study supported by The National Institute on Drug Abuse (NIDA), Grant number R01DA023577.

Documentation

The Opioid teaching dataset contains longitudinal information for 362 individuals who at baseline were age 18 to 23 years, had used non-prescribed pharmaceutical opioids (PO, "pain pills"), were not dependent on POs, and had never used heroin (Carlson et al., 2016). They were interviewed approximately every 6 months for 3 years (a total of seven waves = 0, 1, ..., 7). The dataset contains 1853 observations.

Each row contains the subject and wave identifiers `RANDID` and `wave` and the time variables `START` and `STOP`, which define the time interval (years from initiation of PO use) associated with that row. Time-invariant variables in this dataset include age at initiation of PO use (`age_at_init`) and `sex` – these variables are the same in all rows for the same individual.

Time-varying variables can change between rows and the value in any given row represents that variable's value in the time interval `(START, STOP]` (the interval runs from immediately after `START` up to and including `STOP`). The time-varying variables in the dataset are heroin use (`heroin`), lifetime opioid dependence (`dep_lifetime`) based on DSM-IV criteria (Forman et al., 2004; Hudziak et al., 1993), typically taking POs non-orally in the past 6 months (e.g., snorting, injecting) (`po_nonoral`), ever used POs to self-medicate a health problem in the past 6 months (`self_medicate`), lifetime psychiatric comorbidity (antisocial personality disorder, depression, generalized anxiety disorder, mania, or post-traumatic stress disorder) (`psych_lifetime`) and lifetime cocaine use (`coca_lifetime`).

Interviews asked individuals to report about their experiences and behavior since the last interview (about 6 months), so the temporal ordering of any changes in time-dependent variables since the previous interview is not known. Therefore, other than heroin use, which is the event of interest, all time-dependent predictors were lagged to ensure that they temporally preceded the outcome. While there were up to seven interviews per individual, the lagged dataset has only up to six. The first row in the data for each individual contains their outcome value at their first follow-up visit and their baseline values of the other time-varying variables. Each subsequent row contains an outcome value at a given interview along with the values of time-varying variables at the previous interview.

Teaching Dataset

Any analyses, interpretations, or conclusions reached herein are only for the purpose of illustrating regression methods and are credited to the author, not to NIDA. The author makes no claim or implication that any inferences derived from this teaching dataset are valid.

Creating the Teaching Dataset

To create the teaching dataset, do the following:

- Download `opioid_rmph.rData` from RMPH Resources[39].
- Place this `.Rdata` file in your "Data" folder.

Rows and Columns

This file has the following numbers of rows and columns:

```
load("Data/opioid_rmph.rData")
dim(opioid)
```

```
## [1] 1853   13
```

```
xfun::session_info()
```

[39] https://github.com/rwnahhas/RMPH_Resources

Bibliography

Allison, P. (2010). *Survival Analysis Using SAS: A Practical Guide.* SAS Press, Cary, NC, 2nd edition.

Allison, P. (2012). When can you safely ignore multicollinearity? statisticalhorizons.com/multicollinearity. Online; accessed 23-August-2021.

Alsuwaidi, A. R., Al Hosani, F. I., Al Memari, S., Narchi, H., Abdel Wareth, L., Kamal, H., Al Ketbi, M., Al Baloushi, D., Elfateh, A., Khudair, A., Al Mazrouei, S., Al Humaidan, H. S., Alghaithi, N., Afsh, K., Al Kaabi, N., Altrabulsi, B., Jones, M., Shaban, S., Sheek-Hussein, M., and Zoubeidi, T. (2021). Seroprevalence of COVID-19 infection in the Emirate of Abu Dhabi, United Arab Emirates: A population-based cross-sectional study. *International Journal of Epidemiology*, 50(4):1077–1090.

Altieri, N., Barter, R. L., Duncan, J., Dwivedi, R., Kumbier, K., Li, X., Netzorg, R., Park, B., Singh, C., Tan, Y. S., Tang, T., Wang, Y., Zhang, C., and Yu, B. (2021). Curating a COVID-19 data repository and forecasting county-level death counts in the United States. *Harvard Data Science Review (Special Issue 1).*

Altman, D. G. (1985). Comparability of randomised groups. *Journal of the Royal Statistical Society. Series D (The Statistician)*, 34(1):125–136.

American Medical Association (2020). New AMA policies recognize race as a social, not biological, construct. *AMA Press Releases.*

Babyak, M. A. (2004). What you see may not be what you get: A brief, nontechnical introduction to overfitting in regression-type models. *Psychosomatic Medicine*, 66(3):411–421.

Blakesley, R. E., Mazumdar, S., Dew, M. A., Houck, P. R., Tang, G., Reynolds, C. F., and Butters, M. A. (2009). Comparisons of methods for multiple hypothesis testing in neuropsychological research. *Neuropsychology*, 23(2):255–264.

Box, G. and Cox, D. (1964). An analysis of transformations. *Journal of the Royal Statistical Society. Series B (Methodological)*, 26(2):211–252.

Burris, H. H., Lorch, S. A., Kirpalani, H., Pursley, D. M., Elovitz, M. A., and Clougherty, J. E. (2019). Racial disparities in preterm birth in USA: A biosensor of physical and social environmental exposures. *Archives of Disease in Childhood*, 104(10):931–935.

Carlson, R. G., Nahhas, R. W., Martins, S. S., and Daniulaityte, R. (2016). Predictors of transition to heroin use among initially non-opioid dependent illicit pharmaceutical opioid users: A natural history study. *Drug and Alcohol Dependence*, 160:127–134.

Centers for Disease Control and Prevention (CDC). National Center for Health Statistics (NCHS) (2017). National Health and Nutrition Examination Survey Data. U.S. Department of Health and Human Services, Centers for Disease Control and Prevention, wwwn.cdc.gov/nchs/nhanes/continuousnhanes/default.aspx?BeginYear=2017.

Cook, R. D. (1977). Detection of influential observation in linear regression. *Technometrics*, 19(1):15–18.

Cox, D. R. (1972). Regression models and life-tables. *Journal of the Royal Statistical Society. Series B (Methodological)*, 34(2):187–220.

Dales, L. G. and Ury, H. K. (1978). An improper use of statistical significance testing in studying covariables. *International Journal of Epidemiology*, 7(4):373–375.

Dawber, T. R., Meadors, G. F., and Moore, F. E. (1951). Epidemiological approaches to heart disease: The Framingham Study. *American Journal of Public Health and the Nations Health*, 41(3):279–286.

Degerud, E., Høiseth, G., Mørland, J., Ariansen, I., Graff-Iversen, S., Ystrom, E., Zuccolo, L., Tell, G. S., and Næss, O. (2021). Associations of binge drinking with the risks of ischemic heart disease and stroke: A study of pooled Norwegian Health Surveys. *American Journal of Epidemiology*, 190(8):1592–1603.

Diggle, P., Heagerty, P., Liang, K., and Zeger, S. (2002). *Analysis of Longitudinal Data*. Oxford University Press, Oxford.

Donoghoe, M. W. and Marschner, I. C. (2018). logbin: An R package for relative risk regression using the log-binomial model. *Journal of Statistical Software*, 86:1–22.

Efron, B. (2013). *Large-Scale Inference: Empirical Bayes Methods for Estimation, Testing, and Prediction*. Cambridge University Press, Cambridge, reprint edition.

Efron, B. (2023). *Exponential Families in Theory and Practice*. Cambridge University Press, Cambridge.

Enders, C. K. (2010). *Applied Missing Data Analysis*. Guilford Press, New York, NY.

Errickson, J. (2023). *howManyImputations: Calculate How many Imputations are Needed for Multiple Imputation*. R package version 0.2.4.

Fagerland, M. W. and Hosmer, D. W. (2013). A goodness-of-fit test for the proportional odds regression model. *Statistics in Medicine*, 32(13):2235–2249.

Falissard, B. (2022). *psy: Various Procedures Used in Psychometrics*. R package version 1.2.

Faraway, J. (1992). On the cost of data analysis. *Journal of Computational and Graphical Statistics*, 1(3):213–229.

Faraway, J. J. (2016). *Extending the Linear Model with R: Generalized Linear, Mixed Effects and Nonparametric Regression Models*. Chapman & Hall/CRC Press, Boca Raton, 2nd edition.

Fitzmaurice, G., Laird, N., and Ware, J. (2011). *Applied Longitudinal Analysis*. John Wiley & Sons, Inc., Hoboken, 2nd edition.

Forman, R. F., Svikis, D., Montoya, I. D., and Blaine, J. (2004). Selection of a substance use disorder diagnostic instrument by the National Drug Abuse Treatment Clinical Trials Network. *Journal of Substance Abuse Treatment*, 27(1):1–8.

Fox, J. (2015). *Applied Regression Analysis and Generalized Linear Models*. Sage Publications, Inc., Los Angeles, 3rd edition.

Fox, J. and Monette, G. (1992). Generalized collinearity diagnostics. *Journal of the American Statistical Association*, 87(417):178–183.

Fox, J. and Weisberg, S. (2019). *An R Companion to Applied Regression*. Sage Publications, Inc., Los Angeles, 3rd edition.

Fox, J., Weisberg, S., and Price, B. (2023). *car: Companion to Applied Regression*. R package version 3.1-2, https://CRAN.R-project.org/package=car.

Friedman, J. H., Hastie, T., and Tibshirani, R. (2010). Regularization paths for generalized linear models via coordinate descent. *Journal of Statistical Software*, 33:1–22.

Gail, M. H., Lubin, J. H., and Rubinstein, L. V. (1981). Likelihood calculations for matched case-control studies and survival studies with tied death times. *Biometrika*, 68(3):703–707.

Gohel, D. and Skintzos, P. (2023). *flextable: Functions for Tabular Reporting*. R package version 0.9.4, https://davidgohel.github.io/flextable/.

Grambsch, P. M. and Therneau, T. M. (1994). Proportional hazards tests and diagnostics based on weighted residuals. *Biometrika*, 81(3):515–526.

Grambsch, P. M. and Therneau, T. M. (2000). *Modeling Survival Data: Extending the Cox Model*. Springer, New York.

Greenland, S. (1994). Alternative models for ordinal logistic regression. *Statistics in Medicine*, 13(16):1665–1677.

Greenland, S. (2000). When should epidemiologic regressions use random coefficients? *Biometrics*, 56(3):915–921.

Greenland, S. and Hofman, A. (2019). Multiple comparisons controversies are about context and costs, not frequentism versus Bayesianism. *European Journal of Epidemiology*, 34(9):801–808.

Greenland, S., Mansournia, M. A., and Altman, D. G. (2016a). Sparse data bias: A problem hiding in plain sight. *BMJ (Clinical research ed.)*, 352:i1981.

Greenland, S., Senn, S. J., Rothman, K. J., Carlin, J. B., Poole, C., Goodman, S. N., and Altman, D. G. (2016b). Statistical tests, p-values, confidence intervals, and power: A guide to misinterpretations. *The American Statistician*, 70(2):Online Supplement, 1–12.

Grund, S., Lüdtke, O., and Robitzsch, A. (2016). Pooling ANOVA results from multiply imputed datasets: A simulation study. *Methodology: European Journal of Research Methods for the Behavioral and Social Sciences*, 12(3):75–88.

Grund, S., Robitzsch, A., and Luedtke, O. (2023). *mitml: Tools for Multiple Imputation in Multilevel Modeling*. R package version 0.4-5.

Hahs-Vaughn, D. L., McWayne, C. M., Bulotsky-Shearer, R. J., Wen, X., and Faria, A.-M. (2011). Methodological considerations in using complex survey data: An applied example with the Head Start Family and Child Experiences Survey. *Evaluation Review*, 35(3):269–303.

Harrell, Jr., F. E. (2015). *Regression Modeling Strategies*. Springer International Publishing, Switzerland, 2nd edition.

Harrell, Jr., F. E. (2023). *Hmisc: Harrell Miscellaneous*. R package version 5.1-1.

Harrington, D. P. and Fleming, T. R. (1982). A class of rank test procedures for censored survival data. *Biometrika*, 69(3):553–566.

Hastie, T., Tibshirani, R., and Friedman, J. (2016). *The Elements of Statistical Learning: Data Mining, Inference, and Prediction*. Springer, New York, NY, 2nd edition.

Hawkins, D. and Weisberg, S. (2017). Combining the Box-Cox power and generalised log transformations to accommodate nonpositive responses in linear and mixed-effects linear models. *South African Statistical Journal*, 51(2):317–328.

Hayes, A. F. (2022). *Introduction to Mediation, Moderation, and Conditional Process Analysis: A Regression-Based Approach.* The Guilford Press, New York, NY, 3rd edition.

Hayes, A. F. and Cai, L. (2007). Using heteroskedasticity-consistent standard error estimators in OLS regression: An introduction and software implementation. *Behavior Research Methods*, 39(4):709–722.

Hebbali, A. (2024). *olsrr: Tools for Building OLS Regression Models.* R package version 0.6.0, https://github.com/rsquaredacademy/olsrr.

Heinze, G., Ploner, M., Jiricka, L., and Steiner, G. (2023a). *coxphf: Cox Regression with Firth's Penalized Likelihood.* R package version 1.13.4.

Heinze, G., Ploner, M., Jiricka, L., and Steiner, G. (2023b). *logistf: Firth's Bias-Reduced Logistic Regression.* R package version 1.26.0.

Heinze, G. and Schemper, M. (2002). A solution to the problem of separation in logistic regression. *Statistics in Medicine*, 21(16):2409–2419.

Hommel, G. (1988). A stagewise rejective multiple test procedure based on a modified Bonferroni test. *Biometrika*, 75(2):383–386.

Hosmer, D. W. and Lemeshow, S. (2000). *Applied Logistic Regression.* John Wiley & Sons, New York, 2nd edition.

Hudziak, J. J., Helzer, J. E., Wetzel, M. W., Kessel, K. B., McGee, B., Janca, A., and Przybeck, T. (1993). The use of the DSM-III-R Checklist for initial diagnostic assessments. *Comprehensive Psychiatry*, 34(6):375–383.

Iannone, R., Cheng, J., Schloerke, B., Hughes, E., Lauer, A., and Seo, J. (2023). *gt: Easily Create Presentation-Ready Display Tables.* R package version 0.10.0.

Jackson, C. (2023). *flexsurv: Flexible Parametric Survival and Multi-State Models.* R package version 2.2.2.

James, G., Witten, D., Hastie, T., and Tibshirani, R. (2021). *An Introduction to Statistical Learning: with Applications in R.* Springer, New York, NY, 2nd edition.

Kassambara, A., Kosinski, M., and Biecek, P. (2021). *survminer: Drawing Survival Curves using ggplot2.* R package version 0.4.9.

Katikireddi, S., Lal, S., Carrol, E., Niedzwiedz, C., Khunti, K., Dundas, R., Diderichsen, F., and Barr, B. (2021). Unequal impact of the COVID-19 crisis on minority ethnic groups: A framework for understanding and addressing inequalities. *J Epidemiol Community Health*, 75(10):970–974.

Klein, J. P. and Moeschberger, M. L. (2010). *Survival Analysis: Techniques for Censored and Truncated Data.* Springer, New York, NY.

Kleinman, K. and Horton, N. J. (2014). *SAS and R.* Routledge, Boca Raton, 2nd edition.

Kroenke, K. and Spitzer, R. (2002). The PHQ-9: A new depression diagnostic and severity measure. *Psychiatric Annals*, 32(9):509–515.

Kroenke, K., Spitzer, R. L., and Williams, J. B. W. (2001). The PHQ-9. *Journal of General Internal Medicine*, 16(9):606–613.

Laird, N. and Ware, J. (1982). Random-effects models for longitudinal data. *Biometrics*, 38(4):963–974.

Lele, S. R., Keim, J. L., and Solymos, P. (2023). *ResourceSelection: Resource Selection (Probability) Functions for Use-Availability Data*. R package version 0.3-6.

Li, K. H., Meng, X. L., Raghunathan, T. E., and Rubin, D. B. (1991). Significance levels from repeated p-values with multiply-imputed data. *Statistica Sinica*, 1(1):65–92.

Liang, K. and Zeger, S. (1986). Longitudinal data analysis using generalized linear models. *Biometrika*, 73(1):13–22.

Liang, K. and Zeger, S. (1993). Regression analysis for correlated data. *Annual Review of Public Health*, 14(1):43–68.

Little, R. J. A. and Rubin, D. B. (2019). *Statistical Analysis with Missing Data*. Wiley, Hoboken, NJ, 3rd edition.

Liu, B., Yu, M., Graubard, B. I., Troiano, R. P., and Schenker, N. (2016). Multiple imputation of completely missing repeated measures data within person from a complex sample: Application to accelerometer data in the National Health and Nutrition Examination Survey. *Statistics in Medicine*, 35(28):5170–5188.

Logan, J. A. (1983). A multivariate model for mobility tables. *American Journal of Sociology*, 89(2):324–349.

Lohr, S. L. (2021). *Sampling: Design and Analysis*. Chapman & Hall/CRC Press, Boca Raton, 3rd edition.

Lumley, T. (2004). Analysis of complex survey samples. *Journal of Statistical Software*, 9(1):1–19. R package verson 2.2.

Lumley, T. (2010). *Complex Surveys: A Guide to Analysis Using R: A Guide to Analysis Using R*. John Wiley & Sons, Hoboken.

Lumley, T. (2023). *survey: Analysis of Complex Survey Samples*. R package version 4.2-1.

Lumley, T. and Scott, A. (2013). Partial likelihood ratio tests for the Cox model under complex sampling. *Statistics in Medicine*, 32(1):110–123.

Lumley, T. and Scott, A. (2014). Tests for regression models fitted to survey data. *Australian & New Zealand Journal of Statistics*, 56(1):1–14.

Lumley, T. and Scott, A. (2017). Fitting regression models to survey data. *Statistical Science*, 32(2):265–278.

Lüdecke, D. (2021). *sjmisc: Data and Variable Transformation Functions*. R package version 2.8.9.

Lüdecke, D. (2023). *sjPlot: Data Visualization for Statistics in Social Science*. R package version 2.8.15.

Mahmood, S. S., Levy, D., Vasan, R. S., and Wang, T. J. (2014). The Framingham Heart Study and the epidemiology of cardiovascular diseases: A historical perspective. *Lancet*, 383(9921):999–1008.

Mansournia, M. A., Geroldinger, A., Greenland, S., and Heinze, G. (2018). Separation in logistic regression: Causes, consequences, and control. *American Journal of Epidemiology*, 187(4):864–870.

Matsuzawa, Y. (2006). The metabolic syndrome and adipocytokines. *FEBS Lett*, 580(12):2917–2921.

Meghani, S. and Chittams, J. (2015). Controlling for socioeconomic status in pain disparities research: All-else-equal analysis when "all else" is not equal. *Pain Medicine*, 16(12):2222–2225.

Michalowicz, B. S., Hodges, J. S., DiAngelis, A. J., Lupo, V. R., Novak, M. J., Ferguson, J. E., Buchanan, W., Bofill, J., Papapanou, P. N., Mitchell, D. A., Matseoane, S., Tschida, P. A., and for the OPT Study (2006). Treatment of periodontal disease and the risk of preterm birth. *The New England Journal of Medicine*, 355(18):1885–1894.

Moher, D., Hopewell, S., Schulz, K. F., Montori, V., Gotzsche, P. C., Devereaux, P. J., Elbourne, D., Egger, M., and Altman, D. G. (2010). CONSORT 2010 Explanation and Elaboration: Updated guidelines for reporting parallel group randomised trials. *BMJ*, 340:c869.

Nahhas, R. W. (2024). An Introduction to R for Research. bookdown.org/rwnahhas/IntroToR.

National Center for Health Statistics (2021a). NHANES - About the National Health and Nutrition Examination Survey. www.cdc.gov/nchs/nhanes. Online; accessed 18-November-2021.

National Center for Health Statistics (2021b). NHANES Questionnaires, Datasets, and Related Documentation. wwwn.cdc.gov/nchs/nhanes/Default.aspx. Online; accessed 18-November-2021.

National Center for Health Statistics (2022). Birth Data. www.cdc.gov/nchs/nvss/births.htm. Online; accessed 22-May-2022.

National Heart, Lung and Blood Institute (2022). About BioLINCC. biolincc.nhlbi.nih.gov/about. Online; accessed 5-May-2022.

Olivier, C. B., Struß, L., Sünnen, N., Kaier, K., Heger, L. A., Westermann, D., Meerpohl, J. J., and Mahaffey, K. W. (2024). Accuracy of event rate and effect size estimation in major cardiovascular trials: A systematic review. *JAMA Network Open*, 7(4):e248818.

Pinheiro, J. and Bates, D. (2000). *Mixed-effects models in S and S-PLUS*. Springer-Verlag, New York.

Pinheiro, J., Bates, D., and R Core Team (2023). *nlme: Linear and Nonlinear Mixed Effects Models*. R package version 3.1-163.

Quartagno, M., Carpenter, J. R., and Goldstein, H. (2020). Multiple imputation with survey weights: A multilevel approach. *Journal of Survey Statistics and Methodology*, 8(5):965–989.

R Core Team (2022). R: A language and environment for statistical computing. www.R-project.org. R Foundation for Statistical Computing, Vienna, Austria.

Rafi, Z. and Greenland, S. (2020a). Semantic and cognitive tools to aid statistical science: replace confidence and significance by compatibility and surprise. *BMC Medical Research Methodology*, 20(1):244.

Rafi, Z. and Greenland, S. (2020b). Technical issues in the interpretation of S-values and their relation to other information measures. *arXiv*, stat(arXiv:2008.12991v3).

Rao, J. N. K. and Scott, A. J. (1984). On chi-squared tests for multiway contingency tables with cell proportions estimated from survey data. *The Annals of Statistics*, 12(1):46–60.

Rebora, P., Salim, A., and Reilly, M. (2018). *bshazard: Nonparametric Smoothing of the Hazard Function.* R package version 1.1.

Rich, B. (2023). *table1: Tables of Descriptive Statistics in HTML.* R package version 1.4.3.

Ripley, B. (2023). *MASS: Support Functions and Datasets for Venables and Ripley's MASS.* R package version 7.3-60.

Robitzsch, A., Grund, S., and Henke, T. (2023). *miceadds: Some Additional Multiple Imputation Functions, Especially for mice.* R package version 3.16-18, https://sites.google.com/site/alexanderrobitzsch2/software.

Rogers, N., Blodgett, J., Searle, S., Cooper, R., Davis, D., and Pinto Pereira, S. (2021). Early-life socioeconomic position and the accumulation of health-related deficits by midlife in the 1958 British Birth Cohort Study. *American Journal of Epidemiology*, 190(8):1550–1560.

Rolland, C., Hession, M., and Broom, I. (2011). Effect of weight loss on adipokine levels in obese patients. *Diabetes Metab Syndr Obes*, 4:315–23.

Rothman, K. J. (1990). No adjustments are needed for multiple comparisons. *Epidemiology*, 1(1):43–46.

Rothman, K. J., Greenland, S., and Lash, T. L. (2008). *Modern Epidemiology.* Lippincott Williams & Wilkins, Philadelphia, PA, 3rd edition.

Rubin, D. (1987). *Multiple Imputation for Nonresponse in Surveys.* John Wiley & Sons, Inc., New York.

Rubin, D. B. (1976). Inference and missing data. *Biometrika*, 63(3):581–592.

Rubin, D. B. (1996). Multiple imputation after 18+ years. *J Am Stat Assoc*, 91(434):473–489.

Sakai-Bizmark, R., Kumamaru, H., Estevez, D., Marr, E., Haghnazarian, E., Bedel, L., Mena, L., and Kaplan, M. (2021). Health-care utilization due to suicide attempts among homeless youth in New York State. *American Journal of Epidemiology*, 190(8):1582–1591.

Sarkar, S. K. (1998). Some probability inequalities for ordered MTP 2 random variables: A proof of the Simes conjecture. *The Annals of Statistics*, 26(2):494–504.

Sarkar, S. K. and Chang, C.-K. (1997). The Simes Method for multiple hypothesis testing with positively dependent test statistics. *Journal of the American Statistical Association*, 92(440):1601–1608.

Schmidt, A. and Finan, C. (2018). Linear regression and the normality assumption. *Journal of Clinical Epidemiology*, 98:146–151.

Schnake-Mahl, A. S. and Bilal, U. (2021). Schnake-Mahl and Bilal respond to "Structural racism and COVID-19 mortality in the US". *American Journal of Epidemiology*, 190(8):1447–1451.

Senn, S. (1994). Testing for baseline balance in clinical trials. *Statistics in Medicine*, 13(17):1715–1726.

Sjoberg, D. D., Larmarange, J., Curry, M., Lavery, J., Whiting, K., and Zabor, E. C. (2023). *gtsummary: Presentation-Ready Data Summary and Analytic Result Tables.* R package version 1.7.2, https://www.danieldsjoberg.com/gtsummary/.

Sjoberg, D. D., Whiting, K., Curry, M., Lavery, J. A., and Larmarange, J. (2021). Reproducible summary tables with the gtsummary package. *The R Journal*, 13:570–580.

Skinner, C. and Wakefield, J. (2017). Introduction to the design and analysis of complex survey data. *Statistical Science*, 32(2):165–175.

Steyerberg, E. W. (2009). *Clinical Prediction Models: A Practical Approach to Development, Validation, and Updating.* Springer, New York, NY.

Therneau, T., Crowson, C., and Atkinson, E. (2015). Adjusted surival curves. cran.r-project.org/web/packages/survival/vignettes/adjcurve.pdf. Online; accessed 13-January-2023.

Therneau, T., Crowson, C., and Atkinson, E. (2023). Using time dependent covariates and time dependent coefficients in the Cox model. cran.r-project.org/web/packages/survival/vignettes/timedep.pdf. Online; accessed 6-November-2023.

Therneau, T. M. (2023). *survival: Survival Analysis.* R package version 3.5-7.

United Nations Development Programme (2020). Human development data. http://hdr.undp.org/en/content/copyright-and-terms-use. Online; accessed 8-September-2021.

U.S. Department of Health and Human Services, Substance Abuse and Mental Health Services Administration, Center for Behavioral Health Statistics and Quality (2019). National Survey on Drug Use and Health 2019 (NSDUH-2019-DS0001). www.samhsa.gov/data/data-we-collect/nsduh-national-survey-drug-use-and-health. Online; accessed 22-May-2021.

van Buuren, S. (2018). *Flexible Imputation of Missing Data.* Chapman & Hall/CRC Press, Boca Raton, 2nd edition.

van Buuren, S. and Groothuis-Oudshoorn, K. (2011). mice: Multivariate imputation by chained equations in R. *Journal of Statistical Software*, 45(3):1–67.

van Buuren, S. and Groothuis-Oudshoorn, K. (2023). *mice: Multivariate Imputation by Chained Equations.* R package version 3.16.0.

Vandenbroucke, J. P., von Elm, E., Altman, D. G., Gøtzsche, P. C., Mulrow, C. D., Pocock, S. J., Poole, C., Schlesselman, J. J., Egger, M., and Initiative, f. t. S. (2007). Strengthening the Reporting of Observational Studies in Epidemiology (STROBE): Explanation and Elaboration. *Epidemiology*, 18(6):805–835.

Venables, W. and Ripley, B. (2002). *Modern Applied Statistics with S.* Springer, New York, 4th edition.

von Hippel, P. T. (2009). How to impute interactions, squares, and other transformed variables. *Sociological Methodology*, 39(1):265–291.

von Hippel, P. T. (2019). How many imputations do you need? statisticalhorizons.com/how-many-imputations. Online; accessed 5-January-2024.

von Hippel, P. T. (2020). How many imputations do you need? A two-stage calculation using a quadratic rule. *Sociological Methods & Research*, 49(3):699–718.

Warnes, G. R., Bolker, B., Lumley, T., and Johnson, R. C. (2022). *gmodels: Various R Programming Tools for Model Fitting.* R package version 2.18.1.1.

Wasserstein, R. L. and Lazar, N. A. (2016). The ASA statement on p-values: Context, process, and purpose. *The American Statistician*, 70(2):129–133.

Wasserstein, R. L., Schirm, A. L., and Lazar, N. A. (2019). Moving to a world beyond p < 0.05". *The American Statistician*, 73(sup1):1–19.

Weisberg, S. (1985). *Applied Linear Regression.* Wiley, Hoboken, NJ, 2nd edition.

Weisberg, S. (2014). *Applied Linear Regression.* Wiley, Hoboken, NJ, 4th edition.

Westfall, P. H. and Young, S. S. (1993). *Resampling-Based Multiple Testing, Examples and Methods for p-value Adjustment.* John Wiley & Sons, Inc., New York, NY.

White, H. (1980). A heteroskedasticity-consistent covariance matrix estimator and a direct test for heteroskedasticity. *Econometrica*, 48(4):817–838.

White, I. R. and Royston, P. (2009). Imputing missing covariate values for the Cox model. *Statistics in Medicine*, 28(15):1982–1998.

Wickham, H. (2019). *Advanced R.* Chapman & Hall/CRC Press, Boca Raton, 2nd edition.

Wickham, H. (2023). *tidyverse: Easily Install and Load the Tidyverse.* R package version 2.0.0, https://github.com/tidyverse/tidyverse.

Wickham, H., Çetinkaya Rundel, M., and Grolemund, G. (2023). *R for Data Science: Import, Tidy, Transform, Visualize, and Model Data.* O'Reilly Media, Sebastopol, CA, 2nd edition.

Williamson, T., Eliasziw, M., and Fick, G. H. (2013). Log-binomial models: Exploring failed convergence. *Emerging Themes in Epidemiology*, 10(1):14.

Xie, Y. (2015). *Dynamic Documents with R and knitr.* Chapman & Hall/CRC Press, Boca Raton, 2nd edition.

Xie, Y. (2023). *bookdown: Authoring Books and Technical Documents with R Markdown.* R package version 0.37, https://pkgs.rstudio.com/bookdown/.

Yehia, B., Winegar, A., Fogel, R., Fakih, M., Ottenbacher, A., Jesser, C., Bufalino, A., Huang, R., and Cacchione, J. (2020). Association of race with mortality among patients hospitalized with Coronavirus Disease 2019 (COVID-19) at 92 U.S. hospitals. *JAMA Network Open*, 3(8):e2018039.

Yoshida, K. and Bartel, A. (2022). *tableone: Create Table 1 to Describe Baseline Characteristics with or without Propensity Score Weights.* R package version 0.13.2.

Zalla, L. C., Martin, C. L., Edwards, J. K., Gartner, D. R., and Noppert, G. A. (2021). A geography of risk: Structural racism and Coronavirus Disease 2019 mortality in the United States. *American Journal of Epidemiology*, 190(8):1439–1446.

Zeger, S. and Liang, K. (1986). Longitudinal data analysis for discrete and continuous outcomes. *Biometrics*, 42(1):121–130.

Zeileis, A., Kleiber, C., and Jackman, S. (2008). Regression models for count data in R. *Journal of Statistical Software*, 27:1–25.

Zeola, M. P., Guina, J., and Nahhas, R. W. (2017). Mental health referrals reduce recidivism in first-time juvenile offenders, but how do we determine who is referred? *The Psychiatric Quarterly*, 88(1):167–183.

Index

Printed in the United States
by Baker & Taylor Publisher Services